Green Nanobiotechnology

This book provides a comprehensive exploration of green nanotechnology covering principles, applications, and ethical considerations. The book begins with an introductory exploration of nanotechnology, followed by in-depth discussions on the synthesis of ozone-friendly nanomaterials and the emerging practice of green synthesis. It delves into the diverse applications of green nanoparticles, spanning biomedical applications, tissue engineering, biosensors, antimicrobials, and vaccine development. It explores applications of nanotechnology in environmental sciences including bioremediation, microengineered ceramics for environmental protection, and the modification of advanced nano-polymer composites. The environmental fate and ecotoxicological implications of nanomaterials are thoroughly examined, followed by discussions on the energy-saving potential and sustainable fuel development in the realm of green nanotechnology. The book concludes with a focus on responsible and ethical considerations, addressing the legal, socio-economic, and ethical impacts of nanotechnology, making it an important resource for researchers, academics, and professionals in nanobiotechnology and biomedical sciences.

Key Features:

- Explores fundamental principles and applications of green nanotechnology in biomedical sciences and environmental protection.
- Reviews the use of green nanoparticles in tissue engineering, biosensors, antimicrobials, and vaccine development.
- Highlights the environmental impact of nanomaterials, examining their roles in treatment, bioremediation, and the development of advanced nano-polymer composites.
- Examines ecotoxicological implications of nanomaterials for humans and animals.
- Provides a strong emphasis on responsible and ethical considerations, exploring the legal, socio-economic, and ethical impacts of nanotechnology.

Nanotechnology Advances and Applications

About the Series

This book series explores the broader spectrum of applications of nanoparticles across biomedical, agricultural, and environmental sciences. Focusing on the biomedical applications of nanoparticles, the books cover their role in targeted drug delivery systems, advancements in nanoscale imaging for diagnostics, innovations in regenerative medicine and tissue engineering, and cutting-edge research in immunotherapy and vaccine development. In the realm of agricultural applications of nanoparticles, the series focuses on their use as nano-fertilizers, nano-pesticides, and in soil health monitoring techniques, sustainable farming practices, and the environmental benefits of reducing agricultural waste and chemical usage. The volumes cover the technologies behind antimicrobial packaging, freshness indicators, and the environmental impact of nano-enabled packaging solution. The series also explores applications of nanoparticles in environmental sciences, including advanced filtration systems for air and water purification, efficient removal of heavy metals and pollutants through nano-adsorbents, and the development of sustainable energy solutions with biofuel cells. Each volume also addresses the ethical, safety, and regulatory aspects of nanotechnology, providing a balanced and comprehensive perspective.

TITLES

1. Green Nanobiotechnology (9781032820453)
Edited by Atul Thakur, Preeti Thakur, Deepa Suhag, and S M Paul Khurana

2. Advancements in Nanobiology (9781032750491)
Edited by Atul Thakur, Preeti Thakur, and Deepa Suhag

Green Nanobiotechnology

Edited by
Atul Thakur, Preeti Thakur, Deepa Suhag,
and S M Paul Khurana

CRC Press
Taylor & Francis Group
Boca Raton London New York

CRC Press is an imprint of the
Taylor & Francis Group, an **informa** business

Designed cover image: Shutterstock

First edition published 2025
by CRC Press
2385 NW Executive Center Drive, Suite 320, Boca Raton FL 33431

and by CRC Press
4 Park Square, Milton Park, Abingdon, Oxon, OX14 4RN

CRC Press is an imprint of Taylor & Francis Group, LLC

ISBN: 9781032820453 (hbk)
ISBN: 9781032820460 (pbk)
ISBN: 9781003502692 (ebk)

DOI: 10.1201/9781003502692

Typeset in Times
by codeMantra

Contents

Editors

Atul Thakur is the Director of Amity Centre for Nanotechnology and Amity School of Applied Sciences at Amity University Haryana. He was recognized among the top 2% of researchers by Stanford University ranking. Dr. Thakur received his Post Doctorate from the University of Brest, France, and thereafter from the world-renowned National Taiwan University, Taiwan. He completed his M.Phil. and Ph.D. degrees from Himachal Pradesh University, Shimla. Dr. Thakur has published more than 205 research papers in international, peer-reviewed, and SCI and Scopus-indexed journals, holds fifty patents and has authored four books. His research areas include ferrite nanomaterials for water purification, agricultural applications, microwave and high-frequency applications, radar-absorbing materials, antenna miniaturization, high-density memory storage systems, sensor applications, and metamaterials.

Preeti Thakur is Professor of Physics and the Head of Amity Institute of Nanotechnology at Amity University Haryana, India. She was recognized among the top 2% of researchers by Stanford University ranking. She holds a post-doctoral degree from Taiwan and was a Gold medalist in her Master's program. Dr. Thakur earned her Ph.D. in Physics from Himachal Pradesh University, Shimla, India. She is an Editor for the Journal of Magnetism and Magnetic Materials. Her research areas include synthesis and characterization of magnetic nanoferrites for applications such as radar-absorbing materials, sensors, high frequency, wastewater treatment, and agricultural applications. Dr. Thakur has more than 180 publications in SCI/Scopus-indexed international journals, holds forty-five patents, and has edited books on nanomaterials and published specialized book chapters.

Dr. Deepa Suhag is Assistant Professor in the Department of Nanotechnology, Amity University Haryana. She has published 17 research articles in major international and has filed 6 patents of which 1 has been granted by the Government of India. Patent titled "Method for preparation of highly fluorescent biocompatible Sulphur doped graphene quantum dots from affordable agro-industrial bio-waste cane molasses using hydrothermal synthesis for bioimaging application" was granted on 19.05.2022. Her research area spans from biomaterial engineering to wound healing and tissue regeneration. She currently has a combined research funding of 10 million from various funding Indian government agencies. Dr. Deepa Suhag's major collaborators are from CNCI Kolkata, AIIMS Delhi, IIT Delhi, INST Mohali, RGCB Kerala, and Harvard University, USA.

Dr. S M Paul Khurana obtained M.Sc. (Plant Pathology) in 1965 and Ph. D. (Virus Pathology) in 1968 from DDU University of Gorakhpur. He did 2yrs post-doctoral work on Advanced Plant Virology at Kyushu University, Fukuoka (Japan) and later worked at the University of Minnesota, St. Paul (USA) on immunodiagnostics. Dr Khurana started his scientific career as Junior Scientist (Pathology) at Sugarcane Breeding Institute, Coimbatore. He joined Central Potato Research Institute, Shimla as CSIR Pool Research Officer in 1973 and continued there in different capacities as the AICRP Coordinator and the Director there till November 2004, when he joined as the Vice-Chancellor of Rani Durgavati Vishwavidyalaya, Jabalpur, Madhya Pradesh. Dr. Khurana is a Fellow of National Academy of Agricultural Sciences, National Academy of Biological Sciences, Confederation of Hort Research Institutes, Distinguished Fellow of the Indian Potato Association. He guided one D Sc,16 Ph D and has published over 245 research papers in journals of repute, authored more than 140 reviews/book chapters and has to his credit 25 books.

Contributors

Arti
Amity Institute of Nanotechnology
Amity University Haryana
Gurugram, India

Surendra Kumar Biswal
International PranaGraf Mintech Research Centre
Bhubaneswar, Odisha, India

P. B. Dandge
Department of Biochemistry
Shivaji University
Kolhapur, Maharashtra, India

Sumistha Das
Amity Institute of Biotechnology
Amity University Haryana
Gurugram, India

Lipsa Dash
Department of Entomology, Institute of Agricultural Sciences
SOA University
Bhubaneswar, Odisha, India

Tapan Dash
International PranaGraf Mintech Research Centre
Bhubaneswar, Odisha, India
and
Centurion University of Technology and Management
Odisha, India

Nitai Debnath
Amity Institute of Biotechnology
Amity University Haryana
Gurugram, India

Taranga Dehury
School of Materials Science and Technology
Indian Institute of Technology (BHU)
Varanasi, Uttar Pradesh, India

Neetu Dhanda
Department of Physics
Amity University Haryana
Gurugram, India

Sunita Dhar
Centurion University of Technology and Management
Odisha, India

Irina Edelman
Kirensky Institute of Physics, Federal Research Center KSC
Siberian Branch, Russian Academy of Sciences
Krasnoyarsk, Russia

M. Love Edet
School of Life Sciences
B. S. Abdul Rahman Crescent Institute of Science and Technology
Vandalur, Chennai, India

Ritika Gera
Department of Chemistry
Amity University Haryana
Manesar, Gurgaon, Haryana, India

S. Hemalatha
School of Life Sciences
B. S. Abdul Rahman Crescent Institute of Science and Technology
Vandalur, Chennai, India

Bikash Ranjan Jena
School of Pharmacy and Life Sciences
Centurion University of Technology and Management
Bhubaneswar, Odisha, India

Moni Kharb
Amity Institute of Nanotechnology
Amity University Haryana
Gurugram, India

S M Paul Khurana
Amity Institute of Biotechnology
Amity University Haryana
Gurugram, India

Deok-kee Kim
Department of Electrical Engineering
Sejong University
Seoul, South Korea

P. N. Kirdat
Department of Biotechnology
K.B.P. College Vashi
Navi Mumbai, Maharashtra, India

Arun Kumar
Department of Chemistry, School of Physical Sciences
Doon University
Dehradun, Uttarakhand, India

Praveen Kumar
Department of Chemistry
Doon University
Dehradun, Uttarakhand, India

Chandan Kumar Mandal
Department of Chemistry
Amity University Haryana
Manesar, Gurgaon, Haryana, India

Suchismita Mishra
Department of Floriculture and Landscaping
Institute of Agricultural Sciences
SOA University
Bhubaneswar, Odisha, India

Nibedita Mohanty
GIET University
Gunpur, Odisha, India

Sushree Subhadarshinee Mohapatra
GIET University
Gunpur, Odisha, India

Nancy
Department of Physics
Amity University Haryana
Gurugram, India

Sergey Ovchinnikov
Kirensky Institute of Physics
Russian Academy of Sciences
Krasnoyarsk, Russia

Larissa V. Panina
National University of Science and Technology
MISIS, Moscow, Russia

Abhilash Pathania
Amity Institute of Nanotechnology
Amity University Haryana
Gurugram, India

Tapan Kumar Patnaik
GIET University
Gunpur, Odisha, India

Ch. Niranjan Patra
Roland Institute of Pharmaceutical Sciences
Berhampur, Odisha, India

Gurudutta Pattnaik
Department of Pharmaceutics
Centurion University of Technology and Management
Odisha, India

Ayushi Pradhan
School of Pharmacy and Life Sciences
Centurion University of Technology and Management
Bhubaneswar, Odisha, India

Gyaneshwar Kumar Rao
Department of Chemistry, Biochemistry and Forensic Science
Amity University Haryana
Manesar, Gurgaon, Haryana, India

Chandana Rath
School of Materials Science and Technology
Indian Institute of Technology (BHU)
Varanasi, Uttar Pradesh, India

Raksha Rathore
Amity Institute of Nanotechnology
Amity University Haryana
Gurugram, India

Blaise Ravelo
School of Electronics and Information Engineering
Nanjing University of Information Science & Technology
Nanjing, China

Anand Salvi
Department of Chemistry
Amity University Haryana
Gurugram, India

Manish Shandilya
Department of Chemistry
Amity University Haryana
Gurugram, India

Shivani
Amity Institute of Nanotechnology
Amity University Haryana
Gurugram, India

M. Sai Sindhu
Department of Vegetable Science
College of Horticulture and Forestry
PAU, Ludhiana, India

Amitender Singh
Amity Institute of Nanotechnology
Amity University Haryana
Gurugram, India

Yassine Slimani
Department of Biophysics
Institute for Research and Medical Consultations (IRMC)
Imam Abdulrahman Bin Faisal University
Dammam, Saudi Arabia

Monika Sohlot
Amity Institute of Biotechnology
Amity University Haryana
Gurugram, India

Deepa Suhag
Amity Institute of Nanotechnology
Amity University Haryana
Gurugram, India

An-Cheng Aidan Sun
Department of Chemical Engineering and Materials Science
Yuan Ze University
Taoyuan, Taiwan

Shilpa Taneja
NBGSM College Sohna
Haryana, India

Atul Thakur
Amity Institute of Nanotechnology
Amity University Haryana
Gurugram, India
and
School of Electronics and Information Engineering
Nanjing University of Information Science & Technology
Nanjing, China

Preeti Thakur
Amity Institute of Nanotechnology
Amity University Haryana
Gurugram, India

Prakash Chander Thapliyal
CSIR-Central Building Research Institute
Roorkee, Uttarakhand, India

Barsha Tripathy
Department of Vegetable Science
Institute of Agricultural Sciences
SOA University
Bhubaneswar, Odisha, India

Alex V. Trukhanov
National University of Science and Technology
MISIS, Moscow, Russia

Ritesh Verma
Department of Physics
Amity University Haryana
Gurugram, India

Fayu Wan
School of Electronics and Information Engineering
Nanjing University of Information Science & Technology
Nanjing, China

Hui-Min David Wang
Graduate Institute of Biomedical Engineering
National Chung Hsing University
Taiwan

Kavita Yadav
G.C.W. Gurawara, Rewari, Department of Higher Education Haryana
Rewari, India

Nisha Yadav
Department of Chemistry
Amity University Haryana
Manesar, Gurgaon, Haryana, India

1 Green Synthesis of Nanomaterials
An Eco-Friendly Approach

P. N. Kirdat
K.B.P. College Vashi

P. B. Dandge
Shivaji University

Deok-kee Kim
Sejong University

1.1 INTRODUCTION OF NANOSCIENCE AND NANOTECHNOLOGY

In the past decades, more research interest has developed in the field of nanoscience and nanotechnology because nanoparticles (NPs) act as carriers for small and large molecules. The actual meaning of "nano" in the Latin language is "dwarf." The nanosize means one billionth of a meter ($1\ nm = 10^{-9} m$). Nanotechnology is not an individual scientific discipline, but it also connects other conventional disciplines like physical, chemical, and biological sciences, bringing them together to generate novel technologies [1]. Nanotechnology comprises development at the atomic and molecular levels to generate components at the nano level. The idea of nanotechnology was first given by physicist Richard Feynman during his talk entitled "There's Plenty of Room at the Bottom" at the American Physical Society meeting (1959). A group of chemists in 1985 invented a football-like molecule called buckminsterfullerene containing 60 carbon atoms (also named Fullerene or Buckyball). Fullerene is any component exclusively composed of carbon to form a hollow sphere, ellipsoid, or tube-like structure. The sphere-shaped fullerene is called Buckyball, while the cylindrical molecule is defined as a carbon nanotube (CNT) [2]. In the fourth century AD, NPs and nanostructures were used by Roman peoples. The Lycurgus cup, in the British Museum collection, is one of the ancient synthetic nanomaterials. It is made up of dichroic glass, which exhibits a color-changing effect in certain lighting situations (green and red-purple colors). The combination of gold and silver NPs in late Medieval church windows showed a luminescent red and yellow color. Also, the sixteenth-century Italians produced Renaissance pottery using NPs. In early 2004, researchers discovered a new type of carbon nanomaterial having a size less than 10 nm, named carbon dots (C-dots). The innovation of graphene in 2004 underlined the importance of carbon-based materials in each sector of science and engineering. In 2006, Paul Rothemund constructed the "scaffolded DNA origami" using a one-pot process by increasing the complexity and dimension of self-assembled DNA nanostructures.

Thus, from the ancient to the modern period, nanotechnology became an interdisciplinary research area to discover innovative solutions for upcoming challenges in human life and the environment [3].

DOI: 10.1201/9781003502692-1

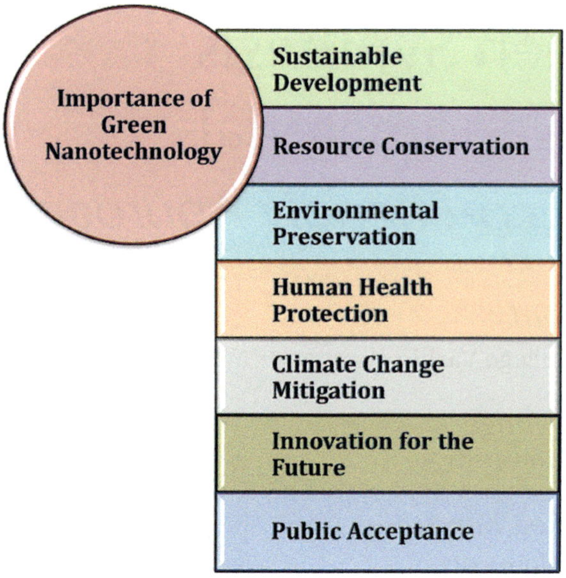

FIGURE 1.1 Importance of green nanotechnology.

1.2 GREEN NANOTECHNOLOGY

Nanotechnology provides a sustainable future due to the continual development of green chemistry and green nanotechnology. Green nanotechnology is the implementation of principles of green chemistry and engineering in the field of nanotechnology to generate nanoscale components in an eco-friendly manner to minimize human health and environmental threats; also it helps to improve environmental sustainability. Green NPs produced from the diverse green nanotechnological methodologies contain a well-defined chemical configuration, size, and applications in various technological fields [4].

Because of environmental sustainability and cost-effectiveness, the green nano-products provide benefits over the chemically manufactured nano-products as they supply direct or indirect solutions to environmental problems at the nano level. The green NPs synthesized via the green method lead to the synthesis of less harmful chemicals, minimize waste formation, and reduce the formation of other unsafe derivatives [5]. Figure 1.1 represents the importance of green nanotechnology. Green NPs synthesized through ecological-friendly procedures have significant importance in the fields of environmental microbiology, medical biology, industrial microbiology, bioremediation, clean technology, and electronics.

1.3 TRADITIONAL METHOD OF NANOPARTICLE SYNTHESIS

There are different physical and chemical traditional methods employed to synthesize NPs using various chemicals. These techniques comprise spray pyrolysis, electrospinning, physical vapor deposition, laser ablation, ion sputtering, solvothermal synthesis, electrochemical synthesis, chemical deposition, chemical vapor deposition, chemical reduction, and sol–gel method. In general, there are two perspectives for nanoparticle synthesis: top-down and bottom-up approaches. In the top-down method, nanosize components were prepared using greater, superficially organized microscopic ingredients, while the bottom-up process leads to the collection of molecular compounds that together form more complex structures. The micro-fabrication procedures were used to prepare nano-objects of specific shape and size in the top-down approach whereas the bottom-up approach uses self-assembled properties of single molecules to form nano-objects of certain specifications [6].

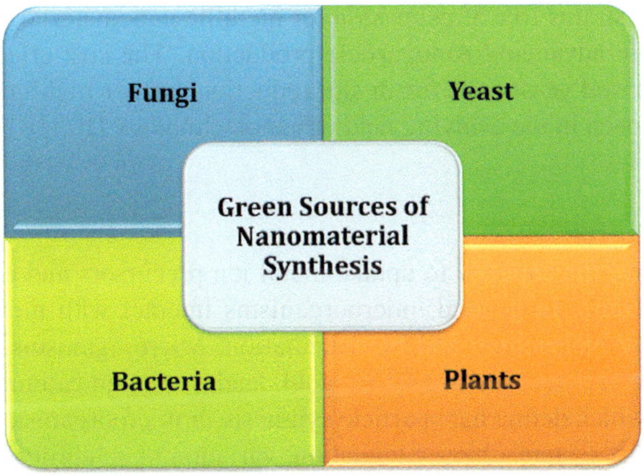

FIGURE 1.2 Green sources of nanomaterial synthesis.

1.4 GREEN SOURCES OF NANOMATERIAL SYNTHESIS

The physical and chemical procedures of nanoparticle synthesis include the utilization of high-energy radiation and poisonous reducing and stabilizing agents, which impact negatively on both living things and the environment (Figure 1.2). On the contrary, green synthesis includes an eco-friendly and cost-effective bio-synthetic method that requires minimum energy and less harmful reducing and stabilizing agents to carry out the reaction [7].

1.4.1 Fungi

Fungi prove to be highly effective in the biosynthesis of metal and metal oxide NPs, yielding monodispersed particles with well-defined morphologies [8]. The diverse array of intracellular enzymes in fungi positions them as superior biological agents for the eco-friendly synthesis of NPs. This is attributed to the presence of reducing constituents on their cell exteriors, enabling fungi to surpass bacterial sources in nanoparticle production [8]. The enzymatic reduction facilitated by reductase enzymes within the fungal cell wall plays a pivotal role in converting metals into nanoparticle form [8]. Various fungi have been employed in the synthesis of metal oxide NPs, such as silver, gold, titanium, and zinc oxide [7]. Fungal proteins and enzymes emerge as efficient catalysts in this process, contributing to the synthesis of NPs with specific capping components. These components play a crucial role in controlling the size and shape of the biosynthesized NPs [9].

1.4.2 Yeast

Yeasts, comprising single-celled eukaryotic microorganisms, boast a diverse array of approximately 1,500 recognized species [10]. Within the realm of nanomaterial synthesis, certain research groups have achieved success in utilizing yeast as a source.

Notably, silver-tolerant yeast strains and the common yeast *Saccharomyces cerevisiae* have been instrumental in the biological synthesis of silver and gold NPs [10]. In particular, the use of deceased *S. cerevisiae* cells has proven effective for the in situ reduction of gold NPs, offering advantages such as reduced production costs and simplified handling procedures [11]. The utilization of *S. cerevisiae* for gold nanoparticle synthesis involves strategic control over various parameters to achieve desired outcomes. By manipulating the concentration of the precursor, pH levels, temperature conditions, and exposure time, researchers can exert influence over the size of the synthesized NPs. This level of control not only enhances the precision of the synthesis process but also provides avenues for tailoring the NPs to specific applications [11]. The vast diversity of yeast species, coupled with the successful utilization of strains like *S. cerevisiae* for metallic nanomaterial synthesis, underscores the potential of yeasts in advancing nanoparticle production. The cost-effective and manageable nature of utilizing deceased yeast cells for in situ reduction further highlights the versatility and practicality of this approach in the evolving field of nanotechnology [10,11].

1.4.3 Bacteria

The microorganisms have the capacity to uptake metal ion precursors and can synthesize NPs by the detoxification procedure. In general, microorganisms interact with metals and can accumulate or extract metals by bioleaching and bioremediation. Microorganisms behave as prominent biological agents for the various NPs like silver, gold, lead, iron, and cadmium. There are several mechanisms proposed which define nanoparticle synthesis in microorganisms. These mechanisms include biosorption, efflux systems, bioaccumulation, variation of solubility and toxicity by oxidation or reduction, bioreduction of enzymes, extracellular precipitation of metals and their transport systems, as well as aldehyde- and ketone-mediated cell wall reduction [12,13]. Few investigations suggested that microbial enzymes, cell walls, and cell wall proteins may have the capability to reduce metal ions into their nano form [11].

1.4.4 Plants

The biosynthesis of metal and metal oxide NPs using different parts of the plant is one of the economically feasible methods and hence used for large-scale production. Several research groups studied the mechanism behind the formation of NPs. It was proposed that various biomolecules (vitamins, amino acids, proteins, phenolic acids, alkaloids, etc.) in plants lead to the bioreduction of NPs.

Phenolic acid is one of the potent antioxidant agents, containing hydroxyl and carboxyl groups which can attach to metals. The active hydrogen in phenolic acid may be responsible for the conversion of metal ions into metal NPs [14]. Different parts of plants like leaves, stems, roots, flowers, and fruits were efficaciously employed for the synthesis of NPs [15]. This synthetic method prevents the use of chemical stabilizers as biomolecules existing in the extract behave as a reducing and stabilizing agent. From ancient times, plants have acted as an efficient biochemical source because of the presence of phytochemicals as naturally occurring antioxidants, which are generally secondary metabolites (phenols, tannins, carbohydrates, saponins, flavonoids, amino acids, proteins, and polysaccharides). These biomolecules inhibit the agglomeration of particles by eco-friendly procedures and have no side consequences in synthesizing NPs. The antioxidants and phytochemicals existing in the plant extract decrease the metal ions in the salt solution to generate the first phase of nanoparticle formation more actively. In the second stage, nucleation results in the growth and stabilization of NPs. Next to nucleation, the various phytochemicals cause the capping and stabilization of generated NPs to form different shapes including cubes, spheres, hexagons, triangles, rods, wires, and pentagons. It was also studied that the quantity of the extract, pH, reaction time, and concentration of secondary metabolites on the size of NPs [16].

1.5 MECHANISM OF NANOMATERIAL SYNTHESIS: GREEN APPROACH

The principal mechanism behind nanoparticle formation is nucleation and growth because this method permits more flexibility and control of the nanoparticle structure and assembly, basically on nanoparticle diameter and randomly arranged molecules. The nucleation process is a critical procedure to understand NPs morphology as well as an important step to develop desired properties in NPs. Nucleation starts the generation of nanostructures (crystalline or amorphous) from its reaction stage (gas, plasma, liquid, solid). Nucleation is the first step in the development of a crystal structure from a solid, liquid, or vapor. In this process, small quantities of ions, atoms, or molecules are arranged in characteristic patterns, and additional particles are deposited in the form of crystals. There are some factors which influence the nucleation and growth of crystals. An increase in temperature decreases the critical supersaturation, resulting in an increase in the nucleation rate. Reaction volume may minimize the nucleation rate; therefore, the volume must be constant for a reaction system. The ionic strength influences ions in the reaction system, which can interfere with crystal interface and growth. The proportion of the ionic species with the crystal unit stoichiometry influences the ionic species adsorption on nuclei and its successive crystal growth.

The extremely supersaturated solutions cause shear-dependent nucleation after agitation or generate nucleation upon an increase in fluctuation. Surface active compounds (surfactants) exhibit a significant capacity to maintain crystal growth and control their shape and size. The aggregation of particles can be reduced or inhibited using an appropriate surfactant or stabilizer (cetyl trimethyl ammonium bromide, polyvinylpyrollidine, citrate, polymers, etc.) [17]. There are three techniques through which living organisms carry out the synthesis of NPs; these are intracellular (endogenous) synthesis, extracellular (exogenous) synthesis, and using biochemicals. The intracellular NP biosynthesis is dependent on the capacity of specific organisms to draw out metals from the growing medium and concentrate them. The synthesis happens inside the cell cytoplasm because it has reductive potential. Integral enzymes and other biomolecules are also involved in this process. Hence, the bioreduction procedure occurs intracellularly, and then the generated NPs are released into the exterior matrix. At the end of the process, the polysaccharides present in the extracellular matrix act as a capping agent to stabilize the formed NPs [18]. The intracellular NPs have some disadvantages, and therefore this method is not suitable at the industrial level. In this, the morphology of NP is difficult to control, a less efficient process that is difficult for isolation and purification [19]. To avoid these limitations, the scientists carried out a study on the development of an extracellular approach for NP synthesis. It was also designated as in vitro NP synthesis. This process is largely dependent on reducing and capping properties of interior biomolecules from plants and microorganisms.

Extracellular approaches include extracts from the source organism, which comprises the overall biocomposition of the organism. Figure 1.3 depicts the general process of NP synthesis from plant extracts. Instead of various biomolecules, a single biomolecule was also utilized for the synthesis of NPs.

The classical green synthesis includes a mixture of a biological extract with a metal salt solution, which provides the precursor for NPs. Secondary metabolites like phenolic acid, flavonoids, terpenoids, and alkaloids carry out the transfer of electrons to metallic ions, which results in a reduction of metal precursors [20–22]. Due to reduction, the reduced ions start to organize in a systematic arrangement to form a crystalline nature nucleus. The reduced ions get settled down on the nuclei surface, leading to the expansion of NPs. The biomolecules which behave as capping agents attach to the surface of particles and stabilize their size, which inhibits further synthesis of NPs [18].

1.5.1 Titanium Dioxide (TiO_2) NPs

Titanium dioxide (TiO_2) is one of the less toxic, inert, and less costly materials with more refractive index and UV absorption ability, which makes it an eco-friendly catalyst. TiO_2 nanoparticle is commonly utilized in sunscreen creams, paints, plastics, papers, inks, food colors, and toothpaste for whiteness and opaqueness because of its better physical strength and less toxicity. It is a well-recognized metal oxide which plays a vital role in various applications. Due to its distinctive antimicrobial potential and chemical stability, it has more value in the fields of chemistry and nanomedicines.

TiO_2 NPs are utilized in cosmetic creams and ointments to slow skin aging and cure sunburn. The efficiency of TiO_2 NPs is related to their greater surface area, which leads to a rise in surface energy, helping to increase their microbial destruction capacity. TiO_2 NPs are more effective against bacterial strains, which are resistant to presently available antibiotics. Living organisms like *Morinda citrifolia*, *Aspergillus flavus*, *Trigonella foenum-graecum*, *Aspergillus niger*, and *Bacillus subtilis* were employed for the green synthesis of TiO_2 NPs [23].

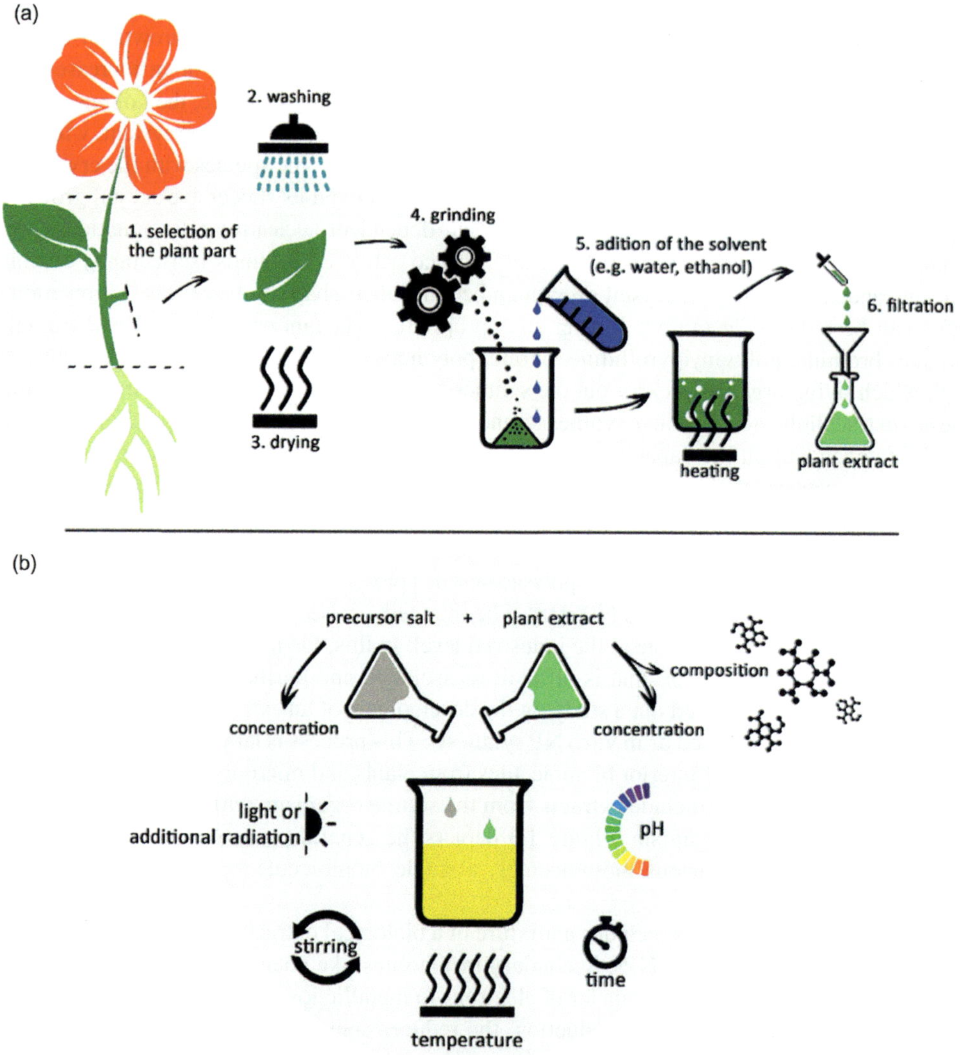

FIGURE 1.3 The typical approach to green synthesis of NPs: (a) preparation of the plant extract; (b) factors affecting the reaction of green synthesis and the characteristics of the resulting NPs [18]. (Open Access Image.)

1.5.2 Silver NPs

Silver NPs (Ag NPs) have been mostly synthesized and studied because of their exclusive physical, chemical, and biological characteristics. In the world, about 55 tons/year of silver NPs are synthesized. The synthetic procedure of AgNP includes the reduction of the silver ions, nucleation, and growth of the NPs. The extracts from various plants act as reducing agents, while silver nitrate ($Ag(NO)_3$) acts as a precursor for synthesis. Researchers described that key components such as proteins, carbohydrates, flavonoids, phenols, and vitamins were involved in the reduction of $Ag(NO)_3$.

Some researchers found that green-synthesized AgNPs can be useful in drug delivery as it has more compatibility with a few pharmaceutical drugs and acts as natural capping agents. Researchers studied that *Chelidonium majus L.* containing alkaloids and flavonoids behaves as a reducing and capping agent for AgNP synthesis. Researchers found that polysaccharide pullulan and the oxidized form of pullulan are beneficial to synthesizing and stabilizing Ag NPs. They suggested that the size of the NPs depends on the concentration of Ag $(NO)_3$ and pullulan [24,25]. The extracts from the root, shoot, fruit, flower, and leaves of various medicinal and non-medicinal plants were successfully utilized for the synthesis of silver NPs.

1.5.3 Iron Oxide NPs

Iron oxide is one of the broadly studied transition metal oxides due to its significant crystal structures, low cost, magnetic characteristics, different oxidation states, and environment-friendly nature. Iron oxide NPs are usually utilized as high-performing anodes in lithium-ion batteries and photo-electrochemical cells as well as to sense carbon monoxide gas.

It is used as a catalyst in the decay of phenols and oxidation of benzyl alcohols, in bioremediation, as a safe labeling agent in the biomedical field, and in targeted drug delivery systems.

Iron oxide NPs also have applications in electronic devices [26]. The synthesis of iron and iron oxide NPs is mainly influenced by the concentration of antioxidant compounds present in various plant extracts. The plants containing more phenolic content exhibit the highest reductive potential. Plants like *Camellia sinensis*, *Caricaya papaya*, *Lawsonia inermis*, and *Gardenia jasminoides* are used to synthesize iron oxide NPs [26].

1.5.4 Copper Oxide NPs (CuO)

The biological synthesis of CuO NPs having antimicrobial potential is the key interest in the biomedical field. These are well-defined particles with controlled size and shape. Some researchers studied the green procedure of extracellular synthesis of copper oxide NPs. Copper NPs exhibit specific physical, chemical, and biological characteristics and have a minimum cost of preparation. Copper NPs can simply oxidize into copper oxide.

To avoid CuO NP from oxidation, these are generally encapsulated with organic or inorganic materials like carbon and silica. Various biological entities like plants, bacteria, fungi, yeast, and actinomycetes were utilized for CuO NP synthesis. They carried out intracellular and extracellular synthesis of NPs [27]. Plants like *Cassia alata*, *Aloe vera*, *Malva sylvestris*, and *Phyllanthus amarus* are exploited for the synthesis of copper oxide NPs with efficient antimicrobial capacity [28].

1.5.5 Gold NPs (Au NP)

Au NPs have gained considerable interest because of their manageable size, shape, and surface characteristics. Due to their distinctive properties, AuNPs have probable applications in various fields including biosensors, genetic engineering, drug delivery systems, molecular biology, and therapeutics. Plants are employed as eco-friendly agents for Au NP synthesis via a green chemistry approach. It is observed that the pH of the reaction medium affects the shape of NPs such as rod, decahedral, icosahedral, and hexagonal. The leaf extracts of *Eucalyptus macrocarpa*, *Psidium guajava*, *A. vera*, and *Coleus aromaticus* can be utilized as sources for gold NP synthesis. From these leaves extract, spherical-size NPs of size around 20–80 nm are produced [29].

1.6 APPLICATORY STUDIES OF GREEN NPS

Recently, NPs have had increasing demand due to their tremendous applications in industries, electronics, environment, energy, and more specifically biomedical sectors. The NPs of gold (AuNP) and silver (AgNP) are usually employed for various applications in the biological field. Basically, plant-derived NPs do not cause any major side effects on living things and the environment. They exhibit a variety of applications in different regions including.

i. Nanomedicine and human health care (antimicrobial, anti-inflammatory, antiparasitic, antiproliferative, pro-apoptotic, antioxidative, etc.)
ii. Agriculture (controlled release of agrochemicals for accurate farming, targeted delivery of biomolecules, effective absorption of nutrients, identification, and diagnosis of plant diseases)
iii. Food science and technology (processing, storage, and packaging) as well as bioengineering (biocatalysts, photocatalysts, biosensors, etc.)
iv. Cosmetics (anti-aging, sunscreen, hair development, bioactive compounds delivery, nano-emulsion, etc.) [30].

1.6.1 Anti-Inflammatory Potential of NPs

In current years, NPs have been modified as anti-inflammatory agents. The surface-to-volume ratio of NPs is larger so it can be utilized as a barricading component associated with inflammation like cytokines and inflammation-assisting enzymes related to other complements. Several metal oxide NPs such as silver, gold, copper, and iron oxide have been studied for their tremendous anti-inflammatory capacities. Swelling or inflammation is the body's immediate reaction to internal damage, infection, hormone imbalance, and failure in the inner structures or exterior features such as an attack by an external component [31].

Anti-inflammation is one of the vital mechanisms in wound healing. It is a cascade reaction that generates immune system-activating substances such as interleukins and cytokines, which can be generated by keratinocytes including T lymphocytes, B lymphocytes, and macrophages. Biosynthesized gold NPs show positive wound-healing mechanisms and tissue regeneration in inflammatory activities. The studies evidenced that biosynthesized gold and platinum NPs provide a substitute source for the treatment of inflammation in a natural way [32].

1.6.2 Treatment of Wastewater

The increasing growth rate of population, industrial development, and unnecessary usage of chemicals result in the contamination of the water environment due to the release of wastewater into the environment.

The natural water reservoirs are not appropriate for consumption because of the presence of different organic (dyes, pesticides, surfactants, etc.), inorganic (fluoride, arsenic, copper, mercury, etc.), biological (algae, bacteria, viruses, etc.), and radiological contaminants (cesium, plutonium, uranium, etc.). Numerous techniques, physical, chemical, and biological, were employed for wastewater treatment. However, recent research has been carried out to develop cost-effective innovative technologies to improve water purification.

At present, nanotechnology provides a novel approach for the elimination of contaminants from wastewater with more efficacy. In the adsorption process, gaseous or liquid particles are attached to the surface of a solid, generating a layer or films of molecules. This method is useful only for the external layer of adsorbents where the adsorbate gathers. The adsorption procedure may be physical or chemical depending on van der forces, covalent interaction, or electrostatic attraction between the absorbent and the adsorbate. Adsorption is the most frequently employed technique for the exclusion of contaminants from water due to its minimum cost, simple procedure, and lack of the development of secondary pollutants. The recent improvement in nanotechnology generates numerous nanostructured components, which are utilized as adsorbents for their probable application in the treatment of industrial water discharges, surface water, drinking water, and groundwater. Nano-adsorbents possess high porosity, more active surface area, and small size; hence, they have greater efficiency and quicker adsorption degree compared to traditional adsorbents. Furthermore, these nano-adsorbents display more reactivity and catalytic proficiency. Some nanomaterials like CNTs, ferric oxide (Fe_3O_4), graphene, titanium oxide (TiO_2), manganese oxide (MnO_2), zinc oxide

(ZnO), and magnesium oxide (MgO) are effectively utilized as absorbents for the exclusion of contaminants like heavy metals, azo dyes, and other pollutants from water. Additionally, various metal oxide NPs act as superparamagnetic, and, hence, they exert an external magnetic field on the reaction mixture to carry out the easy separation of contaminants. Researchers synthesized green magnetite NPs from crude latex of *Jatropha curcas* and leaf extract of *Cinnamomum tamala*, and it was found that the NPs helped to remove methylene blue dye, Cu (II), and Co (II) from aqueous solution. The silver NPs synthesized from the leaf extract of the *Ficus* tree (*Ficus benjamina*) and iron oxide NPs prepared from tangerine peel extract using the co-precipitation method were used to remove cadmium (II) from the contaminated solution. Zinc oxide NPs synthesized from *A. vera* and *Cassava* starch exhibit a high capacity for copper ion removal. These nano-adsorbents proved significant effectiveness in the elimination of pollutants from wastewater. The major limitations of this process are the toxicity of remaining NPs in the wastewater and decreased potential activity due to the use of a large number of NPs in the treatment procedure to reduce the time duration. Filtration is one of the techniques employed to remove contaminants from polluted water or wastewater. For this purpose, nanofiltration is a more effective and efficient method for the exclusion of various types of contaminants (organic, heavy metals, pathogens, etc.) from wastewater, and its removal capacity is mostly reliant on the pore dimension and charge characteristics of the filter membrane. Several studies were carried out on the improvement and usage of a complex membrane generated from the polymeric or inorganic membrane by the addition of NPs into it for wastewater treatment. The combination of metal oxide NPs such as silica, alumina, zeolite, and TiO_2 into polymeric membranes increases the membrane hydrophobicity and permeability. Also, the integration of membrane matrix with antimicrobial NPs like silver helps to inhibit bacterial adhesion and biofilm formation.

Metal and metal oxide NPs are broadly utilized as nano-catalysts in water treatment because of their more surface-to-volume ratio and catalytic properties of the surface; thus, they degrade different contaminants including dyes, pesticides, herbicides, nitro aromatics, and polychlorinated biphenyls and improve the quality of water. These nano-catalysts are electrocatalysts, photocatalysts, and Fenton-based catalysts. The photocatalysis of contaminants occurred due to the photoexcitation of electrons existent in the catalysts. The holes (h^+) and electrons (e^-) generated by the light irradiation process and formed holes (h^+) are trapped by water molecules in an aqueous solution, which later forms the hydroxyl radicals (OH^-). These hydroxyl ions are very reactive and strong oxidizing agents, which cause the oxidation of organic impurities and lead to the formation of degrading end products like water and gaseous compounds. Several studies stated that green-synthesized metal and metal oxides including Ag, Au, Pt, Pd, ZnO, CuO, FeO, NiO, TiO_2, SnO_2, CeO_2, etc. revealed efficient photocatalytic ability for the degradation of various organic dyes and pollutants [33–38].

1.6.3 Food Industry and Cosmetics

Some cosmetic products including makeup accessories, moisturizers, hair care products, and sunscreen are based on the use of NPs and nanomaterials. NPs like TiO_2 and ZnO are mainly used in cosmetics as UV filters. Also, NPs are used as delivery vehicles in the cosmetic industry. Nanomaterials containing solid lipid NPs and lipid nanostructured carriers behave as more prominent delivery vehicles than liposomes. Nanospheres or nanoemulsions have better penetration capacity so they easily carry out skin penetration. In hair treatment products, nanoemulsions support the easy transport of desired medicine to deeper hair shafts. The utilization of zinc and titanium NPs in sunscreen lotions leads to transparent, less-greasy-texture, minimum-odor lotion, which is more absorbed by the skin [39]. In the food sector, the use of this transforming technology, that is nanotechnology, has increased enormously. Various NPs are incorporated into a variety of food products to increase their specific properties and also develop new properties. All over the world, nanotechnology has become an interesting research area for the development and manufacturing of agricultural products, processed foods, and food packaging ingredients on a large scale.

The increase in the demands of consumers regarding quality and hygiene food products results in the generation of nanomaterial-containing food products without disturbing their original nutritional value. These nanomaterials comprise important elements and are non-hazardous and steady at high pressure and temperature [40]. Nanotechnology can provide a variety of advantages at different steps of food products ranging from manufacturing to processing and packaging and possesses the capacity to improve food quality and safety and provides more health benefits. Few researchers and industrial organizations are engaged in the development of new procedures and products, which can be directly used in the food sector. It includes two main groups: nanostructured food ingredients and the nano sensing of food.

Nanostructured food constituents are widely employed in food processing as food additives, antimicrobial agents, carriers for the smooth delivery of nutrients, anti-caking mediators, and fillers that help to improve the mechanical strength and durability of the packaged material, etc. The nano sensing of food is used to attain better quality and safety of food [40]. Numerous studies have proved that nanomaterials are potential agents that help to improve food safety by raising the effectiveness of packaging without affecting their nutritional value. They also increase the efficacy of additives without altering food taste and physical characteristics.

Even though nanotechnology supplies effective benefits to the food sector, it also faces some challenges like the cost-effectiveness of the process, synthesis of edible NPs, formation of efficient nano formulations which are safe for human consumption, and generation of a less toxic delivery system [41,42]. Green synthesis methodology may cover these disadvantages, but it has some controversies in a few cases such as their scientific uncertainty, their negative impact on human health and environment, toxicity, and accumulation of particles in nature. Hence, to overcome these problems, there is a need to carry out large-scale research [43].

1.6.4 Nanomaterials in Therapeutics and Drug Delivery System

NPs are a critical factor useful for innovations in the biomedical and therapeutic field.

Cancer is one of the most harmful and deadly diseases affecting the world population in the twenty-first century. Therefore, there is a need to discover an anticancer medicine for better treatment. Green-synthesized metal oxide NPs have a vital role in photothermal therapy because they are illuminated by the light and can assist the controlled release of the drug at its target site. The NPs like Zn and Ce oxides act as prominent anticancer medicines [31]. The cerium core present in CeO (cerium oxide) NPs is bounded by an oxygen lattice, which possesses better potential as a therapeutic agent [44]. Silver (Ag) NPs synthesized from pulp extract of *Abelmoschus esculentus* (L.) exhibit good therapeutic capacity and the ability to kill Jurkat cells in vitro.

The anticancer capability of AgNP was specifically related to the higher levels of reactive oxygen species (ROS) and damage to the mitochondrial membrane [45,46]. The Punica granatum-mediated silver NPs were found to be more effective against the liver cancer cell line (HepG2). Likewise, AgNPs generated from *Olax scandens* leaf extract exhibited anticancer activities against various cancer cells (B16: mouse melanoma cell line; A549: human lung cancer cell lines; and MCF7: human breast cancer cells) [31]. In current reports, it was found that iron oxide NPs possess double capability; that is they behave as both magnetic and photothermal agents in cancer treatment. This double potential causes total cell death through apoptosis. Also, the combination of iron oxide NPs with laser therapy results in a complete inhibition of tumor cells in vivo [47]. Studies indicated that photothermal treatment using green-synthesized iron oxide NPs, combined with the drug temozolomide under the influence of near-infrared light, kills glioblastoma cancer cells [47]. NPs have been generated for the formulations of eye drops or injectable solutions. Drugs packed with NPs obtain better drug pharmacokinetics, pharmacodynamics, immune-activating capacity, non-specific toxicity, and bio-identification and therefore increase the efficiency of the drugs [48]. Polymeric NPs prepared from chitosan provide a better drug delivery carrier and supply a new way for the development of delivery systems. These NPs can easily pass through biological obstacles and preserve macromolecules like oligonucleotides, peptides, and genes from the degeneration of biological media, which results in the distribution of drugs or macromolecules to the target position followed by their controlled release [49]. NPs offer a favorable approach for the precise delivery of medicines against the human immunodeficiency virus (HIV) named lamivudine, which functions as a strong inhibitor of type 1 and type 2 HIV [50]. Superparamagnetic iron oxide NPs (SPIONs) along with the drug have been employed for targeted drug delivery. The drug can quickly attach to the SPION exterior, and due to the external magnetic field, it reaches its preferred site where NPs can enter the target cell and dispatch the drug [51]. The *Hybanthus enneaspermus* leaf-extract mediated AgNPs were found to be efficient reducing agents, having substantial capacity to activate drug delivery, antibacterial coverings, and wound dressings [31].

1.6.5 NPs as Antimicrobial Agents

Generally, NPs demonstrate high bactericidal capacities against various pathogenic microorganisms (Figure 1.4). Based on particle size, capping process, and NP concentration metal NPs can be classified into bactericidal and bacteriostatic. Through various analyses, it was manifested that NPs exhibit antibacterial action on both gram-positive and gram-negative strains. The gram-negative bacterial cell wall is composed of a thin peptidoglycan polymer layer (~7 to 8 nm) with a negative charge on it. This property is directly related to the antibacterial capacity of NPs. The thin cell wall causes simple penetration of NPs into the bacterial cell wall, while electrostatic interaction between cells is produced by a negatively charged surface. This process leads to the generation of ROS and oxidative stress, which results in the inhibition and destruction of bacterial cells. The gram-positive bacteria contain less negative charge on their surface, which enriches the penetration of NPs, permitting the entry of negatively charged superoxide radical anions and peroxide ions to carry out cell inhibition at a comparatively minimum concentration. The metal and metal oxide NPs are recognized to be an expectant fighting weapon against a wide array of pathogenic strains including multidrug-resistant strains.

The fungicidal and fungistatic potential of NPs is being analyzed to restrict outbreaks produced by pathogenic fungi. The fungicidal ability occurred due to the size and shape of NPs. The small-sized nano discs exhibit the inhibition of ATPase, which forms intracellular acidification and leads to cell destruction. In some instances, NPs exhibit fungicidal action causing the deformation of hyphae and inhibiting it [52].

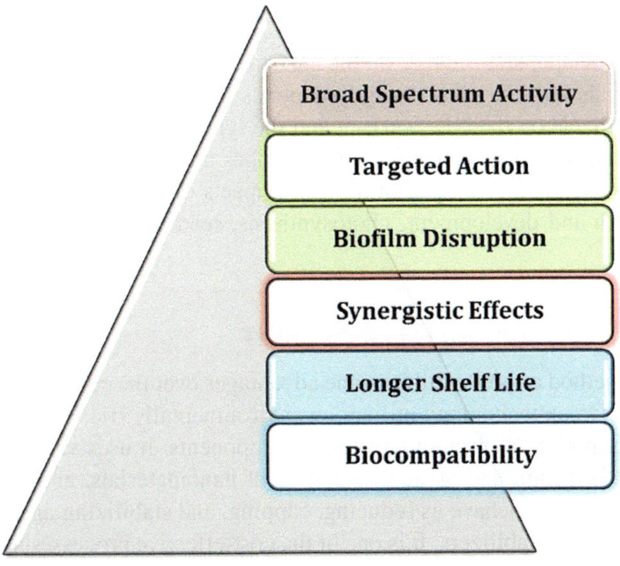

FIGURE 1.4 Nanoparticles as antimicrobial agents.

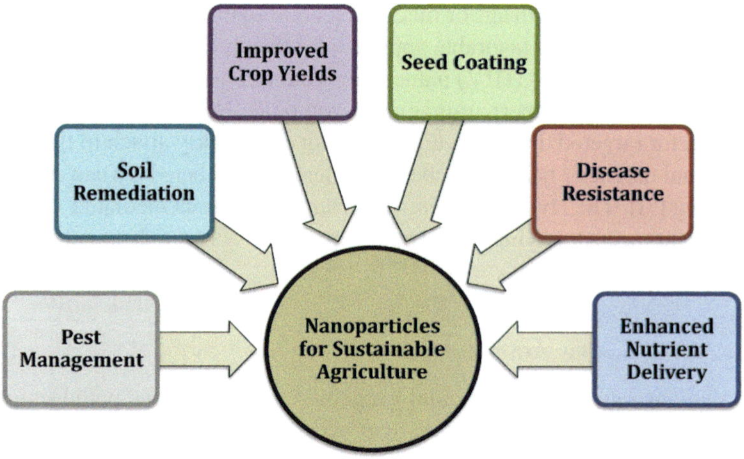

FIGURE 1.5 Nanoparticles for sustainable agriculture.

1.6.6 NPs for Sustainable Agriculture

Agriculture is the primary source of economy for developing countries and supplies nutritional food for life (Figure 1.5). In the present situation, the agriculture sector is facing a wide variety of challenges such as pollution of soil and water, decreased fertility of the soil, unpredictable climate variation, and use of harmful fertilizers and pesticides, and the major problem is increasing food demand of the growing population. New technologies make innovative evolution for the development of sustainable agriculture, and they provide maximum yield from agriculture. Many types of nanomaterials [53] like nanoclays, nanotubes, and nanowires exhibit distinctive surface chemistry and electrical and optical characteristics that offer them superior sensitivity, enhanced detection limits, and quick response times. The main objective of nanomaterials in agriculture is to minimize nutrient losses, improve yield, decrease production costs from agriculture, and increase plant protection. With the help of innovations in nanotechnology, more effective nanoformulations are developed for sustainable agriculture. These newly synthesized materials have the capability to change the composition of plants immediately once entering the complex plant–soil system. Numerous factors influence the uptake and transport of NPs, specifically plant structure and physicochemical properties of NPs like shape, size, charge on the surface, and possible interaction with plants. The uptake of nanomaterials into the plants mainly depends on the size, chemical configuration, and functional groups available on the NPs' surface and the type of covering. The communication and uptake of nanomaterials result in molecular-level changes and influence the complete physiology of plants. The utilization of NPs in agriculture also impacts other important parameters of plants including plant growth and development, photosynthesis, seed viability, as well as the growth of shoots and roots.

1.7 PROS AND CONS OF GREEN APPROACH

The green synthesis method of NPs exhibits some advantages over the commonly used conventional synthetic procedures. Green technology utilizes an environmentally friendly approach and protects the environment from pollution through hazardous components. It uses various biological entities such as plants, bacteria, fungi, and algae as a source of nanomaterials, and the different biomolecules present in these agents behave as reducing, capping, and stabilizing agents. So, ultimately it reduces the use of chemical stabilizers. It is one of the cost-effective processes as it does not require more sophisticated instruments and costly chemicals for synthesis. The NPs synthesized by green strategy are safe and non-toxic for applications in biomedical fields. Also, the process does not release harmful end products into the environment. Besides all these advantages, the green process has some disadvantages, and more research must be carried out to minimize these problems. Green synthesis is a time-consuming process and may produce less nanoparticle yield. Some dangerous atomic weapons can be produced using nanotechnology. As NPs are small enough, they get easily inhaled into the body and can impair heart and vascular functions. The nanosized particles can be mixed in air and water, so there is a possibility of increased environmental pollution and there is a necessity to carry out more research on it.

Hence, there is a need to synthesize more NPs by minimizing its issues for improvements in various applications in the biomedical, environment, agriculture, and electronic industries.

1.8 FUTURE PERSPECTIVES

Nanometer-sized herbal medicines have a bright future in the development of nanomedicines to overcome the issues related to health and the environment and help to manage problems due to chemical components.

This results in the usage of green pharmaceuticals in the delivery of nanodrugs to improve the utilization of herbal drugs and support in treating several disorders. Plant NPs can be employed in a wide range to treat oral diseases, for cancer inhibition and treatment, to decrease oxidative stress, as anti-inflammatory agents, in wastewater treatment, for the bioremediation of harmful dyes, and in various electronic devices as energy storage systems and memory devices [54,55]. The future of innovative technology is connected to the developments in the nanotechnology sector. Clean energy production and the minimization of pollution are related to improvements in nanomaterial-dependent engineering approaches. This new technology provides efficient products for energy storage devices.

1.9 CONCLUSION

In recent times, the utilization of various biological agents has proven to be highly effective in the successful green synthesis of diverse metal and metal oxide NPs. This innovative approach not only minimizes the potential side effects associated with traditional physical and chemical procedures but also significantly reduces the reliance on harmful chemical compounds. By embracing this eco-friendly strategy, there emerges a tangible benefit to both living organisms and the environment at large. Furthermore, the continuous and dedicated research efforts in the field have led to significant advancements in the synthesis and application studies of these green NPs. The commitment to exploring novel methods and materials underscores the potential for sustainable practices in nanotechnology. As a result, the green synthesis of NPs offers a promising avenue for minimizing environmental impact and promoting the overall well-being of ecosystems.

In essence, the ongoing pursuit of knowledge and innovation in the realm of green nanoparticle synthesis provides a fresh perspective for the current generation. This eco-conscious approach not only addresses the limitations of conventional methods but also opens new possibilities for sustainable development and responsible utilization of nanomaterials. As the scientific community continues to explore these avenues, the potential benefits to both human society and the broader environment become increasingly apparent.

REFERENCES

1. M. Rangasamy (2011). Nano technology: a review. *Journal of Applied Pharmaceutical Science*, 1, 08–16.
2. S. K. Subedi (2013). An introduction to nanotechnology and its implications. *The Himalayan Physics*, 4, 4.

3. P. Thakur, A. Thakur (2022). Introduction to nanotechnology. In: Thakur, A., Thakur, P., Khurana, S. P. (eds) *Synthesis and Applications of Nanoparticles*. Springer, Singapore. https://doi.org/10.1007/978-981-16-6819-7_1.
4. S. Suresh, S. Chandran, G. R. Prashob, L. M. Rohini, S. Sasi, U.S. Sudhi (2021). Green nanotechnology: a review. *Indo American Journal of Pharmaceutical Sciences*, 8, 263–270.
5. G. Aarti, B. Sneha (2014). Green nanotechnology. *Bioevolution*, 5(6), 3–4.
6. D. Nath, P. Banerjee (2013). Green nanotechnology – a new hope for medical biology. *Environmental Toxicology and Pharmacology*, 36, 997–1014.
7. J. Singh, T. Dutta, K. H. Kim, M. Rawa, P. Samddar, P. Kumar (2018). Green synthesis of metals and their oxide nanoparticles: applications for environmental remediation. *Journal of Nano Biotechnology*, 16(1), 84.
8. A. Thakur, P. Thakur, S. Baccar (2023). Structural properties of nanoparticles. In: Suhag, D., Thakur, A., Thakur, P. (eds) *Integrated Nanomaterials and their Applications*. Springer, Singapore. https://doi.org/10.1007/978-981-99-6105-4_4
9. D. S. Balaji, S. Basavaraja, R. Deshpande, D. B. Mahesh, B. K. Prabhakar, A. Venkataraman (2009). Extracellular biosynthesis of functionalized silver nanoparticles by strains of cladosporium cladosporioides fungus. *Colloids and Surfaces B: Biointerfaces*, 68(1), 88–92.
10. R. Rathore, D. Suhag, F. Wan, A. Thakur, P. Thakur (2023). Everyday nanotechnology. In: Suhag, D., Thakur, A., Thakur, P. (eds) *Integrated Nanomaterials and their Applications*. Springer, Singapore. https://doi.org/10.1007/978-981-99-6105-4_2
11. T. Thunugunta, A. C. Reddy, D. C. L. Reddy (2015). Green synthesis of nanoparticles: current prospectus. *Nanotechnology Reviews*, 4(4), 303–323.
12. M. Gericke, A. Pinches (2006). Biological synthesis of metal nanoparticles. *Hydrometallurgy*, 83(1), 132–140.
13. M. I. Husseiny, M. A. EI. Aziz, Y. Badr, M. A Mahmoud (2007). Biosynthesis of gold nanoparticles using Pseudomonas aeruginosa. *Spectrochim Acta Part A*, 6(3–4), 1003–1006.
14. S. Kandasamy, R. Sorna Prema (2015). Methods of synthesis of nano particles and its applications. *Journal of Chemical and Pharmaceutical Research*, 7(3), 278–285.
15. L. Krishnia, P. Thakur, A. Thakur (2022). Synthesis of nanoparticles by physical route. In: Thakur, A., Thakur, P., Khurana, S. P. (eds) *Synthesis and Applications of Nanoparticles*. Springer, Singapore. https://doi.org/10.1007/978-981-16-6819-7_3
16. D. Gnanasangeetha, M. Suresh (2020). A review on green synthesis of metal and metal oxide nanoparticles. *Nature Environment and Pollution Technology*, 19(5), 1789–1800.
17. V. Harish, Md. M. Ansari, D. Tewari, M. Gaur, A. Bihari Yadav, M. L. G. Betancourt, F. M. A. Haleem, M. Bechelany, A. Barhoum (2022). Nanoparticle and nanostructure synthesis and controlled growth methods. *Nanomaterials*, 12(18), 3226.
18. B. A. Miu, A. Dinischiotu (2022). New green approaches in nanoparticles synthesis: an overview. *Molecules*, 27(19), 6472.
19. N. Pantidos (2014). Biological synthesis of metallic nanoparticles by bacteria, fungi and plants. *Journal of Nanomedicine and Nanotechnology*, 6(2), 257–262.
20. F. Tasca, R. Antiochia (2020). Biocide activity of green quercetin-mediated synthesized silver nanoparticles. *Nanomaterials*, 10(5), 909.
21. V. C. Thipe, P. Amiri, P. Bloebaum, R. Karikachery, M. Khoobchandani, K. K. Katti, S. S. Jurisson, K. V. Katti (2019). Development of resveratrol-conjugated gold nanoparticles: interrelationship of increased resveratrol corona on anti-tumor efficacy against breast, pancreatic and prostate cancers. *International Journal of Nanomedicine*, 14, 4413–4428.
22. S. Jain, M. S. Mehata (2017). Medicinal plant leaf extract and pure flavonoid mediated green synthesis of silver nanoparticles and their enhanced antibacterial property. *Scientific Reports*, 7, 15867.
23. A. O. Adesina, O. R. Adeoyo, O. M. Bankole, A. G. Olaremu, C. A. Osunla (2022). Green approach to the synthesis of metal oxide nanoparticles used as alternative remedy to multidrug resistance. *Saudi Journal of Engineering and Technology*, 7(6), 270–277.
24. V.C. Karade, S.B. Parit, V.V. Dawkar, R.S. Devan, R.J. Choudhary, V.V. Kedge (2019). A green approach for the synthesis of α-Fe2O3 nanoparticles from Gardenia resinifera plant and it's In vitro hyperthermia application, *Heliyon*, 5(7).
25. L. G. Torresdey (2016). Plant-based green synthesis of metallic nanoparticles: scientific curiosity or a realistic alternative to chemical synthesis? *Nanotechnology for Environmental Engineering*, 1, 1–29.
26. P. N. Kirdat, P. B. Dandge, R. M. Hagwane, A. S. Nikam, S. P. Mahadik, S. T. Jirange (2020). Synthesis and characterization of ginger (Z. officinale) extract mediated iron oxide nanoparticles and its antibacterial activity. *Materials Today: Proceedings*, 43(4), 2826–2831.

27. S. Drummer, T. Madzimbamuto, M. Chowdhury (2021). Green synthesis of transition metals nanoparticle and their oxides: a review. *Materials*, 14(11), 2700.
28. V. Mishra, R. Sharma, N. D. Jasuja, D. K. Gupta (2014). A review on green synthesis of nanoparticles and evaluation of antimicrobial activity. *International Journal of Green and Herbal Chemistry*, 333(82), 81–94.
29. I. Hussain, N. B. Singh, A. Singh, H. Singh, S. C. Singh (2016). Green synthesis of nanoparticles and its potential application. *Biotechnology Letters*, 38, 545–560.
30. A. Verma, S. P. Gautam, K. K. Bansal, N. Prabhakar, J. M. Rosenholm (2019). Green nanotechnology: advancement in phytoformulation research. *Medicines*, 6(1), 39.
31. C. Hano, B. H. Abbasi (2021). Plant based green synthesis of nanoparticles: production, characterization and applications. *Biomolecules*, 12(1), 31.
32. M. S. Samuel, M. Ravikumar, A. John, J. E. Selvarajan, H. Patel, P. S. Chander, J. Soundarya, S. Vuppala, R. Balaji, N. Chandrasekar (2022). A review on green synthesis of nanoparticles and their diverse biomedical and environmental applications. *Catalyst*, 12(5), 459.
33. P. Kuppusamy, M. M. Yusoff, G. P. Maniam, N. Govindan (2016). Biosynthesis of metallic nanoparticles using plant derivatives and their new avenues in pharmacological applications–an updated report. *Saudi Pharmaceutical Journal*, 24(4), 473–484.
34. P. K. Dikshit, J. Kumar, A. K. Das, S. Sadhu, S. Sharma, S. Singh, P. K. Gupta, B. S. Kim (2021). Green synthesis of metallic nanoparticles: applications and limitations. *Catalysts*, 11(8), 902.
35. S. Hemmati, L. Mehrazin, H. Ghorban, S. H. Garakani, T. H. Mobaraki, P. Mohammadi, H. Veisi (2018). Green synthesis of Pd nanoparticles supported on reduced graphene oxide, using the extract of Rosa canina fruit, and their use as recyclable and heterogeneous nanocatalysts for the degradation of dye pollutants in water. *RSC Advances*, 37, 21020–21028.
36. B. Kumar, K. Smita, L. Cumbal (2015). Phytosynthesis of gold nanoparticles using Capsicum baccatum L. *Cogent Chemistry*, 1(1), 1120982.
37. B. Kumar, K. Smita, L. Cumbal, A. Debut (2016). One pot synthesis and characterization of gold nanocatalyst using Sacha inchi (Plukenetia volubilis) oil: green approach. *Journal of Photochemistry and Photobiology B: Biology*, 158, 55–60.
38. M. B. Sumi, A. Devadiga, K. V. Shetty, M. B. Saidutta (2017). Solar photocatalytically active, engineered silver nanoparticle synthesis using aqueous extract of mesocarp of Cocos nucifera (Red Spicata Dwarf). *Journal of Experimental Nanoscience*, 12(1), 14–32.
39. K. Tahir, S. Nazir, B. Li, A. U. Khan, Z. U. H. Khan, A. Ahmad, F. U. Khan (2015). An efficient photo catalytic activity of green synthesized silver nanoparticles using Salvadora persica stem extract. *Separation and Purification Technology*, 150, 316–324.
40. A. Hameed, G. R. Fatima, K. Malik, A. Muqadas, M. Fazalur-Rehman (2019). Scope of nanotechnology in cosmetics: dermatology and skin care products. *Journal of Medicinal Chemistry*, 2(1), 9–16.
41. F. Khajehei, C. Piatti, S. Graeff-Hönninger (2019). Novel food technologies and their acceptance. In: Piatti, C., Graeff-Hönninger, S., Khajehei, F. (eds) *Food Tech Transitions*. Springer, Berlin, 3–22.
42. A. B. Perumal, R. B. Nambiar, P. S. Sellamuthu, E. R. Sadiku (eds) (2019). Application of biosynthesized nanoparticles in food, food packaging and dairy industries. In: *Biological Synthesis of Nanoparticles and Their Applications*, CRC Press, Boca Raton, FL, USA, 145–158.
43. H. Ma, Z. Chen, X. Gao, W. Liu, H. Zhu (2019). 3D hierarchically gold nanoparticle- decorated porous carbon for high-performance supercapacitors. *Scientific Reports*, 9, 17065.
44. W. Du, Z. Zhu, Y. Wang, J. Liu, W. Yang, X. Qian, H. Pang (2014). One-step synthesis of CoNi2S4 nanoparticles for supercapacitor electrodes. *RSC Advances*, 4, 6998–7002.
45. Y. Gao, F. Gao, K. Chen, J.L. Ma (2014). Cerium oxide nanoparticles in cancer. *OncoTargets Therapy*, 7, 835–840.
46. M. M. R. Mollick, D. Rana, S. K. Dash, S. Chattopadhyay, B. Bhowmick, D. Maity, D. Mondal, S. Pattanayak, S. Roy, M. Chakraborty et al. (2019). Studies on green synthesized silver nanoparticles using Abelmoschus esculentus (L.) pulp extract having anticancer (in vitro) and antimicrobial applications. *Arabian Journal of Chemistry*, 12(8), 2572–2584.
47. A. Espinosa, R. Di Corato, J. Kolosnjaj-Tabi, P. Flaud, T. Pellegrino, C. Wilhelm (2016). Duality of iron oxide nanoparticles in cancer therapy: amplification of heating efficiency by magnetic hyperthermia and photothermal bimodal treatment. *ACS Nano*, 10(2), 2436–2446.
48. Y. M. Kwon, J. Je, S. H. Cha, Y. Oh, W. H. Cho (2019). Synergistic combination of chemo-phototherapy based on temozolomide/ICGloaded iron oxide nanoparticles for brain cancer treatment. *Oncology Reports*, 42(5), 1709–1724.
49. D. R. Janagam, L. Wu, T. L. Lowe (2017). Nanoparticles for drug delivery to the anterior segment of the eye. *Advanced Drug Delivery Review*, 122(1), 31–64.

50. M. Rajan, V. Raj (2013). Potential drug delivery applications of chitosan based nanomaterials. *International Review of Chemical Engineering*, 5(2), 145–155.
51. A. Dev, N. S. Binulal, A. Anitha, S. V. Nair, T. Furuike, H. Tamura, R. Jayakumar (2010). Preparation of poly (lactic acid)/chitosan nanoparticles for anti-HIV drug delivery applications. *Carbohydrate Polymer*, 80(3), 833–838.
52. K. S. Siddiqi, A. U. Rahman, A. H. Tajuddin (2016). Biogenic fabrication of iron/iron oxide nanoparticles and their application. *Nanoscale Research Letters*, 11(1), 498.
53. P. Thakur, A. Thakur (2022). Nanomaterials, their types and properties. In: Thakur, A., Thakur, P., Khurana, S. P. (eds) *Synthesis and Applications of Nanoparticles*. Springer, Singapore. https://doi.org/10.1007/978-981-16-6819-7_2
54. D. Mittal, G. Kaur, P. Singh, K. Yadav, S. A. Ali (2020). Nanoparticle based sustainable agriculture and food science: recent advances and future outlook. *Frontiers in Nanotechnology*, 2, 579954.
55. M. Yazdanian, P. Rostamzadeh, M. Rahbar, M. Alam, K. Abbasi, E. Tahmasebi, H. Tebyaniyan, R. Ranjbar, A. Seifalian, A. Yazdanian (2022). The potential application of green-synthesized metal nanoparticles in dentistry: a comprehensive review. *Hindawi, Bioinorganic Chemistry and Application*, 2022(2), 1–27.

2 Synthesis, Types, and Characterization of Nanomaterials

M. Love Edet and S. Hemalatha
B.S. Abdul Rahman Crescent Institute of Science and Technology

2.1 INTRODUCTION

In its etymological roots, the English term "nano" can be traced back to the Greek word "nanos," which means "one billionth"; in simple terms, 1 nm is equal to 1 billion meters. Nanotechnology, which makes use of extremely small dimensions on the nanoscale, has ushered in a new age in the development and implementation of materials across many fields. About 800 nano-based products are available now, and more are constantly being created [1]. Nanomaterials have recently garnered a lot of attention from the scientific community due to their unique optical, physical, electrical, and chemical properties. Nanoparticles (NPs) made of various materials, such as carbon, ceramic, polymer, lipid, magnetic, and metallic, are used in many applications. Metal NPs (including copper, gold, silver, and mercury) are among the most adaptable. NPs can be synthesized through physical, biological, chemical, or hybrid methods. Out of these, physical and chemical methods have a low output of nanomaterials and pose a serious threat to human health and the environment. Because of these issues, a greener, biocompatible, and cheaper alternative based on the use of natural chemicals and microorganisms has been developed. Nanotechnology allows NPs to be manufactured using biological components including yeast, bacteria, fungi, and plants. These synthetic NPs have several potential uses in diverse fields especially in biomedical and agricultural sectors [2].

2.2 BRIEF HISTORY AND CONCEPT

National Nanotechnology Initiative (NNI) defines nanotechnology as "the manipulation of matter with at least one dimension sized between 1 and 100 nm, where distinctive phenomena enable new applications in diverse field ranging from biology, physics, chemistry, medicine, engineering, and electronics" [3,4]. Two prerequisites are identified for nanotechnology in this definition. The first is a matter of scale; in nanotechnology, one can manipulate both the shape and size of a structure by working at a certain nanometer level. The second criterion is that the materials employed in nanotechnology exhibit innovative features as a result of their nanoscale [5]. In 1974, Norio Taniguchi, a Japanese scientist, defined nanotechnology as "the process of separating, consolidating, and deforming materials by one atom or another molecule," 15 years after Richard Feynman first proposed the idea. In 1981, Eric Drexler put forward the concept of nanotechnology, the field that makes use of such accuracy in working with materials on the atomic level. The first book to explore the promising new topic of nanotechnology, *Engines of Creation: The Coming Era of Nanotechnology*, was published under his pen in 1986 [4,6].

Nanotechnology and nanoscience are two phrases that are sometimes used interchangeably but have distinct meanings. "Nanotechnology" refers to the use of novel nanomaterials and nanosized components in goods, whereas "nanoscience" encompasses the study, manipulation, and engineering

DOI: 10.1201/9781003502692-2

of matter, particles, and structure on the nanometer scale [7]. Each concept represents a thriving area of research into improving the reliability, effectiveness, robustness, and accuracy of devices, structures, and systems operating at the nanoscale (the range of sizes between 1 and 100 nm). Democritus, a Greek philosopher from the fifth century BCE, is credited with introducing the concept of nanotechnology to the world [6]. Nanotechnology is one of, if not the most, promising technological developments of the twenty-first century because of its wide range of potential applications in fields as diverse as biology, physics, chemistry, material science, computing, food science, agriculture, and medicine (most recently in the field of cancer) [8].

2.3 NANOPARTICLES

NPs range in material, dimension, shape, and size, but they all act as a physical link between bulk materials and atoms or molecules [9]. Nanodots, graphene, carbon nanotubes, and gold NPs are all examples of NPs with varying numbers of dimensions [10]. The many advantageous properties of NPs have paved the way for their utilization in several industries.

2.3.1 Properties

The attributes of NPs are based on their size and characteristics, such as color, melting temperature, magnetic behavior, and redox potential [11]. The physical and the chemical are the two most important categories of characteristics.

2.3.2 Physical Properties

Examples of physical properties of NPs include light sensitivity, stiffness, conductivity and reluctance, absorption, transmission, reflection, and scattering of visible light, as well as measurements of UV radiation absorption in solution or on the particle's surface, which are all examples of NP optical characteristics. Mechanical properties including tensile strength, ductility, elasticity, and flexibility can be used to enhance NP applications. There is a wide spectrum of magnetic and electrical characteristics in NPs, from resistance to conductivity to semi-conductivity. Additional physical properties that are important for the utilization of nanomaterial in many modern products include hydrophobicity, hydrophilicity, diffusion, melting temperatures, and suspension [12].

2.3.3 Chemical Properties

NPs' reactivity, due to their nanoscale size and high surface-to-volume ratio, is a crucial chemical feature that dictates the sensitivity and stability of their applications, which include targeting environmental factors including light, heat, moisture, and atmosphere. Properties such as antimicrobial, disinfectant, antifungal, and poisonous make NPs useful in a variety of fields, including environmental science and medicine [13,14]. Chemical features including oxidation, flammability, corrosive, anti-corrosive, and reducing properties determine how NPs are used.

2.4 CLASSIFICATION OF NANOMATERIALS

NPs are categorized based on their size, shape, morphology, and chemical characteristics. Because of this, NPs are typically categorized based on their physical and chemical properties [15,16]. NPs can be broken down into two broad categories: inorganic and organic carbon-based (Figure 2.1).

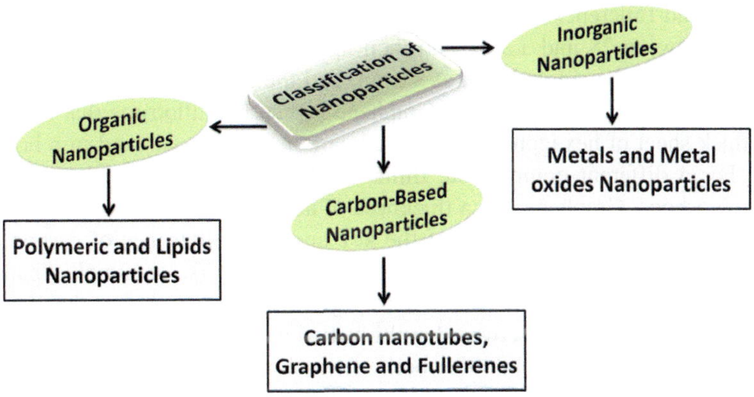

FIGURE 2.1 Types of nanoparticles.

2.4.1 Organic NPs

These are solid NPs with a range size of 10 nm–1 μm in diameter. Nanospheres and nano-capsules are the most common shapes for polymer NPs (also called polymeric NPs or PNPs). Adsorbed molecules cover the nanosphere polymer's surface, giving it a solid texture. The other molecules are protected by the solid mass of nano-capsules [17]. Lipid NPs have a diameter between 10 and 1,000 nm, and their shape is typically spherical. Like PNP, they have a solid, strong lipid nucleus and a lipophilic structure kept together by surfactants or emulsifiers [18].

These NPs are great vectors for medication or DNA delivery because of their well-documented affinity for molecules in the form of protein conjugates in two ways: (a) with the help of an adsorbed drug system and (b) with the help of an entrapped drug system [17,18]. Due to their biodegradability, lack of toxicity, insulation against electricity, and insensitivity to heat and light, they pose no threat to the natural world [19]. NPs of these types include well-known dendrimers, liposomes, micelles, and ferritin, which have proven to be biodegradable and non-toxic [20].

2.4.2 Inorganic NPs

NPs in this category primarily include metal and metal oxide NPs, along with doped variants consisting of metal/metal oxide and metal sulfide-based NPs. These particles, ranging from copper (Cu) to cadmium (Cd), exhibit nanoscale dimensions (10–100 nm), high surface charge density, and varied shapes. Notably, metal NPs, like silver (Ag) and gold (Au), show promise in antibacterial applications due to their high surface-to-volume ratio. Metal oxide NPs, such as zinc oxide (ZnO), derived from metals like iron, aluminum, and titanium, demonstrate enhanced reactivity and performance. Doped metal/metal oxide NPs, chemically altered for stability, efficacy, and environmental safety, have gained attention [21]. For instance, ZnO NPs doped with tantalum (Ta), magnesium (Mg), or antimony (Sb) showed a 5% improvement in antibacterial activity. Diverse doped particles, like Zn/CuO, Mn/ZnO, Cu/TiO, Ag/MgO, Ag and carbon monolith, and Ag and phosphate-based glasses, exhibit unique properties and applications. Metal sulfides, including copper sulfide NPs (CuS NPs), are emerging in forensics for metabolite, food pathogen, and DNA detection, showcasing their potential in biosensing [22].

2.4.3 Carbon-Based NPs

The types of nanomaterials that are of carbon source are called carbon-based NPs or carbon nanomaterials. These nanomaterials vary in shape, size, and function. In this class are carbon nanotubes

(CNTs), graphene, and fullerenes. Carbon fullerenes (C60) are spherical carbon molecules with a molecular weight between 28 and 1,500, and their spherical shape is maintained by sp^2 hybridization. The diameters of the single layer, at 8.2 nm, and the multilayer, at 4–36 nm, make it look like a soccer ball (big, round balls) [23,24]. Graphene is a carbon allotrope that takes the form of a single-atom-thick sheet of hexagonal, sp^2-bonded carbon atoms organized in a honeycomb-like crystal lattice. For a different comparison, think of chicken wire, although much smaller and entirely made of carbons. Graphene has a thickness of about 1 nm and a bond length of 0.142 nm between its carbon atoms. Graphene, one of the key and vital allotropes of carbon, is a crucial component in a wide variety of other carbon compounds. CNTs, or simply "nanotubes," are cylinder-shaped nanostructures formed of rolled-up graphene sheets. Using the Kratschmer and Huffman approach, Sumio Iijima discovered multi-walled nanotubes in 1991, and Donald Bethune discovered single-walled nanotubes in 1993, both of which are made by the addition of transition metals. Other CNT variants include zigzags, chirals, and armchairs, all of which are distinguished by the direction of rolling the graphene sheet [25].

2.5 NP SYNTHESIS

Methods for the synthesis of NPs are classified into two: the bottom-up method and the top-down method (Figure 2.2).

2.5.1 Bottom-Up Method

Using this technique, NPs are formed constructively from the bottom up by molecules or atoms self-recognizing and assembling at the nanoscale (between 1 and 100 nm). The particles synthesized using this process have an ideal crystallographic and surface structure, making them homogeneous, stable, and highly sought after in a variety of fields [26]. The following methods are employed in this synthesis approach.

2.5.1.1 Sol–Gel

Wet-chemical procedures used in the sol–gel process are relatively simple since they include the use of a chemical solution as a precursor, such as metal oxides or chlorides. The NPs are recovered

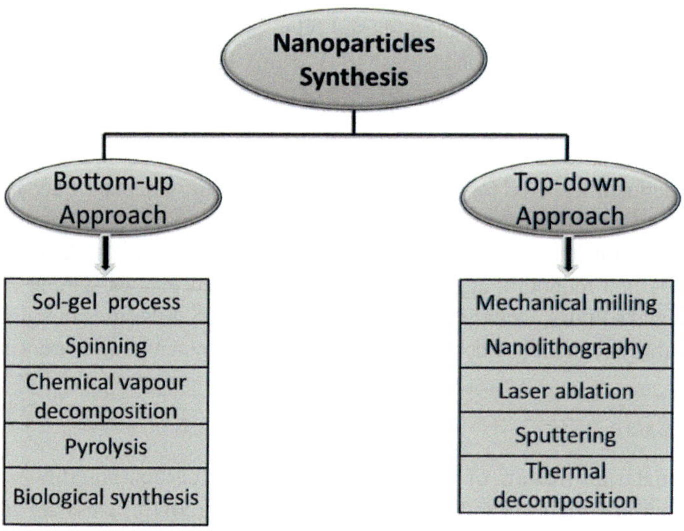

FIGURE 2.2 Different approaches of nanoparticle synthesis.

by first adding the precursor to the host liquid and dispersing it evenly with sonication, stirring, or shaking and then separating the system into liquid and solid phases via sedimentation, centrifugation, or filtration [26,27].

2.5.1.2 Spinning

This is accomplished with the use of a device called a spinning disc reactor, which utilizes the controlled rotation of a disc to alter certain physical properties. The features of the produced NPs are dependent on several factors, some of which are temperature, liquid flow rate, feed location, disc rotation speed, disc surface, and so on. To prevent chemical reactions by depleting oxygen levels, the reactor is first filled with nitrogen or another inert gas, and then the precursor and water are pumped in while the disc is rotated at varying speeds. Molecular or atomic fusion is induced by the spinning, and the resulting precipitate is collected, washed, and dried [28–30].

2.5.1.3 Chemical Vapor Deposition

The term "chemical vapor deposition" (CVD) is used to describe the procedure where molecules of a gas mixture can be deposited onto a substrate in a reaction chamber at room temperature to initiate a chemical reaction and produce a thin layer of the product. In this process, the temperature of the substrate is crucial. This technique can produce pure, uniform, and stable NPs, but its dangerous by-products should not be overlooked [31–33]. Using a reaction chamber and room temperature, molecules from a gas mixture can be deposited onto a substrate to kick off a chemical reaction and produce a thin layer of product. The technical name for this process is CVD. Products that have been found to contain NPs are recycled and returned to consumers. The substrate temperature has a crucial role in determining the success of this method. Although this approach can generate chemically pure, homogeneous, and stable NPs, it does so at the expense of creating several potentially harmful by-products [34].

2.5.1.4 Pyrolysis

To mass-produce NPs, many industries rely on pyrolysis. When the precursor is in liquid or gaseous form, it is delivered through a tiny opening into a high-pressure furnace where it is burned. NPs are separated from the exhaust gases and combustion by-products via air classification. Evaporation is simplified when a laser or plasma is utilized in the furnace rather than a flame [35].

2.5.1.5 Biological Biosynthesis

Biological biosynthesis is a more environmentally friendly way to create NPs because it is non-toxic and biodegradable. This technique uses plant, fungal, and microbial components as an alternative to conventional chemicals. The synthesized NPs we have here are exceptional in several respects [36,37].

2.5.2 Top-Down Method

Destructively breaking down a bulk substance to create a nanoscale particle is at the heart of the top-down method of NP production. This is achieved using the following techniques.

2.5.2.1 Mechanical Milling

Mechanical milling is the most widely used top-down strategy for producing NPs, and it comprises the milling and post-annealing of several materials in an inert atmosphere [38]. The particles are shaped via plastic deformation, their sizes are reduced via fracture, and their sizes are increased via cold welding.

2.5.2.2 Nanolithography

As a technique for fabricating nanoscale structures, nanolithography involves printing the desired structure or form on a light-sensitive substance and then erasing the unwanted shape. Examples of lithographic techniques include electron-beam, optical, scanning probe, multi-photon, and nano-imprint lithography [39,40].

2.5.2.3 Laser Ablation

This is a common method used to produce metal NPs in liquids without the addition of any surfactant or chemical. Simply described, it is the method of making NPs by abrading metal plates in a liquid. In other words, the NPs made with this method are truly unique. Important considerations in this method include the laser's wavelength, power, ablation time, repetition rate, and the solubility of the aqueous solution [41].

2.5.2.4 Sputtering

NPs are deposited on a surface using sputtering, also known as physical vapor deposition, which involves ejecting atoms or molecules from the particle and condensing them onto the surface while maintaining a high vacuum. Most of the time, annealing is done after the deposition stage. Multiple factors, including annealing temperature and time, substrate duration, and layer thickness, allow for the precise manipulation of the resulting NP sizes and shapes [42].

2.5.2.5 Thermal Decomposition

The breakdown of chemical bonds in compounds by applying heat is an example of thermal decomposition, an endothermic chemical decomposition. In this method, NPs are created through the chemical reaction that occurs when metals are heated to their breakdown temperature [43,44].

Strategies for NP synthesis can be categorized into biological, physical, and chemical. Sputtering, pyrolysis, mechanical milling, spinning lithography, and laser ablation are all examples of physical processes used in the creation of NPs. Even though the physical technique is quick, it can be used as a reducing agent, it does not require any harmful chemicals, and the distribution is not uniform. The negatives are low yield, solvent contamination, high energy consumption, and lack of uniform distribution [45].

The chemical approach, while yielding high results, is marked by high costs and the use of hazardous compounds like 2-mercaptoethanol, citrate, thiol-glycerol, and borohydride, posing potential risks [45]. Processes like lithography, sol–gel, laser ablation, CVD, and thermal breakdown are integrated into this method. In contrast, the biological approach, known as the green synthesis method, stands out as an environmentally friendly and safe alternative. This method, which involves fewer steps than conventional synthesis, utilizes plant extracts, fungi, bacteria, and biomolecules like vitamins, amino acids, and proteins. Notably, these biological systems act as both capping and reducing agents for NPs. Starch and D-glucose, for instance, act as reducing agents and a capping molecule in the production of starch–AgNPs [46]. This approach is not only cost-effective but also produces NPs with precisely defined sizes, making it a preferred choice [46].

2.6 SUBSTRATES INVOLVED IN GREEN SYNTHESIS OF NPS

2.6.1 Actinomycetes

NPs have been reported to be synthesized via the use of actinomycetes that carry the enzyme adenine dinucleotide-dependent nitrate reductase, which is responsible for the reduction of metals by electron transfer to form stable NPs. Examples of actinomycetes involved in this synthesis are *Streptomyces* sp. and *Rhodococcus* sp. [47].

2.6.2 Algae

Diverse groups of algae are noted to hold heavy metals and are utilized in the green synthesis of metallic NPs. Some of these algae include *Chlorella vulgaris*, *Tetraselmis kochinensis*, *Sargassum wightii*, brown algae *Fucus vesiculosus*, *Kappaphy cus alvarezii*, *Chondrus crispus*, and green algae *Spirogyra insignis* [47].

2.6.3 Bacteria

Green synthesis of NPs using bacteria has evolved over time. Studies showed that numerous types of bacterial species are used in the synthesis of Au, Ag, titanium, platinum, palladium, magnetite, cadmium sulfide, and titanium dioxide NPs due to their unique ability to reduce metals [48]. Some of these bacteria include *Pseudomonas stutzeri*, *Plectonema boryanum* (cyanobacteria), *Thiobacillus ferrooxidans*, *Rhodopseudomonas capsulate*, *Thiobacillus thiooxidans*, *Sufolobus acido caldarium*, *Staphylococcus aureus*, *Bacillus subtilis*, *Shewanella algae*, *Escherichia coli*, *Pseudomonas aeruginosa*, *Morganella* sp., and many others [48]. Monodispersity, morphology, shape, and size of NPs are majorly affected by the type of bacteria, source of bacteria, concentration of the substrate, temperature, pH, and reaction time.

2.6.4 Fungi

Investigating many types of fungi, such as *Trichoderma asperellum*, *Phoma sp. 3.2883*, *Fusarium* sp., *Aspergillus* sp., and *Phanerochaete chrysosporium* for the synthesis of NPs has captivated scientists because of their many benefits, including increased protein production and increased intracellular metal uptake at the expense of larger NPs than bacteria. Fungi can facilitate the creation of both organic and inorganic nanomaterials [49].

2.6.5 Viruses

The ability of viruses to transport inorganic compounds including cadmium sulfide (CdS), silicon dioxide (SiO_2), zinc sulfide (ZnS), and iron oxide (Fe_2O_3) and transform them into nanosized structures is well-documented. However, this is because the viral capsid proteins form a thick surface, making the virus a reactive surface capable of interacting with metallic ions. Capsid protein molecules can number up to 2,130 on a single tomato mosaic virus particle [50,51].

2.6.6 Yeasts

Yeast is used in the synthesis of nanomaterials due to their capacity for detoxification, biosorption, sequestration, chelation, and bioprecipitation of several hazardous metals found in the environment. Quantum dots of cadmium, gold, lead, and silver are among the nanomaterials shown to be generated by yeast. *Candida glabrata*, *Pichia jadinii*, *Schizosaccharomyces pombe*, *Torulopsis* sp., and MKY3 are just some of the yeasts that have been used for this synthesis [52].

2.6.7 Plants

Utilizing plants for green nanomaterial synthesis stands out as the most environmentally friendly approach due to the abundance of active compounds in various plant parts and extracts. These compounds play a crucial role in reducing metallic ions and stabilizing NPs. The synthesis process is a simple and direct method, involving the mixing of a metal salt solution with a plant extract at room temperature. Notably, this process induces a change in the solution's color, serving as an indicator of NP formation in the reaction mixture [53,54].

2.6.8 Animal Tissues

Different animal tissues utilized for the synthesis of NPs are silkworms, spiders, starfish, and chitosan. Silkworms or silk proteins are natural semi-crystalline biopolymers made up of amino acids and used to produce nanomaterials. Hydroxyapatite (HA) found in starfish and sponges is used to synthesize nano-HA; in addition, earthworms and marine worms have been utilized in the production of AgNPs. The peptide chitosan derived from chitin is another animal substrate used in the green synthesis of nanomaterials like nanofibers and nano-capsules utilized in both medical and industrial sectors [55,56].

2.7 CHARACTERIZATION OF NPS

The behavior, efficacy, bio-distribution, and safety of manufactured NPs are influenced by various factors, including shape, size, surface area, structural composition, surface charge, surface chemistry, crystallography, and concentration. Consequently, a thorough assessment of these factors is crucial [57]. Numerous characterization methods are employed to examine the physiochemical properties of NPs. Below are brief descriptions of some of these methods.

2.7.1 UV–Vis Spectrophotometry

This technique is typically employed to monitor the genesis and persistence of metallic NPs. The absorption peak produced by its NP production occurs in the visible spectrum, with a distinct band of 200–800 nm in wavelength [58]. Several metallic NPs between 2 and 100 nm in size have been studied for their Surface plasmonic resonance (SPR), which is caused by the unrestricted motion of electrons in the appropriate absorption band. This technique is fast and sensitive enough to identify a wide range of NPs [59].

2.7.2 X-ray Diffraction Analysis

Structure analysis, crystallinity measurement, compound identification, quantitative resolution of chemical species, and particle size determination are just some of the many uses for X-ray diffraction (XRD) Researchers can verify the manufacturing of NPs, learn about their physiochemical composition, and establish the purity of the nanomaterials in issue by penetrating them with X-rays and examining the ensuing diffraction pattern. Scherrer's Debye equation is applied to Bragg's law and XRD data to infer particle size [60].

2.7.3 Fourier Transform Infrared Spectroscopy

Fourier Transform Infrared Spectroscopy (FTIR) serves as a valuable technique for examining the surface chemistry of synthesized NPs, particularly the role biomolecules play in their production. By utilizing spectra generated from infrared light passing through a material, FTIR assesses the absorption and transmission properties of NPs. Known for its precision, user-friendliness, and safety, FTIR is the preferred method for identifying functional compounds in NPs [61].

2.7.4 Scanning Electron Microscopy

A high-resolution surface imaging technique, scanning electron microscopy (SEM) employs a beam of powerful electrons to intricately examine objects on the micrometer scale. This method facilitates the determination of NP sizes, surface morphology, distributions, and shapes. While SEM is instrumental in assessing particle morphology and generating histograms, it does have limitations, notably its inability to determine internal structure [62]. Despite this drawback, SEM proves beneficial for evaluating the level of aggregation and the purity of particles.

2.7.5 Transmission Electron Microscopy

The transmission electron microscope (TEM) is a widely recognized and widely utilized tool for gaining information on particle size and shape. TEM's resolution is 1,000 times better than that of SEM's. Magnification is proportional to the square of the distance between the specimen and the objective lens and the square of the distance between the image plane and the objective lens. This allows for a more accurate quantitative assessment of NP characteristics like shape, size, and crystallinity [63,64].

2.7.6 Atomic Force Microscopy

The atomic force microscope (AFM) stands out as a powerful and high-resolution technology utilized for analyzing the physical properties of various surfaces, including polymers, glass, ceramics, composites, and biological samples. Capable of probing both qualitative and quantitative aspects, such as surface texture, morphology, roughness, area, size, and volume distribution, AFM is well-suited for NP characterization [65,66]. Furthermore, it can measure magnetic force, adhesion strength, and the mechanical properties of nanomaterials. Using a laser beam focused with a photodiode, AFM creates a three-dimensional image of the probed subject, with the tip having a diameter of approximately 10 to 20 nm. The two scanning modes available—contact and tapping—require careful selection for accurate sample analysis. However, it's crucial to note that the lateral dimensions of samples may be overstated due to the size of the cantilever. Dynamic light scattering (DLS) significantly aids in the physiochemical characterization of NPs through radiation scattering techniques. Measuring the size and size distributions of submicron to nanometer-sized particles in solution or suspension, DLS relies on light's interaction with subatomic particles. By measuring the intensity of scattered light when a laser passes through a solution or colloid, the hydrodynamic size of a particle can be calculated [67]. Despite its ability to analyze numerous particles simultaneously, DLS has some sample-specific limitations.

2.7.7 Energy-Dispersive X-Ray Spectroscopy

Energy-dispersive X-ray spectroscopy (EDX) is employed to map the internal structure of an NP and calculate its elemental composition near the surface. Typically conducted alongside SEM or TEM, EDX reveals the percentage distribution of elements and their characteristic peaks in the X-ray spectra [68].

2.7.8 Zeta Potential

Zeta potential (ZP) is a technique commonly used to determine the surface charge of uniformly distributed NPs. Predicting the strength of an NP solution, ZP size is assessed by sending a laser beam through the NP solution in an electric field. The Doppler shift in the frequency of scattered light is detected, allowing the determination of ZP. A ZP of +25 or −25 mV generally indicates stability, while lower ZPs may lead to coagulation or agglomeration [69,70].

2.8 CONCLUSION

Nanomaterials play a pivotal role in advancing nanotechnology and nanoscience, employing both top-down and bottom-up methodologies encompassing physical, chemical, and biological processes for synthesis [71,72]. Among these methods, biological approaches have emerged as the most environmentally friendly and safest. The utilization of natural reducing and stabilizing agents in green synthesis techniques offers a sustainable alternative, facilitating the facile production of nanomaterials with applications spanning medicine, cosmetics, and food production. In the realm of nanomaterial characterization, various analytical techniques are employed to discern crucial parameters such as size, shape, surface charge, functional groups, size distribution, aggregation, dispersion, structure, binding hydrodynamics, and confirmation [73,74]. These analytical methodologies are essential for comprehensively understanding the properties of nanomaterials and tailoring them to specific applications. The intricate details of synthesizing diverse nanomaterials and the corresponding characterization methods are meticulously expounded upon in this comprehensive chapter. The integration of green synthesis not only aligns with sustainability goals but also expands the horizons of nanomaterial applications. The exploration of natural sources as reducing and stabilizing agents not only enhances the eco-friendliness of the processes but also broadens the scope for innovative applications in emerging fields. This holistic understanding of nanomaterials, encompassing synthesis strategies and analytical techniques, paves the way for their widespread utilization across various scientific and industrial domains.

REFERENCES

1. Abou El-Nour, K. M., Eftaiha, A. A., Al-Warthan, A., & Ammar, R. A. (2010). Synthesis and applications of silver nanoparticles. *Arabian Journal of Chemistry*, *3*(3), 135–140. https://doi.org/10.1016/j.arabjc.2010.04.008
2. Thakur, P., & Thakur, A. (2022). Introduction to nanotechnology. In: Thakur, A., Thakur, P., & Khurana, S. P. (eds) *Synthesis and Applications of Nanoparticles*. Springer, Singapore. https://doi.org/10.1007/978-981-16-6819-7_1
3. Almatroudi A. (2020). Silver nanoparticles: synthesis, characterisation and biomedical applications. *Open Life Sciences*, *15*(1), 819–839. https://doi.org/10.1515/biol-2020-0094
4. Anu, S. M. E., & Saravanakumar, M. P. (2017). A review on the classification, characterisation, synthesis of nanoparticles and their application. *IOP Conference Series: Materials Science and Engineering*, 263, 032019. https://doi.org/10.1088/1757-899X/263/3/032019
5. Parupudi, A., Mulagapati, S. H. R., & Subramony, J. A. (2022). Chapter 1 - Nanoparticle technologies: recent state of the art and emerging opportunities. In: Kesharwani, P., & Singh, K. K. (eds) *Nanoparticle Therapeutics*. Academic Press, New York, 3–46. https://doi.org/10.1016/B978-0-12-820757-4.00009-0
6. Arakawa, F. S., Shimabuku-Biadola, Q. L., Fernandes Silva, M., & Bergamasco, R. (2020). Development of a new vacuum impregnation method at room atmosphere to produce silver-copper oxide nanoparticles on activated carbon for antibacterial applications. *Environmental Technology*, *41*(18), 2400–2411. https://doi.org/10.1080/09593330.2019.1567607
7. Suhag, D., Thakur, P., & Thakur, A. (2023). Introduction to nanotechnology. In: Suhag, D., Thakur, A., & Thakur, P. (eds) *Integrated Nanomaterials and their Applications*. Springer, Singapore. https://doi.org/10.1007/978-981-99-6105-4_1
8. Bhaviripudi, S., Mile, E., Steiner, S. A., 3rd, Zare, A. T., Dresselhaus, M. S., Belcher, A. M., & Kong, J. (2007). CVD synthesis of single-walled carbon nanotubes from gold nanoparticle catalysts. *Journal of the American Chemical Society*, *129*(6), 1516–1517. https://doi.org/10.1021/ja0673332
9. Cabral, M., Pedrosa, F., Margarido, F., & Nogueira, C. A. (2013). End-of-life Zn-MnO_2 batteries: electrode materials characterization. *Environmental Technology*, *34*(9–12), 1283–1295. https://doi.org/10.1080/09593330.2012.745621
10. Cao, Y. C., Jin, R., & Mirkin, C. A. (2002). Nanoparticles with Raman spectroscopic fingerprints for DNA and RNA detection. *Science (New York, N.Y.)*, *297*(5586), 1536–1540. https://doi.org/10.1126/science.297.5586.1536
11. Thakur, A., Thakur, P., & Baccar, S. (2023). Structural properties of nanoparticles. In: Suhag, D., Thakur, A., & Thakur, P. (eds) *Integrated Nanomaterials and their Applications*. Springer, Singapore. https://doi.org/10.1007/978-981-99-6105-4_4
12. D'Amato, R., Falconieri, M., Gagliardi, S., Popovici, E., Serra, E., Terranova, G., & Borsella, E. (2013). Synthesis of ceramic nanoparticles by laser pyrolysis: from research to applications. *Journal of Analytical and Applied Pyrolysis*, *104*, 461–469. https://doi.org/10.1016/j.jaap.2013.05.026
13. Das, R., Nath, S. S., Chakdar, D., Gope, G., & Bhattacharjee, R. J. J. O. N. (2009). Preparation of silver nanoparticles and their characterization. *Journal of Nanotechnology*, *5*, 1–6.
14. Dey, A., Mukhopadhyay, A. K., Gangadharan, S., Sinha, M. K., & Basu, D. (2009). Characterization of microplasma sprayed hydroxyapatite coating. *Journal of Thermal Spray Technology*, *18*(4), 578–592. https://doi.org/10.1007/s11666-009-9386-2
15. Dhanasekar, M., Jenefer, V., Nambiar, R. B., Babu, S. G., Selvam, S. P., Neppolian, B., & Bhat, S. V. (2018). Ambient light antimicrobial activity of reduced graphene oxide supported metal doped TiO_2 nanoparticles and their PVA based polymer nanocomposite films. *Materials Research Bulletin*, *97*, 238–243. https://doi.org/10.1016/j.materresbull.2017.08.056

16. Djafari, J., McConnell, M. T., Santos, H. M., Capelo, J. L., Bertolo, E., Harvey, S. C., Lodeiro, C., & Fernández-Lodeiro, J. (2018). Synthesis of gold functionalised nanoparticles with the eranthis hyemalis lectin and preliminary toxicological studies on caenorhabditis elegans. *Materials (Basel, Switzerland), 11*(8), 1363. https://doi.org/10.3390/ma11081363
17. Drexler, E. K. (1986). *Engines of Creation: The Coming Era of Nanotechnology.* Anchor Press; Garden City, NY.
18. El-Kassas, H. Y., & Okbah, M. A. E. A. (2017). Phytotoxic effects of seaweed mediated copper nanoparticles against the harmful alga: Lyngbya majuscula. *Journal of Genetic Engineering and Biotechnology, 15*(1), 41–48.
19. Elmer, W., & White, J. C. (2018). The future of nanotechnology in plant pathology. *Annual Review of Phytopathology, 56,* 111–133. https://doi.org/10.1146/annurev-phyto-080417-050108
20. Elsupikhe, R. F., Shameli, K., Ahmad, M. B., Ibrahim, N. A., & Zainudin, N. (2015). Green sonochemical synthesis of silver nanoparticles at varying concentrations of κ-carrageenan. *Nanoscale Research Letters, 10*(1), 916. https://doi.org/10.1186/s11671-015-0916-1
21. Fissan, H., Ristig, S., Kaminski, H., Asbach, C., & Epple, M. (2014). Comparison of different characterization methods for nanoparticle dispersions before and after aerosolization. *Analytical Methods, 6*(18), 7324–7334. https://doi.org/10.1039/C4AY01203H
22. Gmoshinski, I. V., Khotimchenko, S. A. E., Popov, V. O., Dzantiev, B. B., Zherdev, A. V., Demin, V. F., & Buzulukov, Y. P. (2013). Nanomaterials and nanotechnologies: methods of analysis and control. *Russian Chemical Reviews, 82*(1), 48.
23. Gudikandula, K., & Charya Maringanti, S. (2016). Synthesis of silver nanoparticles by chemical and biological methods and their antimicrobial properties. *Journal of Experimental Nanoscience, 11*(9), 714–721. https://doi.org/10.1080/17458080.2016.1139196
24. Guo, B. L., Han, P., Guo, L. C., Cao, Y. Q., Li, A. D., Kong, J. Z., Zhai, H. F., & Wu, D. (2015). The antibacterial activity of Ta-doped ZnO nanoparticles. *Nanoscale Research Letters, 10*(1), 1047. https://doi.org/10.1186/s11671-015-1047-4
25. Hasan, S. (2015). A review on nanoparticles: their synthesis and types. *Research Journal of Recent Sciences, 2277,* 2502.
26. Hilton, A., Handiseni, M., Choi, W., Wang, X., Grauke, L. J., Yu, C., & Jo, Y. K. (2017). Novel phytosanitary treatment of Xylella fastidiosa-infected pecan scions using carbon nanotubes. In *2017 APS Annual Meeting.* APSNET, San Antonio, TX.
27. Inagaki, S., Ghirlando, R., & Grisshammer, R. (2013). Biophysical characterization of membrane proteins in nanodiscs. *Methods (San Diego, Calif.), 59*(3), 287–300. https://doi.org/10.1016/j.ymeth.2012.11.006
28. Ivanisevic I. (2010). Physical stability studies of miscible amorphous solid dispersions. *Journal of pharmaceutical sciences, 99*(9), 4005–4012. https://doi.org/10.1002/jps.22247
29. Kammler, H. K., Mädler, L., & Pratsinis, S. E. (2001). Flame synthesis of nanoparticles. *Chemical Engineering & Technology: Industrial Chemistry-Plant Equipment-Process Engineering-Biotechnology, 24*(6), 583–596. https://doi.org/10.1021/acs.energyfuels.0c04054
30. Khan, I., Saeed, K., & Khan I. (2019). Nanoparticles: properties, applications and toxicities. *Arabian Journal of Chemistry, 12,* 908–931. https://doi.org/10.1016/j.arabjc.2017.05.011
31. Khodashenas, B., & Ghorbani, H. R. (2014). Synthesis of silver nanoparticles with different shapes. *Arabian Journal of Chemistry, 12,* 1823–1838. https://doi.org/10.1016/j.arabjc.2014.12.014
32. Kinnear, C., Moore, T. L., Rodriguez-Lorenzo, L., Rothen-Rutishauser, B., & Petri-Fink, A. (2017). Form follows function: nanoparticle shape and its implications for nanomedicine. *Chemical Reviews, 117*(17), 11476–11521. https://doi.org/10.1021/acs.chemrev.7b00194
33. Korbekandi, H., Iravani, S., & Abbasi, S. (2009). Production of nanoparticles using organisms. *Critical Reviews in Biotechnology, 29*(4), 279–306. https://doi.org/10.3109/07388550903062462
34. Kou, T., Jin, C., Zhang, C., Sun, J., & Zhang, Z. (2012). Nanoporous core–shell Cu@ Cu_2O nanocomposites with superior photocatalytic properties towards the degradation of methyl orange. *RSC Advances, 2*(33), 12636–12643. https://doi.org/10.1039/c2ra21821f
35. Kumar, R., & Lal, S. (2014). Synthesis of organic nanoparticles and their applications in drug delivery and food nanotechnology: a review. *Journal of Nanomaterials & Molecular Nanotechnology, 3,* 4. https://dpi.org/10.4172/2324-8777.1000150
36. Kummara, S., Patil, M. B., & Uriah, T. (2016). Synthesis, characterization, biocompatible and anticancer activity of green and chemically synthesized silver nanoparticles - a comparative study. *Biomedicine & Pharmacotherapy = Biomedecine & Pharmacotherapie, 84,* 10–21. https://doi.org/10.1016/j.biopha.2016.09.003

37. Kuppusamy, P., Yusoff, M. M., Maniam, G. P., & Govindan, N. (2016). Biosynthesis of metallic nanoparticles using plant derivatives and their new avenues in pharmacological applications - an updated report. *Saudi Pharmaceutical Journal: SPJ: The Official Publication of the Saudi Pharmaceutical Society, 24*(4), 473–484. https://doi.org/10.1016/j.jsps.2014.11.013
38. Lee, D. E., Koo, H., Sun, I. C., Ryu, J. H., Kim, K., & Kwon, I. C. (2012). Multifunctional nanoparticles for multimodal imaging and theragnosis. *Chemical Society Reviews, 41*(7), 2656–2672. https://doi.org/10.1039/c2cs15261d
39. Lee, P. Y., & Wong, K. K. (2011). Nanomedicine: a new frontier in cancer therapeutics. *Current Drug Delivery, 8*(3), 245–253. https://doi.org/10.2174/156720111795256110
40. Li, Y., & Chen, S. M. (2012). The electrochemical properties of acetaminophen on bare glassy carbon electrode. *International Journal of Electrochemical Science, 7,* 13. https://doi.org/10.1016/j.jcis.2016.08.028
41. Lin, P. C., Lin, S., Wang, P. C., & Sridhar, R. (2014). Techniques for physicochemical characterization of nanomaterials. *Biotechnology Advances, 32*(4), 711–726. https://doi.org/10.1016/j.biotechadv.2013.11.006
42. Llorens, A., Lloret, E., Picouet, P. A., Trbojevich, R., & Fernandez, A. (2012). Metallic-based micro and nanocomposites in food contact materials and active food packaging. *Trends in Food Science and Technology, 24,* 19–29. https://doi.org/10.1016/j.tifs.2011.10.001
43. Lugscheider, E., Bärwulf, S., Barimani, C., Riester, M., & Hilgers, H. (1998). Magnetron-sputtered hard material coatings on thermoplastic polymers for clean room applications. *Surface and Coatings Technology, 108,* 398–402. https://doi.org/10.1016/S0257-8972(98)00627-6
44. Malik, P., Shankar, R., Malik, V., Sharma, N., & Mukherjee, T. K. (2014). Green chemistry based benign routes for nanoparticle synthesis. *Journal of Nanoparticles,* 2014, 302429. https://doi.org/10.1155/2014/302429.
45. Malka, E., Perelshtein, I., Lipovsky, A., Shalom, Y., Naparstek, L., Perkas, N., Patick, T., Lubart, R., Nitzan, Y., Banin, E., & Gedanken, A. (2013). Eradication of multi-drug resistant bacteria by a novel Zn-doped CuO nanocomposite. *Small (Weinheim an der Bergstrasse, Germany), 9*(23), 4069–4076. https://doi.org/10.1002/smll.201301081
46. Mann, S., Burkett, S. L., Davis, S. A., Fowler, C. E., Mendelson, N. H., Sims, S. D., & Whilton, N. T. (1997). Sol–gel synthesis of organized matter. *Chemistry of Materials, 9*(11), 2300–2310. https://doi.org/10.1021/cm970274u
47. Mansoor, S., Zahoor, I., Baba, T. R., Padder, S. A., Bhat, Z. A., Koul, A. M., & Jiang, L. (2021). Fabrication of silver nanoparticles against fungal pathogens. *Frontiers in Nanotechnology, 3,* 679358. https://doi.org/10.3389/fnano.2021.679358
48. Mansoori, G., & Fauzi Soelaiman, T. (2005). Nanotechnology—an introduction for the standards community. *Journal of ASTM International, 2,* 1–22.
49. Thakur, A., Chahar, D., & Thakur, P. (2022). Synthesis of nanomaterials by biological route. In: Thakur, A., Thakur, P., & Khurana, S. P. (eds) *Synthesis and Applications of Nanoparticles.* Springer, Singapore. https://doi.org/10.1007/978-981-16-6819-7_5
50. Meena, M., Zehra, A., Swapnil, P. H., Marwal, A., Yadav, G., & Sonigra, P. (2021). Endophytic nanotechnology: an approach to study scope and potential applications. *Frontiers in Chemistry, 9,* 613343. https://doi.org/10.3389/fchem.2021.613343
51. Mohammadi, S., Harvey, A., & Boodhoo, K. V. (2014). Synthesis of TiO_2 nanoparticles in a spinning disc reactor. *Chemical Engineering Journal, 258,* 171–184. https://doi.org/10.1016/j.cej.2014.07.042
52. Mukherjee, P., Senapati, S., Mandal, D., Ahmad, A., Khan, M. I., Kumar, R., & Sastry, M. (2002). Extracellular synthesis of gold nanoparticles by the fungus Fusarium oxysporum. *Chembiochem: A European Journal of Chemical Biology, 3*(5), 461–463. https://doi.org/10.1002/1439-7633(20020503)3:5<461::AID-CBIC461>3.0.CO;2-X
53. Narayanan, K. B., & Sakthivel, N. (2010). Biological synthesis of metal nanoparticles by microbes. *Advances in Colloid and Interface Science, 156*(1–2), 1–13. https://doi.org/10.1016/j.cis.2010.02.001
54. Pimpin, A., & Srituravanich, W. (2012). Review on micro-and nanolithography techniques and their applications. *Engineering Journal, 16*(1), 37–56. https://doi.org/10.4186/ej.2012.16.1.37
55. Rajeshkumar S. (2016). Anticancer activity of eco-friendly gold nanoparticles against lung and liver cancer cells. *Journal, Genetic Engineering & Biotechnology, 14*(1), 195–202. https://doi.org/10.1016/j.jgeb.2016.05.007
56. Raveendran, P., Fu, J., & Wallen, S. L. (2006). A simple and "green" method for the synthesis of Au, Ag, and Au–Ag alloy nanoparticles. *Green Chemistry, 8*(1), 34–38. https://doi.org/10.1039/B512540E
57. Rekha, K., Nirmala, M., Nair, M. G., & Anukaliani, A. (2010). Structural, optical, photocatalytic and antibacterial activity of zinc oxide and manganese doped zinc oxide nanoparticles. *Physica B: Condensed Matter, 405,* 3180–3185. https://doi.org/10.1016/j.physb.2010.04.042

58. Rohman, A., & Man, Y. C. (2010). Fourier transform infrared (FTIR) spectroscopy for analysis of extra virgin olive oil adulterated with palm oil. *Food Research International, 43*(3), 886–892. https://doi.org/10.1016/j.foodres.2009.12.006
59. Roy, A., Bulut, O., Some, S., Mandal, A. K., & Yilmaz, M. D. (2019). Green synthesis of silver nanoparticles: biomolecule-nanoparticle organizations targeting antimicrobial activity. *RSC Advances, 9*, 2673–2702. https://doi.org/10.1039/c8ra08982e
60. Sadrolhosseini, A. R., Mahdi, M. A., & Rashid, F. A. (2018). Laser Ablation technique for synthesis of metal nanoparticle in liquid. In: Yufei, M. (eds) *Laser Technology and its Applications*. IntechOpen, London. https://doi.org/10.5772/intechopen.80374
61. Sarlak, N., Taherifar, A., & Salehi, F. (2014). Synthesis of nanopesticides by encapsulating pesticide nanoparticles using functionalized carbon nanotubes and application of new nanocomposite for plant disease treatment. *Journal of Agricultural and Food Chemistry, 62*(21), 4833–4838. https://doi.org/10.1021/jf404720d
62. Shah, M., Fawcett, D., Sharma, S., Tripathy, S. K., & Poinern, G. E. J. (2015). Green synthesis of metallic nanoparticles via biological entities. *Materials (Basel, Switzerland), 8*(11), 7278–7308. https://doi.org/10.3390/ma8115377
63. Shameli, K., Bin Ahmad, M., Zargar, M., Yunus, W. M., Ibrahim, N. A., Shabanzadeh, P., & Moghaddam, M. G. (2011). Synthesis and characterization of silver/montmorillonite/chitosan bionanocomposites by chemical reduction method and their antibacterial activity. *International Journal of Nanomedicine, 6*, 271–284. https://doi.org/10.2147/IJN.S16043
64. Shnoudeh, A. J., Hamad, I., Abdo, R. W., Qadumii, L., Jaber, A. Y., Surchi, H. S., & Alkelany, S. Z. (2019). Synthesis, characterization, and applications of metal nanoparticles. *Biomaterials and Bionanotechnology, 2019*, 527–612. https://doi.org/10.1016/b978-0-12–814427-5.00015-9
65. Singh, D. K., Pandey, D. K., Yadav, R. R., & Singh, D. (2013). A study of ZnO nanoparticles and ZnO-EG nanofluid. *Journal of Experimental Nanoscience, 8*(5), 731–741. https://doi.org/10.1080/17458080.2011.602369
66. Tai, C. Y., Tai, C. T., Chang, M. H., & Liu, H. S. (2007). Synthesis of magnesium hydroxide and oxide nanoparticles using a spinning disk reactor. *Industrial & Engineering Chemistry Research, 46*(17), 5536–5541. https://doi.org/10.1021/ie060869b
67. Thakkar, K. N., Mhatre, S. S., & Parikh, R. Y. (2010). Biological synthesis of metallic nanoparticles. *Nanomedicine: Nanotechnology, Biology, and Medicine, 6*(2), 257–262. https://doi.org/10.1016/j.nano.2009.07.002
68. Tiede, K., Boxall, A. B., Tear, S. P., Lewis, J., David, H., & Hassellov, M. (2008). Detection and characterization of engineered nanoparticles in food and the environment. *Food Additives & Contaminants. Part A, Chemistry, Analysis, Control, Exposure & Risk Assessment, 25*(7), 795–821. https://doi.org/10.1080/02652030802007553
69. Tsuji, M., Hashimoto, M., Nishizawa, Y., Kubokawa, M., & Tsuji, T. (2005). Microwave-assisted synthesis of metallic nanostructures in solution. *Chemistry (Weinheim an der Bergstrasse, Germany), 11*(2), 440–452. https://doi.org/10.1002/chem.200400417
70. Ulbrich, K., Hola, K., Subr, V., Bakandritsos, A., Tucek, J., & Zboril, R. (2016). Targeted drug delivery with polymers and magnetic nanoparticles: covalent and noncovalent approaches, release control, and clinical studies. *Chemical Reviews, 116* (9), 5338–5431. https://doi.org/10.1021/acs.chemrev.5b00589
71. Rathore, R., Suhag, D., Wan, F., Thakur, A., & Thakur, P. (2023). Everyday nanotechnology. In: Suhag, D., Thakur, A., & Thakur, P. (eds) *Integrated Nanomaterials and their Applications*. Springer, Singapore. https://doi.org/10.1007/978-981-99-6105-4_2
72. Wang, H., Li, Y., Zuo, Y., Li, J., Ma, S., & Cheng, L. (2007). Biocompatibility and osteogenesis of biomimetic nano-hydroxyapatite/polyamide composite scaffolds for bone tissue engineering. *Biomaterials, 28*(22), 3338–3348. https://doi.org/10.1016/j.biomaterials.2007.04.014
73. Wang, X., Liu, X., & Han, H. (2013). Evaluation of antibacterial effects of carbon nanomaterials against copper-resistant Ralstonia solanacearum. *Colloids and Surfaces. B, Biointerfaces, 103*, 136–142. https://doi.org/10.1016/j.colsurfb.2012.09.044
74. Waseda, Y., Matsubara, E., & Shinoda, K. (2011). *X-ray Diffraction Crystallography: Introduction, Examples and Solved Problems*. Springer Verlag, Berlin, Germany. https://doi.org/10.1007/978-3-642-16635-8

3 Synthesis of Ozone-Friendly Nanomaterials

Ayushi Pradhan, Bikash Ranjan Jena, and Gurudutta Pattnaik
Centurion University of Technology and Management

Ch. Niranjan Patra
Roland Institute of Pharmaceutical Sciences

3.1 INTRODUCTION

The use of instinctively sourced resources, such as biopolymers, plant-based extracts, microorganisms, and others, has many advantages over the use of hazardous chemicals in the manufacturing process in terms of environmental friendliness and biocompatibility for a variety of medicinal and pharmaceutical applications. In contrast to more conventional methods like physical and chemical processes, plant extract-based synthetic procedures have gained popularity. Due to various benefits, such as non-hazardous, affordable, and feasible processes with numerous potential applications in biomedicine, nanotechnology, and nano-optoelectronics, among others, greener nanomaterial synthesis has grown in prominence. As a result of global climate change, agriculture around the world is struggling with new challenges. Improved crop yields and long-term sustainability can be attained with the use of cutting-edge nano-engineering, an effective strategy towards achieving food security. By maximising the effectiveness of inputs and decreasing relevant losses, nanotechnology aids in raising agricultural output. Nanomaterials provide for increased fertiliser and pesticide-specific surface area. In addition, nanoparticles (NPs) serve as novel transporters of agrochemicals, allowing for the controlled, site-specific application of nutrients while bolstering crop security. Nano-biosensors and other nanotools aid in the advancement of high-tech farms since they may be used for the targeted, and inputs, such as fertilisers, pesticides, and herbicides, need to be carefully managed and controlled. The combination of biology and nanotechnology has made it much easier for nano-sensors to notice and identify changes or problems in the environment. This study talks about the growing need for both food security and environmental sustainability. It also talks about some recent efforts to use nanotechnologies in agriculture in new and interesting ways.

3.2 OZONE AND ITS VITAL COMPONENTS

Ozone is a gas that is made up of three different atoms of oxygen. Its molecular weight is 47.98, it's a highly unstable material that gives off single oxygen atoms rapidly, and it has a very bad reputation. Its half-life is 40 minutes when the temperature is 20°C, but it is about 140 minutes when the temperature is 0°C. Ozone is odourless and colourless, yet it has a very distinctive fragrance. Ozone can be found naturally in the atmosphere, where it forms a layer that shields life from the sun's harmful ultraviolet rays. Since ozone is denser than air, it sinks to the ground in areas where the gas interacts with contaminants, a process that is referred to as the self-cleaning phenomenon [1–3]. When compared to the anti-microbial potential of chloride, it is approximately 1.5 times more effective than this substance. It is an exceptionally powerful oxidant. In the field of medicine, a gas mixture

DOI: 10.1201/9781003502692-3

that consists of between 0.5% and 5% ozone and 95–99.5% oxygen is utilised. Singlet oxygen, also known as nascent oxygen, has been studied for almost a century for its capacity to kill microorganisms like bacteria, fungi, and viruses.

Ozone therapy is finding new uses as a result of advances in medical technology. It has been proven that silver particles work well in adhesives, implants, and prosthetic materials. It is now possible to make silver NPs with regulated shapes and sizes, as well as specialised duties and being the same everywhere [4,5]. They have an electron configuration of [Kr] 4d105s1. Additionally, they work well as components of tissue conditioners [6], as multifunctional building blocks for dental goods [6,7], and in orthodontics [7,11]. Silver particles have antifungal effects on *Candida albicans*, and this is especially clear when the particles are put in silicone-based liners or resins. In terms of killing bacteria, silver particles change how permeable a bacterium's membrane is, which causes the membrane to break. The material is what causes oxidative stress, which leads to the destruction of cellular structures like DNA, lipids, and proteins, and, in the end, the death of the whole bacterial cell. Recent reviews [8,9] have spoken about a wide range of things about silver particles, including how they are made, how they are used, and how dangerous they are. The time it takes to extract all of the Ag+ from a particle is determined by how it dissolves, which is influenced by pH, the amount of dissolved O_2, the surface layer, and the ionic strength of the medium [10]. Silver NPs can also act as a bioactive agents in the agriculture sector [4,5]. They are also useful as synergistic agents with different antibiotics [6] and as multifunctional building blocks for dental materials [6,7] in orthodontics [11]. *C. albicans* is particularly sensitive to the antifungal effects of silver particles when they are introduced to silicone-based coatings or resins. In terms of antibacterial activity, silver particles damage bacterial membranes by altering their permeability. Stimulating oxidative stress is the substance's main function; this in turn causes the breakdown of DNA [12], lipids, proteins, and finally the entire bacterial cell. Recent reviews [8,9] have discussed all facets of silver particle manufacture, application, and toxicology. Dissolution mechanisms that depend on pH, dissolved oxygen content, surface coating, and ionic strength dictate the time needed to liberate Ag+ from a particle [10].

3.3 TYPES OF NANOMATERIALS USED FOR GREEN SYNTHESIS

Copper oxide (CuO) can inhibit the growth of bacteria and fungi [13]. It's cheaper than silver oxide and the most fundamental part of copper compounds. Multiple physical phenomena, such as high-temperature superconductivity, spin dynamics, and electron correlation, have been attributed to Cu compounds. It can improve heat conductivity and fluid viscosity due to its huge surface area and unique crystalline structures. These features suggest that CuO could help cut down on energy usage. NPs of this material are used as additives in a wide variety of products, including polymers, metallic coatings, and lubricants [14]. Due to their unusually large surface areas and unconventional morphologies, CuO NPs boost the shear bond strength of adhesives [15]. This also impacts their antibacterial characteristics, allowing them to inhibit *Escherichia coli* bacilli but not *Salmonella typhimurium*. However, *Streptococcus mutans* cocci are susceptible to CuO, which has an impact similar to that of silver particles [16]. Other studies have found that CuO can prevent biofilm formation by as much as 80% [17]. Surface coatings that include titanium dioxide (TiO_2) are often employed as a self-cleaning and disinfection agent [18]. Due to its high biocompatibility and excellent mechanical qualities, titanium and its alloys are increasingly used in the field of dental implantology. As a sterilising agent in settings like hospitals, they have been employed to filter out viruses and other organic pollutants from the air and water [19]. However, TiO_2 requires low-ultraviolet (400 nm) light to function. TiO_2 has been proven to facilitate localised drug delivery by nanotubes [20], in addition to its antibacterial characteristics.

Creating particles having a diameter of less than 100 nm is necessary for the transformation of metals into NPs. Platinum nanoparticles (PtNPs) have been shown to exhibit antibacterial and anti-inflammatory effects in a number of studies [21,22]. Additionally, PtNPs' integration into resin-based materials may improve biocompatibility [23], since they have been found to be a powerful antioxidant in vitro. Furthermore, the dentin bond strength is boosted by a factor of two when platinum particles are coupled with a 4-methacryloyloxyethyl trimellitic anhydride (4-META)/methyl methacrylate adhesive [24]. Pt particles, when in contact with bacteria, have been shown to kill the germs [25]. Platinum particles have been shown to be non-genotoxic, unlikely to cause an allergic reaction, and potentially useful in therapeutic settings [26]. This article takes a look at how modern dentists are employing nanotechnology and ozone in their practices. Medicines with antibacterial qualities allow for more efficient therapy with fewer adverse effects.

3.4 FACTORS AFFECTING SYNTHESIS OF NPS

Innovations in nanotechnology have various agricultural applications, with a focus on accelerating the development of novel nano-agrochemicals, such as nano-fertilisers and nano-pesticides. It contributes significantly to sustainable agriculture by increasing agricultural yield while reducing the use of inorganic fertilisers, pesticides, and herbicides. The synthesis procedure, pH, temperature, pressure, time, particle size, pore size, environment, and proximity all influence the quality and quantity of synthesised NPs, as well as their characterisation and applications. Similar efforts are being made on a global scale to develop eco-friendly technologies that use green nanotechnology and biotechnological instruments to manufacture eco-friendly, benign products [25–27].

3.5 NANO-FARMING: A RECENT EVOLUTION IN THE DEVELOPMENT OF AGRICULTURAL TECHNOLOGY

One of the latest advances in technology, nanoparticle engineering, has been shown to have both increased strength and innovative, targeted features. Norio Taniguchi, a professor at the Tokyo University of Science, is credited with creating the word "nanotechnology" in 1974 [7]. NPs are an emerging technology with promising applications in agriculture; however, they are still in the experimental stages [6]. Although the word "nanotechnology" has been in use for some time and has been put to use in a wide variety of academic disciplines, the notion that NPs can be helpful for agricultural advancement is a relatively new one. As a result of recent developments in nanoparticle synthesis, nanomaterials are finding increasing application in many different areas, including medicine, the environment, agriculture, and food processing. These changes have nearly universally improved farming throughout human history [25]. A decline in crop output due to biotic and abiotic stresses such as nutrient scarcity and environmental contamination are just two examples of the unprecedented difficulties that modern agriculture must overcome. Nanotechnology's intriguing new applications in precision agriculture (Figure 3.1) offer promising answers to these problems. "Farming" and "precision agriculture" are two terms that have recently become popular. This statement alludes to the development of wireless networking and the miniaturisation of sensors used to track, analyse, and regulate farming methods. It is the practice of overseeing the cultivation of agricultural products from seed to harvest [26,27] and includes both field crops and horticultural crops. Tissue engineering and the targeted delivery of CRISPR (clustered regularly interspaced short palindromic repeats)/Cas (CRISPR-associated protein) mRNA and sgRNA have both advanced in recent years [28–30] in the pursuit of developing genetically modified crops. In addition, nanotechnology offers improved answers to the mounting ecological problems. For instance, increasing plant disease resistance and monitoring environmental stress through the use of nano-sensors offers great potential [31,32]. Due to their emphasis on problem-solving and the development of collaborative strategies for sustainable agricultural expansion, the current advancements in nanotechnology have enormous potential to aid society as a whole and in a fair and equitable manner.

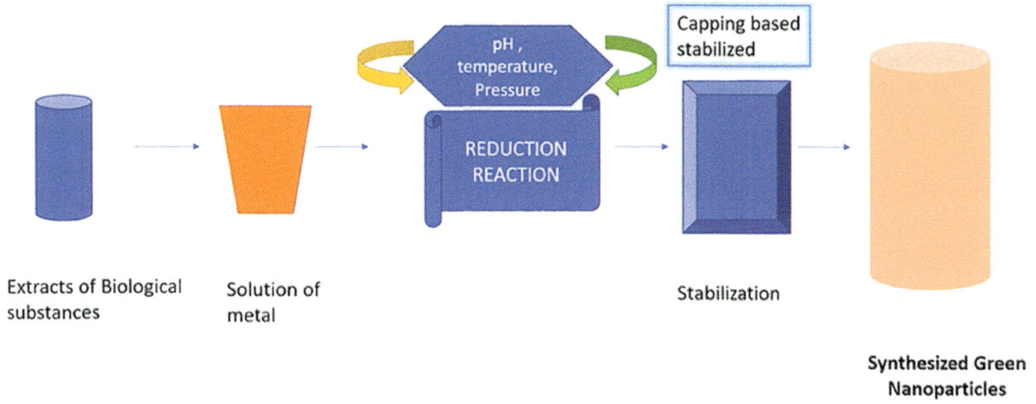

FIGURE 3.1 Schematic diagram elucidating the synthesis of green nanoparticles.

3.5.1 Utilisation of Nanomaterials in Pesticide-Based Plant Protection

The use of nanomaterials in product protection for plants has substantially risen to boost agricultural output. Fungicides, herbicides, and insecticides are typically used extensively and widely in traditional crop protection techniques. More than 90% of pesticides used are either lost in the environment or do not reach the target sites required for efficient pest management [33]. As a result, environmental resources are depleted, driving up the cost of food production. It should be mentioned that the presence of active ingredients (AIs) in a formulation's least effective concentration at the target locations is crucial for increased plant protection against pest invasion and crop loss. In this regard, the study of novel plant protection agents in agriculture has long been a highly conceptualised discipline. One such technique that has fundamentally altered the plant protection industry is nano formulation, sometimes known as pesticide encapsulation. Although other developed nanostructures also exhibit pesticidal characteristics, only a small number of particles are utilised as AIs in pesticide nano formulations [34]. The technique of coating a pesticide's active chemicals with a separate substance that has a variety of sizes in the nanoscale is known as nanoencapsulation. Pesticides are considered the internal phase of encapsulated materials, while coated NPs are considered the external phase of the core material (pesticides) [35]. Pesticide nano formulations or encapsulations enable the persistence or controlled release of active components in root zones or inside plants while maintaining potency. Contrarily, typical pesticide or herbicide formulations not only limit pesticide water solubility but also harm other organisms, resulting in a rise in resistance to the organisms they are intended to control. On the contrary, nano formulations help to get around the aforementioned restrictions [27]. As an illustration, Researchers [36] have shown that pesticide nano formulations increase crop yields by enhancing pesticide efficacy by reducing pesticide transport potential. They discovered that in loamy sand soil saturated with artificial porewater containing Ca^{2+} and Mg^{2+} cations, nano formulations combining polymeric nano capsules and the pyrethroid bifenthrin (nCAP4-BIF) exhibit higher elution over time and enhanced transport capability even after fertiliser addition. This suggests that the application of pyrethroid pesticides via nCAP4 for plant protection may be an option. The improved wettability and dispersion of nano formulations, which stop unfavourable pesticide movement and runoff from organic solvents, may help to explain this. A sustainable agro-environmental system requires materials with increased stiffness, permeability, thermal stability, solubility, crystallinity, and biodegradability, which nanomaterials have [7,34]. More importantly, the integrated pest management approach of the controlled and timed release of active chemicals lowers the total amount of pesticides needed for disease and pest control. Furthermore, little pesticide use is required for sustainable agriculture

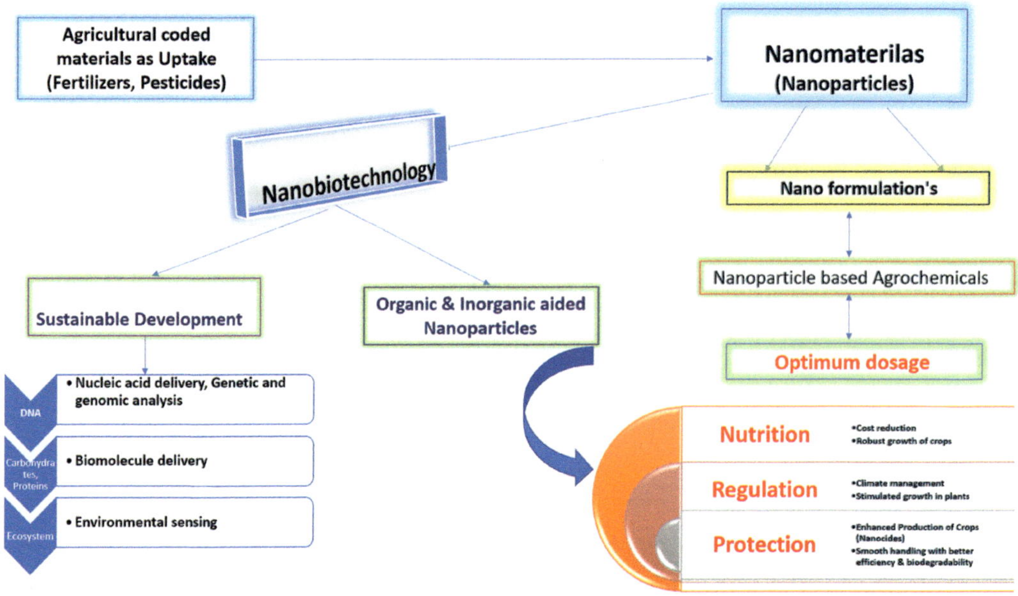

FIGURE 3.2 Nanomaterial-based agrochemicals for sustainable plant growth.

to safeguard the environment and non-target species. Additionally, sparingly using insecticides lowers the cost of agricultural produce. According to estimates, plant diseases, insect pests, and weeds cost agricultural products $2,000 billion annually in losses worldwide, with pathogen control alone costing more than $600 million in the US [35,36]. In these conditions, it has been proposed that the use of NPs is an efficient method for reducing pathogen activity and infection directly, resulting in improved crop development and yield [36]. For instance, clay nanotubes known as halloysites are utilised as low-cost pesticide carriers in agriculture. These nanotubes promise enhanced interaction with little adverse environmental impact in addition to demonstrating a prolonged AI release period [27]. One example of this is nanosilica, which has the hydrophobic feature of instantly killing insects by piercing their cuticle layer on contact [32]. Particularly praiseworthy was De Jorge et al.'s [37,38] examination of the role of nano formulation in AI-controlled release. They investigated whether pheromone nanofiber compositions from *Grapholita molesta* (Lepidoptera: Tortricidae) (Busck) altered mortality over time without releasing AI or having a persistent attract-and-kill effect of pheromone and insecticide.

A schematic diagram elucidating nanomaterial-based agrochemicals for sustainable plant growth regulators in ecosystem management has been depicted in Figure 3.2.

3.6 CONCLUSION

Utilising nanotechnology in agriculture has enhanced both crop quality and yields. To meet anticipated increases in global food consumption, the introduction of engineered nanomaterials and their application within the context of sustainable agriculture have substantially altered the global agricultural landscape. The reduction of pollution is a primary objective of sustainable agriculture trade, and nanomaterials promise improved input management and conservation. By reducing agricultural risks, the promise of nanomaterials encourages a new green revolution. Concerning the ecotoxicity, safe threshold, and absorbency of specific nanomaterials, however, there are numerous unanswered questions. Understanding the behaviour and outcome of altered agricultural inputs and their interactions with biomacromolecules present in living systems and their environments requires urgently more research.

REFERENCES

1. Dwivedi S., Saquib Q., Al-Khedhairy A.A., Musarrat J. Understanding the role of nanomaterials in agriculture. In: Singh D.P., Singh H.B., Prabha R., editors. *Microbial Inoculants in Sustainable Agricultural Productivity*. Springer; New Delhi, India: 2016, pp. 271–288.
2. Kou T.J., Yu W.W., Lam S.K., Chen D.L., Hou Y.P., Li Z.Y. Differential root responses in two cultivars of winter wheat (Triticum aestivum L.) to elevated ozone concentration under fully open-air field conditions. *J. Agron. Crop Sci.* 2018; 204:325–332. doi: 10.1111/jac.12257
3. Xiao M., Song F., Jiao J., Wang X., Xu H., Li H. Identification of the gene Pm47 on chromosome 7BS conferring resistance to powdery mildew in the Chinese wheat landrace Hongyanglazi. *Theor. Appl. Genet.* 2013; 126:1397–1403. doi: 10.1007/s00122-013-2060-6
4. Vermeulen S.J., Aggarwal P.K., Ainslie A., Angelone C., Campbell B.M., Challinor A.J., Hansen J.W., Ingram J.S.I., Jarvis A., Kristjanson P., et al. Options for support to agriculture and food security under climate change. *Environ. Sci. Policy.* 2012; 15:136–144. doi: 10.1016/j.envsci.2011.09.003
5. Panpatte D.G., Jhala Y.K., Shelat H.N., Vyas R.V. Microbial inoculants in sustainable agricultural productivity. In: Singh D., Singh H., Prabha R., editors. *Nanoparticles: The Next Generation Technology for Sustainable Agriculture*. Springer; New Delhi, India: 2016, pp. 289–300.
6. Thakur, A., Thakur, P., Baccar, S. Structural properties of nanoparticles. In: Suhag, D., Thakur, A., Thakur, P., editors. *Integrated Nanomaterials and their Applications*. Springer; Singapore: 2023. doi: 10.1007/978-981-99-6105-4_4
7. Khan M.R., Rizvi T.F. Nanotechnology: scope and application in plant disease management. *Plant Pathol. J.* 2014; 13:214–231. doi: 10.3923/ppj.2014.214.231
8. Afsharinejad A., Davy A., Jennings B., Brennan C. Performance analysis of plant monitoring nanosensor networks at THz frequencies. *IEEE Internet Things J.* 2016; 3:59–69. doi: 10.1109/JIOT.2015.2463685
9. Kwak S.-Y., Wong M.H., Lew T.T.S., Bisker G., Lee M.A., Kaplan A., Dong J., Liu A.T., Koman V.B., Sinclair R. Nanosensor technology applied to living plant systems. *Annu. Rev. Anal. Chem.* 2017; 10:113–140. doi: 10.1146/annurev-anchem-061516-045310
10. Worrall E., Hamid A., Mody K., Mitter N., Pappu H. Nanotechnology for plant disease management. *Agronomy.* 2018; 8:285. doi: 10.3390/agronomy8120285
11. He X., Deng H., Hwang H.-M. The current application of nanotechnology in food and agriculture. *J. Food Drug Anal.* 2018; 27:1–21. doi: 10.1016/j.jfda.2018.12.002
12. Kim D.H., Gopal J., Sivanesan I. Nanomaterials in plant tissue culture: the disclosed and undisclosed. *RSC Adv.* 2017; 7:36492–36505. doi: 10.1039/C7RA07025J
13. Verma S.K., Das A.K., Patel M.K., Shah A., Kumar V., Gantait S. Engineered nanomaterials for plant growth and development: a perspective analysis. *Sci. Total Environ.* 2018; 630:1413–1435. doi: 10.1016/j.scitotenv.2018.02.313
14. Kumar S., Nehra M., Dilbaghi N., Marrazza G., Hassan A.A., Kim K.-H. Nano-based smart pesticide formulations: emerging opportunities for agriculture. *J. Control. Release.* 2018; 294:131–153. doi: 10.1016/j.jconrel.2018.12.012
15. McShane H.V., Sunahara G.I. Environmental perspectives. In: Dolez P.I., editors. *Nanoengineering*. Elsevier; Amsterdam, The Netherlands: 2015, pp. 257–283
16. Love J.C., Estroff L.A., Kriebel J.K., Nuzzo R.G., Whitesides G.M. Self-assembled monolayers of thiolates on metals as a form of nanotechnology. *Chem. Rev.* 2005; 105:1103–1169. doi: 10.1021/cr0300789
17. Dahoumane S., Jeffryes C., Mechouet M., Agathos S. Biosynthesis of inorganic nanoparticles: a fresh look at the control of shape, size and composition. *Bioengineering.* 2017; 4:14. doi: 10.3390/bioengineering4010014
18. Kitching M., Ramani M., Marsili E. Fungal biosynthesis of gold nanoparticles: mechanism and scale up. *Microb. Biotechnol.* 2015; 8:904–917. doi: 10.1111/1751–7915.12151
19. Iravani S. Green synthesis of metal nanoparticles using plants. *Green Chem.* 2011; 13:2638–2650. doi: 10.1039/c1gc15386b
20. Park T.J., Lee K.G., Lee S.Y. Advances in microbial biosynthesis of metal nanoparticles. *Appl. Microbiol. Biotechnol.* 2016; 100:521–534. doi: 10.1007/s00253-015-6904–7
21. Bernhardt E.S., Colman B.P., Hochella M.F., Cardinale B.J., Nisbet R.M., Richardson C.J., Yin L.Y. An ecological perspective on nanomaterial impacts in the environment. *J. Environ. Qual.* 2010; 39:1954–1965. doi: 10.2134/jeq2009.0479
22. Tiwari J.N., Tiwari R.N., Kim K.S. Zero-dimensional, one-dimensional, two-dimensional and three-dimensional nanostructured materials for advanced electrochemical energy devices. *Prog. Mater. Sci.* 2012; 57:724–803. doi: 10.1016/j.pmatsci.2011.08.003

23. Lee W., Kang S.H., Kim J.Y., Kolekar G.B., Sung Y.E., Han S.H. TiO2 nanotubes with a ZnO thin energy barrier for improved current efficiency of CdSe quantum-dot-sensitized solar cells. *Nanotechnology.* 2009; 20:335706. doi: 10.1088/0957–4484/20/33/335706
24. Stouwdam J.W., Janssen R.A.J. Red, green, and blue quantum dot LEDs with solution processable ZnO nanocrystal electron injection layers. *J. Mater. Chem.* 2008; 18:1889–1894. doi: 10.1039/b800028j
25. Suhag, D., Thakur, P., Thakur, A. Introduction to nanotechnology. In: Suhag, D., Thakur, A., Thakur, P., editors. *Integrated Nanomaterials and their Applications.* Springer; Singapore: 2023. doi: 10.1007/978-981-99-6105-4_1
26. Khan M.R., Rizvi T.F. Nanotechnology: scope and application in plant disease management. *Plant Pathol. J.* 2014; 13:214–231. doi: 10.3923/ppj.2014.214.231
27. Rathore, R., Suhag, D., Wan, F., Thakur, A., Thakur, P. Everyday nanotechnology. In: Suhag, D., Thakur, A., Thakur, P., editors. *Integrated Nanomaterials and their Applications.* Springer; Singapore: 2023. doi: 10.1007/978-981-99-6105-4_2
28. Ran Y., Liang Z., Gao C. Current and future editing reagent delivery systems for plant genome editing. *Sci. China Life Sci.* 2017; 60:490–505. doi: 10.1007/s11427-017-9022-1
29. Miller J.B., Zhang S., Kos P., Xiong H., Zhou K., Perelman S.S., Zhu H., Siegwart D.J. Non-viral CRISPR/Cas gene editing in vitro and in vivo enabled by synthetic nanoparticle co-delivery of Cas9 mRNA and sgRNA. *Angew. Chem. Int. Ed.* 2017; 56:1059–1063. doi: 10.1002/anie.201610209
30. Kim D.H., Gopal J., Sivanesan I. Nanomaterials in plant tissue culture: the disclosed and undisclosed. *RSC Adv.* 2017; 7:36492–36505. doi: 10.1039/C7RA07025J
31. Afsharinejad A., Davy A., Jennings B., Brennan C. Performance analysis of plant monitoring nanosensor networks at THz frequencies. *IEEE Internet Things J.* 2016; 3:59–69. doi:10.1109/JIOT.2015.2463685
32. Kwak S.-Y., Wong M.H., Lew T.T.S., Bisker G., Lee M.A., Kaplan A., Dong J., Liu A.T., Koman V.B., Sinclair R. Nanosensor technology applied to living plant systems. *Annu. Rev. Anal. Chem.* 2017; 10:113–140. doi: 10.1146/annurev-anchem-061516-045310
33. Nuruzzaman M., Rahman M.M., Liu Y.J., Naidu R. Nanoencapsulation, nano-guard for pesticides: a new window for safe application. *J. Agric. Food Chem.* 2016; 64:1447–1483. doi: 10.1021/acs.jafc.5b05214
34. Haq I.U., Ijaz S. Use of metallic nanoparticles and nanoformulations as nanofungicides for sustainable disease management in plants. In: Prasad R., Kumar V., Kumar M., Choudhary D., editors. *Nanobiotechnology in Bioformulations.* Springer; Cham, Switzerland: 2019, pp. 289–316.
35. Gonzalez-Fernandez R., Prats E., Jorrin-Novo J.V. Proteomics of plant pathogenic fungi. *J. Biomed. Biotechnol.* 2010; 2010:932527. doi: 10.1155/2010/932527
36. Rai M., Ingle A. Role of nanotechnology in agriculture with special reference to management of insect pests. *Appl. Microbiol. Biotechnol.* 2012; 94:287–293. doi: 10.1007/s00253-012-3969-4
37. Bikash Ranjan J., Gurudutta P. Biomedical applications of nanozymes. In: Seema N., editors. *Emerging Environmental Applications of Nanozymes.* NOVA Science Publisher; Hauppauge, NY: 2023. doi: 10.52305/RFFX4767
38. De Jorge B.C., Bisotto-de-Oliveira R., Pereira C.N., Sant'Ana J. Novel nanoscale pheromone dispenser for more accurate evaluation of Grapholita molesta (Lepidoptera: Tortricidae) attract-and-kill strategies in the laboratory. *Pest Manag. Sci.* 2017; 73:1921–1926. doi: 10.1002/ps.4558

4 Nanomaterials in the Environment

Neetu Dhanda and Preeti Thakur
Amity University Haryana

An-Cheng Aidan Sun
Yuan Ze University

Atul Thakur
Amity University Haryana
Yuan Ze University

4.1 INTRODUCTION

Due to the limits of analytical techniques, nanomaterials (NMs), both organic and inorganic, have mostly gone undiscovered and have the potential to constitute pollutants [1,2]. The size of NMs or nanoparticles (NPs) determines how they are classified; sizes can range up to 1,000 nm but are typically characterized as having one dimension between 1 and 100 nm [3]. According to Turan et al. [4], NMs can be distinguished by their surface charge, surface area, degree of aggregation, surface coating, and particle shape [4,5]. NMs can be produced purposely for commercial purposes, happen naturally, or be an inadvertent byproduct. Although we have only lately learned about their existence and significance, the biosphere is abundant with natural NMs and anthropogenic NPs [6]. Natural-memory NMs that are present in the environment can enter the body of a person and affect their health through direct or indirect interactions. Agriculture, engineering, manufacturing, and medicine are among the potential fields of use for nanotechnology. The goal of current research is to advance and commercialize nanotechnology. By 2022, the electronics, energy, and biomedical industries are predicted to account for 70% of the growth in the market for nanotechnology, which is expected to reach 55 billion US dollars [7]. In the modern world, more attempts are being made to combat environmental pollution. Volatile organic compounds, and nitrogen oxides, are present in the air all over the world. Today's organic and inorganic substances found in fertilizers, sewage water, pesticides, and oil spills, all contribute considerably to soil and water contamination. The sources of inorganic and organic pollutants are shown in Figure 4.1. These pollutants pose risks to human health when they are breathed, swallowed, or come into touch with human skin. Therefore, engineers and scientists need to concentrate more on creating more effective technologies to identify and efficiently handle harmful substances in the environment [8]. Additionally, according to scientists, nanotechnology may represent the future of environmental cleanup and the improvement of older technologies. By limiting the discharge of contaminants into the atmosphere or preventing their creation, innovation is also regarded as an effective strategy for controlling environmental pollution. Nanotechnology has recently combined knowledge from a variety of disciplines, such as physics, informatics, biology, and medicine, and this has helped advance modern science and technology [9–11]. As a result, researchers started using NMs in better systems to make environmental monitoring and cleanup easier. Innovative remediation methods can be advanced by using NMs

DOI: 10.1201/9781003502692-4

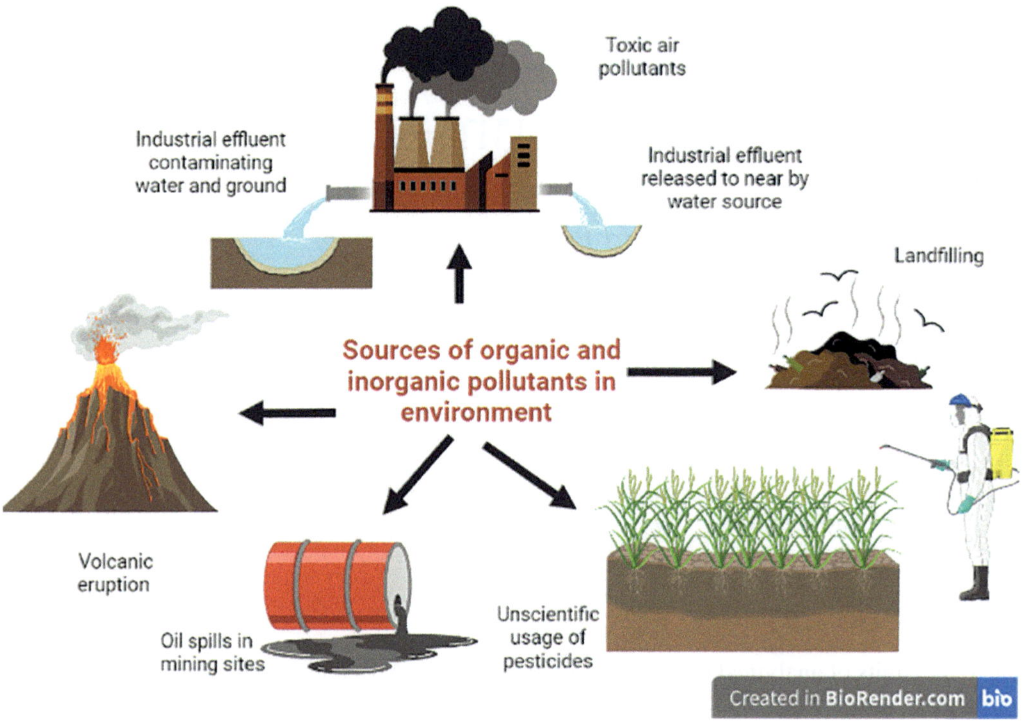

FIGURE 4.1 Sources of inorganic and organic pollutants. (Modified and reproduced from the Ref. [32].)

to create environmental conservation devices that can feel and identify pollutants. Nanomaterials' physical and chemical characteristics, such as their high surface area and reactivations, enhance their ability to absorb substances, making them better as sensors and catalysts. Due to its capacity to enable the advancement of well-established equipment with new characteristics and functions in a variety of disciplines, nanotechnology is now regarded as an emerging technology [12]. Numerous pollutants, such as carbon dioxide, chlorofluorocarbons, and hydrocarbons, have the potential adverse health effects according to the amount of study conducted over the past 10 years on nanomaterials to adsorb the pollutants likewise reflects its rising use. In addition, aqueous sample analysis may not be able to distinguish between natural and synthetic NMs with accuracy [13]. This chapter adopts a novel strategy to incorporate studies on the environmental roles of both natural and synthetic NMs (Table 4.1), their methods of analysis, and their possible effects on human health, as well as commenting on the existing knowledge gaps [14]. It analyzes the reasons behind the presence of NM in the natural world as well as any potential repercussions. Different categories and subgroups of synthetic and natural NMs were summarized in the text [15]. The two more general categories in the citation index of web science, nanotechnology and nanoscience, and environmental science, were the focus of further search refinement [16]. The focus of this chapter will be on how NMs behave in diverse environmental compartments, with a brief explanation of the many analytical techniques used to comprehend NMs in these intricate matrices [17]. Current knowledge gaps will be filled to explore potential future opportunities. Readers in the scientific area will gain a better understanding of the status of information regarding their transit and fate of NMs in diverse natural resources as well as possible exposure and nanotoxicity in humans through this chapter study.

TABLE 4.1
A List of the Numerous Natural and Man-Made (Synthetic) NMs Present in the Environment

Nanomaterials Type	Presence in the Environment	References
	Natural Nanomaterials (NMs)	
Carbon nanotubes	Atmosphere, soot, fires	[18]
Ferrihydrite	Soil, surface water, groundwater	[19]
Mercury nanoparticles	Soil and water resources	[20]
Silicon dioxide	Atmosphere, surface water, groundwater	[21]
Iron oxyhydroxide	Oceans, seas	[22]
Sulfur NMs	Mineral wells, springs	[18]
Manganese based NMs	Oceans, seas	[22]
Silver nanomaterials	Hydrothermal vents, surface water, wastewater	[23]
Nickel, zinc, cadmium, silver, tin, selenium, lead-bismuth	Volcanic eruptions, atmosphere	[24]
Polymeric NMs	Atmosphere	[25]
	Synthetic NMs	
Titanium dioxide	Wastewater, surface runoff, stormwater, surface and groundwater, landfills, atmosphere	[26]
Platinum	Atmosphere	[13]
Iron oxide	Atmosphere, stormwater, surface water, tap water	[17]
Fullerenes (C_{60})	Air, sludge, wastewater	[13]
Cerium oxide	River water	[26]
Zinc oxide	Soil, surface water, crops, landfills	[27]
Nanoplastics	Surface water, sea, ocean	[28]
Zinc sulfide	Stormwater	[29]
Carbon nanotubes	Atmosphere, wastewater, surface water, landfills	[30]
Lead sulfide	Atmosphere	[31]
Silver NMs	Wastewater, landfills, sludge	[32]

4.2 NANOMATERIALS

4.2.1 Natural NMs

Natural nanomaterials (NNMs) are defined as naturally occurring NPs in the earth's crust. These are often created by various biogeochemical processes, shown in Figure 4.2 under the heading "natural nanomaterials". Every year, a million megatons of NNMs with sizes ranging from 1 to a 1,000 nm circle the earth [13,14]. According to Hochella et al. [22], the NNMs atmospheric flux is around 342 megatons per/year [18]. These NNMs can also find their way into natural sources of water. Accepting the function of NNMs in the environment and their involvement in various element cycles is the main goal of the current study. By facilitating biogeochemical interactions, NNMs are recognized to protect a variety of habitats [16]. According to Westerhoff et al. [17] and Griffin et al. [18], NNMs have a wide range of morphologies, organic and inorganic forms, and compositions [17,18]. According to several studies, NNMs can be produced by volcanic eruptions,

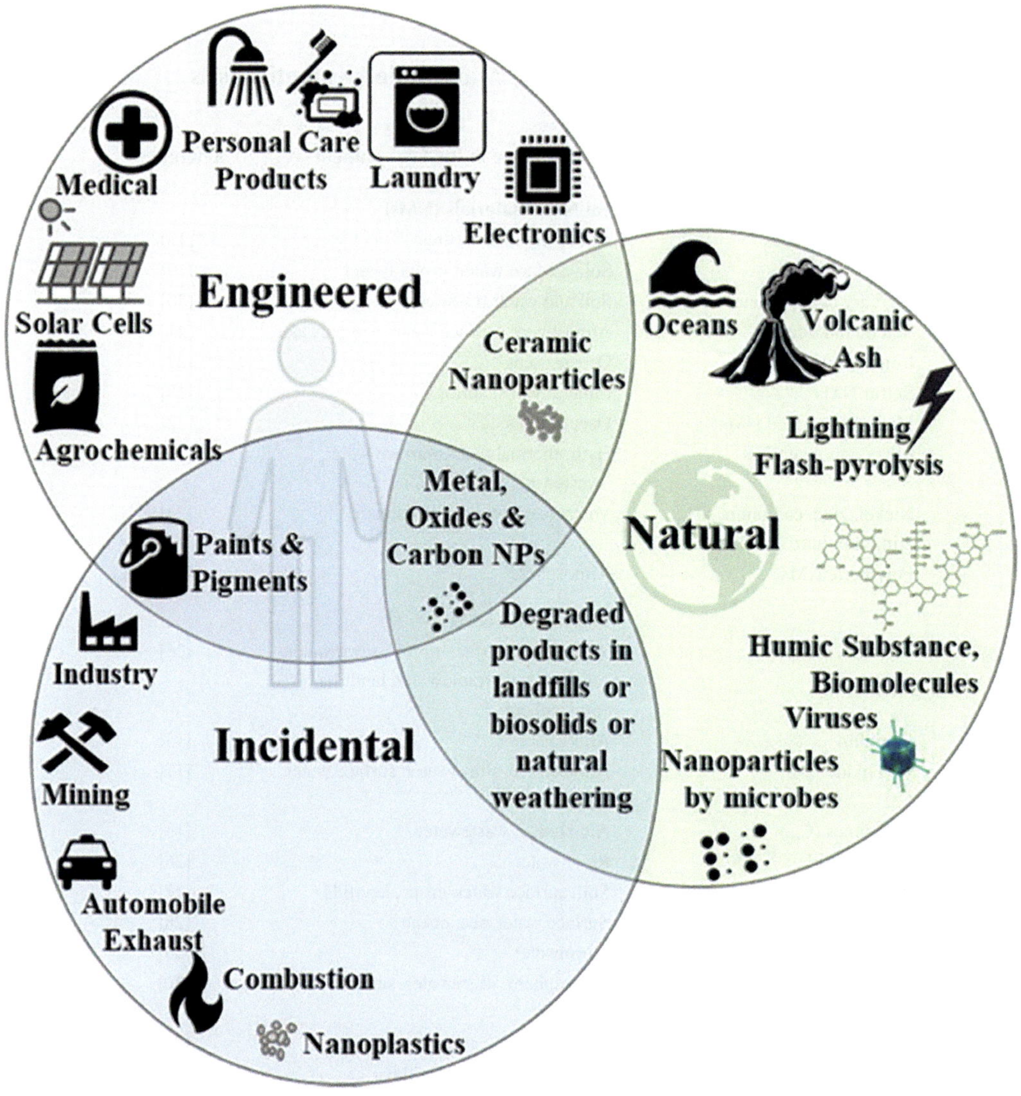

FIGURE 4.2 Different categories of anthropogenic and natural nanoparticles. It demonstrates how these diverse nanomaterials are created, the areas in which they are most prevalent, and the main applications of natural, incidental, and engineered NMs. (Reproduced by permission from the Ref. [31], License number 5550651231934, copyright 2021, Elsevier.)

natural chemical weathering processes, aero-plasma, flash pyrolysis, lightning, metal oxides, and metal nanomaterials formed by abiotic or biotic interactions (such as minerals of selenium, sulfur, and uranium) (Figure 4.2) [33–35].

4.2.1.1 Organic NNMs

The most prevalent form of NNM is humic compounds, which are the portion of the total environmental carbon (organic) pool that can be chemically extracted. According to Grillo et al. [36], humic substances can be a combination of aromatic and aliphatic organic molecules with a molecular weight of low or high and a size that is less than 5 nm [36]. The biomolecules can have a nanoscale size including proteins, peptidoglycans, peptides, and polysaccharides, as well as viruses [37].

4.2.1.2 NNMs from Abiotic, Biotic, and Physical Processes

Tons of NNMs are added to the biosphere each year by active volcanoes. In addition to having numerous components like cadmium, nickel, lead, zinc, and sulfur, NNMs from volcanoes can spread across great distances thanks to the wind [38]. According to Ermolin et al., the concentrations of these metalloids and metals in volcanic ash NNMs can be anywhere between 10 and 500 times greater than their background levels. However, due to their natural origin, NNMs formed from volcanic ash typically have diverse compositions and purity [24]. In space, NMs can be created by temperature, pressure, shock waves, radiation, high-speed physical collisions, and more [39]. According to studies, solid and gas particles are captured in a bolide's Mach cone as they enter the atmosphere and are transformed into a new class of polymeric nanomaterials [40]. Lightning is also known to cause flash pyrolysis that results in the formation of NMs, and lightning strikes on the soil can also directly result in NM production. The distinction between minerals and NNMs is one nuanced but crucial aspect to consider when thinking about NM occurrence. Nanominerals are minerals that occur only at the nanoscale, like ferrihydrite or clay particles [25]. Mineral nanomaterials are defined as minerals as they are known to occur in bulk at nanoscales. Also, the natural aquatic environment can contain a variety of microscopic and nano-polydisperse minerals, mostly calcium and occasionally iron oxides. It is also known that organic and inorganic NNMs can co-exist, with metal oxide and metal NMs being linked to organic NNMs like humic acid [41,42]. The earth's surface is home to a variety of metalorganic NNMs, which are essential for regulating many biogeochemical cycles. NNMs with stable stoichiometry and strong monodispersity can be produced by biotic processes including microbial respiration by bacteria, fungi, and algae [43,44]. According to Joshi et al. [45] and Das et al. [46], microbial respiration can produce inorganic natural NMs like gold, iron, copper, selenium, uranium, and silver, whereas microbial degradation can produce organic NMs such as humic matter. It is also possible to make highly mono-dispersed inorganic natural NMs of silicon, calcium, iron, selenium, and other elements through biomineralization [45,46]. These inorganic NNMs may naturally occur in various microbial processes or can be created by microorganisms [47].

4.2.2 Synthetic NMs

Synthetic NMs are materials at the nanoscale that can be produced by anthropogenic activity, both inadvertent and deliberate. However, the anticipated yearly flux of synthetic nanomaterials into the environment is 10.3 megatons/year to the atmosphere, which is much less than the annual flux of natural NMs [48]. Despite having a lower volume than NNMs, synthetic NMs are regarded as pollutants and a hazard to the environment. In general, synthetic NMs are divided into two categories: engineered NMs (ENMs) and incidental NMs (INMs) (Figure 4.2) [28].

4.2.3 Incidental NMs

According to Baalousha et al. [30], incidental nanomaterials are produced accidentally because of anthropogenic activity. Various sources of accidental NPs are depicted in Figure 4.2, including industrial waste, combustion processes, mining waste, corrosion and wear processes, and car exhaust [30,49,50]. Incidental nanomaterials can be based on carbon, like carbon soot from combustion and metal-based or nano plastics from plastic degradation; caused corrosion in tap water; and can potentially bear humans [51]. Transmission electron microscopy (TEM) and X-ray energy dispersive spectroscopy (X-EDS) were used to analyze the $PM_{2.5}$ air sample that was used in Shanghai, China, on a foggy day, as shown in Figure 4.3, which is taken from Baalousha et al. [30,52,53]. The existence of incidental NMs of lead, iron, and aluminum can be seen in the bright-field TEM micrograph (a) and the X-EDS maps (b–i), which also indicate the iron oxide's presence (c–h), calcium silicate (c–e–g), and galena (f–i), and calcium silicate NMs [54,55].

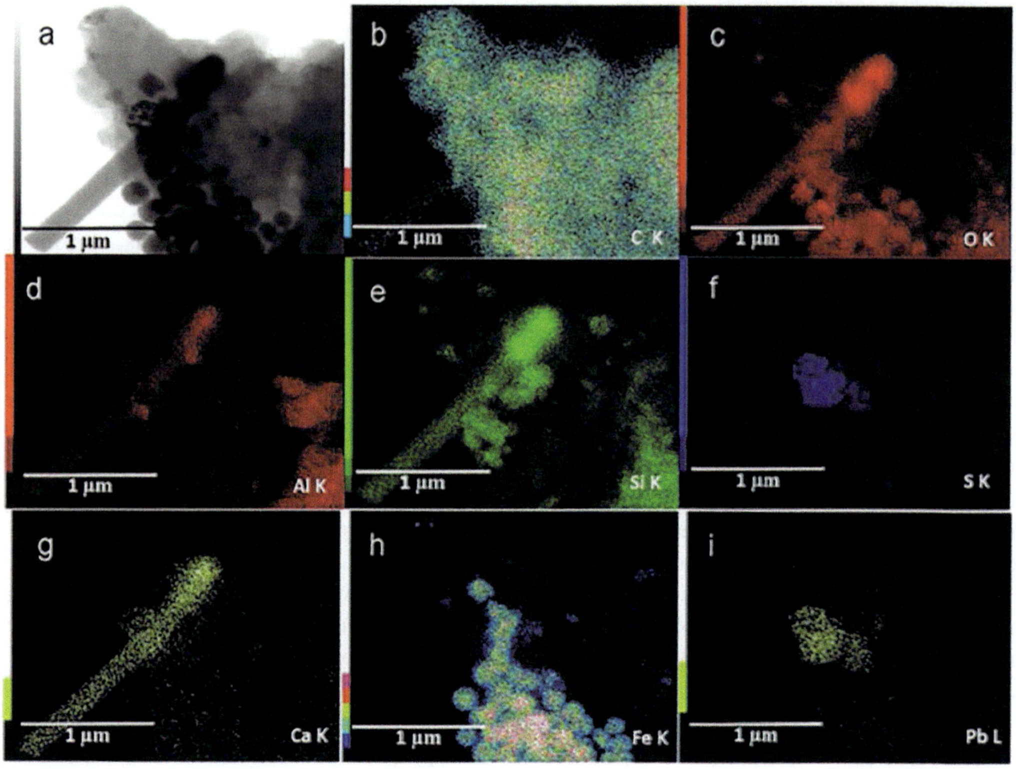

FIGURE 4.3 A typical NM aggregate was identified in a carbon matrix by transmission electron microscopy (TEM) and X-ray energy dispersive spectroscopy (X-EDS) maps of a $PM_{2.5}$ sample taken on a hazy day in Shanghai, China: (a) bright-field micrograph; and (b–i) elemental maps of Al, O, Si, Ca S, Fe, and Pb, respectively. These studies show the presence of nanomaterials such as iron oxide (c and h), galena, PbS (f and i), and Ca_2SiO_4 (c, e, and g). Using their selective area electron diffraction (SAED) patterns, iron oxides were recognized as magnetite NMs. (Reproduced by permission from the Ref. [31], License number 5550651231934, copyright 2021, Elsevier.)

4.2.4 Engineered NMs

Engineered nanomaterials are those that are created for commercial use. The energy industry, computing, telecommunications industry, agrichemicals, and personal care items are all affected by ENMs today. The use of ENMs is growing every day, and more ENMs are entering different water sources [56]. From quantum computing to agriculture, ENMs are applied in practically every technological field. The modern world depends heavily on nanotechnology, which supports numerous sectors. ENMs can be widely categorized based on their morphology, such as carbon- or metal-based nanomaterials, and their composition, such as 0 dimensions (quantum dots), 1 dimension (nanorods), 2 dimensions (graphene), or 3 dimensions (fullerene). These two major categories for ENMs, however, have a lot in common [57–60].

4.3 NMs in the Environment

Natural and manufactured NMs can interact with various environmental compartments via a variety of mechanisms (Table 4.1). NMs are frequently found in the air, and metropolitan areas are likely to have larger concentrations of accidental NMs (INMs). In these environmental compartments, NMs may build up and find their way into various water sources, as well as in the soil near landfills,

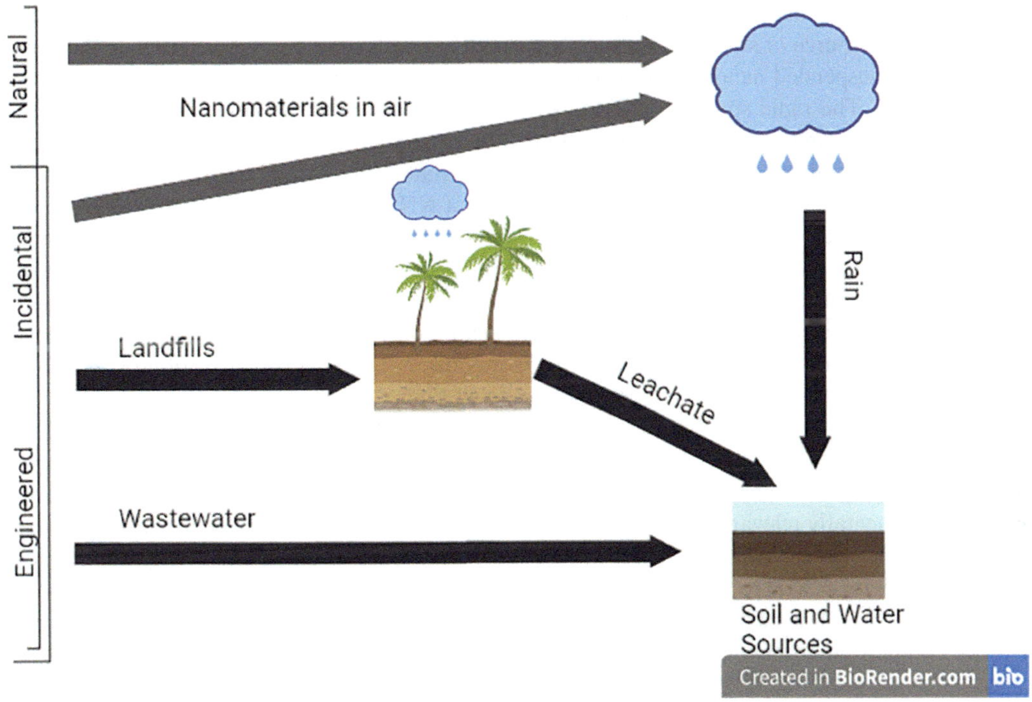

FIGURE 4.4 Routes by which artificial and natural nanoparticles (engineered and incidental) are transported to soil and water sources.

municipal wastewater, and industrial discharges, or by natural processes [20,61,62]. The route that NMs might take to get to various aquatic settings is shown in Figure 4.4. NNMs, which occur in large quantities and are easily accessible and the cosmos, can be recharged into groundwater or end up in precipitation. NNMs can linger in the air as dust for a long time and are probably present in the soil microbial community [63]. Manufacturing emissions, vehicle emissions, and sewage from landfills or wastewater treatment plants, these substances can enter natural water bodies and soil through precipitation from airborne particles or point sources of discharge [64]. Commercially manufactured NMs may be released as waste products or during the production stage of their life cycle into the aquatic environment. These ENMs may gradually travel to surface and groundwater settings or may stay in the soil, where they may later be ingested by plants or animals used as food sources. One of the main vectors for distributing ENMs into the environment is thought to be wastewater [65].

4.3.1 NNMs in the Environment

According to Griffin et al. [18], nature is an expert nanotechnologist, and there are a ton of NNMs found in the natural world. The transformation, fate, and potential toxicity of these substances are typically disregarded [18,66]. The lack of appropriate analytical techniques to assess NNMs in a complex natural setting dependably and reproducibly has been a major factor in their underappreciation. NNMs have contributed to the earth's evolution to its current state, which is another reason that their environmental impact is not well understood. The advancement of nano geosciences has resulted in the introduction of cutting-edge technologies for analyzing the destiny, possible toxicity, and transportation of NNMs as well as for comprehending their complex ecological functions [67]. There is a good chance that there are NNMs in the air that were created by space debris, lightning, forest fires, or volcanic ash. According to Griffin et al. [18], multi-walled carbon nanotubes with a size range between 15 and 70 nm are naturally occurring and are produced by soot. Volcanic eruptions are another source of silicon dioxide NPs that can enter the environment and irritate the eyes [18,68,69]. Air-suspended nanomaterials may be deposited into surface and groundwater sources by precipitation. The rapid spread of NNMs from volcanic ash to various surface water sources (seas, lakes, oceans, and rivers) via rain and winds (Figure 4.4) can enhance the risk of toxicity [70]. Numerous natural water resources, such as aquifers or groundwater, with NNM concentrations in the low mg L1 range, can contain significant amounts of NNMs in a scattered form. NNMs have been found in surface water in a variety of size regimes. The nanoscale size regime can include significant amounts of metal in the ocean [71]. For instance, over 100% of manganese is below 20 nm, while 80% of iron is in the size range of 20–200 nm. These elements also have an impact on biogeochemical cycles (redox based). NNM concentrations, however, can be too low to detect environmental toxicity [23,72–74]. Additionally, it was discovered that the formation of Ag NMs was sped up by iron (Fe (II)/Fe(III)) redox species. In addition to being generated via the reduction of Au ions, in water gold (Au) NNMs are also stabilized by Natural organic matter (NOM). The primary cause of NNMs' toxicity is their tendency to dissolve and release metal ions under anoxic circumstances [22]. This phenomenon is dependent on the mechanics and kinetics of dissolution under environmentally relevant anoxic and oxic conditions. NNMs can, however, also act as a vital source of nutrients. For instance, volcanic ash NMs can support phytoplankton but also raise toxicity levels [75]. It is difficult to study the nanoscale biogeochemical processes that are brought about by different NNMs. It is currently unclear as to how NNMs made of geological materials formed or in the complex environment how they affected various biogeochemical cycles. Recent research suggests that while NNMs may not pose a direct risk as a contaminant, they can regulate the levels of trace elements like arsenic pollution in various water sources [76]. The majority of NNMs have an organic layer on their surface, which creates a net negative charge (surface charge) and improves stabilization through the development of double layers. The NNMs' movement in various water sources can be controlled by the improved colloidal stability, which also regulates the mobility of trace elements bound to the NNMs. Surface charge, Size, and Stokes settling velocity are the three most important variables affecting NNM mobility [77]. The surface charge essentially regulates how NNMs aggregate in surface water. Precipitation can start following aggregation processes by gravitational forces. According to Wagner et al. [78], the mobility of NNMs in groundwater sources depends on size exclusion, pH, particle deposition, redox potential, ionic strength of the bulk solution, and surface charge of NNMs [78]. The general perception is that NNM–metal ion complexation reduces contaminant levels, reduces contaminant levels, and limits elemental mobility, all of which limit the negative impacts of trace elements in various water sources. However, the release of trace elements into water bodies due to the reductive decomposition of NNMs, for instance, can raise their concentration in the water source over the values that are considered to be safe [79]. The tendency of NNMs like metal hydroxides and oxides to form stable complexes with trace elements in the water is what gives them their mobility. The bioavailability of hazardous and necessary metal ions in the soil and natural water sources may be influenced by this characteristic. This complexation may cause the inner-sphere complex to develop between trace metal ions and NNMs. Complex formation is greatly influenced by variables like redox state, pH, ionic strength, the impact of NM interaction, and the content of humic matter relative to NNMs. Understanding the effects of NNMs in the soil can benefit from an understanding of these interactions [80]. At the aqueous–soil organism boundary, NNMs are crucial to many other related activities. NNMs can affect rhizosphere processes and end up in food items if they are present in the soil. The ability to analyze NNM interactions in the environment is severely constrained [81]. Understanding the varied roles of NNMs in microbial–soil–rhizosphere interfaces, climate change, and various natural cycles is now possible thanks to the rapid development of analytical tools to assess NMs. Concerning the chemical makeup of NNMs and their life cycle in natural systems, there is currently a sizable information gap. Further study is needed to understand the fate and behavior of NNMs since they can make it more difficult to comprehend the life cycle of NMs in various natural resources [82].

4.3.2 Synthetic NMs in the Environment

Synthetic NMs are released into the environment in the final stage of their life cycle. Similar to NNMs, it is difficult and complex to trace synthetic nanomaterials in a different natural setting. However, the precise sink and source of synthetic NMs may be able to aid in tracing the natural environment's life cycle. At the point of creation, synthetic NMs, including ENMs and INMs, have the potential to be released into the atmosphere and can hang suspended for a long time. According to Baloushaet al. (2016), road vehicles are the main source of synthetic NMs in densely populated urban areas [30,83]. The galena, calcium silicate, and iron oxide fingerprints are one of the examples of these synthetic NMs discovered embedded in $PM_{2.5}$ samples. The most significant avenue for human exposure to synthetic NMs is thought to be their atmospheric release, whether intentional or unintentional. In samples of water and air taken in the Netherlands, C_{60} was also present [19]. According to Malakar and Snow (2020), synthetic NMs suspended in the air can travel along the same route as NNMs and end up in a variety of water and soil sources. High levels of synthetic NMs can be found in urban run-off, which can be detected in stormwater and traced to wastewater treatment plants where they can further contaminate groundwater and surface runoff can contaminate the surface water sources [84]. ENMs interact with the varied components of various water bodies. The many interactions include precipitation, flocculation, transformation, hetero- and homo-agglomeration, and physical- and chemisorption [85]. After treatment, wastewater discharges a significant number of ENMs, which are then released into water sources (Figure 4.4). Biosolids and waste may also contain ENMs, which can easily contaminate groundwater or seep from landfills to reach surface waters (Figure 4.4). Synthetic NMs can be released into the environment directly or inadvertently via landfills and wastewater discharge. Any of these release phases can cause ENMs to change, which can change their characteristics [86]. According to Bundschuh et al. [86], the indirect release is where these alterations are predominantly served. Different bodies of water may hold onto these released NMs. Also, ENMs can enter the environment during the usage phase, production phase, or after dumping [87]. However, it is anticipated that ENMs, as opposed to INMs, are primarily released during the latter two stages of the life cycle of synthetic NMs. Due to paint peeling and subsequent mobilization, incidental NMs, like TiO_2, used in pigments, the paint industry, and the paper industry are discovered in water sources. The investigations also observed that sludge-treated soils accumulate TiO_2 NMs more frequently than landfills and sediments do [88]. The primary channel for release is wastewater discharge [89]. Almost 90% of all incidental carbon-based NMs, including CNTs, are released during the production phase and can end up in landfills directly. During the production stage, incidental metal NMs like silver may also be discharged into the environment. In addition, landfills, biofilm reactors, and wastewater can all deposit silver NMs [90]. According to studies by Choi et al. [90] and Kaegi et al. [91], silver NMs embedded in fabrics, which make up about 20–100% of the overall particle count, are liberated during laundry and can be seen in the wastewater stream. The main release point for ENMs is a wastewater treatment facility. The dispersion of ENMs to various water bodies is greatly aided by these treatment facilities. The transformation, destiny, behavior, and mobility of ENMs in the intricate wastewater matrix are still largely unknown [91,92]. The use of items based on nanotechnology in the environment is expanding right now. These uses for diverse purposes, such as the application of nano pesticides, nano fertilizers, and groundwater nano remediation, directly release ENMs into the environment. Groundwater remediation using nanoscale zero-valent iron (nZVI) particles [93]. For the point decontamination of groundwater sources, direct injection of nZVI is used. Direct injection into an aquifer may have unintended effects since nZVI can interact with microbes and negatively affect the biogeochemical cycles that are already in place in the subsurface, which can harm groundwater bacteria. The post-treatment ENM residues are assumed to be consumables, yet ENMs in water can be harmful to people due to their mobility, reactivity, and relative stability. Therefore, it is advised to conduct further research on the toxicity caused by direct intake [94]. In five distinct water sources, including groundwater and freshwater, a study is attempted to determine what happened to ENMs utilized in water treatment. After the final membrane filtration step, harmful levels of ENMs that are frequently employed in water treatment operations, such as Ag, TiO_2, and ZnO ENMs, were discovered. By releasing zinc ions through dissolution, ZnO ENMs produce reactive oxygen species. Depending on the pH of the solution, different amounts of ZnO nanomaterials can move across saturated porous media at varying speeds. If not eliminated in the finished product, nano adsorbents used to purify drinking water may have negative health impacts on people and other living things. ENMs can interact with NOMs such as humic matter to become stable in water. Due to the surface charge reversal, ZnO ENMs exhibit greater mobility in the presence of humic compounds [95]. Different aqueous media can stabilize ENMs with the help of humic chemicals. However, other factors, such as the presence of other NNMs, pH, and redox conditions, also have a big impact on how stable, mobile, and transformable they are in these complex systems. Accepting the life cycle of ENMs in complex ecosystems is difficult due to the multi-particle architecture of the natural environment.

Since ENMs have the potential to be harmful, they could be pollutants in many water sources. However, there aren't enough monitoring systems and comprehensive datasets to track nanomaterial levels in drinking water [96]. The most prevalent sources of raw drinking water in the majority of developed countries are surface water sources that have the greatest potential to be contaminated by nanomaterials. To forecast the likelihood of ENMs in drinking water, a variety of treatment facilities, point-of-use sites, and surface water sources were investigated using field data and a material-flow model [97]. Another study on surface water from the Meuse and Ijssel rivers in the Netherlands verified the presence of cerium oxide and Ag ENMs and TiO_2 microparticles. Our knowledge of the risks that synthetic NMs pose to human health in different environmental contexts is mostly insufficient. The impact of synthetic nanomaterials as a pollutant in soils, various water bodies, and the atmosphere cannot be fully understood due to several factors, including their nanoscale size, ephemeral nature, and lack of appropriate monitoring technologies [98]. To understand the exposure and release mechanisms of synthetic nanomaterials and their enduring effects on the environment, these information gaps must be filled.

4.4 ANALYSIS AND SAMPLING OF NMS IN THE ENVIRONMENT

NMs behave differently in their surroundings than ionic species do, and they have quite different properties from their bulk counterparts. The life cycle of nanomaterials in intricate environmental compartments made up of soil, water, sediment, biota, and air is studied at the nanoscale. Because of their specific characteristics, NMs must be studied using specialist analytical methods [99]. The use of nanotechnology and nanogeosciences for environmental research has recently attracted attention and support. Modern analytical nanoscience apparatus offers state-of-the-art technologies that have aided in defining new insights into NMs. To examine the potential impact and occurrence of NMs, environmental scientists have developed specialized sampling approaches that can be augmented by cutting-edge analytical technologies [100]. Additionally, controlled laboratory settings can be used to mimic the natural environment and investigate particular nanomaterials without any outside intervention, which can improve knowledge of their fate, transformation, and environmental impact [101]. A vital tool for studying NMs directly within complicated matrices is electron microscopy, like scanning electron microscopy (SEM) and TEM. The shape, size, and aggregation condition of the relevant NMs can be clarified with the aid of visualization. Additionally, X-EDS in combination with electron microscopy can be used to comprehend the elemental composition of the NM both qualitatively and quantitatively. Environmental samples are ideal for advanced techniques including environmental SEM, scanning TEM, and cryo-TEM. However, using these methods to distinguish between natural and artificial NMs remains difficult. To analyze NMs in biological materials, optical imaging methods like confocal fluorescence microscopy and hyperspectral-enhanced dark field microscopy can be used [102]. Optical microscopy, however, might only apply to fluorescent NMs. One of the most reliable methods for quantifying nanomaterials in various matrices, particularly in aqueous samples, is isotope and atomic spectrometry, more specifically inductively coupled plasma mass spectroscopy (ICP-MS). ICP-MS is a method of choice because of its low detection limits, and single-particle ICP-MS (spICP-MS) and other technologies are constantly being developed to pair with it for simpler NM detection. In this case, a lower concentration of nanomaterials in solution is provided such as one nanomaterial is comparable to a single pulse, and the intensity of each ion can reveal the information of particle size. As a new analytical method, spICP-MS will need to undergo extensive development to become more practical and affordable. To address the drawbacks of spICP-MS, other methods including time of flight and multi-collector are being developed [29,103]. Low NM concentrations in complicated environmental matrices necessitate repurification, which involves separating and purifying individual NMs before examination. The detection and characterization of NMs are made more accessible and accurate through enrichment. To extract and separate NMs of low size from water, field flow fractionation and ICP-MS analysis have been employed extensively [104]. Hydrodynamic chromatography, capillary electrophoresis, and size exclusion chromatography are further separation methods. These methods, nevertheless, might be prone to clogging and less sensitive at smaller dimensions. Dynamic light scattering and NP tracking analysis, two low-cost methods for assessing particle size distribution in liquid, are examples of light scattering techniques. These methods cannot reveal information regarding elemental composition or shape, making them potentially useless for polydisperse, heterogeneous materials [105]. According to Baalousha et al. (2016), filter materials can trap and concentrate NMs in the atmosphere, which can then be studied using the methods previously mentioned. Another continuing investigation of NMs in the air makes use of particle counters to measure the number concentration of NMs in air samples over time and with the potential for size resolution. High-resolution particle analyzers with continuous measurement capabilities and low detection limits include scanning mobility particle sizers with a range of 2.5–1,000 nm and fast mobility particle sizers with a range of 5.6–560 nm [30,106]. However, none of these modern analytical methods can analyze NMs in a complex multi-particle environment with great repeatability and sensitivity. The main criteria for improving the capability of these analytical techniques are reproducibility, sensitivity, and cost-effectiveness [107].

4.5 CONCLUSION

Critical attention must be paid to knowing the life cycle of both manufactured and naturally manufactured NMs. To create plans to protect the environment, it is important to understand the underlying biogeochemical implications of NMs on various natural resources. It is clear and alarming that nanomaterials are frequently found in the air, soil, and water used for food crops. A significant barrier to understanding NMs' fate in the environment is their nanoscale size. To better forecast long-term effects as more NMs, notably ENMs, are released into the environment, it is necessary to better understand their potential toxicity. By taking preventative steps, nanomaterials can be kept out of the environment and have less of an impact on the ecosystem and public health. Natural, inadvertent, and engineered NM exposure and its toxicological implications are not well understood. Modern analytical methods are not difficult enough to analyze nanomaterials in complex biological and ecological environments. The interaction of NMs with cells, tissues, and organs has become better understood because of recent research endeavors. Multiple exposure pathways should be considered, as with any cluster of harmful contaminants, and additional research is required to assess the impact of NMs on human health.

REFERENCES

1. Sharma, V.K., Filip, J., Zboril, R., Varma, R.S., 2015. Natural inorganic nanoparticlesformation, fate, and toxicity in the environment. *Chem. Soc. Rev.* 44, 8410–8423. https://doi.org/10.1039/c5cs00236b
2. Alizadeh, S., Ghoshal, S., Comeau, Y., 2019. Fate and inhibitory effect of silver nanoparticles in high rate moving bed biofilm reactors. *Sci. Total Environ.* 647, 1199–1210. https://doi.org/10.1016/jscitotenv.2018.08.073

3. Allouni, Z.E., Gjerdet, N.R., Cimpan, M.R., Høl, P.J., 2015. The effect of blood protein adsorption on cellular uptake of anatase TiO_2 nanoparticles. *Int. J. Nanomed.* 10, 687–695. https://doi.org/10.2147/IJN.S72726
4. Turan, N.B., Erkan, H.S., Engin, G.O., Bilgili, M.S., 2019. Nanoparticles in the aquatic environment: usage, properties, transformation and toxicity—a review. *Process. Saf. Environ. Prot.* 130, 238–249. https://doi.org/10.1016/jsep.2019.08.014
5. Voegelin, A., Senn, A.C., Kaegi, R., Hug, S.J., 2019. Reductive dissolution of As(V)- bearing Fe(III)-precipitates formed by Fe(II) oxidation in aqueous solutions. *Geochem. Trans.* 20, 2. https://doi.org/10.1186/s12932-019-0062-2
6. Baalousha, M., Stoll, S., Motelica-Heino, M., Guigues, N., Braibant, G., Huneau, F., Le Coustumer, P., 2019. Suspended particulate matter determines physical speciation of Fe, Mn, and trace metals in surface waters of Loire watershed. *Environ. Sci. Pollut. Res.* 26, 5251–5266. https://doi.org/10.1007/s11356-018-1416-5
7. Inshakova, E., Inshakov, O., 2017. World market for nanomaterials: structure and trends. *MATEC Web Conf.* 129, 02013. https://doi.org/10.1051/matecconf/201712902013
8. Wille, G., Hellal, J., Ollivier, P., Richard, A., Burel, A., Jolly, L., Crampon, M., Michel, C., 2017. Cryo-scanning electron microscopy (SEM) and scanning transmission electron microscopy (STEM)-in-SEM for bio- and organo-mineral interface characterization in the environment. *Microsc. Microanal.* 23, 1159–1172. https://doi.org/10.1017/S143192761701265X
9. Dhanda, N., Thakur, P., Kumar, R., Fatima, T., Hameed, S., Slimani, Y., Sun, A.-C.A., Thakur, A., 2023. Green-synthesis of Ni-Co nanoferrites using aloe vera extract: structural, optical, magnetic, and antimicrobial studies. *Appl. Organomet. Chem.* 37(7), e7110. https://doi.org/10.1002/aoc.7110
10. Barton, L.E., Auffan, M., Durenkamp, M., McGrath, S., Bottero, J.Y., Wiesner, M.R., 2015. Monte Carlo simulations of the transformation and removal of Ag, TiO_2, and ZnO nanoparticles in wastewater treatment and land application of biosolids. *Sci. Total Environ.* 511, 535–543. https://doi.org/10.1016/jscitotenv.2014.12.056
11. Thakur, A., Mathur, P., Singh, M. 2007. Controlling the properties of manganese-zinc ferrites by substituting In^{3+} and Al^{3+} ions. *Zeitschrift Fur Phys. Chemie.* 221, 837–845. https://doi.org/10.1524/zpch.2007.221.6.837
12. Dhanda, N., Thakur, P., Thakur, A. 2023. Green synthesis of cobalt ferrite: a study of structural and optical properties. *Mater. Today: Proceed.* 73(2), 237–240. https://doi.org/10.1016/jmatpr.2022.07.202
13. Bäuerlein, P.S., Emke, E., Tromp, P., Hofman, J.A.M.H., Carboni, A., Schooneman, F., de Voogt, P., van Wezel, A.P., 2017. Is there evidence for man-made nanoparticles in the Dutch environment? *Sci. Total Environ.* 576, 273–283. https://doi.org/10.1016/j.scitotenv.2016.09.206
14. Thakur, P., Gahlawat, N., Punia, P. et al. 2022. Cobalt nanoferrites: a review on synthesis, characterization, and applications. *J. Supercond. Nov. Magn.* 35, 2639–2669. https://doi.org/10.1007/s10948-022-06334-1
15. Hochella, M.F., Spencer, M.G., Jones, K.L., 2015. Nanotechnology: nature's gift or scientists' brainchild? *Environ. Sci. Nano.* 2, 114–119. https://doi.org/10.1039/c4en00145a
16. Dhanda, N., Thakur, P., Sun, A.C.A., Thakur, A., 2023. Structural, optical and magnetic properties along with antifungal activity of Ag-doped Ni-Co nanoferrites synthesized by eco-friendly route. *J. Magn. Magn. Mater.* 572, 170598. https://doi.org/10.1016/jmmm.2023.170598
17. Westerhoff, P., Atkinson, A., Fortner, J., Wong, M.S., Zimmerman, J., Gardea-Torresdey, J., Ranville, J., Herckes, P., 2018. Low risk posed by engineered and incidental nanoparticles in drinking water. *Nat. Nanotechnol.* 13, 661–669. https://doi.org/10.1038/s41565-018-0217–9
18. Griffin, S., Masood, M.I., Nasim, M.J., Sarfraz, M., Ebokaiwe, A.P., Schäfer, K.H., Keck, C.M., Jacob, C., 2018. Natural nanoparticles: a particular matter inspired by nature. *Antioxidants.* 7, 3. https://doi.org/10.3390/antiox7010003
19. Malakar, A., Snow, D.D., Ray, C., 2019. Irrigation water quality-a contemporary perspective. *Water* 11, 1482. https://doi.org/10.3390/w11071482
20. Ghoshdastidar, A.J., Ariya, P.A., 2019. The existence of airbornemercury nanoparticles. *Sci. Rep.* 9, 10733. https://doi.org/10.1038/s41598-019-47086-8
21. Lungu, M., Neculae, A., Bunoiu, M., Biris, C., 2015. *Nanoparticles' Promises and Risks: Characterization, Manipulation, and Potential Hazards to Humanity and the Environment.* Springer International Publishing, Cham. https://doi.org/10.1007/978-3-319-11728-7
22. Hochella, M.F., Mogk, D.W., Ranville, J., Allen, I.C., Luther, G.W., Marr, L.C., McGrail, B.P., Murayama, M., Qafoku, N.P., Rosso, K.M., Sahai, N., Schroeder, P.A., Vikesland, P., Westerhoff, P., Yang, Y., 2019. Natural, incidental, and engineered nanomaterials and their impacts on the earth system. *Science* 363 (6434), 8299. https://doi.org/10.1126/science.aau8299

23. Sharma, V.K., Zboril, R., 2017. Silver nanoparticles in natural environment: formation, fate, and toxicity. In: Yan, B., Zhou, H., GardeaTorresdey, J. L. (Eds.), *Bioactivity of Engineered Nanoparticles, Nanomedicine and Nanotoxicology*. Springer-Verlag, Singapore, pp. 239–258. https://doi.org/10.1007/978-981-10-5864-6_10
24. Ermolin, M.S., Fedotov, P.S.,Malik, N.A., Karandashev, V.K., 2018. Nanoparticles of volcanic ash as a carrier for toxic elements on the global scale. *Chemosphere*. 200, 16–22. https://doi.org/10.1016/j.chemosphere.2018.02.089
25. Courty, M.-A., Martinez, J.-M., 2015. Terrestrial carbonaceous debris tracing atmospheric hypervelocity-shock aeroplasma processes. *Proc. Eng*. 103, 81–88. https://doi.org/10.1016/j.proeng.2015.04.012
26. Peters, R.J.B., van Bemmel, G., Milani, N.B.L., den Hertog, G.C.T., Undas, A.K., van der Lee, M., Bouwmeester, H., 2018. Detection of nanoparticles in Dutch surface waters. *Sci. Total Environ*. 621, 210–218. https://doi.org/10.1016/j.scitotenv.2017.11.238
27. Durenkamp, M., Pawlett, M., Ritz, K., Harris, J.A., Neal, A.L., McGrath, S.P., 2016. Nanoparticles withinWWTP sludges haveminimal impact on leachate quality and soilmicrobial community structure and function. *Environ. Pollut*. 211, 399–405. https://doi.org/10.1016/j.envpol.2015.12.063
28. Lehner, R., Weder, C., Petri-Fink, A., Rothen-Rutishauser, B., 2019. Emergence of nanoplastic in the environment and possible impact on human health. *Environ. Sci. Technol*. 53, 1748–1765. https://doi.org/10.1021/acs.est.8b05512
29. Baalousha, M., Yang, Y., Vance, M.E., Colman, B.P., McNeal, S., Xu, J., Blaszczak, J., Steele, M., Bernhardt, E., Hochella, M.F., 2016. Outdoor urban nanomaterials: the emergence of a new, integrated, and critical field of study. *Sci. Total Environ*. 557–558, 740–753. https://doi.org/10.1016/j.scitotenv.2016.03.132
30. Sun, T.Y., Bornhöft, N.A., Hungerbühler, K., Nowack, B., 2016. Dynamic probabilistic modeling of environmental emissions of engineered nanomaterials. *Environ. Sci. Technol*. 50, 4701–4711. https://doi.org/10.1021/acs.est.5b05828
31. Malakar, A. et.al., 2021. Nanomaterials in the environment, human exposure pathway, and health effects: a review. *Sci. Total Environ*. 759, 143470. https://doi.org10.1016/j.scitotenv.2020.143470.
32. Bhavya, G. et.al., 2021. Remediation of emerging environmental pollutants: a review based on advances in the uses of eco-friendly biofabricated nanomaterials. *Chemosphere* 275, 129975. https://doi.org/10.1016/j.chemosphere.2021.129975
33. Dhanda, N., Kumari, S., Kumar, R., Kumar, D., Sun, A.C.A., Thakur, P., Thakur, A., 2023. Influence of Ni over magnetically benign Co ferrite system and study of its structural, optical, and magnetic behavior. *Inorg. Chem. Commun*. 151, 110569. https://doi.org/10.1016/jinoche.2023.110569
34. Boyes, W.K., van Thriel, C., 2020. Neurotoxicology of nanomaterials. *Chem. Res. Toxicol*. 33, 1121–1144. https://doi.org/10.1021/acs.chemrestox.0c00050
35. Mathur, P., Thakur, A., Singh, M., 2008. A study of nano-structured Zn-Mn soft spinel ferrites by the citrate precursor method. *Phys. Scripta*. 77, 045701. https://doi.org/10.1088/0031–8949/77/4/045701
36. Grillo, R., Rosa, A.H., Fraceto, L.F., 2015. Engineered nanoparticles and organicmatter: a review of the state-of-the-art. *Chemosphere*. 119, 608–619. https://doi.org/10.1016/j.chemosphere.2014.07.049
37. Cai, J., Zang, X., Wu, Z., Liu, J., Wang, D., 2019. Translocation of transition metal oxide nanoparticles to breast milk and offspring: the necessity of bridging mother offspring- integration toxicological assessments. *Environ. Int*. 133, 105153. https://doi.org/10.1016/jenvint.2019.105153
38. Chen, L., Li, J., Chen, Z., Gu, Z., Yan, L., Zhao, F., Zhang, A., 2019. Toxicological evaluation of graphene-family nanomaterials. *J. Nanosci. Nanotechnol*. 20, 1993–2006. https://doi.org/10.1166/jnn.2020.17364
39. Christou, A., Stec, A.A., Ahmed, W., Aschberger, K., Amenta, V., 2016. A review of exposure and toxicological aspects of carbon nanotubes, and as additives to fire retardants in polymers. *Crit. Rev. Toxicol*. 46, 74–95. https://doi.org/10.3109/10408444.2015.1082972
40. Coman, V., Oprea, I., Leopold, L.F., Vodnar, D.C., Coman, C., 2019. Soybean interaction with engineered nanomaterials: a literature review of recent data. *Nanomaterials*. 9, 1248. https://doi.org/10.3390/nano9091248
41. Courty, M.-A., Allue, E., Henry, A., 2020. Forming mechanisms of vitrified charcoals in archaeological firing-assemblages. *J. Archaeol. Sci. Rep*. 30, 102215. https://doi.org/10.1016/j.jasrep.2020.102215
42. Crampon, M., Joulian, C., Ollivier, P., Charron, M., Hellal, J., 2019. Shift in natural groundwater bacterial community structure due to zero-valent iron nanoparticles (nZVI). *Front. Microbiol*. 10, 533. https://doi.org/10.3389/fmicb.2019.00533
43. Cronin, J.G., Jones, N., Thornton, C.A., Jenkins, G.J.S., Doak, S.H., Clift, M.J.D., 2020. Nanomaterials and innate immunity: a perspective of the current status in nanosafety. *Chem. Res. Toxicol*. 33, 1061–1073. https://doi.org/10.1021/acs.chemrestox.0c00051

44. De Matteis, V., 2017. Exposure to inorganic nanoparticles: routes of entry, immune response, biodistribution and in vitro/in vivo toxicity evaluation. *Toxics*. 5, 29. https://doi.org/10.3390/toxics5040029
45. Joshi, N., Filip, J., Coker, V.S., Sadhukhan, J., Safarik, I., Bagshaw, H., Lloyd, J.R., 2018. Microbial reduction of natural Fe(III) minerals; toward the sustainable production of functional magnetic nanoparticles. *Front. Environ. Sci.* 6, 127. https://doi.org/10.3389/fenvs.2018.00127
46. Das, R.K., Pachapur, V.L., Lonappan, L., Naghdi, M., Pulicharla, R., Maiti, S., Cledon, M., Dalila, L.M.A., Sarma, S.J., Brar, S.K., 2017. Biological synthesis of metallic nanoparticles: plants, animals and microbial aspects. *Nanotechnol. Environ. Eng.* 2, 18. https://doi.org/10.1007/s41204-017-0029-4
47. De Matteis, V., Rinaldi, R., 2018. Toxicity assessment in the nanoparticle era. *Adv. Exp. Med. Biol.* 1048, 1–19. https://doi.org/10.1007/978-3-319-72041-8_1
48. Ding, G., Zhang, N.,Wang, C., Li, X., Zhang, J., Li, W., Li, R., Yang, Z., 2018. Effect of the size on the aggregation and sedimentation of graphene oxide in seawaters with different salinities. *J. Nanopart. Res.* 20, 313. https://doi.org/10.1007/s11051-018-4421-1
49. Eivazzadeh-Keihan, R., Maleki, A., de la Guardia, M., Bani, M.S., Chenab, K.K., Pashazadeh- Panahi, P., Baradaran, B., Mokhtarzadeh, A., Hamblin, M.R., 2019. Carbon based nanomaterials for tissue engineering of bone: building new bone on small black scaffolds: a review. *J. Adv. Res.* 18, 185–201. https://doi.org/10.1016/j.jare.2019.03.011
50. Erbs, J.J., Berquó, T.S., Reinsch, B.C., Lowry, G.V., Banerjee, S.K., Penn, R.L., 2010. Reductive dissolution of arsenic-bearing ferrihydrite. *Geochim. Cosmochim. Acta* 74, 3382–3395. https://doi.org/10.1016/j.gca.2010.01.033
51. Ermolin, M.S., Fedotov, P.S., 2016. Separation and characterization of environmental nano- and submicron particles. *Rev. Anal. Chem.* 35(4), 185–199. https://doi.org/10.1515/revac-2016-0006
52. Fang, Q., Chen, B., 2012. Adsorption of perchlorate onto raw and oxidized carbon nanotubes in aqueous solution. *Carbon* 50(6), 2209–2219. https://doi.org/10.1016/j.carbon.2012.01.036
53. Fang, Q., Shen, Y., Chen, B., 2015. Synthesis, decoration and properties of threedimensional graphene-based macrostructures: a review. *Chem. Eng. J.* 264, 753–771. https://doi.org/10.1016/j.cej.2014.12.001
54. Faulstich, L., Griffin, S., Nasim, M.J., Masood, M.I., Ali, W., Alhamound, S., Omran, Y., Kim, H., Kharma, A., Schäfer, K.H., Lilischkis, R.,Montenarh, M., Keck, C., Jacob, C., 2017. Nature's hat-trick: can we use sulfur springs as ecological source for materials with agricultural and medical applications? *Int. Biodeterior. Biodegrad.* 119, 678–686. https://doi.org/10.1016/j.ibiod.2016.08.020
55. Rana, K., Thakur, P., Tomar, M., Gupta, V., Thakur, A., 2016. Structural and magnetic properties of Ni-Zn doped BaM nanocomposite via citrate precursor. *AIP Conf. Proc.* 1731, 1–4. https://doi.org/10.1063/1.4947806
56. Gad, G.M.A., Hegazy, M.A., 2019. Optoelectronic properties of gold nanoparticles synthesized by using wet chemical method. *Mater. Res. Express* 6, 085024. https://doi.org/10.1088/2053-1591/ab1bb8
57. Ganguly, P., Breen, A., Pillai, S.C., 2018. Toxicity of nanomaterials: exposure, pathways, assessment, and recent advances. *ACS Biomater. Sci. Eng.* 4, 2237–2275. https://doi.org/10.1021/acsbiomaterials.8b00068
58. Thakur, A., Thakur, P., Hsu, J.H., 2011. Enhancement in dielectric and magnetic properties of In^{3+} substituted Ni-Zn nano-ferrites by coprecipitation method. *IEEE Trans. Magn.* 47(10), 4336–4339. https://doi.org/10.1109/TMAG.2011.2156394
59. Geiser, M., Jeannet, N., Fierz, M., Burtscher, H., 2017. Evaluating adverse effects of inhaled nanoparticles by realistic in vitro technology. *Nanomaterials* 7, 49. https://doi.org/10.3390/nano7020049
60. Ghosh, S., Pradhan, N.R., Mashayekhi, H., Zhang, Q., Pan, B., Xing, B., 2016. Colloidal aggregation and structural assembly of aspect ratio variant goethite (α-FeOOH) with nC60 fullerene in environmental media. *Environ. Pollut.* 219, 1049–1059. https://doi.org/10.1016/j.envpol.2016.09.005
61. Gogos, A.,Wielinski, J., Voegelin, A., Emerich, H., Kaegi, R., 2019. Transformation of cerium dioxide nanoparticles during sewage sludge incineration. *Environ. Sci. Nano* 6, 1765–1776. https://doi.org/10.1039/c9en00281b
62. González-Gálvez, D., Janer, G., Vilar, G., Vílchez, A., Vázquez-Campos, S., 2017. The life cycle of engineered nanoparticles. *Adv. Exp. Med. Biol.* 947, 41–69. https://doi.org/10.1007/978-3-319-47754-1_3
63. Good, K.D., Bergman, L.E., Klara, S.S., Leitch, M.E., VanBriesen, J.M., 2016. Implications of engineered nanomaterials in drinking water sources. *J. Am. Water Works Assoc.* 108, E1–E17. https://doi.org/10.5942/jawwa.2016.108.0013
64. Goswami, L., Kim, K.H., Deep, A., Das, P., Bhattacharya, S.S., Kumar, S., Adelodun, A.A., 2017. Engineered nano particles: nature, behavior, and effect on the environment. *J. Environ. Manag.* 196, 297–315. https://doi.org/10.1016/j.jenvman.2017.01.011

65. Gupta, R., Xie, H., 2018. Nanoparticles in daily life: applications, toxicity and regulations. *J. Environ. Pathol. Toxicol. Oncol.* 37, 209–230. https://doi.org/10.1615/JEnvironPatholToxicolOncol.2018026009
66. Han, J., Zhao, D., Li, D., Wang, X., Jin, Z., Zhao, K., 2018. Polymer-based nanomaterials and applications for vaccines and drugs. *Polymers (Basel)* 10, 31. https://doi.org/10.3390/polym10010031
67. Hartland, A., Lead, J.R., Slaveykova, V.I., 2013. The environmental significance of natural nanoparticles. *Nat. Educ. Knowl.* 4, 7; Hashem, N.M., Gonzalez-Bulnes, A., 2020. State-of-the-art and prospective of nanotechnologies for smart reproductive management of farm animals. *Animals* 10, 840. https://doi.org/10.3390/ani10050840
68. Hashem, N.M., Sallam, S.M., 2020. Reproductive performance of goats treated with free gonadorelin or nanoconjugated gonadorelin at estrus. *Domest. Anim. Endocrinol.* 71, 106390. https://doi.org/10.1016/j.domaniend.2019.106390
69. Hauser, M., Li, G., Nowack, B., 2019. Environmental hazard assessment for polymeric and inorganic nanobiomaterials used in drug delivery. *J. Nanobiotechnol.* 17, 56. https://doi.org/10.1186/s12951-019-0489-8
70. He, C.S., He, P.P., Yang, H.Y., Li, L.L., Lin, Y., Mu, Y., Yu, H.Q., 2017. Impact of zero-valent iron nanoparticles on the activity of anaerobic granular sludge: from macroscopic to microcosmic investigation. *Water Res.* 127, 32–40. https://doi.org/10.1016/j.watres.2017.09.061
71. Hochella, M.F., 2008. Nanogeoscience: from origin to cutting-edge applications. *Elements* 4, 373–379. https://doi.org/10.2113/gselements.4.6.373
72. Yin, Y., Yu, S., Liu, J., Jiang, G., 2014. Thermal and photoinduced reduction of ionic Au(III) to elemental Au nanoparticles by dissolved organic matter in water: possible source of naturally occurring Au nanoparticles. *Environ. Sci. Technol.* 48, 2671–2679. https://doi.org/10.1021/es404195r
73. Adegboyega, N.F., Sharma, V.K., Cizmas, L., Sayes, C.M., 2016. UV light induces Ag nanoparticle formation: roles of natural organic matter, iron, and oxygen. *Environ. Chem. Lett.* 14, 353–357. https://doi.org/10.1007/s10311-016-0577-z
74. Zhang, Z., Yang, X., Shen, M., Yin, Y., Liu, J., 2015. Sunlight-driven reduction of silver ion to silver nanoparticle by organic matter mitigates the acute toxicity of silver to Daphnia magna. *J. Environ. Sci.* 35, 62–68. https://doi.org/10.1016/j.jes.2015.03.007
75. Hoggan, J.L., Sabatini, D.A., Kibbey, T.C.G., 2016. Transport and retention of TiO_2 and polystyrene nanoparticles during drainage from tall heterogeneous layered columns. *J. Contam. Hydrol.* 194, 30–35. https://doi.org/10.1016/j.jconhyd.2016.10.003
76. Huang, Y.W., Cambre, M., Lee, H.J., 2017. The toxicity of nanoparticles depends on multiple molecular and physicochemical mechanisms. *Int. J. Mol. Sci.* 18, 2702. https://doi.org/10.3390/ijms18122702
77. Iavicoli, I., Leso, V., Beezhold, D.H., Shvedova, A.A., 2017. Nanotechnology in agriculture: opportunities, toxicological implications, and occupational risks. *Toxicol. Appl. Pharmacol.* 329, 96–111. https://doi.org/10.1016/j.taap.2017.05.025
78. Wagner, S., Gondikas, A., Neubauer, E., Hofmann, T., Von Der Kammer, F., 2014. Spot the difference: engineered and natural nanoparticles in the environmentrelease, behavior, and fate. *Angew. Chem. Int. Ed.* 53, 12398–12419. https://doi. org/10.1002/anie.201405050; Jain, R., Jordan, N., Tsushima, S., Hübner, R., Weiss, S., Lens, P.N.L., 2017. Shape change of biogenic elemental seleniumnanomaterials fromnanospheres to nanorods decreases their colloidal stability. *Environ. Sci. Nano* 4, 1054–1063. https://doi.org/10.1039/c7en00145b
79. Jawed, A., Saxena, V., Pandey, L.M., 2020. Engineered nanomaterials and their surface functionalization for the removal of heavy metals: a review. *J. Water Process Eng.* 33, 101009. https://doi.org/10.1016/j.jwpe.2019.101009
80. Jeevanandam, J., Barhoum, A., Chan, Y.S., Dufresne, A., Danquah, M.K., 2018. Review on nanoparticles and nanostructured materials: history, sources, toxicity and regulations. *Beilstein J. Nanotechnol.* 9, 1050–1074. https://doi.org/10.3762/bjnano.9.98
81. Jesus, S., Schmutz, M., Som, C., Borchard, G.,Wick, P., Borges, O., 2019. Hazard assessment of polymeric nanobiomaterials for drug delivery: what can we learn from literature so far. *Front. Bioeng. Biotechnol.* 7, 261. https://doi.org/10.3389/fbioe.2019.00261
82. Jiang, D., Zeng, G., Huang, D., Chen, M., Zhang, C., Huang, C.,Wan, J., 2018. Remediation of contaminated soils by enhanced nanoscale zero valent iron. *Environ. Res.* 163, 217–227. https://doi.org/10.1016/j.envres.2018.01.030
83. Jiang, L.,Wang, Y., Liu, Z., Ma, C., Yan, H., Xu, N., Gang, F.,Wang, X., Zhao, L., Sun, X., 2019. Three-dimensional printing and injectable conductive hydrogels for tissue engineering application. *Tissue Eng. B Rev.* 25, 398–411. https://doi.org/10.1089/ten.teb.2019.0100

84. Kah, M., Kookana, R.S., Gogos, A., Bucheli, T.D., 2018. A critical evaluation of nanopesticides and nanofertilizers against their conventional analogues. *Nat. Nanotechnol.* 13, 677–684. https://doi.org/10.1038/s41565-018-0131-1
85. Kanel, S.R., Al-Abed, S.R., 2011. Influence of pH on the transport of nanoscale zinc oxide in saturated porous media. *J. Nanopart. Res.* 13, 4035–4047. https://doi.org/10.1007/s11051-011-0345-8
86. Bundschuh, M., Filser, J., Lüderwald, S., McKee, M.S., Metreveli, G., Schaumann, G.E., Schulz, R., Wagner, S., 2018. Nanoparticles in the environment: where do we come from, where do we go to? *Environ. Sci. Eur.* 30, 6. https://doi.org/10.1186/s12302-018-0132-6
87. Kanel, S.R., Choi, H., Kim, J.Y., Vigneswaran, S., Shim, W.G., 2006a. Removal of arsenic(III) from groundwater using low-cost industrial by-products - blast furnace slag. *Water Qual. Res. J. Can.* 41, 130–139. https://doi.org/10.2166/wqrj.2006.015
88. Kanel, S.R., Greneche, J.M., Choi, H., 2006b. Arsenic(V) removal from groundwater using nano scale zero-valent iron as a colloidal reactive barrier material. *Environ. Sci. Technol.* 40, 2045–2050. https://doi.org/10.1021/es0520924
89. Kanel, S.R., Goswami, R.R., Clement, T.P., Barnett, M.O., Zhao, D., 2008. Two dimensional transport characteristics of surface stabilized zero-valent iron nanoparticles in porous media. *Environ. Sci. Technol.* 42, 896–900. https://doi. org/10.1021/es071774j
90. Choi, S., Johnston, M.V., Wang, G.S., Huang, C.P., 2017. Looking for engineered nanoparticles (ENPs) in wastewater treatment systems: qualification and quantification aspects. *Sci. Total Environ.* 590–591, 809–817. https://doi.org/10.1016/j.scitotenv.2017.03.061
91. Kaegi, R., Voegelin, A., Sinnet, B., Zuleeg, S., Siegrist, H., Burkhardt, M., 2015. Transformation of AgCl nanoparticles in a sewer system - a field study. *Sci. Total Environ.* 535, 20–27. https://doi.org/10.1016/j.scitotenv.2014.12.075
92. Khan, I., Saeed, K., Khan, I., 2019. Nanoparticles: properties, applications and toxicities. *Arab. J. Chem.* 12, 908–931. https://doi.org/10.1016/j.arabjc.2017.05.011
93. Kirkegaard, P., Hansen, S.F., Rygaard, M., 2015. Potential exposure and treatment efficiency of nanoparticles in water supplies based on wastewater reclamation. *Environ. Sci. Nano* 2, 191–202. https://doi.org/10.1039/c4en00192c
94. Koelmans, A.A., Diepens, N.J., Velzeboer, I., Besseling, E., Quik, J.T.K., van de Meent, D., 2015a. Guidance for the prognostic risk assessment of nanomaterials in aquatic ecosystems. *Sci. Total Environ.* 535, 141–149. https://doi.org/10.1016/j.scitotenv.2015.02.032
95. Koelmans, A.A., Quik, J.T.K., Velzeboer, I., 2015b. Lake retention ofmanufactured nanoparticles. *Environ. Pollut.* 196, 171–175. https://doi.org/10.1016/j.envpol.2014.09.025
96. Korin, N., Kanapathipillai, M., Ingber, D.E., 2013. Shear-responsive platelet mimetics for targeted drug delivery. *Isr. J. Chem.* 53, 610–615. https://doi.org/10.1002/ijch.201300052
97. Kretzschmar, R., Sticher, H., 1998. Colloid transport in natural porous media: influence of surface chemistry and flow velocity. *Phys. Chem. Earth* 23, 133–139. https://doi.org/10.1016/S0079-1946(98)00003-2
98. Kunhikrishnan, A., Shon, H.K., Bolan, N.S., El Saliby, I., Vigneswaran, S., 2015. Sources, distribution, environmental fate, and ecological effects of nanomaterials in wastewater streams. *Crit. Rev. Environ. Sci. Technol.* 45, 277–318. https://doi.org/10.1080/10643389.2013.852407
99. Laborda, F., Bolea, E., Cepriá, G., Gómez, M.T., Jiménez, M.S., Pérez-Arantegui, J., Castillo, J.R., 2016. Detection, characterization and quantification of inorganic engineered nanomaterials: a review of techniques and methodological approaches for the analysis of complex samples. *Anal. Chim. Acta* 904, 10–32. https://doi.org/10.1016/j.aca.2015.11.008
100. Larese Filon, F., Mauro, M., Adami, G., Bovenzi, M., Crosera, M., 2015. Nanoparticles skin absorption: new aspects for a safety profile evaluation. *Regul Toxicol. Pharmacol.* 72, 310–322. https://doi.org/10.1016/j.yrtph.2015.05.005
101. Lawrence, J.R., Swerhone, G.D.W., Dynes, J.J., Hitchcock, A.P., Korber, D.R., 2016. Complex organic corona formation on carbon nanotubes reduces microbial toxicity by suppressing reactive oxygen species production. *Environ. Sci. Nano* 3, 181–189. https://doi.org/10.1039/c5en00229j
102. Lead, J.R., Batley, G.E., Alvarez, P.J.J., Croteau, M.N., Handy, R.D., McLaughlin, M.J., Judy, J.D., Schirmer, K., 2018. Nanomaterials in the environment: behavior, fate, bioavailability, and effects—an updated review. *Environ. Toxicol. Chem.* 37, 2029–2063. https://doi.org/10.1002/etc.4147
103. Lei, C., Zhang, L., Yang, K., Zhu, L., Lin, D., 2016. Toxicity of iron-based nanoparticles to green algae: effects of particle size, crystal phase, oxidation state and environmental aging. *Environ. Pollut.* 218, 505–512. https://doi.org/10.1016/j.envpol.2016.07.030

104. Lewis, R.W., Bertsch, P.M., McNear, D.H., 2019. Nanotoxicity of engineered nanomaterials (ENMs) to environmentally relevant beneficial soil bacteria – a critical review. *Nanotoxicology* 13, 392–428. https://doi.org/10.1080/17435390.2018.1530391
105. Liu, J., Zhu, R., Xu, T., Laipan, M., Zhu, Y., Zhou, Q., Zhu, J., He, H., 2018. Interaction of polyhydroxy fullerenes with ferrihydrite: adsorption and aggregation. *J. Environ. Sci. (China)* 64, 1–9. https://doi.org/10.1016/j.jes.2017.06.016
106. Lowry, G.V., Gregory, K.B., Apte, S.C., Lead, J.R., 2012. Transformations of nanomaterials in the environment. *Environ. Sci. Technol.* 46(13), 6893–6899. https://doi.org/10.1021/es300839e
107. Maters, E.C., Delmelle, P., Bonneville, S., 2016. Atmospheric processing of volcanic glass: effects on iron solubility and redox speciation. *Environ. Sci. Technol.* 50, 5033–5040. https://doi.org/10.1021/acs.est.5b06281

5 Nanotech Advancements

Cultivating Sustainable Agriculture for the Future

Barsha Tripathy, Lipsa Dash, and Suchismita Mishra
SOA University

M. Sai Sindhu
College of Horticulture and Forestry

Gurudutta Pattnaik
Centurion University of Technology and Management

5.1 INTRODUCTION

Agriculture is the backbone of most of the developing countries in which a major part of their income comes from the agriculture sector and more than half of the population depends on it for their livelihood. The population of the world is increasing rapidly. It is expected to reach nearly 6 billion by the end of 2050 [1]. The productivity of the crops is decreasing because of environmental impacts, biotic stresses, climatic changes, and water unavailability. To overcome all such barriers, there is a need to develop crop varieties which are resistant to both biotic and abiotic stresses and give more yield in a short period. To overcome all these drawbacks, a smarter way, i.e., nanotechnology (NT), can be one of the sources. NT provides a great scope of novel applications in the plant nutrition fields to achieve the future requests of the rising population because nanoparticles have exclusive physicochemical characteristics, i.e., high surface area, high reactivity, and tunable pore size. NT is one of the rapidly active technologies having the potential for sustainable growth in agriculture. Nanomaterials possess effective properties, which makes them suitable in the agriculture field.

"Nano" is a Greek word meaning one-billionth of something. One-billionth of a meter is termed as one nanometer. A nanometer is 1/80,000 the diameter of a human hair or approximately ten hydrogen atoms wide. The term "nanotechnology" was first defined in 1974 by Norio Taniguchi of the Tokyo Science University. NT is a science that deals with the alteration and generation of materials in the size of one to a hundred nanometers. Joseph and Morrison (2006) defined NT as the manipulation or self-assembly of individual atoms, molecules, or molecular clusters into structures to create material devices with new or vastly different properties. One nanometer is 10^{-9} meters. It combines solid-state physics, chemistry, chemical engineering, biochemistry, biophysics, and materials science. Control of atoms or matter at this nanoscale can give a better understanding of physical, chemical, and biological processes, and the formation of better material or structure has a positive impact. The smaller size of nanomaterial possesses a large surface area and becomes more active. It can also alter atoms, which makes it more reactive in various sectors. NT could be described as an application which involves the ability to see and control every single atom or matter of any structure and manipulate any structure at the atomic level. Because of this unique property NT will be able to solve many inherent problems in agriculture. The application of NT to the agricultural and food industries was first addressed by a US Department of Agriculture roadmap

DOI: 10.1201/9781003502692-5

published in September 2003. It exhibits multiple applications and gives various types of knowledge on sustainable agriculture, food security, and environmental impact. Over-dependency on chemically based fertilizers and pesticides has generated serious problems in sustainability, health issues, and environmental impact. After this, agro-chemicals came into existence, which are environment friendly, but they have also generated some issues like poor shelf-life, non-effective under changing environmental conditions. As a result, nanoparticles came into existence to solve these issues and help in enhancing crop production, food security and sustainability. Nanotechnology likely affects all spheres of agriculture field preparation, crop production to post cooking and food serving. The key focus areas for NT in agricultural research are:

- Capable of improving the soil quality by having efficient nutrients which enhance greater productivity of the soil.
- Facilitating the appropriate process to ensure the availability of nutrients to plants that aid their growth process.
- Influencing the improvement of genetic traits of plants delivery of genes, DNA molecules, and drug molecules at a particular site in plants for the development of pest-repellent properties.
- Tools like quantum dots based on NT are also used properly for monitoring pathogens on a daily basis to aid heathy growth process of plants.
- Smart sensors and smart delivery systems will help the agricultural industry combat viruses and other crop pathogens.
- Nano encapsulation reduces the leaching and evaporation of harmful substances, which play an important role in the protection of the environment.

Nanomaterials can be prepared in two methods: the top-down and bottom-up approaches [2]. Various physical and chemical treatments like sonication, milling, and high-pressure homogenization were involved in the top-down approach to nanoparticles. In the bottom-up approach, the building blocks of the nanoparticles are formed first and assembled to produce the final particles [3]. Nanomaterials are classified into two, namely, organic nanomaterials and inorganic nanomaterials.

5.2 PROPERTIES OF THE NANOPARTICLES

Nanoparticle is defined based on the size at which fundamental characters are different from those of the corresponding bulk material. The properties of the nanoparticles show better results than the bulk particles like increments in the surface area and physical strength. Nanoparticles overlap in size with colloids, which range from 1 nm to 1 mm in diameter [4]; also, the physical properties of nanoparticles are different from the properties of the bulk material [5]. The change in properties is due to the reduced molecular size and also because of changed interactions between molecules. The properties and possibilities of NT, which have great interest in the agricultural revolution, are high reactivity, enhanced bioavailability and bioactivity, adherence effects, and surface effects of nanoparticles [6]. Two principal factors cause the properties of nanomaterials to differ significantly from other materials: increased relative surface area and quantum effects. Morphology-aspect ratio/size, hydrophobicity, solubility-release of toxic species, surface area/roughness, surface species contaminations/adsorption, during synthesis/history, reactive oxygen species (ROS) O_2/H_2O, capacity to produce ROS, structure/composition, competitive binding sites with receptor, and dispersion/aggregation are the important properties of nanoparticles [7]. The model nanomaterials for agricultural applications are supposed to have the following properties:

1. Providing actual concentration and controlled release of fertilizers or pesticides in response to certain conditions (TiO_2 nanoparticles used as plant fertilizer for mung bean to enhance crop production) [8]

2. Improved targeted activity [9]
3. Less eco-harmful with safe and relaxed transport

5.3 NT APPLICATIONS IN AGRICULTURE AND FOOD PRODUCTION

NT will transform the entire food industry, changing the way food is produced, processed, packaged, transported, and consumed (Figure 5.1). The key aspects of these transformations highlight current research in the agrifood industry and what future impacts these may have.

5.3.1 Nano Genetic Manipulation of Agricultural Crops

Nanobiotechnology offers a new set of tools to manipulate genes using nanoparticles, nanofibers, and nano capsules. Properly functionalized nanomaterial serves as vehicles and could carry a larger number of genes and substances that are able to trigger gene expression or control the release of genetic material throughout time in plants [10,11]. Proponents say NT is heading toward taking the genetic engineering of agriculture to the next level down atomic engineering. Atomic engineering could enable the DNA of seeds to be rearranged to obtain different plant properties including color, growth season, and yield [12]. Nanofiber arrays with potential applications in drug delivery, crop engineering, and environmental monitoring can deliver genetic material to cells quickly and efficiently. Controlled biochemical manipulations in cells have been achieved through the integration of carbon nanofibers which are surface-modified with plasmid DNA [13]. Chitosan nanoparticles are quite versatile, as well as their transfection efficiency can be modified; they can be PEGylated to control the release of genetic material as time goes by. The application of fluorescence-labeled starch nanoparticles as plant transgenic vehicle was reported in which the nanoparticle biomaterial was designed in such a way that it binds and transports genes across the cell wall of plant cells by inducing instantaneous pore channels in the cell wall, cell membrane, and nuclear membrane with the help of ultrasound. Argo nano connects the dots in the industrial food chain and goes one

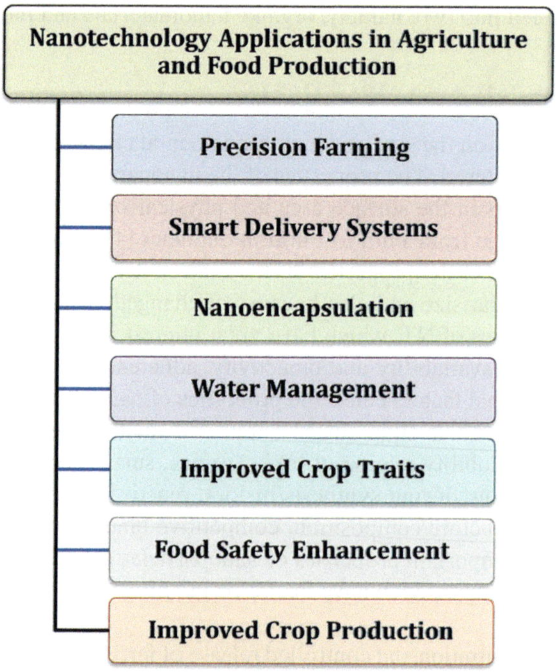

FIGURE 5.1 Various nanotechnology applications in agriculture and food production.

step further down. With new nanoscale techniques of mixing and harnessing genes, GM plants become atomically modified plants. DNA-coated silver nanoparticle (AgNPs coated with plasmid) treatments have been demonstrated to penetrate isolated protoplast of petunia and carry plasmatic DNA into the nucleus by incubation along with ethylene glycol [14]. Nowadays gene gun or particle bombardment is being used for direct delivery of DNA into intact plant cells. Particles used for bombardment are typically made of gold since they readily adsorb DNA and are non-toxic to cells. Experiments showed that the plasmid DNA transferred by gene gun method using gold-capped nanoparticles was successfully expressed in intact tobacco and maize tissues. The major advantage is the simultaneous delivery of both DNA and effector molecules to the specific sites resulting in site-targeted delivery and the expression of chemicals and genes, respectively [14]. The other interesting application of bio-nano sensors is to reduce pollen contamination in wind-pollinated crops. Detecting pollen load that will cause contamination is a sure method to ensure genetic purity. The use of bio-nano sensors specific to contaminating pollen can help alert the possible contamination and thus reduce contamination. The same method can also be used to prevent pollen from genetically modified crops from contaminating field crops. Meanwhile, small-scale farmers who grow tropical agricultural commodities such as rubber, cocoa, coffee, and cotton will find themselves quaint and irrelevant in a new nano economy of "flexible matter" in which the properties of industrial nanoparticles can be adjusted to create cheaper, "smarter" replacements. There are new challenges in this sector including a growing demand for healthy, safe food; an increasing risk of disease; and threats to agricultural and fishery production from changing weather patterns. However, creating a bioeconomy is a challenging and complex process involving the convergence of different branches of science. Nanotechnology has the potential to revolutionize the agricultural and food industry with new tools for the molecular treatment of diseases, rapid disease detection, enhancing the ability of plants to absorb nutrients, etc. [15].

5.3.2 Precision Farming

Precision agricultural techniques might be used to promote crop yields while minimizing input (i.e., fertilizers, pesticides, herbicides, etc.), but not damaging soil and water and decreasing nutrient loss due to leaching and emissions, in addition to enhancing long-term incorporation of nutrients by soil microorganisms (Figure 5.2). Precision farming makes use of computers, global satellite positioning systems, and remote sensing devices to measure highly localized environmental conditions, thus determining whether crops are growing at maximum efficiency or precisely identifying the nature and location of problems. By using centralized data to determine soil conditions and plant development, seeding, fertilizer, chemical, and water use can be fine-tuned to lower production costs and potentially increase production, all benefiting the farmer. In the Erosion, Technology, and Concentration (ETC) group down to the farm, it is described that precision farming can also help reduce agricultural waste and thus keep environmental pollution to a minimum. Although not fully implemented yet, tiny sensors and monitoring systems enabled by NT will have a large impact on future precision farming methodologies. One of the major roles of NT-enabled devices will be the increased use of autonomous sensors linked to a GPS for real-time monitoring. These nano sensors could be distributed throughout the field where they can monitor soil conditions and crop growth. The union of biotechnology and NT in sensors will create equipment of increased sensitivity, allowing an earlier response to environmental changes. Ultimately, precision farming, with the help of smart sensors, will allow enhanced productivity in agriculture by providing accurate information, thus helping farmers to make better decisions. For example:

- Nano sensors utilizing carbon nanotubes 12 or nano cantilevers 13 are small enough to trap and measure individual proteins or even small molecules.
- Nanoparticles or nano surfaces can be engineered to trigger an electrical or chemical signal in the presence of a contaminant such as bacteria.

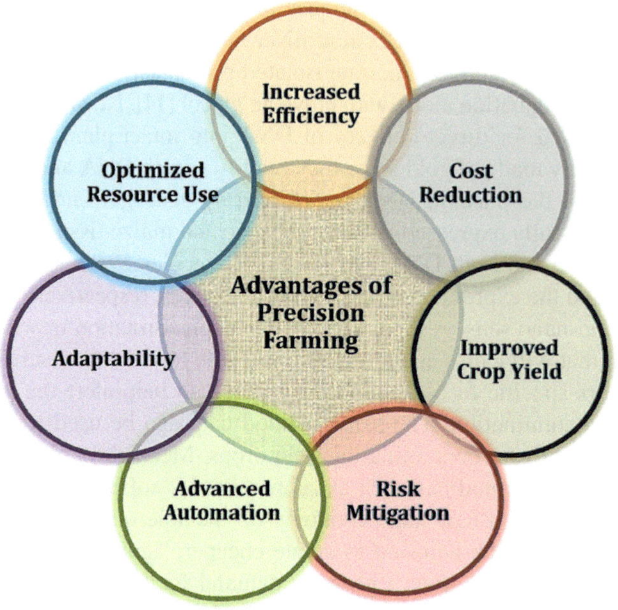

FIGURE 5.2 Advantages of precision farming.

- Other nano sensors work by triggering an enzymatic reaction or by using nanoengineered branching molecules called dendrimers as probes to bind to target chemicals and proteins. Ultimately, precision farming, with the help of smart sensors, will allow enhanced productivity in agriculture by providing accurate information, thus helping farmers to make better decisions.

5.3.3 Nano-Bio Farming

NT can enhance crop yield and nutritional values and can add value to crops or environmental remediation. Particle farming is one such field, which yields nanoparticles for industrial use by growing plants in defined soil. The nanoparticles can be mechanically separated from the plant tissue following harvest [16]. This process opens new opportunities in the recycling of waste and could be useful in areas such as cosmetics, food, or medicine. The most up-to-date research in this field is centered on the production of gold and silver nanoparticles with diverse plants: *Medicago sativa*, *Vigna radiata*, *Arachis hypogaea*, *Cyamopsis tetragonoloba*, *Zea mays*, *Pennisetum glaucum*, *Sorghum vulgare*, *Brassica juncea* extracts from *B. juncea* and *M. sativa*, *Memecylon edule*, and *Allium sativum* L. Depending on the nanoparticle's nature, species of plant, or tissues in which they are stored, metal nanoparticles of diverse shapes and sizes can be obtained. However, all these processes share the advantages of being simple, cost-effective, and environmentally friendly [17].

5.3.4 Nano Sensors

Nano sensors are emerging as promising tools for applications in agriculture and food production (Figure 5.3). Nano sensors can be used for the determination of microbes, contaminants, pollutants, and freshness of the food [18]. NT applications are also being developed to improve soil fertility and crop production. Nano sensors could also monitor crop and animal health, and magnetic nanoparticles could remove soil contaminants. NT is also used to determine the pollutant levels in the environment and the quantity of air dust by using nano-smart dust and gas sensors [19]. "Lab on

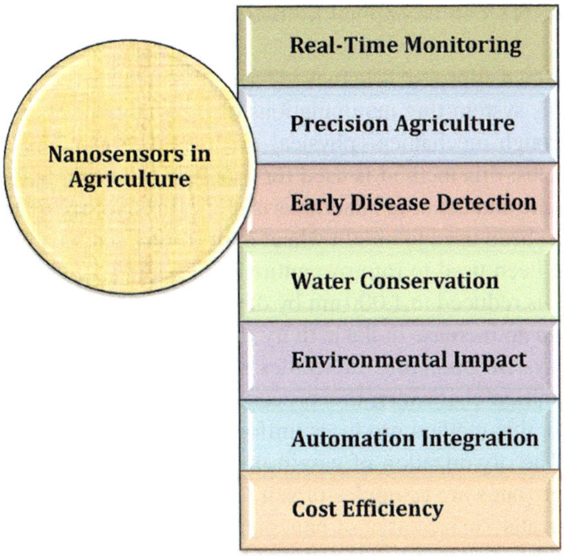

FIGURE 5.3 Advantages of nano sensors in agriculture.

a chip" technology could also have significant impacts on developing nations. Nano sensors for plant pathogen and pesticide detection and nanoparticles for soil conservation or remediation are other areas in agriculture that can benefit from NT. These nano sensors use nanoscale devices to identify and sense conditions that are physically, chemically, or biologically suitable for weed control and translate that response into signals and outputs in a useful form and then transmit it to users [20]. With the help of portable nano sensors, the presence of insects, diseases, pathogens, chemicals, and soil contaminants are easily detected in the field [7]. Similarly, nano-gold-based immune sensors can detect Karnal bunt (*Tilletia indica*) in wheat plants using a surface plasmon resonance sensor [21]. Enzyme immobilization for nano biosensors using nanomaterials involves the high-value low-volume application of enzymes. These nano sensors can also be used to reduce pollen contamination from genetically modified (GM) crops to conventional agronomic field crops [2].

5.3.5 NT and Food Systems

The definition of nanofood is that NT techniques or tools are used during the cultivation, production, processing, or packaging of the food. Since food systems encompass food availability, access, and utilization, the scope of applications of NT for enhancing food security must encompass entire agricultural production–consumption systems. Further, in a rapidly globalizing economy, increasing access to food and its utilization in rural areas will be determined primarily by an increase in rural incomes. Agriculture is the backbone of most developing countries, with more than 60% of the population reliant on it for their livelihood. The primary source of increasing rural incomes has been recognized as value addition across the different links in the agricultural production–consumption chain. These links include farm inputs, farm production systems, postharvest management and processing, and finally markets and consumers. From the food security perspective, it is therefore necessary that the application of NT be not limited to the farm production level but be extended across all the links of the agricultural value chain to increase agricultural productivity, product quality, consumer acceptance, and resource use efficiencies. This will help to reduce farm costs, raise the value of production, increase rural incomes, and enhance the quality of the natural resource base of agricultural production systems [22]. In doing so, it is important to view NT as an enabling technology that can complement conventional technologies and biotechnology [23].

For food applications, NT can be applied by two different approaches, either "bottom-up" or "top-down".

Top-down: The top-down approach is achieved basically by means of the physical processing of the food materials. In this system tiny manipulations of little number of atoms or molecules fashion elegant patterns, through mechanical–physical methods like grinding, milling, and crushing for producing nanoparticles; this method is used for producing nano composites and nano-grained bulk materials like metallic and ceramic nanomaterials in extensive size distribution (10–1,000 nm) [24]. This technology has been used to obtain wheat flour of fine size that has a high water-binding capacity [14]. It has also been used to improve antioxidant activity in green tea powder where the powder size of green tea is reduced to 1,000 nm by dry milling; the high ratio of nutrient digestion and absorption resulted in an increase in the activity of an oxygen-eliminating enzyme [25].

Bottom-Up: In this system numerous molecules self-assemble in parallel steps, as a function of their molecular recognition characters; this processing produces more complex structures from atoms or molecules; also, this method produces uniform controlling sizes, shapes, and size ranges of nanomaterials [24]. The organization of casein micelles or starch and the folding of globular proteins and protein aggregates are examples of self-assembly structures that create stable entities [26,27]. Owing to the greater surface area of nanoparticles per mass unit, they are expected to be more biologically active than larger-sized particles of the same chemical composition, which offers several perspectives for food applications. Nanoparticles can, for instance, be used as bioactive compounds in functional foods [9].

5.3.5.1 Food Processing

The nanoscale is not new to the food and beverage sector, with various phenomena already being witnessed and exploited in nutraceutical and functional food formulation, manufacturing, and processing. An array of food and beverages contain components that are nanoscale in size and processing (e.g., dairy), and the manipulation of naturally occurring nanoparticles is involved [27]. Potential applications include food that can alter its color, flavor, or nutrients to suit each consumer's preference or health requirements; filters that can take out toxins or modify flavors by sifting through certain molecules based on their shape instead of size; and packaging that can detect when its contents are spoiling and change color to warn consumers [28]. Foods can be enriched with fruits and vegetables through NT to deliver higher nutrient density in such foods. This technology also seems to be useful in dissolving additives, such as vitamins, minerals, antioxidants, phytochemicals, and nutritious oils that are not normally soluble [27].

5.3.5.2 Food Packaging

The food packaging industry already incorporates NT in products like polymeric films for food packaging with high antibacterial properties. NT is already being used worldwide to produce antimicrobial food contact materials commercially available as packaging or as coatings on an ever-increasing number of products such as food containers, chopping boards, and refrigerators [29]. Antimicrobial agents inhibit the growth of microbes either by destroying their cellular structure or by inhibiting their metabolic pathway [30]. Nanoparticles contribute to the generation of ROS that inhibits DNA replication and ATP formation resulting in cell damage. Nanomaterials and edible coatings added with nanoparticles are more advantageous than conventional packaging materials in providing better preservation and quality maintenance of food products [14]. Various organic, inorganic, and combined nanoparticles are used in the development of effective food packaging to prevent quantitative and qualitative losses of food products. They improve the mechanical and barrier properties of packaging materials and help in increasing the shelf-life of food products [14].

Considering the concerns on biosafety and consumer acceptance emerging after agribiotechnology-based products have entered the marketplace during the past two decades, it is also essential that new technologies like NT in agricultural and food systems be integrated and deployed after understanding the various societal and environmental implications [31]. The nanotechnologists are more optimistic about the potential to change the existing system of food processing and to ensure the safety of food products, creating a healthy food culture, and they are also hopeful of enhancing the nutritional quality of food through selected additives and improvements to the way the body digests and absorbs food.

5.4 ASSESSING NT FOR ENHANCED FOOD SECURITY IN INDIA

Food security is a primary policy concern of India. Food systems encompass three components: (a) food availability (production, distribution, and exchange); (b) food access (affordability, allocation, and preference); and (c) food utilization (nutritional value, social value, and food safety) [32]. Assessment of emerging technologies like NT is difficult because historical data is not available for impact assessment and much of the work is at the basic research stage with the future promise of a range of applications. In such situations, bibliometrics and patent analysis can be used to both assess the status and trends in technology development and classify and map them to relevant application areas for strategic planning [33]. The framework and databases were used to gauge the type of NT research currently in progress and to assess them from the perspective of food security [34]. A general premise is that basic research is found largely in journals, whereas potential commercial applications are found in patents. More than 60% of records from both databases were on R&D efforts to enhance plant/animal productivity, followed by research in food processing and food packaging which address the other two components of food security systems, namely, food availability and food utilization. Patent documents are also well structured to provide standardized information about citations, issue date, inventors, institutions and their locations technology field classification, etc. Such structured documentation makes them suitable for assessing technology developments in various areas. Bibliometric data, on the contrary, is less precisely structured but amenable to formal key word searches and more intensive text mining approaches for technology assessment. Then both of these databases were sorted and organized to carry out the assessment. This was carried out with respect to the factor, viz, NT product (nanoparticles, nanotubes, quantum dots, etc.). A holistic system framework was developed for patent and bibliometric analysis for the assessment of the potential of NT for enhancing food systems security in India. The framework was developed in two stages: (a) mapping NT to agri-food thematic areas across all the links of the agricultural value chain (farm inputs, production systems, postharvest management including storing and processing, markets, and consumption); (b) mapping NT to the determinants of food security (productivity, soil health, water security, and food quality).

Coming nanotechnologies in the agricultural field seem quite promising. However, the potential risks of using nanoparticles in agriculture are no different than those in any other industry. The environmental group ETC is deeply concerned with the implications and regulation of NT used in food. In a publication [35], the ETC stated that "the merger of nanotech and biotech has unknown consequences for health, biodiversity and the environment".

Emerging technologies like NT can be focused on primary determinants to catalyze research and develop a sustainable food security system. There is a need for investments in capacity building and development of an agri-NT infrastructure in India and for ex-ante assessment of its implications for society.

5.5 NANOPARTICLE-BASED SMART DELIVERY SYSTEMS

Nanoscale devices have the capability to detect and treat an infection, nutrient deficiency, or other health problem, long before symptoms are evident at the macro-scale. Potential applications of NT in crop protection include the controlled release of encapsulated pesticides, fertilizers, and other agrochemicals in protection against pests and pathogens and early detection of plant disease and pollutants including pesticide residues by using nano sensors. Smart delivery systems have the capacity to monitor the effects of the delivery of pharmaceuticals, nutraceuticals, nutrients, food supplements, bioactive compounds, probiotics, chemicals, insecticides, fungicides, and vaccinations. Delivery systems in agriculture are important for the application of pesticides and fertilizers as well as during genetic material-mediated plant improvement. It gives an opportunity to develop improved systems for monitoring environmental conditions and delivering nutrients or pesticides in an appropriate manner, improving our understanding of the biology of different crops and thus potentially enhancing yields or nutritional values. There is a huge scope for applying nanoparticles and nano capsules to plants for agricultural use [36]. Application systems for pesticides need to focus on efficacy enhancement and spray drift management, while fertilizers face problems of bioavailability due to soil chelation, over-application, and runoffs. A viable alternative for these problems is provided by controlled delivery systems for pesticide and fertilizer applications. The controlled delivery technique aims toward the measured release of necessary and enough agrochemicals over a period, to obtain the fullest biological efficacy and to minimize the harmful effects [37]. For this purpose, micro- and sub-micro particles were explored as agrochemical delivery vehicles. In comparison to micronic particles (1,000 nm), nanoparticles (100 nm) offered the advantage of effective loading due to the larger surface area, easy attachment, and fast mass transfer. Controlled release of the active ingredient is achieved due to the slow-release characteristics of the nanomaterial, the bonding of the ingredients to the material, and the environmental conditions.

5.5.1 Delivery of Pesticides/Biopesticides/Herbicides/Fertilizers

NT has the potential for efficient delivery of chemical and biological pesticides using nanosized preparations or nanomaterial-based agrochemical formulations. Improvement in the efficacy of pesticides or herbicides using NT resulted in greater production of crops [38]. The benefits of nanomaterial-based formulations are the improvement of efficacy due to higher surface area, higher solubility, induction of systemic activity due to smaller particle size and higher mobility, and lower toxicity due to the elimination of organic solvents in comparison to conventionally used pesticides and their formulations [39]. These nano-herbicides contain many trillions of active ingredient particles per liter and create an extra surface area by the reduction of particle size, which boosts potency, accelerates plant uptake, increases tank-mix solubility, and reduces settling and separation risks [3]. Infield, the application of NPs for the delivery of pesticides and biopesticides faces several challenges such as multiple environmental perturbations, large areas under spray coverage, and finally cost-effectiveness. In the usual spraying regime, the whole crop is sprayed with the chemical for ease of application involving a high-volume, low-value preparation, whereas nanomaterial-based preparations are expected to involve low-volume, high-value applications. Such controlled nanoparticulate delivery systems will require a targeted delivery approach focused on using the knowledge of the life cycle and the behaviors of the pathogen or pest. The development of nano-herbicides or nano-formulated adjuvants will address the problems in perennial weed management and will help in exhausting weed seedbanks. This technology will decrease or eliminate the effect of excess toxins on the environment and allow HT crop farmers to increase their yields at lower use and costs of herbicides [3].

Slavin et al. [40] reported that the application of large amounts of fertilizers, in the form of ammonium salts, urea, and nitrate or phosphate compounds, is harmful. About 40–70% of nitrogen, 80–90% of phosphorus, and 50–70% of potassium of the applied normal fertilizers are lost to the environment and cannot be absorbed by plants, causing not only substantial economic and resource losses but also very serious environmental pollution. Nanocoatings, or surface coatings of nanomaterials, on fertilizer particles hold the material more strongly from the plant due to higher surface tension than conventional surfaces. Moreover, nanocoatings provide surface protection for larger particles [6,41]. So, there is a need to evolve nano-based fertilizer to address issues of low fertilizer use efficiency, imbalanced fertilization, multi-nutrient deficiencies, and decline of soil organic matter.

5.5.2 Field Application of Nano-Pesticides and Nano-Fertilizers

The mode of pesticide and fertilizer application influences their efficiency and environmental impact [36,42,43]. Conventional fertilizers are generally applied on the crops by either spraying or broadcasting. However, one of the major factors that decide the mode of application is the final concentration of the fertilizers reaching the plant. In the practical scenario, very little concentration to the targeted site is done due to the leaching of chemicals, drift, runoff, evaporation, hydrolysis by soil moisture, and photolytic and microbial degradation. These nanomaterials are used for the preparation of various forms of nano-based agricultural tools to preserve the dynamic nature of the soil and to improve crop productivity by nano-fertilizers, which have made an impeccable impact on the sustainable improvement of agricultural research. Nanomaterials for the improvement and sustenance of soil and improvement of crops are generally either of organic or inorganic or both hybrid origins. Researchers reported that nano-fertilizer refers to a product in a nanometer regime that delivers nutrients to crops. Delivery of agrochemical substances such as fertilizer supplying macro and micronutrients to the plants is an important aspect of the application of NT in agriculture. The choice of fertilizer application method mainly depends on the soil, crop, irrigation type, and the nutrient applied. Current practices involving broadcasting, banding, side dressing, and dusting face problems of runoff due to dissolution in soil moisture and leaching. In the broadcast method of urea application, nitrogen fixation was inhibited while the growth of green algae was favored. In contrast, deep placement of urea granules (1–2 g) did not suppress the growth of nitrogen-fixing blue green algae [44]. NP-coated fertilizers may contribute to the slow release of the fertilizer, preventing the rapid dissolution and therefore harm to the environment.

Currently, the spraying of pesticides involves either knapsacks that deliver large droplets (9–66 μm) associated with splash loss or ultralight volume sprayers for controlled droplet application, with smaller droplets (3–28 μm) causing spray drift. Constraints due to droplet size may be overcome by using NP-encapsulated or nanosized pesticides that will contribute to efficient spraying and reduction of spray drift and splash losses [45]. Another practical problem faced during pesticide application in the field is the settlement of formulation components in the spray tank and the clogging of spray nozzles. The recent nanosized fungicide prevented spray tank filters from clogging, did not require mixing, and did not settle down in the spray tank due to the smaller-sized particles [46].

The potential application of nanocides in different agricultural applications needs further research investigation with respect to synthesis, toxicology, and its effective application at field level [47].

5.6 NT in Crop Nutrition

Plants absorb nutrients from fertilizers, but most conventional fertilizers have low nutrient use and uptake efficiency. Nano-fertilizers are, therefore, engineered to be target-oriented and not easily lost. Plant nutrients could be enriched by applying nanoparticulate nutrients, which are easily absorbed by the plant. The use of chemical fertilizers is an age-long practice and has tremendously increased crop yields. However, they lead to soil mineral imbalance and destroy the soil structure, soil fertility, and general ecosystem, which are serious impediments in the long term. They can improve seed water content when applied moderately. Plants remain the primary source of nourishment for humans, and food quality determines the health of the majority of the people. Consumption of diverse food sources, although recommended as a sustainable solution, is unaffordable to the poor populace, who are at risk of malnutrition. The use of industries for the fortification of food nutrients has not been very successful, except for iodized salt. Biofortification is the concept of increasing the nutrient content of food crops during their cultivation. It enhances the nutritional content of crops, and these strategies include agronomic and breeding methods. A typical example is the development of a new variety of sweet potatoes rich in vitamin A. Although biofortified staple crops deliver low levels of essential nutrients and vitamins per day compared to supplements or industrially enriched foods, they can satisfy the individual daily requirement of micronutrients. To improve the micronutrient content of crops using agronomic biofortification, the use of phytoavailable micronutrient fertilizers, routine correction of the soil alkalinity, crop rotational methods of planting, and strategic introduction of symbiotic soil microorganisms are required. Although there are some toxicity issues associated with the use of nanoparticles in crops, biologically synthesized nanoparticles may be preferred for agricultural purposes.

5.7 POTENTIAL RISKS OF NT

There are some negative effects of nanomaterials on biological systems and the environment caused by nanoparticles, like chemical hazards on edible plants after treatment with high concentrations of nano silver and also, in some cases, nanomaterial-generated free radicals in living tissue, leading to DNA damage; therefore, NT should be carefully evaluated before increasing the use of the nano agro materials [24].

5.7.1 Future Prospects

NT applications have the huge potential to change agricultural production by allowing better scientific management and conservation efforts to plant production. It can enhance agricultural productivity by using:

- Nanoporous zeolites for controlled release and efficient amount of water, fertilizer, etc.
- Nano capsules for delivering herbicide, vector, and management of pests
- Nano sensors for detecting aquatic toxins and pests
- Nanoscale biopolymer (proteins and carbohydrates)-based nanoparticles with few properties such as low impact on human health and the environment for the disinfection and recycling of heavy metals
- Nanostructured metals for the decomposition of harmful organics at room temperature
- Smart particles for effective environmental monitoring and purification processes
- Nanoparticles as a novel photocatalyst

Thus, NT will transform agricultural practices including advanced pest management in the future. Over the next 20 years, the green revolution would be hastened by means of nanoscience. Nanomaterials would be beneficial in the development and formulation of next-generation pesticides, insecticides, and insect repellents. Thus, NT is considered one of the best possible solutions to the problems present in the food and agriculture sectors.

5.8 CONCLUSION

Still, the full potential of NT in the agricultural and food industry is yet to be realized and is gradually moving from theoretical knowledge toward the application regime. Considering the great challenges we will be facing due to a growing global population and climate change, the application of nanotechnologies and the introduction of nanomaterials in agriculture have greatly contributed to addressing the issue of sustainability. The use of NT could permit rapid advances in agricultural research, such as reproductive science and technology which will produce a large number of seeds and fruits unaffected by season and period, early detection of stresses, alleviating stress effects, and disease prevention and treatment in plants. A main contribution anticipated is the application of nanoparticles to stabilize biocontrol preparations that will go a long way in reducing environmental hazards. In fact, the efficient use of fertilizers and pesticides can be enhanced using nanoscale carriers and compounds, reducing the amount to be applied without impairing productivity. NT can endeavor to provide and fundamentally streamline the technologies currently used in environmental detection, sensing, and remediation. In the future, nanoscale devices could be used to make agricultural systems smart. Apart from the potential benefits of NT in the agricultural sector, it involves some risks. It cannot be claimed with certainty whether those nanotechnologies are fully safe for health or they are harmful. Risks associated with chronic exposure of farmers to nanomaterial, unknown life cycles, interactions with the biotic or abiotic environment, and their possible amplified bioaccumulation effects have not been accounted for, and these should be seriously considered before these applications move from laboratories to the field. The common challenges related to commercializing NT are high processing costs, problems in the scalability of R&D for prototype and industrial production, and concerns about public perception of environment, health, and safety issues. Governments across the world should form common and strict norms and monitoring before commercialization and bulk use of these nanomaterials. Therefore, the fate of nanomaterials and their impact on the environment must be understood clearly. For this, the following points should be taken care of:

1. Addressing the environmental issues, potential risks, toxicity, and consequences of nanoparticles is very much essential.
2. Collaborative studies had to be carried out among various institutions for developing multifunctional, efficient, cost-effective, environment-friendly, and stable nanomaterials.
3. Bioremediation of nanoparticles should be explored for developing an integrated remediation strategy. Specific difficulties while incorporating nanotechnologies in food have to be concerned more with the betterment of farmers and society.

REFERENCES

1. Acharya A, Pal PK. Agriculture nanotechnology: translating research outcome to field applications by influencing environmental sustainability. *NanoImpact*, 2020; 19:100232.
2. Agrawal S, Rathore P. Nanotechnology pros and cons to agriculture: a review. *International Journal of Current Microbiology and Applied Sciences*, 2014; 3:43–55.
3. Peerzada AM, O'Donnell C, Adkins S. Optimizing herbicide use in herbicidetolerant crops: challenges, opportunities, and recommendations. *Agronomic Crops*, 2019; 2:283–316.
4. Bharti MK, Chalia S, Thakur P, Sridhara SN, Thakur A, Sharma PB. Nano ferrites heterogeneous catalysts for biodiesel production from soybean and canola oil: a review. *Environmental Chemistry Letters*, 2021; 19(5):3727–3746.
5. Brady NR, Weil RR. *The Nature and Properties of Soils*. Hoboken, NJ: Prentice Hall, 1999; pp. 415–473.
6. Brock DA, Douglas TE, Queller DC, Strassmann JE. Primitive agriculture in a social amoeba. *Nature* 2011; 469:393–396.
7. Buffle J. The key role of environmental colloids/nanoparticles for the sustainability of life. *Environmental Chemistry* 2006; 3(3):155–158.
8. Chau CF, Wu SH, Yen GC. The development of regulations for food nanotechnology. *Trends in Food Science and Technology*, 2007; 18:269–280.
9. Chhipa H. Nanofertilizers and nanopesticides for agriculture. *Environmental Chemistry Letters*, 2017; 15(1):15–22
10. Christou P, McCabe DE, Swain WF. Stable transformation of soybean callus by DNA-coated gold particles. *Plant Physiology*, 1988; 87:671–674.
11. Corradini E, Moura MR, Mattoso LHC. A preliminary study of the incorporation of NP fertilizer into chitosan nanoparticles express. *Polymer Letter*, 2010; 4:509–515.
12. Taneja S, Punia P, Thakur P, Thakur A. Synthesis of nanomaterials by chemical route. In *Synthesis and Applications of Nanoparticles*. Singapore: Springer Nature Singapore, 2022; pp. 61–76.
13. Degant O, Schwechten D. Wheat flour with increased water binding capacity and process and equipment for its manufacture. German Patent DE10107885A1, 2002.
14. DeRosa MC, Monreal C, Schnitzer M, Walsh R, Sultan Y. Nanotechnology in fertilizers. *Nature Nanotechnology*, 2010; 5(2):91.
15. Dhillon N, Mukhopadhyay S. Nanotechnology and allelopathy: synergism in action. *Journal of Crop and Weed*, 2015; 11:187–191.
16. Garrard. Brainy food: academe, industry sink their teeth into edible nano. *Small Times*, 2004; 21:2–20.

17. Hoffmann WC, Walker TW, Smith VI, Martin DE, Fritz BK. Droplet-size characterization of hand-held atomization equipment typically used in vector control. *Journal of American Mosquito Control Association*, 2007; 23:315–320.
18. Kabiri S, Degryse F, Tran DNH, Da-Silva RC, Mclaughlin MJ, Losic, D. Graphene oxide; a new carrier for slow release of plant micronutrients. *ACS Applied Materials & Interfaces*, 2017; 9:43325.
19. Shibata, T. Method for producing green tea in microfine powder. United States Patent US6416803B1, 2002.
20. Down on the farm-published by ETC group (2004) www.etcgroup.org/documents/ETC_DOTFarm2004.pdf.
21. Somasundaran P, Fang X, Ponnurangam S, Li B. Nanoparticles: characteristics, mechanisms and modulation of bio toxicity. *Kona Powder and Particle Journal*, 2010; 28:38–49.
22. Sastry KR, Rashmi HB, Rao NH. Nanotechnology patents as R&D indicators for disease management strategies in agriculture. *Journal Intelligence Property Rights*, 2010; 1(5):197–205.
23. Khan SM, Ali S, Nawaz A, Bukhari SAH, Ejaz S, Ahmad S. Integrated pest and disease management for better agronomic crop production. In: Hasanuzzaman M. (eds.) *Agronomic Crops*. Singapore: Springer, 2019; pp. 385–428.
24. White PJ, Broadley MR. Physiological limits to zinc biofortification of edible crops. *Frontiers in Plant Science*, 2011; 2:1–11.
25. Singh S, Singh M, Agrawal VV, Kumar A. An attempt to develop surface plasmon resonance based immunosensor for Karnal bunt (Tilletia indica) diagnosis based on the experience of nano-gold based lateral flow immunodipstick test. *Thin Solid Films*, 2010; 519:1156–1159.
26. Thakur P, Verma Y, Thakur A. Toxicity of nanomaterials: an overview. *Synthesis and Applications of Nanoparticles*, 2022; 12:535–544.
27. Lu CM., et al. Research on the effect of nanometer materials on germination and growth enhancement of Glycine max and its mechanism. *Soybean Science*, 21(3); (2002):168–171.
28. Morales-Díaz AB, Ortega-Ortíz H, Juárez-Maldonado A, Cadenas-Pliego G, González-Morales S, Benavides-Mendoza A. Application of nano-elements in plant nutrition and its impact in ecosystems. *Advances in Natural Sciences: Nanoscience and Nanotechnology*, 2017; 8:013001.
29. Stein AJ, Nestel P, Meenakshi JV, Qaim M, Sachdev HPS, Bhutta ZA. Plant breeding to control zinc deficiency in India: how cost-effective is biofortification? *Public Health Nutrition*, 2007; 10:492–501.
30. Matthews GA, Thomas N. Working towards more efficient application of Pesticides. *Pest Management Science*, 2000; 56:974–976.
31. Sastry RK, Rashmi HB, Rao NH. Nanotechnology for enhancing food security in India. *Food Policy*, 2011; 36:391–400.
32. Gutiérrez FJ, Mussons ML, Gatón P, Rojo R 2011 Nanotechnology and Food Industry. In: Benjamin V. (eds.) *Scientific, Health and Social Aspects of the Food Industry*, Croatia Book Chapter. New York: In Tech.
33. Hussain T. Nanocides: smart delivery system in agriculture and horticultural crops. *Advances in Plants & Agriculture Research* 2017; 6(6):175. https://doi.org/10.15406/apar.2017.06.00233
34. Scott N, Chen H. Nanoscale science and engineering for agriculture and food systems. A report submitted to cooperative state research, education and extension service, USDA, National Planning Workshop, Washington, 2003.
35. Duhan JS, Kumar R, Kumar N, Kaur P, Nehra K, Duhan S. Nanotechnology: the new perspective in precision agriculture. *Biotechnol Report*, 2017; 15:11–23
36. Joseph T, Morrison M. Nanotechnology in agriculture and food institute of nanotechnology. A *Nano Forum Report*, 2006; 1:414–419.
37. Vijayakumar MD, Surendhar GJ, Natrayan L, Patil PP, Bupathi Ram PM, Paramasivam P. Evolution and recent scenario of nanotechnology in agriculture and food industries. *Journal of Nanomaterials*, 2022; 2022:1–17.
38. Rad SJ, Naderi R, Alizadeh H, Yaraghi AS. Silver-nanoparticle as a vector in gene delivery by incubation. *IRJALS*, 2013; 02:21–33.
39. Sastry KR, Rao NH, Cahoon R, Tucker T. Can nanotechnology provide the innovations for a second green revolution in Indian agriculture? In: *Paper presented in NSF Nanoscale Science and Engineering Grantees Conference*, 2007.
40. Slavin YN, Asnis J, H¨ afeli UO, Bach H. Metal nanoparticles: understanding the mechanisms behind antibacterial activity. *Journal of Nanobiotechnology*, 2017; 15(1):65.

41. Sasson Y, Levy G, Toledano O, Ishaaya I. Nanosuspensions: emerging novel agrochemical formulations, In: Ishaaya I, Nauen R, Horowitz AR. (eds.) *Insecticides Design Using Advanced Technologies*. Netherlands: Springer-Verlag, 2007; pp. 1–32.
42. Matthews GA. Developments in application technology. *Environmentalist*, 2008; 28:19–24.
43. Miller G, Kinnear S. Nanotechnology the new threat to food. *Clean Food Organic*, 2007; 4:17.
44. Sahab AF, Waly AI, Sabbour MM, Lubna SN. Synthesis, antifungal and insecticidal potential of chitosan (CS)-g-poly (acrylic acid) (PAA) nanoparticles against some seed borne fungi and insects of soybean. *International Journal of ChemTech Research*, 2015; 8:589–598.
45. hsan M, Mahmood A, Mian MA, Cheema NM. Effect of different methods of fertilizer application to wheat after germination under rainfed conditions. *Journal of Agricultural Research*, 2007; 45:277–281.
46. Roger PA, Kulasooriya SA, Tirol AC, Craswell ET. Deep placement a method of nitrogen fertilizer application compatible with algal nitrogen fixation in wetland rice soils. *Plant and Soil*, 1980; 57:137–142.
47. Ihsan M, Mahmood A, Mian MA, Cheema NM. Effect of different methods of fertilizer application to wheat after germination under rainfed conditions. *Journal of Agricultural Research*, 2007; 45:277–281.

6 Nanotechnology for Green Fuels

Deepa Suhag and Raksha Rathore
Amity University Haryana

Larissa V. Panina and Alex V. Trukhanov
National University of Science and Technology

Preeti Thakur and Atul Thakur
Amity University Haryana

6.1 INTRODUCTION

Biofuels are becoming increasingly important, as seen by the continually rising research activity in both academia and business, as well as the rise in cooperative ventures involving multiple institutions. Because of its evident socioeconomic relevance, public interest in this highly dynamic topic is thus growing together with its volume of publications. To produce biofuels, a wide range of biomass types based on vegetable organic materials can be used. "Energy crops" created for non-nutritional consumption, such as willow, short-rotation trees, or grasses, can be employed. Food and feed crops include wheat, other cereals, rapeseed, sugar cane or sugar beetroot, and wheat. Low-value resources including agricultural residues and wastes, such as straw, bark, reclaimed wood, bagasse, or wastepaper, are widely viewed as the preferred resource to reduce rivalry with traditional applications [1]. However, value chain considerations credit value-added byproducts such as high-energy proteins in the case of some food and feed crops. In 2005, the amount of biomass that might be utilised to make biofuels in Europe was estimated to be 95 million metric tonnes of oil equivalent (Mtoe), with this figure expected to climb to between 112 and 172 Mtoe in 2020 and 243–316 Mtoe in 2,030. In comparison, the transport industry is expected to consume 416.3 Mtoe of energy in 2020 and 437.2 Mtoe of energy in 2030 [2]. Under average conditions, willow biomass outputs are predicted to be around 10 mto (10 metric tonnes) of dry biomass per hectare and year. All types of fuels derived from biomass that are used in the transportation industry are referred to as biofuels. Most biofuels can be used either in place of or in combination with popular fossil fuels like petrol and diesel.

Engines may need to be modified, like in the case of flexible fuel vehicles. Other biofuels, like biomass-based natural gas, necessitate significant adjustments to both vehicle design and the infrastructure for distributing fuel. Due to historical factors, biofuels are divided into three primary categories: so-called first- and second-generation biofuels, as well as the more recent third-generation biofuels [3–5]. Table 6.1 gives a comparison of petroleum fuel, first-generation fuel, and second-generation fuel.

6.2 FIRST-GENERATION BIOFUELS

First-generation biofuels are produced at industrial scales from cereal and oil crops using proven technology. Usually, their reduction in CO_2 emissions is around half that of fossil fuels. The production of crops and fuels, which requires a lot of energy, is one factor in this. These biofuels include

DOI: 10.1201/9781003502692-6

TABLE 6.1
Comparison of Petroleum Fuel, First-Generation Fuel, and Second-Generation Fuel

Property	Petroleum Fuel	First-Generation Fuel	Second-Generation Fuel
Source material	Crude oil (fossil)	Food crops (e.g., corn)	Non-food biomass
Production process	Refining	Fermentation/distillation	Advanced biochemical processes
Feedstock competition	None	Potential with food crops	Reduced (non-food)
Greenhouse gas emissions	High	Moderate	Varies (lower potential)
Energy content	High	Moderate	Moderate to high
Environmental impact	High (fossil origin)	Moderate (land use)	Potential lower impact
Sustainability	Limited (depleting)	Concerns (food vs. fuel)	Improved sustainability

ethanol, biogas, ethyl-tertiary-butyl ether, biodiesel, and pure vegetable oil. The usage of wood gas as an alternative fuel may be discussed for historical reasons. Because of the constraints on easy and affordable access to oil during the Second World War, this was extremely popular in many European and Asian countries both during and after the conflict [6–8]. Wood gas is produced by thermal gasifying wood, which results in a mixture of around 49 vol.% CO_2, 34 vol.% CO, 13 vol.% CH_4, 2 vol.% ethylene, and 2 vol.% H_2 [9].

6.2.1 Bioethanol

The practice of using ethanol as gasoline dates back approximately 150 years and is not a recent invention: To create a market for ethanol's mass manufacture, Eugen Langen's sugar firm financed Nikolaus August Otto while he created his spark-ignition engine prototype in the 1860s [10]. "To design a vehicle inexpensive for the working family and powered by a fuel that will promote the rural farm economy," was Henry Ford's stated goal. His 1908 Model T's original fuel specification called for only 100% ethanol to operate. In the 1930s, Rhenania-Ossag, Monheim (later Deutsche Shell AG) sold dynamin, a petroleum gas blend with 45% benzene and up to 10% ethanol from potatoes [11–13]. To meet the requirements for biocomponents in the petrol pool, it is usual practice to blend bioethanol directly into the fuel (low and high blends up to 95% ethanol, or E95), which boosts the octane ratings of the low blends (Figure 6.1). The use of ethanol as a raw material for ethers like ethyl-tertiary-butyl ether, which can be easily blended with petrol, provides an alternate option [11]. Reduced greenhouse gas (GHG) emissions can be attained using bioethanol, a type of biofuel, in the transportation sector [12–14]. The total amount of ethanol produced in the world in 2007 was 64,126 million litres, and 49,531 million litres of it was sold for the fuel market [15].

6.2.2 Lipid-Based Biofuels

The biggest source of nutritional energy in humans is lipids, which are found in adipose tissue triacylglycerol, intramuscular triglyceride, and dietary-derived fatty acids from plasma triacylglycerol and very low-density lipoproteins. Fat has an advantage over carbohydrates as a stored energy source because it has a higher energy density while having a lower relative weight. Compared to glucose, fatty acids produce more adenosine triphosphate per molecule. In the end, more energy may be obtained from a gramme of fat (9 kcal/gm) than from a gramme of protein (4 kcal/gm) or a gramme of carbohydrate (4 kcal/gm). Although there is a certain amount of fat that may be oxidised during exercise, dieters and endurance athletes are keen to burn more fat when exercising. Sport does really cause both men and women to burn more energy [16].

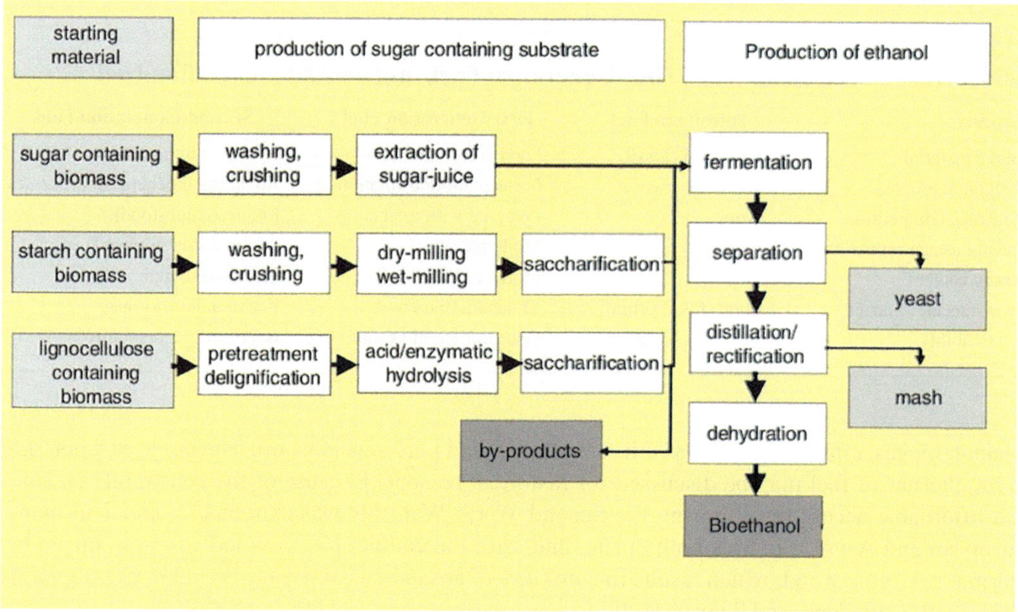

FIGURE 6.1 Bioethanol production.

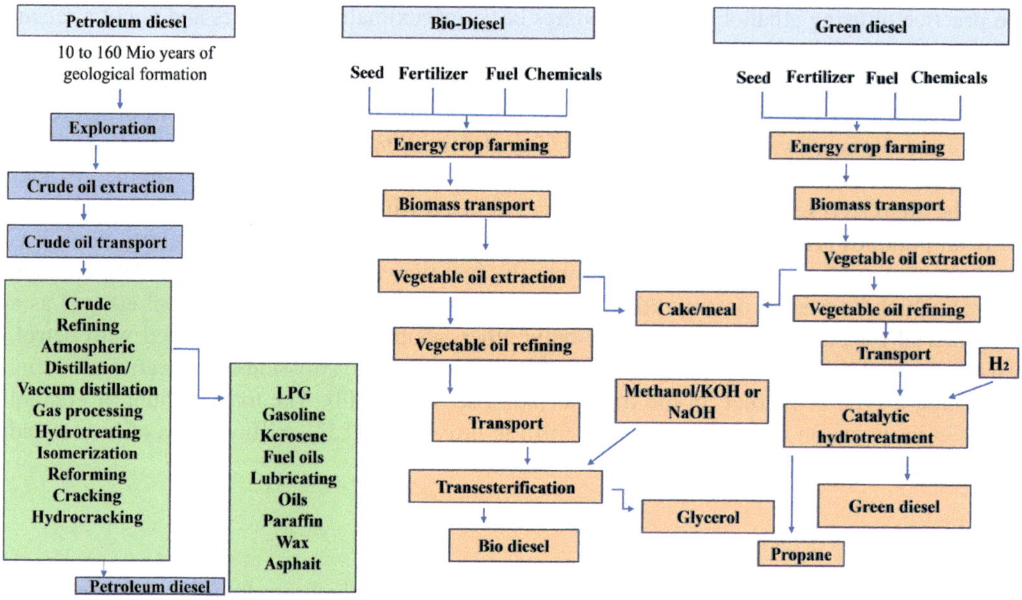

FIGURE 6.2 Lipid-based fuel production.

Lipids are also appropriate as an "alternative" diesel fuel source due to their high energy density, liquid nature, and structural similarity to hexadecane (cetane, the high-quality benchmark of the ignition quality of diesel fuel, i.e. cetane number 14) [17].

The fact that lipids may be used as a source of energy for human nutrition and as a fuel for transportation highlights the crucial function that they play while also highlighting the tension over their sustainability in terms of availability for both food and fuel uses. Figure 6.2 gives a flowchart for lipid-based fuel production.

6.2.3 Methane via Anaerobic Digestion

Anaerobic digestion or fermentation of biomass, manure, or biodegradable wastes, such as sewage sludge, produces biomethane (biogas). Methane and carbon dioxide make up most of the gas that results. Biomethane can be cleaned and compressed to be used as fuel for combustion engines or fuel cells, as well as for the generation of power or warmth [18,19].

More and more, it aims to supply biomethane as green gas to the natural gas infrastructure. To achieve the strict requirements, thorough cleaning is required. In this context, more than 5 million methane-fuelled cars are in use worldwide today. Biomethane is also employed as a transportation fuel. Biomethane has the largest production of gasoline equivalent per area compared to all other biofuels.

6.3 SECOND-GENERATION BIOFUELS

Lignocellulosic resources, such as low-value crops and forestry or agricultural residues [20] or purpose-grown plants from short-rotation forestry, are used to make second-generation biofuels. Extended raw material alternatives provide a much higher production capacity and CO_2 reduction potential than first-generation biofuels. Such a feedstock is also potentially more environmentally friendly and does not directly compete with food.

Advanced production techniques are used since they are currently accessible, even on a commercial level. The most significant second-generation biofuels include hydrogen, synthetic natural gas (SNG), butanol, hydrothermal upgrading (HTU) diesel, bioethanol from lignocellulosic raw materials, and the broad class of biomass to liquid (BtL) fuels based on synthesis gas (syngas). The latter fuel type consists of BtL petrol, BtL diesel, dimethyl ether, methanol, and ethanol.

6.3.1 Through Biomass Gasification of Hydrogen

As long as the gasification process is optimised to create a high hydrogen content in the resultant gas mixture, hydrogen can be produced by gassy biomass [21]. To increase hydrogen yields and decrease carbon monoxide contents, a downstream water gas shift process can be implemented. Dark fermentation and supercritical water gasification are other possibilities for producing hydrogen from biomass [22,23]. Crude biohydrogen can be utilised as fuel for compatible combustion engines or fuel cells after being cleaned, compressed, liquefied, or stored in appropriate storage media.

6.3.2 SNG Produced by Gasification of Biomass

In addition to anaerobic biomass digestion, biomass gasification is another method for producing methane, and, compared to the former, the latter technique allows for the utilisation of a greater variety of biomass types. Methane produced in this way, also known as SNG (bio-SNG), has a high-octane rating and can be utilised as fuel for adapted spark-ignition engines after being cleaned, compressed, or liquefied. Supercritical water gasification of biomass is an alternative method for making SNG [24,25]. Since wet biomass may be used in such a process, it comes as no surprise that it is quite appealing.

6.3.3 Biobutanol

From sugars acquired from food crops or sugars obtained from the cellulose degradation of *Clostridium* bacteria (Acetone, Butanol, Ethanol (ABE) fermentation), acetone, ethanol, and butanol can be generated in a ratio of around 3:6:1. Another appealing alternative is the use of butanol/diesel and butanol/gasoline blends in unaltered or lightly modified engines. However, much like with ethanol, it is necessary to optimise the conversion of cellulose to fermentable sugars and the subsequent fermentation [26].

6.3.4 Diesel HTU

The so-called HTU process converts biomass at a temperature of 300–350°C and a pressure of 120–180 bar to produce a mixture of hydrocarbons, carbon dioxide, water, and dissolved organics. This mixture can then be further processed in a catalytic hydro de-oxygenation step to produce diesel with properties resembling those of fossil diesel [27]. Wet biomass feedstocks can be used without drying, which is a key benefit. In contrast, water works as a solvent and a reactant simultaneously in hydrothermal conditions, producing a product with less oxygen than biocrude produced via pyrolysis.

6.4 OIL PYROLYSIS

Biomass is thermally cracked to create gaseous, liquid, and solid byproducts in an oxygen-free environment during pyrolysis (Figure 6.3). The pyrolysis reactor may be internally heated using the burnable gas and the char. Coke made from fossil fuels can be replaced with char as well. Fast or flash pyrolysis produces large amounts of liquid condensates up to 60% of weight on a base free of water and ash, which are primarily suitable for use as fuel. However, because it contains 400 or more different chemical compounds, including significant amounts of water (15 weight percent water created during fast pyrolysis reaction plus the humidity of the biomass feedstock), and oxygenated chemicals that further reduce the heating value, pyrolysis oil is not a viable fuel [27,28].

6.5 BIOFUELS BASED ON SYNGAS

Almost any sort of biomass, organic residue, or waste can be used to create so-called (bio)synfuels using BtL technology. Depending on the particular procedure, they can be a direct replacement for conventional petrol or diesel made from petroleum and be extremely similar to, or even better, preserving engine types and fuel distribution infrastructure. The technologies work by gassifying suitable biomass-derived materials to syngas, which is analogous to existing established gas-to-liquid or coal-to-liquid processes. The gas is then cleaned, and depending on the desired product, the

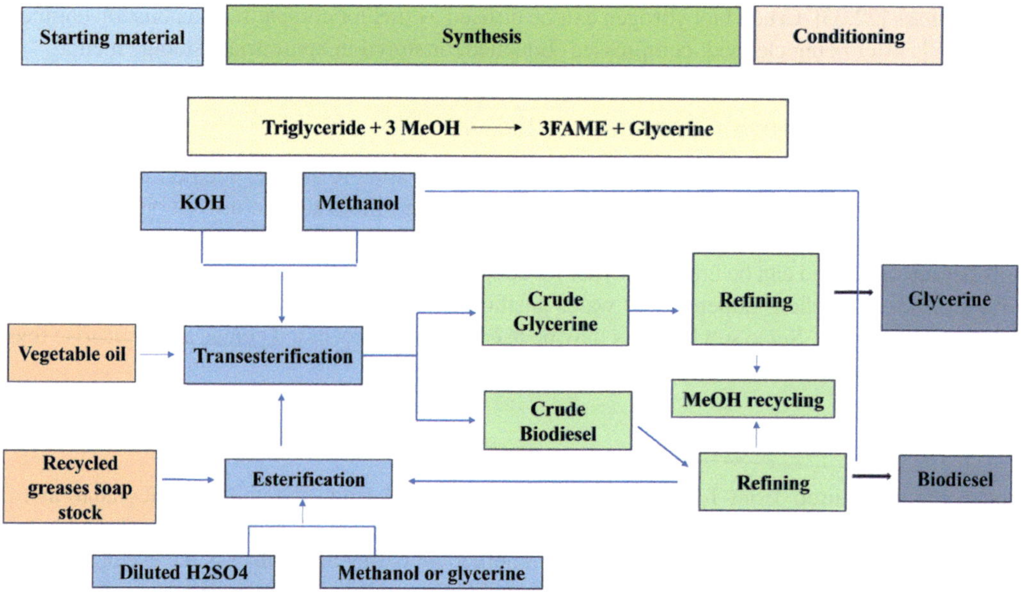

FIGURE 6.3 Biodiesel production.

hydrogen/carbon monoxide ratio is firmly adjusted to the appropriate specifications. The resulting syngas is a great starting point for the production of various products [29,30].

Syngas-produced diesel or petrol can more easily be adapted to meet changing demands compared to fuels made from crude oil. The development of advanced fuels for improved motor concepts, such as combined combustion system engines that combine the benefits of diesel compression ignition and Otto spark-ignition technologies, as well as meeting new emission standards, can both benefit from this [31,32].

6.6 BIOFUEL PRODUCTION WITH GREEN NANOTECHNOLOGIES

The lack of technology for effectively converting biomass into liquid fuels is the main barrier to the production of lignocellulosic-derived biofuels. Therefore, there is a pressing need for the creation of effective technologies to address this issue. In this context, developments in nanotechnology portend the future creation of effective methods for producing biofuels from biomass because they have made it feasible to comprehend and control chemistry at the molecular level. Due to their special qualities, several nanomaterials, including TiO_2, Fe_3O_4, SnO_2, ZnO, carbon, graphene, and fullerene, have been used in the manufacturing of biofuels. Additionally, the manufacture of bioethanol has benefited from the immobilisation of several enzymes, including cellulases and hemicelluloses, on a variety of nanomaterials [33]. By allowing enzyme recycling, this tactic lowers the cost of the procedure. The environment and human health are, however, seriously threatened by the release of nanoparticles (NPs) into the environment [34]. Therefore, a life cycle approach must be used to identify and reduce any environmental effects that may result from the manufacture of nanomaterials [35].

There aren't many studies on the synthesis of biofuels from lignocellulose biomass using green nanotechnology because it is still in its early stages. Tables 6.2 and 6.3 summarise recent studies on the application of green nanotechnology to the production of biofuels from lignocellulosic biomass. Wasted tea (solid waste) was converted into biofuels by researchers via nano-catalytic gasification, followed by biodiesel trans-esterification of the liquid component and fermentation of the solid waste by *Aspergillus niger* [36]. Figure 6.3 gives a flowchart for the synthesis of biodiesel.

To increase the scope of worldwide market penetration of biofuels, the production of biofuels must be separated from food crops. One of the challenges this technology faces is the inherent reluctance of cellulosic feedstocks to conversion into simpler sugars that can be fermented into ethanol. The promise here is in using NPs as immobilising beds for expensive enzymes, which can then be used to break down long-chain cellulose polymers into smaller fermentable sugars [3].

Louisiana Tech University is one of several institutions around the world practising this [37]. The purpose is to generate ethanol from non-edible cellulosic biomass such as wood, grass, and stalks. When compared to fossil fuels, this method of ethanol production can reduce GHG emissions by around 86%.

TABLE 6.2
Biocatalysts in Feedstock Conversion

Feedstock	Biocatalyst	Product
Sugarcane	Yeast	Ethanol
Algae	Algal enzymes	Biodiesel
Corn starch	Bacteria	Lactic acid
Agricultural residues	Fungi	Cellulase for bioethanol
Palm oil	Lipases	Biodiesel and oleochemicals
Woody biomass	Acidic pretreatment	Sugars for bio-based chemicals
Methane (biogas)	Methanogenic bacteria	Methane and digestate

TABLE 6.3
Catalysing Feedstock Transformation Using Biocatalysts

Feedstock	Biocatalyst	Product
Spent tea (*Camellia sinensis*)	Enzymes (e.g., cellulase)	Bioethanol, biogas
Microcrystalline cellulose	Microorganisms	Biofuels, biochemicals
Jackfruit waste	Enzymes	Bioethanol, organic fertilisers
Sugarcane leaves	Bacteria, enzymes	Bioethanol, bioplastics
Sesbania aculeate	Algae, bacteria	Biogas, soil conditioners

The field of nanotechnology research into the manufacture of biofuels from biomass is rapidly developing. For example, in 2007, the oil firm BP awarded the University of California at Berkeley and the University of Illinois $500 million in research money to examine the conversion of maize, plant material, algae, and switch grass into fuel [38].

Berkeley has already investigated low-cost solar panels using nanotechnology [39]. However, Berkeley's newly founded Energy Biosciences Institute will focus on creating fuel with as little negative environmental repercussions as feasible.

A three-pronged plan is being implemented, beginning with technology for more efficient crop production, improving feedstock processing, and developing novel biofuels.

6.6.1 Nanotech in Liquid Additives

The utilisation of solid NPs as catalysts to make biofuels from biomass, landfill methane, and algae was discussed in all previous presentations and debates. The following sections will look at the applications of liquid NPs or droplets [40]. Consider multifunctional surface-active liquid additives, which improve lubricity by causing a monolayer to form on surfaces that come into touch with additive fuel [37]. The treatment rate for lubricity is determined by the adsorption saturation concentration. Speculate that the development of nano-emulsions is what results in the increased detergency and water co-solvency. Also, assume that the behaviour of microdroplets is what leads to more complete combustion and the ensuing increase in fuel efficiency. These nanodroplets are the product of the additive's surfactant effect.

According to researchers [41], fuel nano-emulsions with water and surfactant are thermodynamically stable, microscopically isotropic, and nano-structured (thus, nano-emulsions). Their research found that using these nano-structures with fuel, water, and surfactant can break the normal trade-off between reducing soot and NO_x emissions by doing both at the same time. Many researchers have filed patent applications for micro-emulsions used as fuel. The behaviour of stable nano-emulsions of water, surfactant, and diesel (and most likely biodiesel) is explained as follows.

Surfactant components (oleic acid) and nitrogen-containing compounds (amines) dissolve quickly in diesel fuel (and maybe biodiesel fuel) and bind water to it without being stirred. Water droplets as small as a nanometer in size help to stabilise the emulsion [42]. When this fuel composition is used, it nearly eliminates soot and reduces nitrogen oxide emissions by up to 80%. It also burns cleanly, emitting only water, carbon dioxide, and nitrogen. The result is a "liquid sponge" that, like conventional diesel fuel, may be kept indefinitely without fear of phase separation.

6.6.2 Metallic NPs for Biogas Production Enhancements

Anthropogenic and natural resources have both been used to obtain NPs. A very high concentration of NPs may have built up in waste sludge. However, there is still much research needed on the toxicity of NPs and their effects on the sludge treatment stream [42,43].

Nguyen investigated the effects of ZnO NPs and CeO_2 NPs on sludge toxicity to bacteria and plants, and sludge dewatering. Concentration is critical for evaluating an NP's role in methane and biogas production. Not all NPs activate the anaerobic digestion mechanism; rather, some drastically reduce output when compared to a control sample. NP concentration and kind are crucial factors [44–46].

Luna del Risco [37] described the AD process of cattle manure and the impacts of metal oxide on methane and biogas generation. CuO NPs influenced the experiment more than the other test substances. The 15 mg/L Cu concentration resulted in a 30% decrease in total biogas output on day 14. CuO inhibited biogas generation by 19% and 60% at concentrations of 120 and 240 mg/L, respectively, although this inhibition was mitigated by the addition of Cu microparticles. The differences between the bulk particles and the NPs of CuO were supported statistically (p, 0.05). NP toxicity to bacteria, according to researchers, disrupts the cell membrane and hinders the generation of biogas by releasing metal ions. While comparing test samples containing ZnO NPs to ZnO bulk particles, ZnO NP doses of 120 and 240 mg/L inhibited the maximum yield by 74% and 43%, respectively. At day 14, test bottles containing bulk ZnO showed an 18% and 72% decrease in biogas emission, respectively. ZnO, both in bulk and as NPs, had little effect on biogas inhibition [47].

The facts presented above led to the conclusion that the concentration and size of CuO and ZnO NPs influence biogas yield. CeO_2 can boost biogas generation by 11% even at a low concentration of 10 mg. A bacterial toxicity test confirmed the positive benefits of CeO_2 at a similar dose. Instead of occurring naturally, the toxicity levels of both NPs decreased when they were added to the sludge. Bacterial toxicity decreased once the AD process was completed. The time required to dewater the digested sludge was directly related to the NP exposure concentration. When the AD operations were done, the toxicity of the sludge was also greatly reduced. Bacterial viability was reduced by 47.5% when the sludge was subjected to 1,000 mg/L CeO_2 NPs, whereas it was reduced by 30.4% following the AD process.

Because of the existence of Fe_3C and non-toxic Fe_3C ions, nano iron oxide (Fe_3O_4 NPs) can improve methane production. For 60 days, an anaerobic waste digester was fed 100 ppm of 7 nm-sized Fe_3O_4 NPs at 37°C. The most significant increase in biogas output was reported when NPs were used, with a 234% increase in methane production and an 18% increase in biogas production [47]. Increased organic matter processing and methane generation can be accomplished using a Fe_3O_4 (magnetite) NP delivery system due to an improvement in AD. The FeC_2/FeC_3 ions in the reactor are responsible for the performance boost. These ions are introduced as NPs, much like controlled drug delivery systems. Methane and hydrogen production rates increase because Fe is important in electron transport, and bacterial growth is enhanced by boosting enzymatic activities [47]. Fe_3O_4 NPs are the most used materials because they are more biocompatible and have lower toxicity. In a study utilising fresh raw manure, another researcher [45] found that Ni NPs with a concentration of 2 mg/L and a particle size of 17 nm produced the greatest average methane and biogas yields of 78.53% and 116.76%, respectively, after 40 days of use. While Fe_3O_4 enhanced methane production by 115.66% and biogas output by 73% at a concentration of 20 mg/L and particle size of 7 mm, biogas production increased by 71% and methane production increased by 45.92% when Co was used, which has a particle size of 28 nm and a concentration of 1 mg/L. A concentration of 20 mg/L of Fe NPs with a particle size of 9 nm boosted biogas output by 47.7%.

Researchers found that adding nZVI (10 mg/g) and Fe_2O_3 (100 mg/g) NPs to waste-activated sludge increased methane generation by 120% and 117%, respectively. These findings suggest that lower concentrations of the NPs increase microbial populations of bacteria and archaea.

6.7 CONCLUSION

By lowering GHG emissions and the consumption of non-renewable resources, higher usage of biofuels would promote sustainable development. In place of conventional feedstocks (starch crops), lignocellulosic biomass, such as agricultural and forestry waste, may prove to be a suitable low-cost and widely available source of sugar for fermentation into transportation fuels. Several variables contribute to cellulose in the ability of biomass to resist hydrolysis, including cellulose crystallinity, accessible surface area, lignin-provided protection, and hemicellulose-provided sheathing.

The intrinsic structure and preparation of the biomass are the fundamental reasons for the subsequent hydrolysis. The conditions utilised in the preferred pretreatment process will affect a number of substrate features, which will control the substrate's susceptibility to hydrolysis and subsequent fermentation of the freed sugars. Despite efforts to improve energy efficiency and diversify energy systems, such as increased use of electricity in transportation, rising demand for energy services in the next decades will necessitate a greater supply of liquid fuels. Biofuels are critical in this environment. However, future biofuel supplies must be large enough to make extensive use of non-food feedstocks and novel technologies. Nanotechnologies are the most likely contenders to play a large role in this energy future. They will aid in the commercialisation of liquid biofuels such as biogas, fatty esters, and renewable hydrocarbons generated from algae, carbohydrates, and sugars. Nanotechnologies will improve the efficiency of using liquid fuels, particularly biofuels, by improving nanodroplet combustion. Although there are risks involved with any new technology, the world now looks to be significantly better prepared to analysing risks and taking necessary action. Nanotechnologies for biofuels can be created without jeopardising environmental safety, public health, or security. However, nanotechnologies have a far-reaching and profound impact beyond biofuels, providing hope in a wide range of sectors, most notably human health.

REFERENCES

1. Bhandari, G., Dhasmana, A., Chaudhary, P., Gupta, S., Gangola, S., Gupta, A., Rustagi, S., Shende, S., Rajput, V. D., Minkina, T., Malik, S., & Slama, P. (2023). A perspective review on green nanotechnology in agro-ecosystems: opportunities for sustainable agricultural practices & environmental remediation. *Agriculture*, 13(3), 668.
2. Ramsurn, H., & Gupta, R. B. (2013). Nanotechnology in solar and biofuels. *ACS Sustainable Chemistry & Engineering*, 1(7), 779–797.
3. Sharma, Y., Manro, B., Thakur, P., Khera, A., Dey, P., Gupta, D., Khajuria, A., Jaiswal, P. K., Barnwal, R. P., & Singh, G. (2022). Nanotechnology as a vital science in accelerating biofuel production, a boon or bane. *Biofuels, Bioproducts and Biorefining*, 17(3), 616–663.
4. Negi, P., Singh, Y., Yadav, A. K., Singh, N. K., Sharma, A., & Singh, R. (2022c). Challenges and future opportunities of nanoparticle applications to various biofuel generation processes – a review. *Proceedings of the Institution of Mechanical Engineers, Part E: Journal of Process Mechanical Engineering*, 237(3), 095440892211454.
5. Kanakdande, A. P., & Mane, R. S. (2023). Nanotechnology for Bioenergy and Biofuel Production. In: Mane, R. S., Sharma, R. P., Kanakdande, A. P., editors. *Nanomaterials for Sustainable Development*. Springer; New Delhi, India, pp. 283–296.
6. Neupane, D. (2023). Biofuels from renewable sources, a potential option for biodiesel production. *Bioengineering*, 10(1), 29.
7. Maheshwari, P., Haider, M. B., Yusuf, M., Klemeš, J. J., Bokhari, A., Beg, M., & Jaiswal, A. K. (2022). A review on latest trends in cleaner biodiesel production: role of feedstock, production methods, and catalysts. *Journal of Cleaner Production*, 355, 131588.
8. Jeswani, H. K., Chilvers, A., & Azapagic, A. (2020). Environmental sustainability of biofuels: a review. *Proceedings of the Royal Society A*, 476(2243), 20200351.
9. Tayari, S., Abedi, R., & Rahi, A. (2020). Comparative assessment of engine performance and emissions fueled with three different biodiesel generations. *Renewable Energy*, 147, 1058–1069.
10. Duque, A., Álvarez, C., Doménech, P., Manzanares, P., & Moreno, A. D. (2021). Advanced bioethanol production: from novel raw materials to integrated biorefineries. *Processes*, 9(2), 206.
11. Vasić, K., Knez, Ž., & Leitgeb, M. (2021). Bioethanol production by enzymatic hydrolysis from different lignocellulosic sources. *Molecules*, 26(3), 753.
12. Zabed, H., Sahu, J. N., Suely, A., Boyce, A. N., & Faruq, G. (2017). Bioethanol production from renewable sources: current perspectives and technological progress. *Renewable and Sustainable Energy Reviews*, 71, 475–501.

13. Rastogi, M., & Shrivastava, S. (2017). Recent advances in second generation bioethanol production: an insight to pretreatment, saccharification and fermentation processes. *Renewable & Sustainable Energy Reviews*, 80, 330–340.
14. Gupta, A., & Verma, J. P. (2015). Sustainable bio-ethanol production from agro-residues: a review. *Renewable & Sustainable Energy Reviews*, 41, 550–567.
15. Mohanty, S. K., & Swain, M. R. (2019). *Bioethanol Production from Corn and Wheat: Food, Fuel, and Future*. Elsevier; Amsterdam, The Netherlands, pp. 45–59.
16. Babu, S. S., Gondi, R., Vincent, G. S., John Samuel, G. C., & Jeyakumar, R. B. (2022). Microalgae biomass and lipids as feedstock for biofuels: sustainable biotechnology strategies. *Sustainability*, 14(22), 15070.
17. Alishah Aratboni, H., Rafiei, N., Garcia-Granados, R., Alemzadeh, A., & Morones-Ramírez, J. R. (2019). Biomass and lipid induction strategies in microalgae for biofuel production and other applications. *Microbial Cell Factories*, 18, 1–17.
18. Klassen, V., Blifernez-Klassen, O., Wibberg, D., Winkler, A. M., Kalinowski, J., Posten, C., & Kruse, O. (2017). Highly efficient methane generation from untreated microalgae biomass. *Biotechnology for Biofuels*, 10(1), 186.
19. Weiland, P. (2010). Biogas production: current state and perspectives. *Applied Microbiology and Biotechnology*, 85, 849–860.
20. Bhuiya, M. M. K., Rasul, M. G., Khan, M. M. K., Ashwath, N., Azad, A. K., & Hazrat, M. A. (2014). Second generation biodiesel: potential alternative to-edible oil-derived biodiesel. *Energy Procedia*, 61, 1969–1972.
21. Cao, L., Yu, I. K., Xiong, X., Tsang, D. C., Zhang, S., Clark, J. H., Hu, C., Ng, Y. H., Shang, J., & Ok, Y. S. (2020). Biorenewable hydrogen production through biomass gasification: a review and future prospects. *Environmental Research*, 186, 109547.
22. Barbuzza, E., Buceti, G., Pozio, A., Santarelli, M., & Tosti, S. (2019). Gasification of wood biomass with renewable hydrogen for the production of synthetic natural gas. *Fuel*, 242, 520–531.
23. Mohseni, F., Görling, M., & Alvfors, P. (2011). Synergy Effects on Combining Hydrogen and Gasification for Synthetic Biogas. World Renewable Energy Congress, Linköping, Sweden.
24. Molino, A., & Braccio, G. (2015). Synthetic natural gas SNG production from biomass gasification – thermodynamics and processing aspects. *Fuel*, 139, 425–429.
25. Van Der Meijden, C., Veringa, H., & Rabou, L. (2010). The production of synthetic natural gas (SNG): a comparison of three wood gasification systems for energy balance and overall efficiency. *Biomass & Bioenergy*, 34(3), 302–311.
26. Amiri, H., & Karimi, K. (2019). *Biobutanol Production*. Elsevier; Amsterdam, The Netherlands, pp. 109–133.
27. Tuli, D. K., & Kasture, S. (2022). *Biodiesel and Green Diesel*. Elsevier; Amsterdam, The Netherlands, pp. 119–133.
28. Ağbulut, Ü., Sirohi, R., Lichtfouse, E., Chen, W., Len, C., Show, P. L., Le, A., Nguyen, X. P., & Hoang, A. T. (2023). Microalgae bio-oil production by pyrolysis and hydrothermal liquefaction: mechanism and characteristics. *Bioresource Technology*, 376, 128860.
29. Alherbawi, M., Parthasarathy, P., Al-Ansari, T., Mackey, H. R., & McKay, G. (2021). Potential of drop-in biofuel production from camel manure by hydrothermal liquefaction and biocrude upgrading: a Qatar case study. *Energy*, 232, 121027.
30. Ghazi, F. M. G., Abbaspour, M., & Rahimpour, M. R. (2023). *Biofuel Production from Syngas*. Elsevier; Amsterdam, The Netherlands, pp. 271–286.
31. Yasin, M., Cha, M., Chang, I. S., Atiyeh, H. K., Munasinghe, P. C. M., & Khanal, S. K. (2019). *Syngas Fermentation into Biofuels and Biochemicals*. Elsevier; Amsterdam, The Netherlands, pp. 301–327.
32. Bermejo, S., Lebrero, R., & Muñoz, R. (2022). Syngas biomethanation: current state and future perspectives. *Bioresource Technology*, 358, 127436.
33. Al-Timimi, B. A., & Yaakob, Z. (2022). *Catalysts for the Simultaneous Production of Syngas and Carbon Nanofilaments Via Catalytic Decomposition of Biogas*. IntechOpen; London, UK.
34. Thakur, P., Verma, Y., & Thakur, A. (2022). *Toxicity of Nanomaterials: An Overview In Synthesis and Applications ofNanoparticles*. Springer Nature; Singapore, pp. 535–544. https://doi.org/10.1007/978-981-16-6819-7_25
35. Thakur, P., & Thakur, A. (2022). Introduction to nanotechnology. In: Thakur, A., Thakur, P., Khurana, S.P., editors. *Synthesis and Applications of Nanoparticles*. Springer; Singapore. https://doi.org/10.1007/978-981-16-6819-7

36. Bhuyan, N., Dutta, A., Mohan, R., Bora, N., & Kataki, R. (2021). Advances in nanotechnology for biofuel production. In: Praveen Kumar R., Bharathiraja B., editors. *Nanomaterials*. Elsevier; Amsterdam, The Netherlands, pp. 533–562.
37. Luna-delRisco, M., Orupõld, K., & Dubourguier, H.C. (2011). Particle-size effect of CuO and ZnO on biogas and methane production during anaerobic digestion. *Journal of hazardous materials*, 189(1–2), 603–608.
38. Anto, S., Mukherjee, S. S., Muthappa, R., Mathimani, T., Garlapati, D., Kumar, S. S., Verma, T. N., & Pugazhendhi, A. (2020). Algae as green energy reserve: technological outlook on biofuel production. *Chemosphere*, 242, 125079. https://doi.org/10.1016/j.chemosphere.2019.125079
39. Eggert, H., & Greaker, M. (2014). Promoting second generation biofuels: does the first generation pave the road?. *Energies*, 7(7), 4430–4445.
40. Trindade, S. C. (2011). *Nanotech Biofuels and Fuel Additives*. Tech eBooks; Birmingham, UK.
41. Guo, J., Cheng, J., Tan, H., Sun, Q., Yang, J., & Liu, W. (2020). Constructing a novel and high-performance liquid nanoparticle additive from a Ga-based liquid metal. *Nanoscale*, 12(16), 9208–9218.
42. Alrawashdeh, K. A. B., Al-Zboon, K. K., Rabadi, S. A., Gul, E., Al-Samrraie, L. A., Ali, R., & Al-Tabbal, J. A. (2022). Impact of Iron oxide nanoparticles on sustainable production of biogas through anaerobic co-digestion of chicken waste and wastewater. *Frontiers in Chemical Engineering*, 4, 5844.
43. Amo-Duodu, G., Rathilal, S., Chollom, M. N., & Tetteh, E. K. (2021). Application of metallic nanoparticles for biogas enhancement using the biomethane potential test. *Scientific African*, 12, e00728.
44. Thakur, A., Thakur, P., & Khurana, S. P. (eds). (2022). *Synthesis and Applications of Nanoparticles*, Springer Nature; Singapore, pp 1–544. https://doi.org/10.1007/978-981-16-6819-7
45. Abdelsalam, E., Samer, M., Attia, Y. A., Abdel-Hadi, M. A., Hassan, H. E., & Badr, Y. (2017). Effects of Co and Ni nanoparticles on biogas and methane production from anaerobic digestion of slurry. *Energy Conversion and Management*, 141, 108–119.
46. Amen, T. W., Eljamal, O., Khalil, A. M., Sugihara, Y., & Matsunaga, N. (2018). Methane yield enhancement by the addition of new novel of iron and copper-iron bimetallic nanoparticles. *Chemical Engineering and Processing-Process Intensification*, 130, 253–261.
47. Faisal, S., Yusuf Hafeez, F., Zafar, Y., Majeed, S., Leng, X., Zhao, S., Saif, I., et al. (2018). A review on nanoparticles as boon for biogas producers—nano fuels and biosensing monitoring. *Applied Sciences*, 9(1), 59.

7 Nanomaterials for Biodiesels

Abhilash Pathania
Amity University Haryana

Fayu Wan
Nanjing University of Information Science and Technology

Preeti Thakur
Amity University Haryana

Atul Thakur
Amity University Haryana
Nanjing University of Information Science and Technology

7.1 INTRODUCTION

Petroleum, natural gas, and coal serve as global primary energy sources, with an escalating demand fueled by rapid population growth and industrialization [1]. Approximately 80% of the annual worldwide energy demand, which stands at around 580 TJ, is met through the combustion of conventional fossil fuels, leading to detrimental greenhouse gas emissions and subsequent global warming [1]. Notably, 120 countries have committed to achieving zero carbon emissions by 2050 or earlier, aligning with the net-zero future goal outlined by the UNFCCC [1]. The depletion and limited supply of fossil fuels indicate an imminent exhaustion of reserves, prompting concerted efforts by engineers and scientists to develop renewable energy sources [1]. Projections suggest that, by 2040, half of the global energy requirement will be fulfilled by renewable sources, resulting in a substantial 70% reduction in greenhouse gas emissions [1]. The combustion of fossil fuels releases an alarming 27 billion tons of CO_2 annually [1]. Figure 7.1 provides a visual representation of the energy requirement fulfilled by various fuels on a global scale. Presently, only 18% of the total global energy demand is met by renewable sources, encompassing geothermal, biofuels, solar, wind, and hydro, among others [2]. Biomass, hydro, wind, solar, thermal, and geothermal sources contribute 13%, 3%, 0.9%, 0.7%, 0.2%, and 0.23%, respectively [2]. Recent decades have witnessed increased attention toward biofuel production, leveraging low-cost sources like waste animal fats

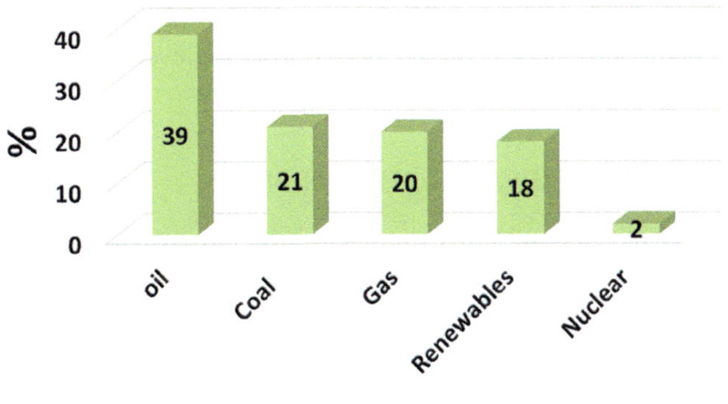

FIGURE 7.1 Contribution of different energy sources for meeting global energy demands.

DOI: 10.1201/9781003502692-7

and algal lipids. Biodiesel, a product of the photosynthesis process, boasts advantages such as excellent lubricating properties, low CO emissions, high flash point, low toxicity, biodegradability, and the absence of sulfur-containing compounds [2]. The biofuel production process involves converting biomass into fluid fuels like biogas, biohydrogen, and biodiesel [2].

Biodiesel, derived from diverse feedstocks like plant and animal oils or triacylglycerides (TAGs), serves as a sustainable energy source, offering a viable alternative to petroleum fuels [3]. Its versatility allows direct use or blending with regular diesel in any proportion, providing an affordable and reliable replacement for traditional fuels [3]. However, commercial biodiesel production faces challenges related to production costs, feedstock availability, and process complexity [3]. Transesterification is the predominant technique employed for biodiesel generation [3]. In the realm of biodiesel manufacturing, nanoparticles play a crucial role, particularly in the synthesis of alternative biodiesel and other bio-based products through the adaptive bio-refinery platform [3]. Nano catalysts, including conventional nanomaterials like nano-silicon, magnetic nanoparticles (MNPs), and nano metal particles, as well as nanostructured materials such as carbon nanotubes (CNTs), nanofibers, and metal-organic frameworks, address issues like mass transfer resistance, quick deactivation, and inefficiency due to their extensive reaction surfaces and robust catalytic activity [3]. This holds significance not only for biodiesel production but also for the broader spectrum of manufacturing cellulosic fuels and renewable chemical applications [3].

In summary, biodiesel from non-food sources, produced using transesterification, is a sustainable energy option, and the integration of nanoparticles, including various nano catalysts, plays a pivotal role in overcoming challenges and enhancing efficiency in biodiesel production and related bio-based processes [3].

7.2 HISTORICAL BACKGROUND OF BIODIESEL

The term "biodiesel" encompasses mono-alkyl esters of long-chain fatty acids derived from animal and vegetable fats, produced with or without catalysts [4]. Various feedstocks can be utilized in the biodiesel production process, with a recent emphasis on vegetable oils. However, concerns over food scarcity and environmental issues, such as deforestation, have led to a shift toward non-edible oils to address these challenges [4]. The diesel engine, invented by German scientist Rudolf Diesel, originally explored alternative energy sources, including the use of peanut oil as fuel, foreseeing the significance of vegetable oil as a non-petroleum fuel equivalent to petroleum and coal tar products [3]. Although Diesel's engine was initially adapted to run on petroleum, his vision laid the foundation for a clean, recyclable, and locally produced fuel. In 1940, France made the first attempt to create biodiesel based on vegetable oil, but widespread adoption as a petroleum diesel substitute occurred slowly until the late 1970s and 1980s, driven by concerns about high petroleum prices [3].

Amid rising petroleum consumption and environmental considerations, scientists are keenly interested in biodiesel synthesis. Four commonly employed techniques for biodiesel synthesis include dilution with hydrocarbon blending [5], microemulsions [6], pyrolysis [7], and transesterification [8]. Each method contributes to the ongoing efforts to develop sustainable alternatives to traditional petroleum-based fuels [5–8].

7.3 FEEDSTOCK FOR BIODIESEL PRODUCTION

The commercial viability of biodiesel hinges on its affordability, environmental friendliness, and accessibility. Biodiesel production costs are heavily influenced by feedstock prices, constituting approximately 75% of the total production expenses, alongside the chosen synthesis technique [9]. Lignocellulosic biomass, the fourth major energy source after petroleum, coal, and natural gas, plays a pivotal role. Biomass categories encompass wood, agricultural biomass, solid waste, landfill biomass, and biomass used in alcoholic beverage production [10]. Compared to fossil fuels, biomass-derived fuel exhibits lower sulfur and nitrogen levels, resulting in reduced environmental

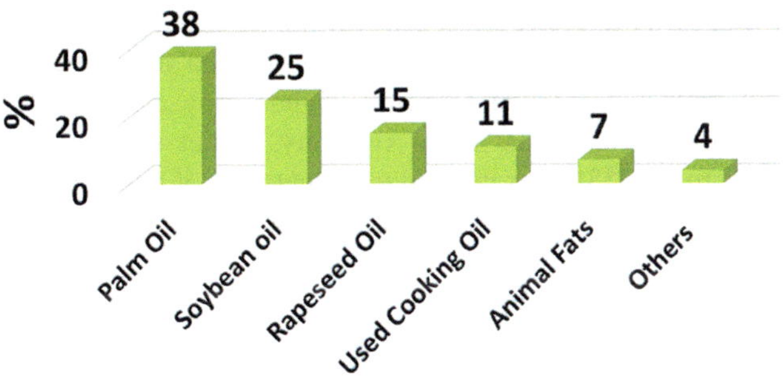

FIGURE 7.2 The distribution of feedstock shares in global biodiesel production in 2020.

pollution. Fatty acid methyl esters, with 9% more oxygen, facilitate complete combustion, aiding in the reduction of carbon dioxide emissions when biodiesel is derived from waste biomass-based oil. Furthermore, the closed carbon cycle, involving the natural absorption of carbon dioxide by biomass plants, helps limit atmospheric carbon dioxide levels. Various feedstocks for biodiesel synthesis have been evaluated and classified into biomass waste oils, animal fats, edible oils, and non-edible oils. The use of edible plant oils has fluctuated due to their dual role in the food and biodiesel industries. Despite 95% of global biodiesel production relying on food oils like sunflower, palm, and soybean oil, their continued use has contributed to increased starvation, heightened food costs, and deforestation. Environmental factors influencing the growth of plants producing edible oils make it impractical to use them as a primary raw material for fatty acid methyl ester production. Consequently, the search for low-cost biodiesel feedstocks persists. Figure 7.2 depicts the distribution of feedstock shares in global biodiesel production in 2020.

Soybean oil, a prominent global source of edible oils, is derived from the most widely grown plant on Earth. Pimental and Patzek noted that 5,546 kg of soybeans yield 1,000 kg of oil [11]. While soybean oil synthesis is expected to further improve, non-edible oils from lignocellulosic biomass and waste products of agricultural and forestry activities are considered superior feedstocks for biodiesel due to their avoidance of potential food crises. These non-edible oils are produced in large quantities to address energy concerns, and cellulosic biomasses exhibit adaptability to diverse environmental conditions, allowing for widespread cultivation. Polanga oil, originating in Malaysia, Indonesia, India, and the Philippines, offers an alternative. With a dark green color and a composition of 72.65% oleic, linoleic, and linolenic acids, it presents a viable biodiesel source [12]. Comprehensive engine testing indicates that polanga-based biodiesel can substitute for diesel fuel without major hardware modifications, exhibiting characteristics akin to diesel fuel.

In India, Madhuca Indica oil is abundant, with the tree thriving in sandy soil and its seeds containing approximately 50% oil content. Waste frying oil emerges as the most cost-effective biodiesel feedstock, priced at half the cost of vegetable oil. However, high cooking temperatures elevate the concentrations of free fatty acids (FFAs) and triglycerides hydrolysis in waste frying oil, potentially impeding biodiesel production and reducing economic reliability, impacting the transesterification reaction rate.

7.4 BIODIESEL SYNTHESIS TECHNIQUES

The innovative exploration of derivatives from vegetable oils has addressed challenges associated with diesel fuels, particularly issues like low volatility, high viscosity, and polyunsaturation in crude vegetable oils [1]. One approach to overcoming these challenges is through dilution with hydrocarbon blending [13]. The dilution process involves blending vegetable oil with petroleum fuel and solvents, enhancing the cetane number and reducing viscosity for improved use in automotive engines [13]. For instance, the viscosity of sunflower oil can be significantly decreased from 31 to 4.47 cP at 40 through this dilution process [13]. However, it's worth noting that the use of mixed vegetable oil in fuel injection engines may lead to problems such as coking and sticking of injector nozzles [13]. Advantages of dilution with hydrocarbon blending include easy availability, liquid nature, and recyclability, while drawbacks encompass lower volatility, higher viscosity, and the reactivity of unsaturated hydrocarbon chains [13]. This method represents one of four strategies, alongside pyrolysis, microemulsion, and transesterification, to address the challenges associated with vegetable oil-based diesel fuels [1].

7.4.1 Pyrolysis

Pyrolysis, a transformative process achieved through heating or heating with a catalyst, involves breaking chemical bonds in the absence of oxygen and converting materials into smaller molecules [14]. The temperature range for pyrolyzing vegetable oil typically falls between 250°C and 350°C [14]. There are three modes of pyrolysis: flash pyrolysis (500°C), fast pyrolysis (500°C), and slow pyrolysis (400°C) [14]. In the refinement of crude oil through pyrolysis, diesel oil is obtained, along with other by-products like gasoline and kerosene oil [14]. For instance, 68 kg of saponified tung oil yields approximately 50 L of crude oil [14]. Notably, jojoba oil proves more advantageous than soybean oil in generating low-molecular-weight hydrocarbons and fatty acids during pyrolysis [14]. Catalytic thermal cracking, utilizing bauxite as a catalyst, accelerates the secondary cracking during soybean oil pyrolysis, leading to the controlled synthesis of various hydrocarbons [14]. Different concentrations of bauxite (10%, 20%, and 30% weight percent) were employed, and specific temperature ranges (380–400°C) facilitated thermal cracking [14]. Simple cracking produced products containing extended chain fatty acids, while catalytic thermal cracking enhanced secondary cracking, yielding diverse hydrocarbons like alkanes, alkenes, and aromatics [14].

The thermal cracking technique, although utilized for low-cost feedstocks, involves complex reaction mechanisms, making it challenging to achieve desired outcomes [15].

7.4.2 Microemulsion

Microemulsion involves the spontaneous formation of two typically incompatible liquids, like low-molecular-weight alcohol and vegetable oil, into optically isotropic fluid microstructures ranging from 1 to 150 nm in size [16]. By utilizing microemulsions containing butanol and hexanol, the viscosity of vegetable oil, particularly peanut oil, can be decreased, making it a viable fuel source for diesel with improved cetane number and spray characteristics [16]. However, prolonged use of microemulsified diesel in engines can lead to issues like injector needle sticking, carbon deposit accumulation, and inadequate ignition [16].

7.4.3 Transesterification

Transesterification, a widely used process for biodiesel production in both industrial and laboratory settings, is preferred due to its ability to produce significant volumes of fuel compared to conventional techniques like pyrolysis, blending, and emulsification [17]. In transesterification, edible or non-edible oil, mixed with or without a catalyst, undergoes a reaction where three moles of alcohol react with one mole of oil to produce biodiesel [17]. This process is illustrated in the catalytic transesterification reaction of *Brassica nigra* oil in Figure 7.3 [17].

Transesterification is a process wherein triglycerides, commonly found in oils derived from organic biomass, are combined with alcohols, often methanol, to produce fatty acid esters, specifically fatty acid methyl esters (FAMEs) (Figure 7.4) [18]. The reaction, accelerated by a catalyst, aims to remove the glycerol backbone from triglycerides, yielding FAME chains with combustion characteristics akin to regular diesel [18]. The transesterification process involves the progression from triacylglycerol to di- and monoacylglycerols, ultimately resulting in glycerol elimination.

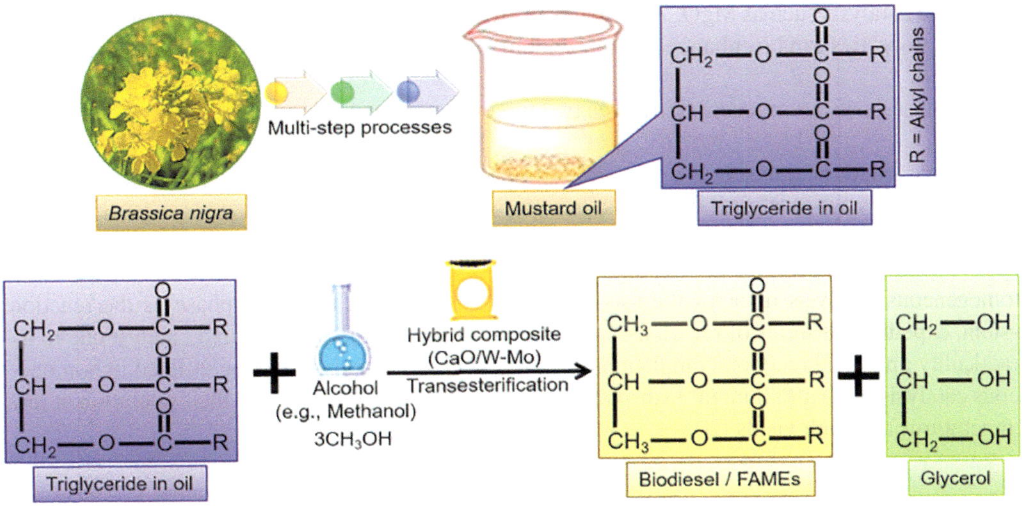

FIGURE 7.3 The catalytic transesterification reaction of *Brassica nigra* oil into biodiesel. (Reprinted from Ref. [18] with permission from Springer Nature.)

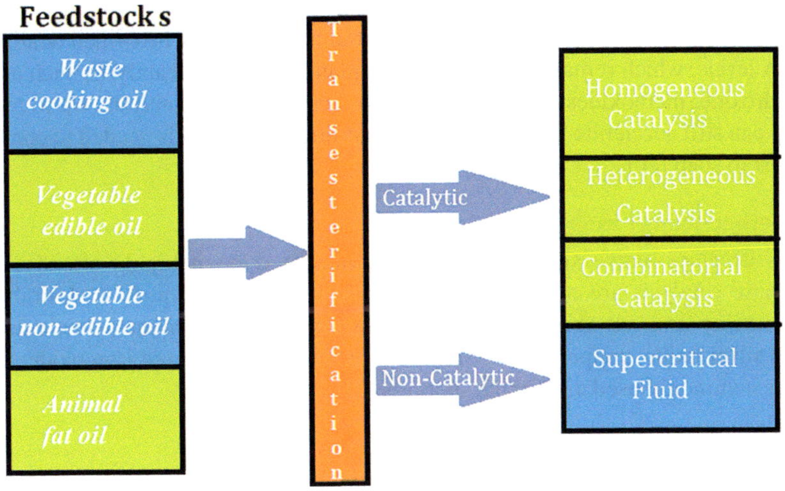

FIGURE 7.4 The transesterification processes.

At each stage, methyl ester is produced, leading to the creation of FAMEs (biodiesel) and glycerol when fatty acids (oils) react with alcohol in the presence of a suitable catalyst [18]. Strong bases such as sodium hydroxide, potassium hydroxide, or sodium methoxide, as well as acids like hydrochloric acid or sulfuric acid, can serve as catalysts [18]. The catalyst plays a crucial role in initiating the reaction and acts as an alcohol solubilizer, as alcohol is weakly soluble in the oil phase, and uncatalyzed reactions are notably slow [18].

The catalyst not only expedites alcohol solubility but also plays a crucial role in maintaining a manageable reaction rate [19]. The distinctive surface area of the catalyst stands out as a key variable in biodiesel production [19]. Nano catalysts, with their increased specific surface area, prove advantageous in the transesterification reaction. Calcium oxide (CaO), a cost-effective nano catalyst, is widely used for biodiesel production due to its robust catalytic activity [19]. CaO can be employed either independently or in combination with other components for biodiesel synthesis.

Other nano catalysts such as MgO, ZnO, and SrO are also utilized in the transesterification process, and their properties can be enhanced by incorporating materials like Fe_3O_4, $CuFe_2O_4$, and activated carbon [19].

The transesterification reaction involves two types of catalysts:

i. Homogeneous catalysis
ii. Heterogeneous catalysis [19]

7.4.3.1 Homogeneous Catalysis

Homogeneous catalysis involves the use of a chemical that is in the same phase as the reaction system, expediting a series of reactions. In biodiesel synthesis, it is the preferred catalyst due to its practicality and rapid reaction completion. Both acid and base catalysts can be utilized in homogeneous catalysis, with alkali metal hydroxides and alkoxides being favored for their operation at low temperatures and high yields [20].

7.4.3.1.1 Acid Catalyst

In acid catalysis, reactions are catalyzed by Bronsted acids like hydrochloric acid, sulfonic acid, and sulfuric acid. Acid catalysis is versatile, able to catalyze both transesterification and esterification processes, and not affected by the presence of FFAs [20].

7.4.3.1.2 Base Catalyst

Base catalysts effectively convert oils into biodiesel, but a limitation arises when dealing with oils with high FFA levels, which remain as soap in large quantities. Alkali catalysts, such as sodium or potassium hydroxide, methoxides, and carbonates, are common base catalysts. Up to 5% FFAs, an alkali catalyst can still accelerate the process, but an additional catalyst is needed to compensate for the catalyst lost to soap [20].

7.4.3.2 Heterogeneous Catalysis

This catalysis develops active sites during the reaction, making it highly effective and easily recyclable. These catalysts are non-corrosive, are cost-effective, and possess desirable qualities such as recyclability and reusability. Various nanomaterials like titanium oxide, calcium oxide, cerium-doped silver oxide, copper oxide, zirconium oxide, gold-supported iron oxide, and magnesium oxide are commonly used as enzyme supports [21].

7.4.3.2.1 Alkali-Based Catalyst

Alkali-based heterogeneous catalysts can be modified for improved activity, selectivity, and extended life. Despite advantages, challenges such as low FFA levels, diffusional restrictions, high molar alcohol-to-oil ratio, high cost compared to conventional processes, and catalyst siphoning still impede commercialization [21].

7.4.3.2.2 Acid Catalyst

Inorganic polymeric materials are often used in heterogeneous acid catalysis because they enable easy separation from the reaction media, reusability, and the promotion of green technologies. Heterogeneous acid catalysts are less toxic and corrosive and have fewer negative effects on the environment. These catalysts may contain many acid sites with variable degrees of Bronsted and Lewis acidity. Due to the frequent poor catalytic activity of microporous inorganic oxides, the application of these catalysts is restricted at some crucial places. High solubility, poor thermal stability, little surface area, difficulties in synthesizing chemically antagonistic features, and a dearth of reactive sites are some of these characteristics [22].

7.4.3.2.3 Enzyme-Based Catalyst

Enzyme-based catalysts, also known as biocatalysts, derive from living organisms and facilitate chemical reactions while maintaining their own chemical composition stability [23]. In biodiesel production, two main categories of enzyme-based biocatalysts, extracellular and intracellular lipases, are frequently employed. Extracellular lipases are obtained from microbial broth and subsequently purified, while intracellular lipases are found within cells or in cell walls that divide to form new cells. Biocatalysts, particularly enzymes, reduce bottlenecks in the purification of glycerol and di- and monoacylglycerols, diminishing the production of waste products like soaps and colors [23]. The efficiency of the bio-catalyzed transesterification process relies on the enzyme source and various process variables. Unlike chemical catalysts, biocatalysts can be applied to a wide range of triglyceride sources, even with FFAs ranging from 0.5% to 80%. However, the use of extracellular enzymes as catalysts faces challenges due to difficulties in separation and purification techniques, along with associated costs [23].

7.5 NANOMATERIALS IN TRANSESTERIFICATION

Nanomaterials have garnered recent attention in biodiesel production due to their beneficial morphological and physicochemical characteristics, such as a high surface-to-volume ratio, increased reactivity, photocatalytic activity, and low cytotoxicity [24]. In transesterification processes, nano catalysts have proven highly effective, outperforming conventional catalysts due to their larger surface area, greater saponification resistance, enhanced stability, and increased reusability (Figure 7.5) [25]. Designing nano catalysts for stability at high temperatures is crucial for influencing reaction conditions and increasing their recyclability and reusability. Various catalysts, encompassing organic and inorganic substances like ion exchange resin, enzymes, oxides, and metal salts, are being developed. Recent research has explored nano catalysts made of polymers, zeolite, and carbon-based nanomaterials, primarily chosen for their high catalytic activity, rapid reaction, and low reaction temperature [26]. The creation and modification of nano catalysts hold the potential

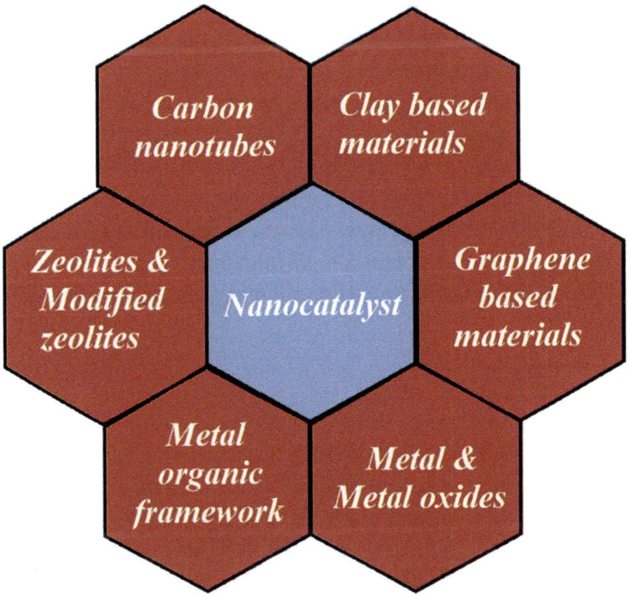

FIGURE 7.5 Different types of nano catalysts used for biodiesel production.

to significantly boost biofuel production. Various methods, including chemical vapor deposition, impregnation, precipitation, and electrochemical deposition, can be employed in nano catalyst synthesis. Noteworthy nano catalysts, such as metal oxide nano catalysts (CaO, ZnO), zeolites, hydrotalcites, and nano biocatalysts, have been utilized in transesterification for biodiesel production [26].

7.5.1 Carbon Nano Catalysts

Nano catalysts made from carbon materials, including CNTs, graphene, and reduced graphene oxides, exhibit improved physicochemical properties, reflecting the diverse shapes and sizes of the created nanocomposite materials [27]. CNTs, owing to their diverse structural, mechanical, thermal, and biocompatibility qualities, are anticipated to serve as excellent catalysts with applications in fuel cells and other electrocatalytical devices. CNTs are often derived from wound graphite sheets, forming cylindrical structures with high surface areas measured in nanometers. Immobilized or conjugated CNTs, particularly functionalized multi-walled carbon nanotubes (MWNTs), have been shown to enhance the activity and stability of enzymes, such as lipases. For instance, MWNT-immobilized lipase demonstrated increased levels of aldehydes, carboxylates, and peroxides during ester hydrolysis compared to free lipase in an aqueous medium over time [28]. Researchers have replaced CNTs with lipase enzymes for biodiesel synthesis due to their superior thermal strength and recyclable nature [29]. Surface-linked dendrimers have further improved the catalytic activity of these catalysts, achieving a remarkable yield of 92.8% overall and notable recyclability for almost 20 additional cycles. Nano catalysts composed of biochar are effective in catalyzing transesterification and are typically produced through the pyrolysis and carbonization of biomass from various feedstocks. Biochar-based nano catalysts, while extending the reaction time, exhibit poor heat stability. Graphene oxide, with its increased surface area and distinctive functional groups, also demonstrates enhanced fatty acid conversion [27].

7.5.2 Magnetic Nano Catalysts

The use of materials such as metal oxides, iron, cobalt, nickel, and platinum alloys is common in the synthesis of magnetic nano catalysts. MNPs offer various advantages, including higher molecular weight, smaller size, quantum properties, and the capability to transport other chemicals, making them valuable in applications like drug delivery. In the biofuel industry, MNPs find widespread usage, particularly in the production of sugars from lignocellulosic materials through the use of immobilized enzymes. MNPs prove to be versatile nanoparticles, as they can be attached to other catalytic nanoparticles or utilized as an immobilization support for enzymes, enabling repeated use. MNPs with enzymes can be recycled multiple times. Additionally, these nano catalysts exhibit applications in hydrogenation, photooxidation, and inductive heating through a high-frequency magnetic field. Furthermore, MNPs serve as carriers for enzymes, offering enhanced control under a magnetic field [30].

7.5.3 Zeolite Nano Catalysts

The potential catalytic activity of nano zeolites is attributed to their strong acidity sites, large surface area, specificity, and unique molecular sieving characteristics. Nano zeolites, with their large exterior surface areas and inherent hydrophobic nature, facilitate enzyme access to substrates, exhibiting high solubility in both polar and non-polar solvents. While natural zeolite materials are less commonly used in commercial industries, commercially available synthetic zeolites, including ZSM-5, X, Y, and beta, are extensively employed in biodiesel production. Transesterification catalyzed by ZSM-based nano catalysts, such as Zbio, exhibits higher activation energy due to increased acid active sites, resulting in a biodiesel production efficiency of 87%. These zeolite nano catalysts open new opportunities for scaling up biodiesel synthesis through the transesterification of various feedstock materials [31].

7.6 FACTORS AFFECTING BIODIESEL PRODUCTION

The factors that must be considered to optimize the transesterification stage affecting the synthesis of biodiesel using nanomaterials for the economic viability of biodiesel production are:

i. Methanol-to-oil ratio
ii. Concentration of nanomaterial
iii. Water content
iv. Temperature

7.6.1 Methanol-to-Oil Ratio

A key parameter in the biodiesel transesterification reaction is the stoichiometric ratio of methanol to oil. Due to interactions between the FFAs and the catalysts, soap is produced when the molar ratio of methanol to oil is not taken correctly. The methanol/oil ratio determines the biomass content. Depending on the feedstock used, the climax point of the methanol/oil molar ratio varies. An increase in the methanol-to-oil molar ratio helped to boost the conversion of biodiesel. The conversion yield tends to decrease as the ratio of methanol to oil is over its maximum value. This behavior could be explained by an excess of methanol, which would lower the catalyst concentration and increase the yield of the reverse transesterification reaction [32].

7.6.2 Water Content

The water content of oils and fats is another parameter of biodiesel generation. The presence of water has an impact on the biodiesel yield, reaction speed, enzyme stability, and activity. It is well known that an enzyme needs some water to remain active in a non-aqueous environment. The reaction that produces biodiesel has a wide range of acceptable water concentrations. TAGs are hydrolyzed in the presence of water to produce FFAs. High levels of FFAs in a fundamental homogeneous transesterification process boost soap formation and reduce the efficiency of TAG to ester conversion. Base-catalyzed transesterification allows the use of water-free and low-acid feedstocks. The cost of the entire process will increase if the feedstock contains more alkaline catalyst to neutralize the FFAs than if it does an acid. Because water promotes the creation of soap and foaming increases viscosity, its presence has a more detrimental effect than the presence of FFAs. The catalyst is employed to create soap and FFA content when there is water present, hence reducing the biodiesel concentration [33].

7.6.3 Concentration of Nanomaterial

The production rate of biodiesel is positively impacted by the nano catalyst concentration. However, the studies reveal that each catalyst has a maximum limit beyond which the conversion yield tends to decline. Additionally, the type of nanomaterial used affects the yield of biodiesel conversion. The excessive catalyst loading decreased the concentration of FAMEs, which may be related to problems with mass transfer, product desorption restrictions, or interference interactions between active sites and reactants. Additionally, if the catalyst loading is too high, the base catalyst may convert the fatty acid methyl ester product into other by-products. Therefore, the correct quantity of chicken manure catalyst is essential for maximizing biodiesel synthesis [34].

7.6.4 Temperature

Temperature is a critical factor influencing the reaction of nano catalysts in biodiesel synthesis, impacting the catalytic activity and, consequently, the reaction rate. In the case of solid calcium di-glyceroxide for biodiesel generation, the research involved varying reaction temperatures from 50°C to 70°C in 5°C increments. It was observed that increasing the temperature from 50°C to 65°C led to a significant rise in conversion, but further increasing it to 70°C resulted in a slight decrease in conversion. The occurrence of cavitation, induced by evaporation and the phase change from liquid to vapor, can strain the system and disrupt the process. In a study investigating the effects of supercritical CO_2 on lipid extraction from nano-chloropsis gaditana, it was noted that increasing temperature and CO_2 density enhanced lipid and carotenoid yields. Temperature increases up to the optimum value accelerated the reaction rate, improved oil solubility in the alcohol phase, and lowered oil viscosity. However, exceeding the optimum temperature led to material evaporation and reduced yield [35].

7.7 TYPE OF ALCOHOL

One mole of feedstock oil can be transesterified into three moles of alcohol. Due to their economic cost, methanol and ethanol are the most widely used alcohols during transesterification reactions. FAME is the biodiesel product that is produced when methanol is used, whereas fatty acid ethyl ester (FAEE) is produced when ethanol is used. Additionally, FAME is superior to FAEE as it gives maximum engine performance. Methanol is less expensive, more flammable, and reactive than FAEE. FAEE is thicker than FAME. Methanol is relatively cheap and performs transesterification reactions most effectively. Hence, producing biodiesel with it offers more economic advantages over producing it with other alcohols. However, the ideal amount of methanol should be utilized [36]. A slower reaction rate will result from using less methanol; however, using more methanol will make it difficult to separate the by-product (glycerin)

7.8 TYPE OF NANOMATERIALS FOR BIODIESEL

Metal-based nanoparticles like titanium, cerium, aluminum, iron, silver, copper, CNTs, and magnesium enhance the catalytic activity of biodiesel, resulting in complete fuel combustion and improved engine performance [37]. Magnetic nano catalysts are cost-effective, easily isolated from reaction media, and can be reused for large-scale biodiesel synthesis [24]. For instance, activated carbon/$CuFe_2O_4$@CaO nano catalyst exhibited a high-quality diesel yield of 95.6% after a 4-hour reaction at 65°C with a methanol/oil molar ratio of 12:1 [38]. In the transesterification of chicken fat, MgO and MgO@Na_2O catalysts yielded 95.17% and 95.22%, respectively, after a 4-hour reaction at 65°C with an alcohol-to-oil ratio of 24:1 [39]. Nano CaO catalysts resulted in a biodiesel production yield of 94.4% at a temperature of 65°C after a 5-hour reaction with an alcohol-to-oil ratio of 1:9 and a catalyst quantity of 1 wt% [40]. Two different nano catalysts, CaO/$CuFe_2O_4$ and AC/$CuFe_2O_4$@CaO, were employed in chicken oil biodiesel synthesis, yielding 94.52% and 95.6%, respectively [41].

Green nanoparticles of MgO, produced with *Parthenium hysterophorus* leaf extract, accelerated the transesterification of *Cichorium intybus* seed oil, achieving a 95% yield of FAMEs under optimized conditions [42]. TiO_2 nano catalysts were utilized in the transesterification of *Carthamus tinctorius* L. for biodiesel synthesis, reaching an optimum yield of 95% with an oil-to-methanol ratio of 1:10, using 0.25 g of TiO_2 catalyst for a 120-minute reaction at 65°C [43].

7.9 CHALLENGES AND PROSPECTS

Nanomaterials, with their small size, ease of synthesis, and wide range of applications, are highly valuable; however, certain metal oxide nanoparticles have been associated with cytotoxic effects in humans. Most notably, titanium nanoparticles have been found to bio-aggregate in the liver, lungs, and brain of exposed rodents, leading to neurobehavioral abnormalities and organ damage. The extensive use or release of nanoparticles in the environment raises concerns, and the long-term repercussions, both positive and negative, especially in large-scale applications like ethanol and biodiesel production, require thorough research. There is a need for experts globally to suggest and study the long-term impacts of nanomaterial use while finding ways to minimize harmful effects. Strategies include developing plans for utilizing non-edible feedstock oils and increasing biomass production from algae. Multifunctional nano catalysts with stable active sites are being developed, and carbonaceous and zeolite-based catalysts are functionalized with polymers and nanomaterials to enhance stability, specificity, and catalytic activity. Structural modification of nano-based materials based on size and form improves the kinetics of biodiesel synthesis processes. Statistical techniques are being developed for the effective optimization of reaction conditions on a large scale. Technological advancements are crucial to scaling up the reaction process, reducing mass transfer resistance and energy consumption, and optimizing by-product utilization, thereby increasing biodiesel production cost-effectively [43].

7.10 CONCLUSION

The production of biodiesel holds significant potential to revolutionize global energy usage, inspiring ideas for sustainable recycling and contributing to the concept of a circular economy. While biodiesel production can promote environmentally beneficial consumption patterns, scaling up challenges like poor yield, the food vs. feed paradox, and a lack of statistical expertise for maximizing biofuel potential. Transesterification, known for its economic viability, sustainability, and high output, is a frequently used method for biodiesel manufacturing. A notable advancement is the utilization of nanoparticles as catalysts in biodiesel synthesis. Nano catalysts, with features like a high surface-to-volume ratio, variable shape, improved activity, and specificity, play a crucial role in enhancing transesterification yields. Functionalizing nano catalysts with various groups can alter their properties, making them more acidic, reducing the leaching of active metals, and increasing active sites for catalysis. Properly using nano catalysts in different ratios can improve the quality and yield of biodiesel. Future research should focus on scaling up biodiesel production using feedstocks from waste and biomass. Exploring the synthesis and functionalization of nano catalysts with minimal leaching and disintegration is also a crucial area of investigation. Technical development is necessary to scale up the reaction process, reduce mass transfer resistance and energy consumption, and optimize by-product utilization. Innovative technological approaches are essential for comprehending and harnessing the utility of biodiesel to meet global energy demands sustainably and environmentally soundly [43].

REFERENCES

1. Martchamadol J, Kumar S. Thailand's energy security indicators. *Renew Sustain Energy Rev* 2012;16:6103–22. https://doi.org/10.1016/j.rser.2012.06.021
2. Prokopowicz A, Zaciera M, Sobczak A, Bielaczyc P, Woodburn J. The effects of neat biodiesel and biodiesel and HVO blends in diesel fuel on exhaust emissions from a light duty vehicle with a diesel engine. *Environ Sci Technol* 2015;49:7473–82. https://doi.org/10.1021/acs.est.5b00648
3. Thakur A, Thakur P, Baccar S. Structural properties of nanoparticles. In: Suhag D, Thakur A, Thakur P. (eds) *Integrated Nanomaterials and their Applications.* Springer, Singapore, 2023. https://doi.org/10.1007/978-981-99-6105-4_4
4. Sahani S, Roy T, Sharma YC. Smart waste management of waste cooking oil for large-scale high-quality biodiesel production using Sr-Ti mixed metal oxide as solid catalyst: optimization and E-metrics studies. *Waste Manage* 2020;108:189–201.
5. Silva MAA, Correa RA, Tavares MGdO, Antoniosi Filho NR. A new spectrophotometric method for determination of biodiesel content in biodiesel/diesel blends. *Fuel* 2015;143:16–20.
6. Amais RS, Garcia EE, Monteiro MR, Nogueira ARA, Nobrega JA. Direct analysis of biodiesel microemulsions using an inductively coupled plasma mass spectrometry. *Microchem J* 2010;96(1):146–50.
7. Laksmono N, Paraschiv M, Loubar K, Tazerout M. Biodiesel production from biomass gasification tar via thermal/catalytic cracking. *Fuel Process Technol* 2013;106:776–83.

8. Yan S, Lu H, Liang B. Supported CaO catalysts used in the transesterification of rapeseed oil for the purpose of biodiesel production. *Energy Fuels* 2008;22(1):646–51.
9. Azcan N, Yilmaz O. Microwave assisted transesterification of waste frying oil and concentrate methyl ester content of biodiesel by molecular distillation. *Fuel* 2013;104:614–9.
10. Perez S, Del Molino E, Barrio VL. Modeling and testing of a milli-structured reactor for carbon dioxide methanation. *Int J Chem Reactor Eng* 2019;17(11):1–12.
11. Pimentel D, Patzek TW. Ethanol production using corn, switchgrass, and wood; biodiesel production using soybean and sunflower. *Nat Resour Res* 2005;14(1):65–76.
12. Sahoo PK, Das LM, Babu MKG, Naik SN. Biodiesel development from high acid value polanga seed oil and performance evaluation in a CI engine. *Fuel* 2007;86(3):448–54.
13. Mishra VK, Goswami R. A review of production, properties and advantages of biodiesel. *Biofuels* 2018;9(2):273–89.
14. Prado CMR, Antoniosi Filho NR. Production and characterization of the biofuels obtained by thermal cracking and thermal catalytic cracking of vegetable oils. *J Anal Appl Pyrol* 2009;86(2):338–47.
15. Luo Y, Ahmed I, Kubatova A, Sťavova, Aulich T, Sadrameli SM, et al. The thermal cracking of soybean/canola oils and their methyl esters. *Fuel Process Technol* 2010;91(6):613–7.
16. Gashaw A, Teshita A. Production of biodiesel from waste cooking oil and factors affecting its formation: a review. *Int J Renew Sustain Energy* 2014;3:92–8.
17. Fatima U, Ahmad F, Ramzan M, Aziz S, Tariq M, Iqbal H, Imran M. Catalytic transformation of *Brassica nigra* oil into biodiesel using in-house engineered green catalyst: development and characterization. *Clean Technol Environ Policy* 2021;4:1–11. https://doi.org/10.1007/s10098-021-02170-4
18. Kamonsuangkasem K, Therdthianwong S, Therdthianwong A, Thammajak N. Remarkable activity and stability of Ni catalyst supported on CeO_2-Al_2O_3 via $CeAlO_3$ perovskite towards glycerol steam reforming for hydrogen production. *Appl Catal B* 2017;218:650–63.
19. Sharma YC, Singh B, Korstad J. Latest developments on application of heterogenous basic catalysts for an efficient and eco-friendly synthesis of biodiesel: a review. *Fuel* 2011;90(4):1309–24.
20. Nayab R, Imran M, Ramzan M, et al. Sustainable biodiesel production via catalytic and non-catalytic transesterification of feedstock materials – a review. *Fuel* 2022;328:125254.
21. Wan F, Thakur A, Thakur P. Classification of nanomaterials (carbon, metals, polymers, bio-ceramics). In: Suhag D, Thakur A, Thakur P. (eds) *Integrated Nanomaterials and their Applications*. Springer, Singapore, 2023. https://doi.org/10.1007/978-981-99-6105-4_3
22. Touqeer T, Mumtaz MW, Mukhtar H, Irfan A, Akram S, Shabbir A, et al. Fe_3O_4-PDA-Lipase as surface functionalized nano biocatalyst for the production of biodiesel using waste cooking oil as feedstock: characterization and process optimization. *Energies* 2020;13(1):177.
23. Hama S, Noda H, Kondo A. How lipase technology contributes to evolution of biodiesel production using multiple feedstocks. *Curr Opin Biotechnol* 2018;50:57–64. https://doi.org/10.1016/j.copbio.2017.11.001
24. Gardy J, Osatiashtiani A, Cespedes O, Hassanpour A, Lai X, Lee AF, et al. A magnetically separable SO_4/Fe-Al-TiO_2 solid acid catalyst for biodiesel production from waste cooking oil. *Appl Catal B* 2018;234:268–78.
25. Qiu F, Li Y, Yang D, Li X, Sun P. Heterogeneous solid base nanocatalyst: preparation, characterization and application in biodiesel production. *Bioresour Technol* 2011;102:4150–56.
26. Thakur P, Thakur A. Nanomaterials, their types and properties. In: Thakur A, Thakur P, Khurana SP. (eds) *Synthesis and Applications of Nanoparticles*. Springer, Singapore, 2022. https://doi.org/10.1007/978-981-16-6819-7_2
27. Rasouli H, Esmaeili H. Characterization of MgO nanocatalyst to produce biodiesel from goat fat using transesterification process. *3 Biotech* 2019;9:429. https://doi.org/10.1007/s13205-019-1963-6
28. Nizami AS, Rehan M. Towards nanotechnology-based biofuel industry. *Biofuel Res J* 2018;5:798–9. https://doi.org/10.18331/BRJ2018.5.2.2
29. Rai M, dos Santos JC, Soler MF, Marcelino PRF, Brumano LP, Ingle AP, Gaikwad SAG, da Silva SS. Strategic role of nanotechnology for production of bioethanol and biodiesel. *Nanotechnol Rev* 2016;5:231–50. https://doi.org/10.1515/ntrev-2015-0069
30. Ehsan M, Suhag D, Rathore R, Thakur A, Thakur P. Metal nanoparticles in the field of medicine and pharmacology. In: Suhag D, Thakur A, Thakur P. (eds) *Integrated Nanomaterials and their Applications*. Springer, Singapore, 2023. https://doi.org/10.1007/978-981-99-6105-4_7
31. Zhang Z, Yuan X, Miao S, Li H, Shan W, Jia M, Zhang C. Effect of Fe additives on the catalytic performance of ion-exchanged CsX zeolites for side-chain alkylation of toluene with methanol. *Catalysts* 2019;9:829. https://doi.org/10.3390/catal9100829

32. Mansir N, Teo SH, Rashid U, Taufiq-Yap YH. Efficient waste Gallus domesticus shell derived calcium-based catalyst for biodiesel production. *Fuel* 2018;211: 67–75.
33. García MT, Martín JFG. Production of biofuels and numerical modelling of chemical combustion systems. *Processes* 2021;9(5): 829. https://doi.org/10.3390/pr9050829
34. Li F, Hülsey MJ, Yan N, Dai Y, Wang CH. Co-transesterification of waste cooking oil, algal oil and dimethyl carbonate over sustainable nanoparticle catalysts. *Chem Eng J* 2021;405:127036. https://doi.org/10.1016/j.cej.2020.127036
35. Millao S, Uquiche E. Extraction of oil and carotenoids from pelletized microalgae using supercritical carbon dioxide. *J Supercrit Fluids* 2016;116:223–31. https://doi.org/10.1016/j.supflu.2016.05.049
36. Gebremariam SN, Marchetti JM. Biodiesel production technologies: review. *AIMS Energy* 2017;5(3):425–57.
37. Krishnia L, Thakur P, Thakur A. Synthesis of nanoparticles by physical route. In: Thakur A, Thakur P, Khurana, SP. (eds) *Synthesis and Applications of Nanoparticles*. Springer, Singapore, 2022. https://doi.org/10.1007/978-981-16-6819-7_3
38. Seffati K, Honarvar B, Esmaeili H, Esfandiari N. Enhanced biodiesel production from chicken fat using $CaO/CuFe_2O_4$ nanocatalyst and its combination with diesel to improve fuel properties. *Fuel* 2019;235:1238–44. https://doi.org/10.1016/j.fuel.2018.08.118
39. Bahador F, Foroutan R, Nourafkan E, Peighambardoust SJ, Esmaeili H. Enhancement of biodiesel production from chicken fat using MgO and $MgO@Na_2O$ nanocatalysts. *Chem Eng Technol* 2021;44(1):77–84. https://doi.org/10.1002/ceat.202000511
40. Keihani M, Esmaeili H, & Rouhi P. Biodiesel production from chicken fat using nano-calcium oxide catalyst and improving the fuel properties via blending with diesel. *Phys Chem Res* 2018;6(3):521–9. https://doi.org/10.22036/PCR.2018.114565.1453
41. Seffati K, Esmaeili H, Honarvar B, Esfandiari N. $AC/CuFe_2O_4@$ CaO as a novel nanocatalyst to produce biodiesel from chicken fat. *Renew Energy* 2020;147:25–34. https://doi.org/10.1016/j.renene.2019.08.105
42. Rozina, MA, Zafar M. Synthesis of green and non-toxic biodiesel from non-edible seed oil of Cichorium intybus using recyclable nanoparticles of MgO, *Materials Today Communications* 2023;35:105611.
43. Jan HA, Saqib NU, Khusro A et al. Synthesis of biodiesel from Carthamus tinctorius L. oil using TiO_2 nanoparticles as a catalyst, *Journal of King Saud University – Science* 2022;34:102317.

8 Nano Sensor for Environmental Pollution Detection

Amitender Singh
Amity University Haryana

Kavita Yadav
G.C.W. Gurawara, Rewari

Fayu Wan
Nanjing University of Information Science & Technology

Preeti Thakur
Amity University Haryana

Atul Thakur
Amity University Haryana
Nanjing University of Information Science & Technology

8.1 INTRODUCTION

With the advent of the digital era, the craving and necessity to comprehend the environment, particularly the quality of air, soil, and water, has rapidly grown. The main barrier to comprehending the environment is collecting sufficient data on a wide range of pollutants of environmental significance and storing all the obtained data. Several methods have emerged over the past many years for the highly accurate and precise detection of environmental pollutants, and nanomaterial-enabled sensors are one of them. The potential to statistically grasp nature in a systematized way will shortly be a reality because of the promise of simple, affordable, field-deployable technology.

8.2 ENVIRONMENTAL POLLUTION PROBLEM

The escalating issue of environmental pollution is a global concern, stemming from rapid urbanization, population growth, and various contributors such as traffic expansion, industrialization, and energy utilization [1]. Air pollution, characterized by the presence of organic, inorganic, and hybrid pollutants, adversely impacts air, soil, and water quality, posing threats to biodiversity and human health [1]. The release of hazardous chemicals into the atmosphere from sources like industrial emissions and transportation not only harms ecosystems but also contributes to severe health conditions, particularly heart and lung diseases linked to long-term exposure to particulate matter, gases, and heavy metals [1]. The World Health Organization (WHO) reports millions of annual deaths attributable to air pollution [1]. Notable air pollutants encompass carbon monoxide, nitrogen oxide, sulphur dioxide, industrial and transport smoke, and volatile organic compounds (VOCs), all stemming from human activities, including the burning of fossil fuels and industrial processes [1]. Non-biodegradable heavy metals such as lead, mercury, arsenic, and chromium further contaminate the environment, emphasizing the urgent need for advanced technologies to monitor and detect pollution levels before surpassing acceptable limits [2].

 DOI: 10.1201/9781003502692-8

8.2.1 Nanotechnology for Environmental Pollution

Nanomaterial-enabled nano sensors are an interesting new technology that offers precise environmental contamination detection at precisely low levels [2,3]. These sensors are appealing because they could enable simple, on-site contamination detection without the use of pricey lab apparatus. There was a need to deliberately discuss the detection of specific pollutants since environmental scientists, researchers, and engineers are frequently interested in establishing whether a specific contamination exists at a field site and whether its concentration is beyond the regulatory limit. By measuring physical properties like velocity, volume, pressure, temperature, concentration, electrical forces, and magnetic forces, nanotechnology can identify pollutants at extremely low levels to monitor the environment. Before looking at the use of nano sensors for the detection of various groups of environmental contaminants, we discuss nano sensor design, nanomaterials, developments in sensing methods, important potentials for nano sensors, and environmentally friendly nanostructure production and outline obstacles for future development. This chapter's goal is to thoroughly highlight the use of nanoparticles as sensing materials in nano sensors for environmental pollution detection. Here, we have addressed how several nanomaterials have recently been used to detect and sense a number of environmental contaminants. Before discussing current advancements in nanomaterial-assisted detection, we briefly review the fundamental ideas underpinning a nano-enabled sensor. There are an almost infinite number of compounds that pose a threat to the environment, and while it would be impossible to list them all, the examples given in this chapter show how the basic nano sensor designs work.

8.2.2 Role of Nanotechnology and Nano Sensors in the Mitigation of Environmental Pollution

The need for nano sensors has increased during the past few decades on a global scale. Nano sensors have been creatively used in devices for environmental pollution detection. It can be used on a broad scale and is also used in the design of the constructions. This helps environmental scientists in developing more economical, effective, and selective nanomaterials [4,5]. The nanoworld is an interesting place where a lot of things happen that are very different from the macro world. Due to the necessity of nano sensors for carrying out many daily tasks, this technology is successfully applied in a variety of industries including environment trackers to meet ongoing demands [6]. The relationship between the improvement of human health and the environment makes environmental control one of the choices on a worldwide scale. Traditional methods that rely on old analytical tools have significant disadvantages. Different chromatographic techniques are among the few methods used for the measurements of environmental pollutants, but they come at a high expense in terms of equipment, reagents, and time-consuming sample pre-treatment [7,8]. The machinery is large, bulky, difficult to use, energy-intensive, and expensive to operate and maintain. When a decision must be made quickly, as in the case of the present environmental concern, real-time measurements are used. If real-time measurements cannot be made, various methods and models of environmental pollutant distribution measurement are used, and their significance is based on the accuracy of these methods and models. Thus, even though stations accurately identify air contaminants, their spatial–temporal resolution is insufficient to capture the spatial–temporal evolution of air pollution. Nanotechnology is a cutting-edge technology that offers a viable way to develop novel nanomaterials with special features utilized for long-term pollution control. Greater catalytic performance, strong electrical conductivity, better hardness and strength, large active surface area, increase in electrochemical signals, and extension of exploring instruments are a few of these nanomaterial qualities [8].

Nanotechnology offers a great chance to measure, monitor, control, and lessen environmental pollution. Nanomaterials have also been employed to address possible problems with the photocatalytic destruction of several organic pollutants in water [9,10]. The removal of air pollutants by nanomaterials has also been studied. For example, TiO_2/Ag tissue nanocomposites are effective at removing toluene and airborne pathogens from vapours [11]. In addition, researchers have shown that hollow doped platinum nanostructures can degrade iso-propanol in the gas phase when exposed to sunlight [12]. To monitor hydrogen gas in the environment, Jung et al. [13] created solid-state potentiometric sensors based on a Nafion electrolyte. These sensors have a long-term stability of more than 3 months, good sensitivity, a rapid reaction time, and a wide linear range. Using a Cu hollow sphere and monooxygenase enzyme, Mohamed and Awad described a unique method for creating nanomaterials that was used in the elimination of gas contaminants under visual irradiation [11]. However, effective control systems that can quickly identify and quantify the pollutant sources are crucially needed to prevent or reduce the harm brought on by atmospheric pollution. The development of nano sensors for the detection of air pollution and sensing of gases (such as CO_2, SO_2, O_2, O_3, H_2, and different organic vapours) will eventually be crucial for reducing industrial and automobile emissions, improving human health, and maintaining environmental control.

8.3 NANO SENSORS

Nano sensors are recently being used on devices used in nearly all spheres of technology and have been tested and incorporated intensively [14,15]. The nanomaterial surface properties have a significant impact on the analytical performance of nano sensor systems [16]. Recently, nano sensors with electromagnetic and chemical properties are gaining importance.

8.3.1 Definition of Nano Sensor

A sensor functions as an apparatus that electronically recognizes a changeable quantity, transforming measurements into specific signals. In contrast, a nano sensor, with at least one dimension at 100 nm, serves as a nanoscale sensing device for data collection and analysis. Essential in nanotechnology, nano sensors contribute to spotting physical and chemical changes, evaluating harmful substances in various environments, and monitoring biomolecules and biochemical alterations. Prioritizing sensitivity, accuracy, selectivity, and stability, nano sensors play a crucial role in detecting microscopic particles, monitoring processes in hard-to-reach areas, and analysing water, soil, and atmospheric elements. They are instrumental in assessing air and water quality for environmental purposes.

The escalating concern over chemical gaseous pollutants causing air pollution underscores their negative impact on plant, animal, and human health. Detecting pollution becomes a pivotal step in pollution control, with the potential to enhance human health and environmental sustainability through efficient and affordable pollution level sensors (Figure 8.1). Advanced pollution detection technology

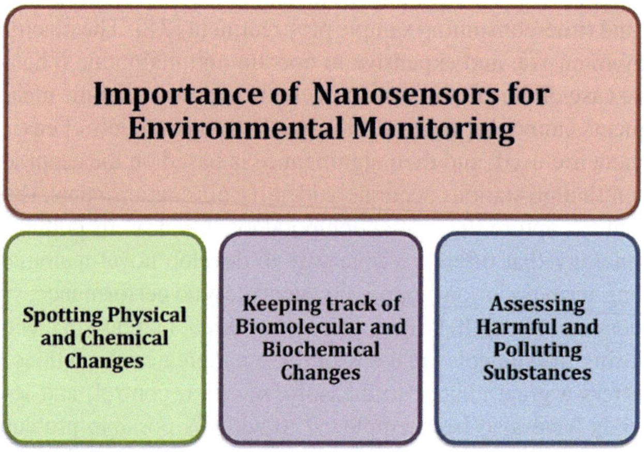

FIGURE 8.1 Importance of nano sensors for environmental monitoring.

holds the promise of improving ecosystem process control and environmental decision-making. Continuous monitoring instruments that rapidly provide information on contaminants are deemed highly efficient, addressing the urgent need to manage pollutants affecting ecosystems and human health. Emerging technologies focus on cost-effective, sensitive, user-friendly, quick, and portable sensing techniques, emphasizing environmental monitoring and pollution control. In the realm of nanotechnology for environmental applications, short-term research outcomes aim at creating innovative, smart, and improved sensors for the rapid detection of chemical, physical, and biological pollutants.

Nano sensors have emerged to furnish insights into nanoparticles utilized in chemical and biological sensory applications. Witnessing substantial growth, the nano sensor market anticipates further advancements, driven by reduced detection times, enhanced diagnostics, and cost-effective solutions. The escalating demand for high-quality, low-cost products is expected to parallel the increased cost-efficiency of nano sensors. With their diminutive size, economical nature, minimal power consumption, and heightened reliability, nano sensors are poised to expand possibilities. The market's trajectory foresees a surge in demand for nano sensors, attributed to their distinctive features. Recent advancements in electromagnetic nano sensors have given rise to modern nanoelectronics components, such as nanobatteries, nanoantenna, nanoscale logical circuits, and nano memories [17]. This progression opens avenues for innovative applications in various fields, particularly environmental sectors. However, the use of nanomaterials for sensing environmental contaminants raises pertinent questions and challenges that warrant exploration.

- How can a sensor tell if a pollutant is present, specifically?
- What are the various types of nano sensors for environmental pollutants?
- Which nanomaterials are employed to create sensors that can identify gaseous, pathogenic, and chemical pollutants?
- What are the characteristics of a nanomaterial-based nano sensor for environmental pollutants?
- What are the advantages and limitations of environmental nano sensors?

We will answer these questions in subsequent discussion in this chapter.

8.3.2 Classification of Nano Sensors

Nanotechnology is interested in the physical and chemical characteristics of materials at the nanoscale. The properties of materials are very different at the nanoscale compared to their properties at larger scales. These phenomena can be utilized by nano sensors. To detect and measure properties at the nanoscale, nano sensors may not necessarily be small enough; instead, they may be larger devices that make use of the properties of nanomaterials. They are capable of simple nanoscale interactions and the detection of processes that are not observable at the macroscale. Nano sensors are playing an important role and have a profound effect on environment monitoring. There are several different categories of nano sensors. The potential categorizations of nano sensors depending upon the stimuli are depicted in Figure 8.2. Currently, fluorescence is the primary transduction method due to its great sensitivity and relatively simple operation. Physical sensors, which operate on the nanoscale, convert signals from physical qualities like force, pressure, flow temperature, and other factors into detectable and measurable signals. In the examination of environmental sample residues, the creation of new medications, the monitoring of pollution, and the assay of organophosphorus chemicals, chemical nano sensors are frequently used to detect a variety of substances.

- **Mechanical nano sensors**: For the measurements of mechanical properties at the nanoscale, mechanical nano sensors have comparative benefits over optical and electromagnetic nano sensors. There are numerous varieties of mechanical nano sensors, including nanomechanical cantilever sensors based on carbon nanotubes (CNTs). The first

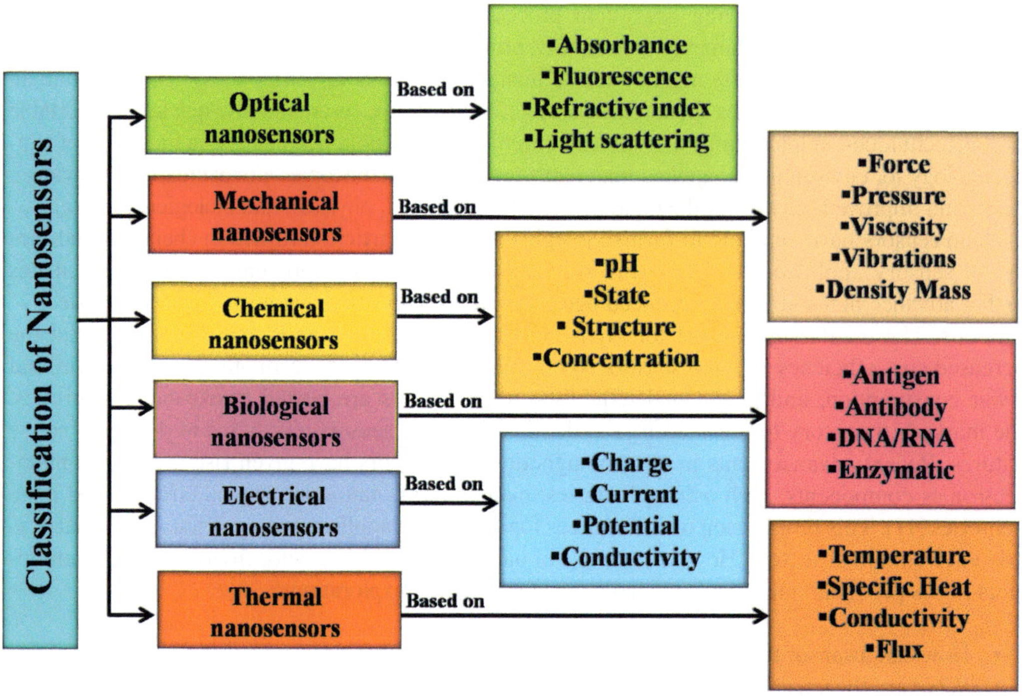

FIGURE 8.2 Classification of nano sensors depending upon stimuli.

mechanical nano sensor was proposed by Binh et al. to measure the elastic and vibrational characteristics of a nanosphere linked to a tapered cantilever [18]. For use in nanodevice components in microelectronic devices for environment monitoring, mechanical sensors play a critical role.

- **Optical nano sensors:** Chemical analysis can be monitored by optical sensors. They are reliant on the optical characteristics of nanomaterials. They can be used in a variety of fields, including the environmental sciences, the chemical industry, and the pharmaceutical industry. The first optical nano sensor to be described was used to monitor pH. Fluorescent sensors are essentially particles with one or more photoactive units and a binding component [19]. The optical nano sensor also has the benefit of achieving the lowest amount of invasion.
- **Electromagnetic nano sensors:** Electromagnetic nano sensors can be divided into two categories based on their monitoring of variables for the detection of objects: (a) Monitoring by measuring electrical current, and (b) Monitoring via measuring magnetism.
- **Measuring electrical current:** Researchers have studied the gas sensing of hydrogen sulphide gas with Au nanoparticles. Gold and chromium electrodes serve as the source and drain in each sensor cell. The two electrodes have a gap that is typically between 40 nm and 60 nm wide. Randomly distributed Au nanoparticles are applied to the gap region. The hopping phenomenon is prevented by the creation of sulphide shells. The Label-free process used in this strategy has an advantage over the use of dyes.
- **Magnetism measurement:** These magnetic nano sensors have been developed to recognize biomolecules, including pathogens (such as viruses). Iron oxide magnetic nanoparticles make up magnetic nano sensors. These magnetic nanoparticles create stable nano-assemblies when they bind to the desired chemical target. This causes the surrounding water molecules' spin-spin relaxation time to decrease in a comparable manner, which can then be picked up by magnetic resonance (NMR/MRI) methods [20].

- **Electrometers:** The mechanical element is coupled with charge using a mechanical resonator, a detecting electrode, and a gate electrode.
- **Chemical nano sensors:** These can be used to study a single molecule or a single chemical. Some features, such as pH and various ion concentrations, were measured using a variety of optical chemical nano sensors.
- **Biosensors:** These sensors use biological molecules like enzymes, antibodies, or living organisms to detect the presence of specific chemicals. Because the transduction techniques are comparable to chemical sensors, the biosensor is typically thought of as a subset of chemical sensors.

According to their energy source, structure, and uses, nano sensors can also be categorized.

- Nano sensors can be classified into two distinct categories depending on their energy source. Active nano sensors, exemplified by thermistors, necessitate an external energy source, while passive nano sensors, such as thermocouples and piezoelectric sensors, operate without an external power supply.
- Another classification criterion is the structural characteristics of nano sensors, leading to four categories: optical nano sensors, electromagnetic nano sensors, and mechanical/vibrational nano sensors.
- Further categorization based on application reveals four key types of nano sensors: chemical sensors, deployable nano sensors, electrometers, and biosensors. This diverse classification system highlights the versatility of nano sensors in various fields, allowing for specialized applications in chemical analysis, deployable scenarios, electrical measurements, and biological sensing.

8.3.3 Detection of Pollutants (Gas/Chemical/Pathogens etc.) by the Nano Sensor

A sensor comprises three primary components: a nanomaterial (or materials), a recognition component providing specificity, and a signal transduction technique, as illustrated in Figure 8.3. While these components may not always be distinctly separate entities within a sensor, each nano sensor can be categorized based on these three divisions. Multiplex detection, allowing a sensor to identify

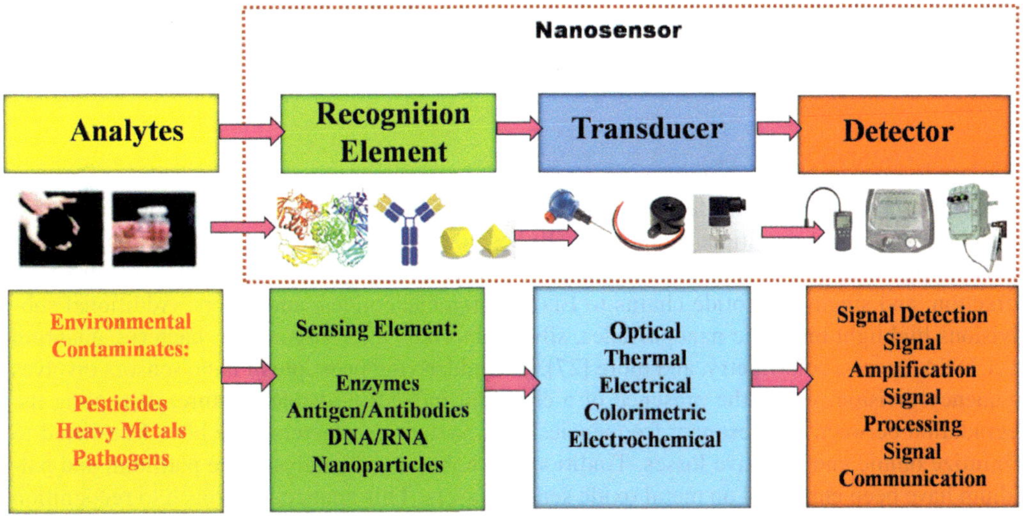

FIGURE 8.3 Various components of nano sensors for the detection of environmental pollution.

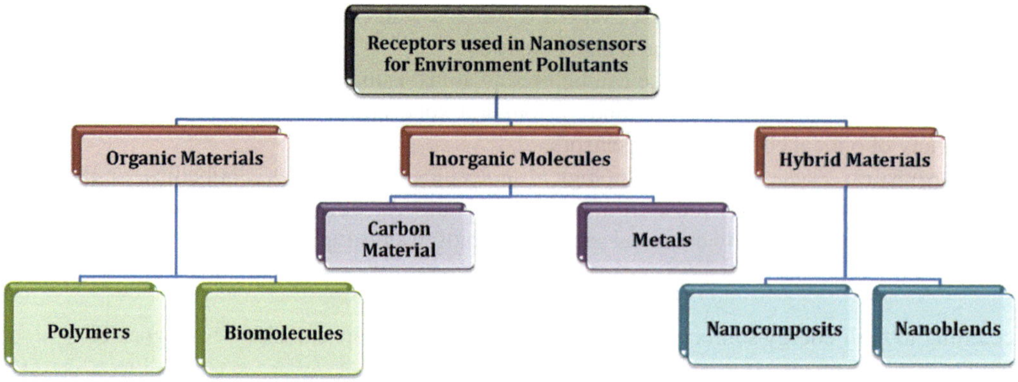

FIGURE 8.4 Various types of receptor materials used in nano sensors for environmental pollution monitoring.

multiple analytes, is a feasible capability. In the context of air pollutants, the interaction with the receptor initiates a reaction. Subsequently, the microelectronic signal processor amplifies the converted signal to a measurable form in the transducer section of the system [21]. This records the response from the receptor through various transducers. The core components of sensors involve substances responsive to changes in chemical or physical properties, translated into electrical signals by transducers [22]. The imperative need to enhance the efficiency of pollutant detection underscores the ongoing urgency in this domain.

Recently, the use of nanotechnology has shown promise in the development of sensors that overcome some of the drawbacks of gas sensors, such as limited sensitivity, high power consumption, and unstable performance [23,24]. The development of more modern sensors using novel transduction materials made possible by nanotechnology can be explained by the demand for intelligent sensing devices that are portable, selective, highly sensitive, affordable, rapid to respond, and extremely reliable. Examining the current varieties of nanomaterials used as a receptor in chemical sensors to find air contaminants is momentarily taken into consideration [24]. By assisting in the cheaper and more precise collection of various contaminants, nano sensors enhance exposure assessment. Metal and metal oxide semiconductors, catalytic materials, polymers, and other materials are examples of nanomaterial receptors. Numerous types of nanomaterials, such as inorganic (carbon and metal products), organic (polymers, biomolecules, etc.), and hybrid nanomaterials, are utilized as receptors in nano sensors as shown in Figure 8.4 [25,26].

8.3.4 Recognition Elements for Nano Sensors

Selectivity stands as a pivotal factor in the effective development of biosensors, and Figure 8.5 delineates various recognition components linked to nano sensors for practical applications. These components encompass a spectrum of categories, ranging from antibodies, electrostatic attraction, and single nucleotide chains to EDTA, oligonucleotides, and others. Additional subdivisions highlight magnetic nanoparticles, silver nanorods, microorganisms, bacteria, quantum dots, DNA, spores, colorimetry, and more [27]. The underlying principle of nano sensors involves frequency division, where the presence of a chemical induces a visible fracture, disrupting the degradation between two resonant frequencies [28]. System parameters have been fine-tuned to minimize ohmic and radiative losses. Traditionally, noble metal catalysts like platinum and palladium have been employed on metal oxide sensing layers. This nuanced approach to recognition elements and system design underscores the intricate considerations involved in optimizing nano sensor performance.

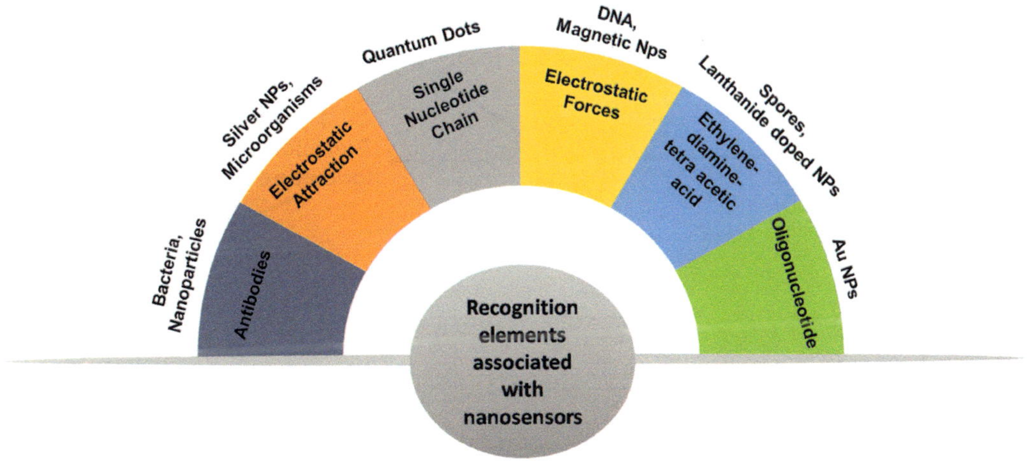

FIGURE 8.5 Recognition elements for nano sensors.

8.3.5 Signal Transduction

Signal transduction methods convert recognition events into computable signals that can be further processed to produce the relevant data. Optical, electrochemical, and magnetic signal transduction are the three main signal transduction techniques used in nano-enabled sensors. For widespread public use, optical methods are preferred, especially optical sensors that provide results in the visible range. High specificity, simplicity, and ease of miniaturization are all characteristics of electrochemical sensing techniques. Magnetic transduction techniques have a lower background signal than optical and electrochemical techniques, making them the best choice for low-concentration samples. Other sensor methods pre-concentrate the analyte using magnetic materials before using an optical or electrochemical transduction approach. Figure 8.6 shows the various parts of the nano sensor working in a co-ordinating manner for monitoring agroecosystem as part of the environmental monitoring of soil.

8.4 NANOMATERIALS FOR ENVIRONMENTAL POLLUTION MONITORING

Nanostructured materials utilized in nano sensor production encompass thin films, nanoscale wires, CNTs, metal and metal oxide nanoparticles, polymers, and biomaterials. The heightened sensitivity of nano sensors stems from the similarity in size between nanomaterials (such as metal ions, pathogens, proteins, antibodies, and DNA) and the analyte of interest, enabling the exploration of previously inaccessible matrices. Examples of nanostructured materials employed in nano sensor fabrication include nanoscale wires, CNTs, thin films, metal and metal oxide nanoparticles, polymers, and biomaterials. In recent chemoreceptive sensing applications, graphene nanostructures have garnered significant attention. Carbon-based nanomaterials offer high conductivity, stability, cost-effectiveness, and ease of surface functionalization. CNTs, along with graphene and nano- and mesoporous carbon, find use in various electro-analytical applications. Their nanostructures expose surface groups for effective analyte-transduction material binding, facilitating the efficient detection of environmental pollutants.

Metal nanoparticles, with unique physical and chemical characteristics, are widely employed in diverse applications. Metals like Au, Pt, Pd, Ag, Cu, and Co feature prominently in sensing applications. Metal nanoparticle-based sensors hold promise for heightened sensitivity and selectivity, particularly with specialized signal amplifications. The evolution of nano sensor research is fuelled by advancements in metal nanoparticles, bio-functionalized nanoparticles, and nanocomposites.

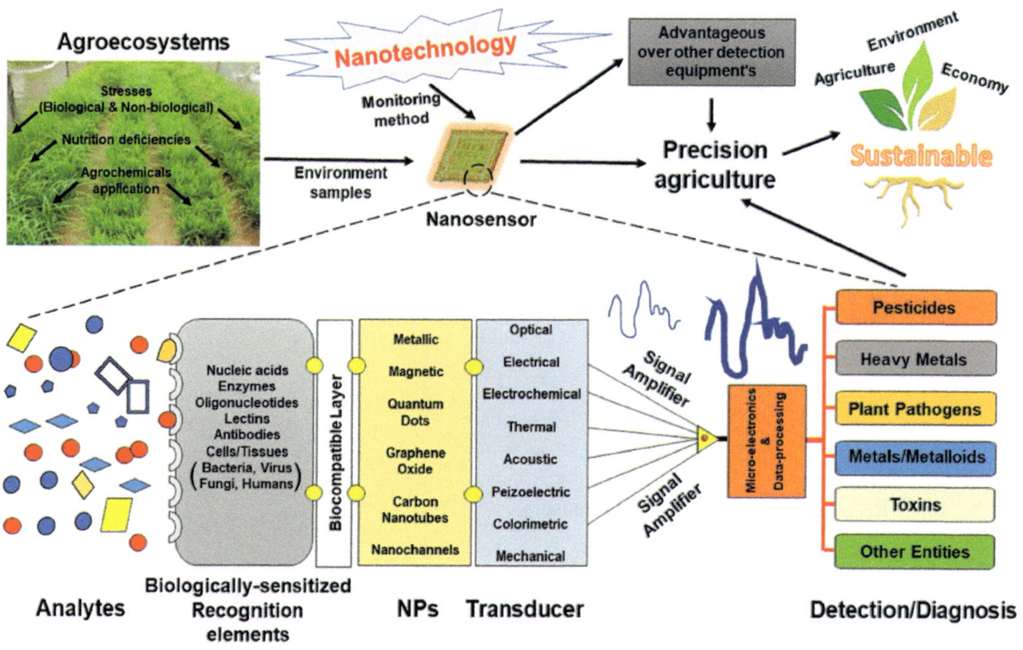

FIGURE 8.6 Various components of nano sensors for monitoring ergo-ecosystems as part of environment monitoring. (Sharma, P., Pandey, V., Sharma, M.M.M. et al. *Nanoscale Res. Lett.* 16, 136 (2021). https://doi.org/10.1186/s11671-021-03593-0.)

8.4.1 Metal and Noble Metals Nanomaterials

Metal nanoparticles have special physical, chemical, and electrical characteristics that are used in a wide variety of applications. For sensing, a variety of metals including Au, Pt, Ag, Cu, Co, and other metals have been used [29]. The use of controlled signal amplifications to increase the sensitivity and selectivity of metal nanoparticle-based sensors has significant potential. Applications for environmental monitoring and food safety have led to the development of various sophisticated analytical techniques. The noble metals, like rhodium (Rh), ruthenium (Ru), silver (Ag), osmium (Os), iridium (Ir), platinum (Pt), and gold (Au), have exceptional resistance to corrosion and oxidation even at high temperatures [30].

- **Gold nanoparticles:** The development of nano sensors utilizing gold (Au) has garnered significant attention in environmental applications. The exceptional stability and complete recovery in chemical redox processes make Au nanoparticles efficient electrocatalysts in various electrochemical reactions [31]. Electrodes based on Au nanoparticles offer several advantages, including an increased signal-to-noise ratio, excellent selectivity, enhanced catalytic activity, and improved diffusion of electroactive species. Researchers presented a cost-effective method for fabricating electrochemical gold-based nano sensors, highlighting the impact of nanoparticle shape on sensitivity. Gold nanoparticles, owing to their intriguing chemical, optical, and catalytic properties, find frequent use in biological and chemical sensors. Some research indicates the utilization of gold nanoparticles as nano-catalysts for electrochemical protein detection. Yuan et al. introduced Au nanoparticles ultrasonically into a carbonitride/graphene composite, creating Au nanoparticles/carbon nitride/graphene composite-based electrochemical sensors [32]. The synergy between Au nanoparticles and GN/C3N4 facilitates charge transfer and efficient catalytic actions and ultimately enhances sensitivity.

- **Silver nanoparticles:** Silver nanoparticles have become a common choice in sensors due to their superior surface-enhanced Raman scattering and catalytic activity. Kariuki and colleagues developed an electrochemical Ag nano sensor for nitrobenzene detection, showcasing rapid response, excellent sensitivity, and specificity in immunosensors based on silver nanoparticles [33]. Silver nanoparticles, coupled with various matrices like metal oxides, silicate networks, polymers, graphene, and fibres, exhibit high sensing efficiency and stability. The diverse production methods for silver nanoparticles, including chemical reduction, template induction, photo-irradiation, and electrochemical deposition, offer flexibility in tailoring their properties for specific sensor applications. The integration of Ag nanoparticles with Si nanocolumns through simple immersion procedures enhances gas sensor performance, demonstrating improved recovery time, sensitivity, and response.
- **Platinum nanoparticles:** Platinum nanoparticles have been employed in electrochemical studies and show good catalytic characteristics. Lebegue et al., for instance, have created a very sensitive H_2O_2 sensor [34]. Compared to a platinum bulk electrode, the modified electrode that included platinum nanoparticles showed a sensitive reaction to H_2O_2. Platinum nanoparticle-based electrode materials can be made using a variety of manufacturing processes, including chemical reduction, electrochemical deposition, and photochemical deposition. The development of Pt nanoparticles on multiple electrode surfaces, with exceptional chemical inertness, good stability, low background current, and strong catalytic and sensing capabilities, depends significantly on the production method choice. Pt nanoparticles' utility depends on interatomic bond lengths, melting temperatures, chemical reactivity, and also optical and electrical properties, all of which can be adjusted morphologically. Additionally, essential characteristics of Pt nanoparticles include their chemical composition and surface quality, which can have an impact on the electron transport pathways.

For detecting Hg^{2+}, Mahmoudian et al. described the production and characterization of platinum-based nano spherical [35]. The linear range of the electrochemical sensor was 5–500 nM, while the Hg^{2+} detection limit was 0.27 nM. Ag^{+}, Fe^{2+}, Pb^{2+}, Mn^{2+}, K^{+}, Ni^{2+}, Pd^{2+}, Cu^{2+}, Sn^{2+}, and Zn^{2+} were the least disruptive ions, and their interference with the detection of Hg^{2+} was minimal. A potentiometric hydrogen sensor with high sensitivity, rapid response, and recovery time; a broad linear range; and greater stability has been created by Jung et al. [13].

- **Palladium nanoparticles:** Palladium nanoparticles have numerous applications as catalysts and sensors for gases and dangerous poisonous compounds including biomolecules. Pd is more plentiful than other noble metals like Au and Pt, making it a less expensive alternative when building different electrochemical sensors. This provided an opportunity for electron tunneling, which made it possible for electrons to go from the active site to the electrode and improved the efficiency of electrochemical sensing [36]. Palladium–graphene nanocomposites were electrodeposited on indium tin oxide in a single step, as described by He et al., to create a sensitive hydrazine sensor [37]. Electrochemical experiments showed that the current was linearly dependent on N_2H_4 concentrations in the range of 0.1–2.5 mM under ideal conditions. According to Lupan et al., the high H_2 solubility of Pd results in a significant increase in the room temperature catalytic activity when Pd nanoparticles are present on ZnO [38]. As a result, there are more clusters (catalytic centres), and the pace at which the response and recovery processes are saturating is decreased. Incorporating these metals into semiconducting oxide micro and nanostructures has been suggested as a means to most effectively enhance their UV and gas-sensing capabilities. Adsorption, doping, surface functionalization, and composition are the techniques that have the greatest research behind them.

- **Copper:** Due to its outstanding electrical conductivity, electrocatalytic capabilities, and inexpensive cost when compared to noble metals like platinum, gold, and silver, copper has drawn the attention of numerous researchers as an ideal sensing material. Numerous special characteristics of copper nanostructures include their fast mass-transport rate, high surface-to-volume ratio, and better S/N ratio in electro-analytical investigations. Copper nanoclusters were created by Li et al. using a straightforward one-step electro-deposition procedure [39].

8.4.2 Metal Oxide Nanoparticles

Metal oxide nanoparticles offer distinct advantages such as ultrahigh surface area, cost-effectiveness, and unique characteristics. These ceramic-based nanomaterials have found wide-spread application in developing highly efficient nano sensors for environmental and process monitoring. Areas like combustion and emissions monitoring, petroleum refinery oversight, and renewable energy technologies benefit from the use of metal oxide nanoparticles. Commercial solid-state chemical sensors, predominantly based on well-structured and doped metal oxides like SnO_2 and ZnO, provide high sensitivity, stability, and cost-efficiency, enabling the detection of various gases [40]. The fundamental sensing principle of metal oxide-based gas sensors revolves around changes in electrical conductivity resulting from charge transfer between interacting molecules and surface complexes. These nanoparticles play a crucial role in the synthesis of electrochemical sensors, offering a substantial surface area, high adsorptive capacity, distinctive electrochemical activity, and durability. The final sensor's performance is influenced by factors such as morphology, surface area, particle size, and surface functionality when using metal oxide nanomaterials. One-dimensional nanostructures prove to be an excellent system for electrochemical sensing of environmental contaminants. Resistive gas sensors based on nanostructured metal oxide semiconductors play a critical role in detecting environmental pollutants like toxic gases and VOCs. The sensors operate on the principle of resistance variation caused by the interaction of gas molecules on the electrode surface under test.

- **Tin oxide:** Tin oxide (SnO_2) nanoparticles are particularly noteworthy as sensing components for gas sensors. Nanostructures, especially nanowires, exhibit higher sensitivity to gases like ethanol compared to their non-nanostructured counterparts. Researchers have explored the interaction of SnO_2 for the simultaneous and selective electrochemical detection of minute heavy metal ions in drinking water, further showcasing the versatile applications of these metal oxide nanoparticles.
- **Zinc oxide (ZnO):** Nanostructures exhibit an exceptional electron transfer rate, facilitating the direct electrochemistry of biomolecules and unveiling their inherent electrochemical capabilities. Compared to its bulk material, ZnO nanostructures offer key advantages, including a high surface-to-volume ratio, non-toxicity, low cost, chemical stability, environmental friendliness, and superior electron communication properties [41]. In gas-sensing applications, ZnO proves advantageous due to its high conductivity, excellent biocompatibility, superior oxidation resistance, and thermal/chemical stability. Operating as an n-type semiconductor, ZnO functions effectively at elevated temperatures between 200°C and 450°C, boasting a significant band gap energy of 3.37 eV [42].

The morphology of sensing materials, particularly in ZnO-based sensors, significantly influences their gas-sensing properties. Nanostructures like wires, rods, tubes, and flower-shaped ZnO formations can be developed at low temperatures, showcasing diverse morphologies. Various techniques, including oxidation, reduction, decomposition, and electrode positioning, have been employed to create ZnO nanostructures, leveraging their low density, extensive active surface area, and surface permeability [43].

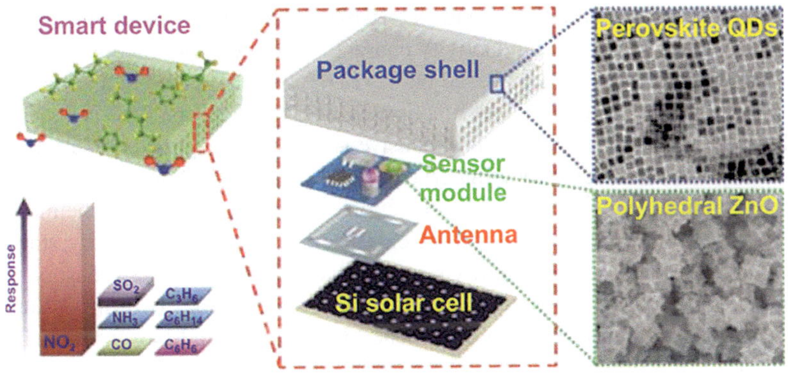

FIGURE 8.7 Nano sensor-based tele-monitoring system used for remote gas molecular tracking using QDs and ZnO. (Jin, H., Yu, J., Cui, D. et al. *Nano-Micro Lett.* 13, 32 (2021). https://doi.org/10.1007/s40820-020-00551-w.)

Multiple methods, such as vapour phase transfer and wet chemical procedures, exist for crafting ZnO nanostructures, with microwave-assisted synthesis emerging as a straightforward and rapid approach. Additives play a crucial role in influencing sensitivity and selectivity, considering the optical, electrical, and magnetic properties of ZnO. Researchers, like Akshaya Kumar et al. [44], have employed hydrothermal methods to generate ZnO nanorods for applications like pH sensing, providing a simple and affordable means to determine water pH. Additionally, Schottky diodes based on AlGaN/GaN heterostructures functionalized with ZnO nanorods enable ammonia detection. Figure 8.7 illustrates a nano sensor-based tele-monitoring system for remote gas molecular tracking, employing quantum dots and ZnO.

- **Nickel oxide:** Model semiconductors for p-type conductivity are NiO nanostructures. They have a wide range of uses, including catalysis, battery electrodes, and gas sensors. NiO's flower-like shape may increase the electrode's electrochemical activity and create a greater contact area between the active material and the electrolyte [44]. Compared to traditional materials, they have superior electrochemical characteristics. The experimental findings showed that rose-like NiO nanoparticles are highly sensitive to formaldehyde gas [29].
- **Titanium oxides:** Electrochemical and optical biosensors were successfully created using this material. TiO_2 nanostructures can also be utilized in electrochemical sensors for pharmaceutical, chemical, and medical purposes. Electrodes made of nanoporous TiO_2 film were used to immobilize a variety of proteins. To measure levodopa, Mazloum-Ardakani et al. modified carbon paste electrodes by adding TiO_2 nanoparticles [45]. The modified electrodes' efficient electrocatalytic activity in reducing the anodic overpotential was demonstrated by the differential pulse voltammetry study technique.

8.4.3 Carbon-Based Nanomaterials

Excellent characteristics of carbon-based nanomaterials include outstanding conductivity, high stability, low cost, wide application windows, and simple surface functionalization, making them a preferential material for nano sensor applications. For diverse electro-analytical applications, CNTs, graphene, and nano and mesoporous carbon were utilized. Their nanostructures effectively expose surface groups, resulting in good environmental pollutant detection capability.

- **Carbon nanotubes:** Since its discovery by Sumio Iijima in 1991, CNTs have become one of the most significant materials due to their distinct electrical, chemical, and mechanical properties. CNTs can be produced using a variety of ways, including chemical vapour deposition processes, laser ablation, and electrical arc discharge. Depending on the level of chirality, CNTs can have the conductivity characteristics of either metals or semiconductors. Environmental nano sensors based on CNTs have been created in a variety of forms, including composite, paste, film, and functionalized CNTs sensors. This is a result of their special qualities, which include their expansive surface area, quick charge transfers, and compatibility with other electrode materials. Single-walled CNT nano sensors have been created by Maduraiveeran et al. for the detection of hazardous phenolic chemicals, which are frequently found in aqueous and biological systems [46]. High sensitivity, stable performance, and remarkable reproducibility were all characteristics of these sensors.

- **Graphene:** Fast electron transit is made possible by the special two-dimensional nanostructure known as graphene. Both electrochemical sensors and biosensors may be used in the future. It has about 260 times more surface area than graphite and twice as much as CNTs, with a potential surface area of 2,630 m^2/g. Additionally, graphene has outstanding mechanical and thermal properties. As a result, graphene significantly enhances the surface area of the materials, increasing their electrochemical catalytic activity. The Hummers technique, electrochemical reduction, and chemical vapour deposition are only a few examples of the several affordable and high-yield processes available for the synthesis of graphene. Graphene is the perfect material for environmental sensing due to its shape and electrochemical characteristics. The remarkable performance of graphene nanoribbon-based electrodes for the detection of explosives has been concluded by Goh et al. [47]. Metal oxide nanostructures can be deposited on graphene using a variety of methods, including chemical synthesis, electro-deposition process, and hydrothermal processes. Nanocomposites made of graphene and cobalt hexacyanoferrate have been created by Luo et al. for the monitoring of hazardous hydrazine and nitrite [48]. Nanomaterials based on graphene have also been utilized to detect heavy metal contamination and gaseous pollutants. High-sensitivity electrodes consisting of Pd-decorated rGO channels and graphene that were synthesized using chemical vapour deposition have been created by the researchers. They were used to detect nitric oxide gas.
- **Porous carbon:** High surface area, easily accessible surface chemistry, and a condensed electron and mass transfer channel are the characteristics of porous carbon. In the world of electrochemical sensors, it has garnered a lot of interest. Porous materials can be divided into three categories based on their pore diameters, according to the International Union of Pure and Applied Chemistry classification: microporous <2 nm, mesoporous 2–50 nm, and macroporous>50 nm. Using macro-/mesoporous carbon materials, Ma et al. have devised a method for the electrochemical detection of NB [49]. Screen-printed electrodes based on bismuth porous carbon nanocomposite have been created by Niu et al. for the detection of heavy metals [50]. An integrated sol–gel and pyrolysis technique was used to create the nanocomposite, which was then milled to a precise particle size distribution. The resultant electrodes demonstrated high sensitivity towards the detection of different ions in drinking water. To monitor harmful metal ions, Veerakumar and colleagues created Pd nanoparticles distributed on porous activated carbons [51]. The porous activated carbons serve as reliable support for the dispersion of Pd NPs with good results. They have enormous pore volumes, high porosity, and high surface areas. Their nano-molar detection limits make them excellent for use as nano sensors for the simultaneous detection of Pb^{2+} and Hg^{2+} metal ions.

8.4.4 Polymers and Bio-Nanomaterials

The polymeric and biological materials-based nanostructured electrochemical sensors and biosensors showed excellent performance with quick response and selectivity. Determining the necessary sensing qualities based on the structural and functional complexity of polymeric and biomaterials is particularly challenging. Electrochemical sensors can be made using polymeric and bio-nanomaterials by combining cutting-edge analytical and scientific techniques, such as nano-fabrication and micro-fluidics.

- **Polymer nanomaterial:** For the detection of environmental contaminants, numerous efforts have been made to develop the technology of polymeric nanoparticles. Numerous environmental applications require the detection and identification of chemically and biologically harmful contaminations in gases and liquids. Polymeric nanoparticles offer a variety of analytical techniques for this purpose. The creation of nanocomposites with a variety of additives, including CNTs, further enhances the sensing capabilities of polymeric nanomaterials. Enhancing the selectivity and superior sensitivity requires the contributions of both the matrix and the nanofiller. The bentonite nanohybrid-modified polyaniline nanofibers were created for gas sensor applications that analyse harmful chemicals such as acetone, benzene, ethanol, and toluene. Indium tin oxide electrodes modified with poly(dopamine) have recently been created by Liu et al. for the detection of hydrazine [52].
- **Bio-nanomaterials:** Numerous options for nano sensors are made possible by the combination of the catalytic activity of biomolecules and the unique properties of nanoscale materials. The self-organization of biological molecules can result in a precise nanostructure with biomaterials. Sensors for the detection of *E. coli* have been created by Li et al. [53]. This could result in a transportable biosensor technique for regular foodborne pathogen surveillance.

8.5 SENSING OF VARIOUS ENVIRONMENTAL POLLUTANTS

Figure 8.8 illustrates various sensing methods used by nano sensors. Such sensing methods have been categorized into four classes: mechanical, fluorescence and Surface-enhanced Raman spectroscopy (SERS), optical localized surface plasmon resonance (LSPR), and colorimetric.

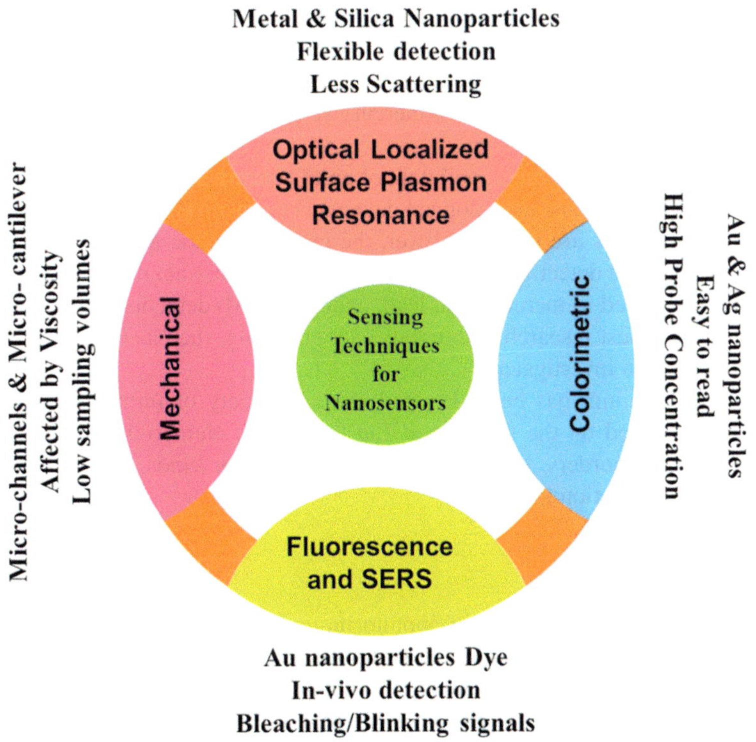

FIGURE 8.8 Sensing methods used by nano sensors for the detection of environment pollutants.

TABLE 8.1
Summary of Main Pesticide Classes with Examples

S.No.	Class of Pesticides	Types
1	Neonicotinoids	Insecticide
2	Pyrethroids	Insecticide
3	Carbamates	Fungicide, insecticide, acaricide
4	Organochlorines	Insecticide, acaricide, fungicide
5	Organophosphates	Insecticide, acaricide
6	Phenoxy	Herbicide
7	Triazines	Herbicides

8.5.1 Pesticides

Given their extensive usage, toxicity, and propensity for bioaccumulation, pesticides are of tremendous relevance for detection. Commercial pesticides currently contain many chemical classes; Table 8.1 provides a summary of the main pesticide classes [54]. The focus is currently on the classes carbamates, neonicotinoids, and triazines for pesticide detection.

8.5.2 Heavy Metals

Mercury, lead, and chromium detection are some of the heavy metals for which nano-enabled sensors have been effectively produced. These environmentally significant contaminants are detected using a wide range of transducers and nanoparticles with the goal of creating sensitive and selective sensors.

Mercury: Researchers are currently working on the geochemical cycling and detection of mercury because of its detrimental neurological effects when exposed to humans [55]. Assays like the mercury sandwich test developed by Liu et al. are only one example of the numerous nano-elements being used to create an increasing number of mercury sensors (Figure 8.9) [56].

- **Lead:** Lead (Pb) is a heavy metal pollutant of great concern since it has been linked to an increased risk of cancer and neurological defects. For sensitive Pb detection, labelled and label-free nano sensors are used. However, the graphene oxide nanocomposite could be employed for multiple detection because each metal has a distinct peak.
- **Cadmium:** Compared to mercury and lead, cadmium (Cd) detection using nano-enabled sensors has less robust research. Numerous nanomaterials, such as Quantum Dots (QDs) and CNTs have been investigated.
- **Chromium:** Chromium has been detected using a variety of immunoassays, although all of them are based on the findings of Liu et al. [57]. Numerous illnesses, including fibro-proliferative disorders, lung cancer, and other tumour kinds, can be brought on by high chromium absorption [58].

8.5.3 Pathogens

Pathogen detection as part of water quality monitoring is becoming essential due to the shortage of potable water across the globe. Waterborne pathogen detection has been a major topic of study since John Snow discovered in 1854 that cholera was spread by drinking tainted water. According to Table 8.2, the WHO has identified various bacteria, viruses, protozoa, and helminths as pathogens of significance in drinking water supplies [59]. Methods for detecting pathogens often concentrate on one of the following: (a) cell detection; (b) detection of genetic material; or (c) detection of a pathogenic product (for example, a toxin).

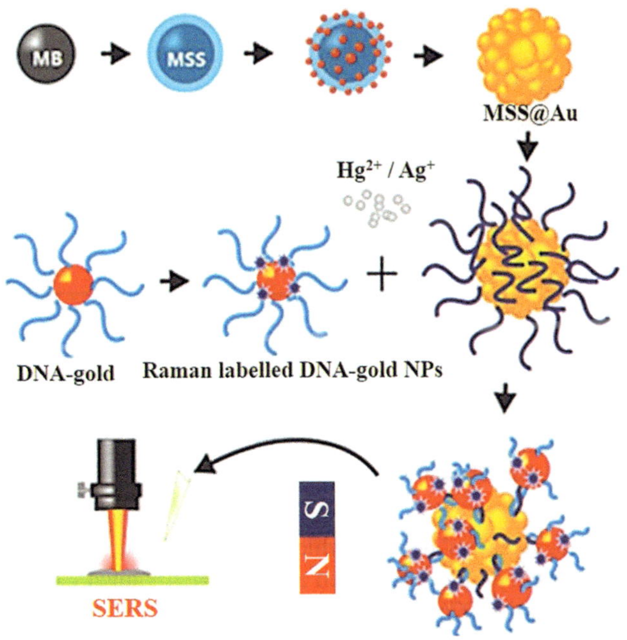

FIGURE 8.9 Schematic illustration of SERS-active system for Hg+ ions. (Min Liu, Zhuyuan Wang, Shenfei Zong, Hui Chen, Dan Zhu, Lei Wu, Guohua Hu, Yiping Cui. *ACS Appl. Mater. Interfaces*. 6(10), 7371–7379 (2014). https://doi.org/10.1021/am5006282.)

TABLE 8.2
Waterborne Pathogens Adapted from WHO Table 7.1 [59]

Pathogen	Examples
Bacteria	*Vibrio cholera*, *Yersinia enterocolitica*, *Salmonella typhi*, *Campylobacter jejuni*, *Campylobacter coli*, *Pseudomonas aeruginosa*, *Escherichia coli*, non-tuberculous mycobacteria
Virus	Adenoviruses, Astroviruses, Enteroviruses, Hepatitis A virus, Sapoviruses, Noroviruses, Rotavirus
Protozoa	*Giardia intestinalis*, *Entamoeba histolytica*, *Acanthamoeba* spp., *Toxoplasma gondii*, *Cyclospora cayetanensis*, *Naegleria fowleri*
Helminth	*Schistosoma* spp., *Dracunculus medinensis*

Here, we have discussed the detection of a few pathogens with the help of nanomaterials: *Legionella pneumophila*, which caused more than 50% of the outbreaks of waterborne diseases between 2011 and 2012 and *Pseudomonas aeruginosa* [60,61].

- ***Vibrio cholerae* and cholera toxin:** It continues to be a major health concern worldwide with an estimated million cases and thousands of deaths per year as a global burden [62]. Acute diarrheal illness called cholera is brought on by consuming tainted water or food that contains the *V. cholerae* bacterium. The cholera toxin, which starts the disease symptoms, is secreted by the bacteria once they colonize the mucosa in the intestines. Both *V. cholerae* and cholera toxin can be detected using nano sensors. In particular, the cholera toxin attaches to the gold nanoparticles that have been modified with galactose, preventing the quantum dots from adhering.

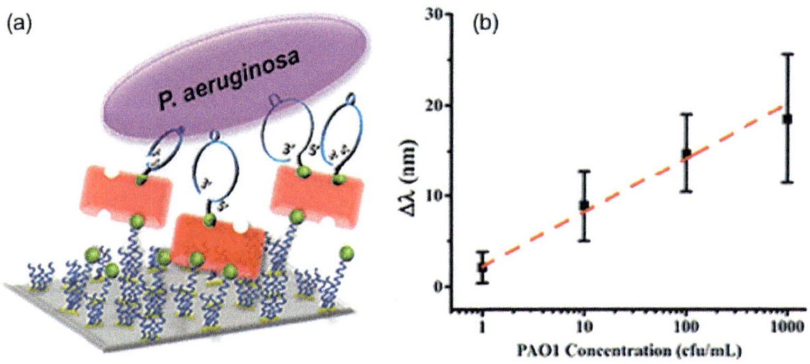

FIGURE 8.10 (a) Schematic illustration of *Pseudomonas aeruginosa* localized surface plasmon resonance (LSPR) sensor chip; (b) sensor calibration curve with error bar representing the standard deviation at a specific bacterial concentration [65]. (Jiayun Hu, Kaiyu Fu, Paul W. Bohn. *Anal. Chem.* 90(3), 2326–2332 (2018). https://doi.org/10.1021/acs.analchem.7b04800.)

- ***L. pneumophila*:** Legionnaires' disease, which resembles pneumonia and is brought on by the bacterium *L. pneumophila*, was named after the well-known outbreak that occurred at the American Legion in 1976. Under specific circumstances, the bacterium can spread throughout a structure (premise) and infect people who breathe in aerosols containing the infectious agent. To find nano-enabled *Legionella*, two methods have been proposed in the literature: whole organism detection and DNA detection [63]. A whole organism sensor was created by Martin et al. by fusing amperometric transduction with a sandwich immunoassay for bacterial capture [64].
- ***P. aeruginosa*:** *P. aeruginosa* is a pathogen that can be found in sources like faecal waste, soil, water, and sewage. The most significant way to get exposed is by dermal contact with a contaminated source or equipment. *P. aeruginosa* has been linked to nosocomial infection outbreaks in hospitals. LSPR chips were created using the fabrication of nano-textured substrates by Hu et al. As shown in Figure 8.10 [65], Researchers used a three-step production process that involved first depositing gold on a glass slide, then depositing silica nanoparticles, and finally depositing a second layer of gold.

8.6 CHARACTERISTICS OF NANOMATERIALS FOR ENVIRONMENTAL POLLUTION DETECTION

- Achieve the best electrochemical active sites by shaping and manipulating the size of nanomaterials.
- Increase the stability and specificity by the enhancement of bi-metallic or tri-metallic nanoparticles.
- Offer highly specific analyte identification of useful compounds.
- Create innovative nanomaterials by correlating the surface functionality, composition, and structure of nanomaterials.
- Enhanced electrochemical qualities by the identification of substrate.
- Enhancing the sensitivity and stability of the nano sensors [29].

8.7 ADVANTAGES AND LIMITATIONS OF NANO SENSORS IN ENVIRONMENTAL POLLUTION MONITORING

As environmental pollution becomes an increasingly serious problem, scientists are particularly searching for cutting-edge methods to monitor potential pollutants. Recent years have seen significant advancements in this area. Nanomaterials' brittle structure drives things to an extreme degree. They have a larger surface area, which makes them more effective and durable. Some nanoparticles, including palladium, platinum, gold, and silver nanoparticles, are used in nano sensors. Nanomaterials can also be easily modified; their composition, size, and structure all affect and combine these components' features [66,67].

8.7.1 Advantages of Nano Sensors in Pollution Monitoring

The substantial surface area of nano sensors translates to enhanced exposure and detection capabilities for target molecules, especially at low concentrations and amid electromagnetic wave variations, accounting for quantum events. Cutting-edge nano sensor devices excel in detecting multiple gases present in the environment through the selective integration of diverse nano sensors. Combining sound waves with bioinspired synthetic polymers and CNTs enables the precise identification of small-molecule targets. These highly accurate sensors offer clear imaging of nanostructures, detect microscopic signals, and pinpoint minute concentrations of specific chemical compounds. Piezoresistive nanomechanical sensors, employed as surface stress sensors, exhibit remarkable sensitivity in detecting gaseous components. Serving as the core of artificial nose sensor systems, these sensors hold significant promise across various industries, including environmental, food, medical, chemical, and pharmaceutical. The evolution of our understanding and interaction with the environment and ourselves will progressively shape advancements in nano sensor technology. Chemical sensors, crucial for environmental monitoring, utilize a receptor layer on the silicone membrane of nanomechanical sensors. Upon interaction with the analyte, surface tension is produced, mechanically deforming small beams that support the membrane. The resulting piezoresistive changes in resistance enable the precise detection of target molecules. Nano sensors, crafted from nanomaterials, exhibit superior precision and sensitivity compared to their conventional counterparts, owing to high nanomaterial surface-to-volume ratios and unique physical properties. The continuous development of nano sensors implies the potential for exponential integration with nanodevices. The data collected by these advanced sensors holds immense transformative potential for the environmental sector, presenting significant opportunities for service evolution.

8.7.2 Limitations of Nano Sensors in Pollution Monitoring

A key constraint of this sensor lies in its dependence on an external power source and a monitoring instrument capable of comparing particle resonance frequencies. This challenge can be addressed through collaboration and communication among nano sensors. For instance, nano sensors emitting identical particles can achieve this at significantly reduced power levels, detecting the radiated energy at multiple frequencies. The resulting data from this collaborative process can be processed and transmitted to a data sink, mitigating the limitations associated with individual sensor power requirements and frequency monitoring capabilities.

8.8 NANO SENSORS FOR INDUSTRIAL PERSPECTIVES OF ENVIRONMENTAL MONITORING

Nano sensors have become integral components of state-of-the-art devices, leveraging the unique properties of materials at nanoscales. This advancement has not only led to the conceptualization of groundbreaking devices but also facilitated the fabrication of nanostructured materials with programmable properties on a global scale, particularly for environmental monitoring. Nano sensor-based trace detectors play a vital role in accurately screening diverse substances across various environments, contributing to the prevention of safety violations. These nanoscale sensors and devices offer continuous and cost-effective monitoring of structural integrities. Furthermore, nanoscale sensors, along with other nano-electronic-enabled innovations, support an efficient and advanced mechanism for collision prevention and the adjustment of travel routes

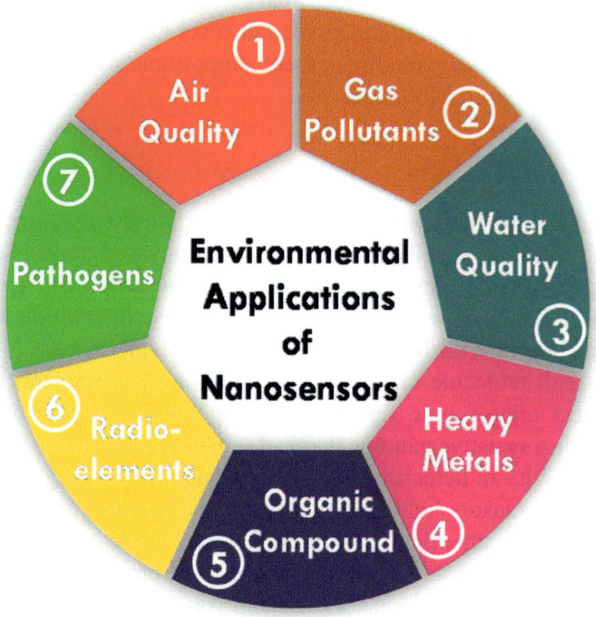

FIGURE 8.11 Applications of nano sensors for the detection of various environmental pollutants.

to alleviate congestion and prevent environmental pollution. Despite these benefits, challenges related to the deployment of expensive technology pose significant obstacles, restraining the broader market expansion for nano sensors. Anticipated advancements in autonomous nanotechnology devices are expected to unlock lucrative opportunities in the market for environmental monitoring, forecasting a positive outlook for nano sensors [68].

Recognizing that miniaturization expands the range of applications, micro and nanoscale sensors are frequently integrated with other sensor technologies. This seamless integration enhances the efficiency of nano sensors, making them instrumental in detecting various environmental pollutants, as illustrated in Figure 8.11.

8.8.1 Significant Potential of Nano Sensors in Environmental Monitoring

Nano sensors play a pivotal role in practical technology, particularly in areas like environmental pollution detection and monitoring. They offer several advantages over traditional equipment, such as lower power consumption, heightened sensitivity, lower analyte detection thresholds, and a closer proximity between the item and the sensor. This enables the detection of chemical vapours even in small quantities [69–71]. The significant potentials of nano sensors are elaborated in Table 8.3. In intelligent environmentally friendly buildings, nano sensors are employed for self-sensing and self-monitoring, contributing to enhanced efficiency and sustainability. With their capability to connect to wireless networks, nano sensors have also found applications in monitoring traffic and environmental pollutants [72]. These versatile sensors serve both mechanical and chemical purposes, adept at identifying chemical substances and gases for detection and pollution monitoring. Nano sensors can function as accelerometers or monitor various physical properties, including temperature, displacement, pressure, and flow [73–75].

TABLE 8.3
Significant Potentials of Nano Sensors in Environmental Monitoring

S.No.	Potential	Description
1	Monitoring soil, water, air, and environmental changes	Nano sensors simplify real-time monitoring of environmental, soil, and water changes, identifying potential risks in nanoscale quantities of solids, liquids, and gases. They prove beneficial in monitoring pesticides and fertilizers across entire farms, allowing farmers to optimize usage based on actual needs.
2	Recognize minute particles	Advancements in nanotechnology empower nano sensors to detect minute components of environmental pollutants, evaluating gases and chemicals in real time to identify pollutants.
3	Information transfer through nanoparticles	Nano sensors play a vital role in the development of silicon computer chips and nanorobotic devices, expanding their applications in various sensor technology fields.
		Nano sensors are becoming more prevalent and are being used in more sensor technology applications. Nanoscience and nano sensor research are concerned with nanoparticles and devices that are applicable to all branches of science, including chemical, environmental, and material science.
4	Verification and measurement of water contents and macronutrient.	Precision farm nanotechnology sensors contribute to assessing water content and optimizing macronutrients for crops, ensuring optimal ripening, fertilization, and crop production. They leverage nanotubes to measure ion concentrations in farm samples, introducing real-time trash detection and improving crop production.
5	Health information of living organisms	In healthcare, nano sensors facilitate the collection of health information, enabling the categorization of specific groups based on genetics. They contribute to continuous monitoring of minute physiological changes, improving disease treatment and overall well-being.
6	Identification and monitoring of environmental gases	Air pollution poses a growing concern, and nano sensors, thanks to nanotechnology advancements, can now detect minute concentrations of various gases in the environment. Metal oxide-based nanostructures integrated with existing technologies enable the identification of gas composition for environmental, industrial, and domestic applications.
7	Monitoring biological environmental pollution	Nano sensors have shown a remarkable ability to track biological cells and creatures and to acquire exact spatial and temporal measurements even at the molecule level. These detect and measure the presence of numerous infectious agents, such as viruses and harmful bacteria, by taking advantage of the special properties of nanomaterials and nanoparticles. To prevent environmental contamination by biological elements, nano sensors are employed to detect pathogens. This technique is an effective tool for applications involving sensing. This method is updated to identify numerous samples of soil and water fertilizers, food pathogens, and other contaminants.
8	Intelligent tracing and tracking of the environment	Nano sensors exhibit a remarkable ability to track biological cells and organisms, providing precise spatial and temporal measurements at the molecular level. They detect and measure the presence of infectious agents, preventing environmental contamination. This technique proves effective in applications such as sensing soil and water fertilizers, detecting food pathogens, and identifying contaminants.

8.9 CONCLUSION

A safe environment with increased health awareness is increasingly in demand all around the world. Nanotechnology examines exciting transdisciplinary scientific information at the nanoscale to collect fundamental distinctions on nano sensors. In the past 20 years, sophisticated nanotechnology has offered a potential opportunity to provide a quick, low-cost method for selecting and accurately identifying environmental pollutants and toxins. The development of nano sensors for environmental contaminants is accelerating, and this chapter has shown novel and inventive combinations of nanomaterials and recognition agents being developed constantly. Additionally, we briefly covered recent developments in advanced nanotechnology applications for reducing ecological pollution, which have the potential to lower the high cost. We also discussed numerous nanostructured materials used in nano sensors including metal, metal oxide, CNTs, graphene, polymers, and biomaterials. To monitor environmental pollution, water quality, and air quality, nanoparticles can be used to create gas/vapour sensors, ultrasonic stamp-sized sensors, and laboratory-on-a-chip sensors. Due to their small size and special optical, magnetic, catalytic, and mechanical properties, nano sensors and optical nanotechnology have made tremendous strides in development and application. Nanomaterials significantly improved pollution detection and tracking. Innovative materials and goods made possible by nanotechnology will increase our ability to identify and manage environmental pollution. Among these nanomaterials, sensors offer the benefit of being able to identify contaminants easily in the field without the use of costly lab apparatus. The drawbacks of earlier sensing technologies are being addressed by current breakthroughs in sensor design. There are still concerns about the selectivity and sensitivity of assays in complicated environmental matrices; however, more research is going on in this field to improve and show the stability and selectivity of their sensors. Also, there are still several obstacles in the way of using plant-based green nanosystems to their full potential. Future nanotechnology industries should encourage the development of safety regulations, such as those relating to the bio-source development related to nanostructures, for better advancement of this technology. One could anticipate more developments in the field of nano sensors for environmental pollution detection and monitoring through the development of newer nano sensors with innovative mechanisms for a healthy environment.

REFERENCES

1. E. F. Mohamed. Nanotechnology: future of environmental air pollution control, *Environ. Manag. Sustain. Dev.* 6 (2017) 429–454.
2. A. Thakur, P. Punia, R. Dhar, R. K. Aggarwal, P. Thakur. Separation of cadmium and chromium heavy metals from industrial wastewater by using Ni-Zn nanoferrites, *Adv. Nano Res.* 12 (2022) 457–465.
3. F-G. Bănică. Nanomaterial applications in optical transduction. In: Florinel-Gabriel Bănică (ed) *Chemical Sensors and Biosensors.* Chichester: Wiley; (2012) 454–472.
4. T. Yang, T. V. Duncan. Challenges and potential solutions for nanosensors intended for use with foods, *Nat. Nanotechnol.* 16 (3) (2021) 251–265.
5. Y. Cui, Q. Wei, H. Park, C. M. Lieber. Nanowire nanosensors for highly sensitive and selective detection of biological and chemical species, *Science* 293 (5533) (2001) 1289–1292.
6. A. Pathania, P. Thakur, A. V. Trukhanov, S. V. Trukhanov, L. V. Panina, U. Lüders, A. Thakur. Development of tungsten doped Ni-Zn nano-ferrites with fast response and recovery time for hydrogen gas sensing application, *Results Phys.* 15 (2019) 102531.
7. S. Hassani, S. Momtaz, F. Vakhshiteh, A. S. Maghsoudi, M. R. Ganjali, P. Norouzi, M. Abdollahi. Biosensors and their applications in detection of organo phosphorus pesticides in the environment, *Arch. Toxicol.* 91 (2017) 109–130.
8. C. I. L. Justino, A. C. Duarte, T.A.P. Rocha-Santos. Review recent progress in biosensors for environmental monitoring: a review, *Sensors* 17 (2017) 1–25.
9. P. Punia, M. K. Bharti, R. Dhar, P. Thakur, A. Thakur. Recent advances in detection and removal of heavy metals from contaminated water, *ChemBioEng Rev.* 9 (2022) 351–369.
10. P. Punia, R. K. Aggarwal, R. Kumar, R. Dhar, P. Thakur, A. Thakur. Adsorption of Cd and Cr ions from industrial wastewater using Ca doped Ni–Zn nanoferrites: synthesis, characterization and isotherm analysis, *Ceram. Int.* 48 (2022) 18048–18056.

11. E. F. Mohamed, G. Awad. Photodegradation of gaseous sstoluene and disinfection of air-borne microorganisms from polluted air using immobilized TiO2 nanoparticle photocatalyst–based filter, *Environ. Sci. Pollut. Res.* 27 (2020) 24507–24517.
12. E. F. Mohamed, T.-O. Do. Synthesis of new hollow nanocomposite photocatalysts: sunlight applications for removal of gaseous organic pollutants, *J. Taiwan Ins. Chem. E.* 2020 (2020) 6–10.
13. S.-W. Jung, E. K. Lee, and S.-Y. Le. Communication-concentration-cell-type nafion-based potentiometric hydrogen sensors, *ECS J. Solid State Sci. Technol.* 7 (2018) Q239.
14. N. Dhanda, S. Kumari, R. Kumar, D. Kumar, A-C. A. Sun, P. Thakur, A. Thakur. Influence of Ni over magnetically benign Co ferrite system and study of its structural, optical, and magnetic behavior, *Inorg. Chem. Commun.* 151 (2023) 110569.
15. A. Baysal, H. Saygin. Smart nanosensors and methods for detection of nanoparticles and their potential toxicity in air. In: Abdeltif Amrane, Aymen Amine Assadi, Phuong Nguyen-Tri, Tuan Anh Nguyen, Sami Rtimi (eds) *Nanomaterials for Air Remediation.* Amsterdam, The Netherlands: Elsevier; (2020) 33–59.
16. A. Thakur, P. Thakur, S. P. Khurana. *Synthesis and Applications of Nanoparticles.* Singapore: Springer; (2022) 1–17.
17. J. M. Dubach, D. I. Harjes, H. A. Clark. Fluorescent ion-selective nanosensors for intracellular analysis with improved lifetime and size, *Nano Lett.* 7 (6) (2007) 1827–1831.
18. V. T. Binh, N. Garcia, A. L. Levanuyk. A mechanical nanosensor in the gigahertz range: where mechanics meets electronics, *Surf. Sci.* 301 (1994) L224.
19. S. Kulmala, J. Suomi. Current status of modern analytical luminescence methods, *Anal. Chim. Acta* 500 (2003) 21.
20. P. Thakur, S. Taneja, D. Chahar, B. Ravelo, A. Thakur. Recent advances on synthesis, characterization and high frequency applications of Ni-Zn ferrite nanoparticles, *J. Magn. Magn. Mater.* 530 (2021) 167925.
21. G. W. Hunter, Z. S Heikh Akbar, S. Bhansali, M. Daniele, P. D. Erb, K. Johnson, C.-C. Liu, D. Miller, O. Oralkan, P. J. Hesketh, P. Manickam, R. L. V. Wal. Editors' choice-critical review-a critical review of solid state gas sensors, *J. Electrochem. Soc.* 167 (037570) (2020) 1–31.
22. R. M. White, P. I. Frederick, F. Doering, W. Cascio, P. Solomon, L. A. Gundel. Sensors and 'Apps' for community-based atmospheric monitoring, *Air Waste Manag. Assoc.* 5 (2012) 36–40.
23. H. Nazemi, A. J oseph, J. Park, A. Emadi. Advanced micro and nano gas sensor technology: a review, *Sensors (Basel)* 19 (2019) 1–23.
24. M. R. Willner, P. J. Vikesl. Nanomaterial enabled sensors for environmental contaminants, *J. Nanobiotechnol.* 16 (2018) 1–16.
25. V. Schroeder, S. Savagatrup, M. He, S. Lin, T. M. Swager. Carbon nanotube chemical sensors, *Chem. Rev.* 119 (2019) 599–663.
26. R. Abdel-Karim, Y. Reda, A. Abdel-Fattah. Review—nanostructured materials based nanosensors, *J. Electrochem. Soc.* 167 (2020), 037554.
27. A. M. Graboski, J. Martinazzo, S. C. Ballen, J. Steffens, C. Steffens. Nanosensors for water quality control. In: *Nanotechnology in the Beverage Industry.* Amsterdam, The Netherlands: Elsevier; (2020) 115–128.
28. P. Thakur, D. Chahar, S. Taneja, N. Bhalla, A. Thakur. A review on MnZn ferrites: synthesis, characterization and applications, *Ceram. Inter.* 46 (2020) 15740–15763.
29. G. Maduraiveeran, W. Jin. Nanomaterials based electrochemical sensor and biosensor platforms for environmental applications, *Trends Environ. Anal. Chem.* 13 (2017) 10.
30. M. Azharuddin, G. H. Zhu, D. Das, E. Ozgur, L. Uzun, A. P. F. Turner, H.K. Patra. A repertoire of biomedical applications of noble metal nanoparticles, *Chem. Commun.* 55 (2019) 6964.
31. N. Ratner, D. Mandler. Electrochemical detection of low concentrations of mercury in water using gold nanoparticles, *Anal. Chem.* 87 (2015) 5148.
32. Y. Yuan, F. Zhang, H. Wang, L. Gao, Z. Wang. A sensor based on Au nanoparticles/carbon nitride/graphene composites for the detection of chloramphenicol and ciprofloxacin, *ECS J. Solid State Sci. Technol.* 7 (2018) M201.
33. V. M. Kariuki, S. A. Fasih-Ahmad, F. J. Osonga, O. A. Sadik. An electrochemical sensor for nitrobenzene using π-conjugated polymer-embedded nanosilver, *Analyst* 141 (2016) 2259.
34. E. Lebegue, C. M. Anderson, J. E. Dick, L. J. Webb, A. J. Bard. Electrochemical detection of single phospholipid vesicle collisions at a Pt ultramicroelectrode, *Langmuir* 31 (2015) 11734.
35. M. R. Mahmoudian, W. J. Basirun, Y. Alias. A sensitive electrochemical Hg2+ ions sensor based on polypyrrole coated nanospherical platinum, *RSC Adv.* 6 (2016) 36459.

36. R. Xi, S.-H. Zhang, L. Zhang, C. Wang, L.-J. Wang, J.-H. Yan, G.-B. Pan. Electrodeposition of Pd-Pt nanocomposites on porous GaN for electrochemical nitrite sensing, *Sensors* 19 (2019) 606.
37. Y. He, X. H. Yang, Y. Huo, Q. Han, J. Zheng. *Open Access J. Chem.* 2 (1) (2018) 15.
38. O. Lupan, V. Postica, R. Adelung, F. Labat, I. Ciofini, U. Schürmann, L. Kienle,L. Chow, B. Viana, T. Pauporte. Functionalized Pd/ZnO nanowires for nanosensors, *Physica Status Solidi RRL.* 12 (2018) 1700321.
39. Y. Li, J. Z. Sun, C. Bian, J. H. Tong, H. P. Dong, H. Zhang, and S. H. Xia. Copper nano-clusters prepared by one-step electrodeposition and its application on nitrate sensing, *AIP Adv.* 5 (2015) 041312.
40. B. Zhang, P.-X. Gao. *Front. Mater.* 6 (2019) 1.
41. M. Luqman, M. Napi, S. M. Sultan, R. Ismail, K. W. How, M. K. Ahmad. Electrochemical-based biosensors on different zinc oxide nanostructures: a review, *Materials* 12 (2019) 2985.
42. G. Zhao, J. Xuan, X. Liu, F. Jia, Y. Sun, M. Sun, G. Yin, B. Liu. Low-cost and high-performance ZnO nanoclusters gas sensor based on new-type FTO electrode for the low-concentration H_2S gas detection, *Nanomaterials* 9 (2019) 435.
43. S. Jung, K. H. Baik, F. Ren, S. J. Pearton, S. Jang. AlGaN/GaN heterostructure based Schottky diode sensors with ZnO nanorods for environmental ammonia monitoring applications, *ECS J. Solid State Sci. Technol.* 7 (2018) Q3020.
44. A. A. Kumar, S. K. N. Kumar, A. A. Almaw, E. F. Renny, B. Shekhar. Hydrothermal growth of zinc oxide (ZnO) nanorods (NRs) on screen printed IDEs for pH measurement application, *J. Electrochem. Soc.* 166 (2019) B3264.
45. M. Mazloum-Ardakani, Z. Taleat, A. Khoshroo, H. Beitollahi, H. Dehghani. Electrocatalytic oxidation and voltammetric determination of levodopa in the presence of carbidopa at the surface of a nanostructure based electrochemical sensor, *Biosens. Bioelectron.* 35 (2012) 75.
46. G. Maduraiveeran, B. Adhikari, A. Chen. Electrochemical sensor based on carbon nanotubes for the simultaneous detection of phenolic pollutants, *Electroanalysis* 27 (2015) 902.
47. M. S. Goh. Graphene-based electrochemical sensor for detection of 2, 4, 6-trinitrotoluene (TNT) in seawater: the comparison of single-, few-, and multilayer graphene nanoribbons and graphite microparticles, *Anal. Bioanal, Chem.* 399 (2011) 127.
48. X. Luo, K. Pan, Y. Yu, A. Zhong, S. Wei, J. Li, J. Shi, X. Li. An electrochemical sensor for hydrazine and nitrite based on graphene–cobalt hexacyanoferrate nanocomposite: Toward environment and food detection, *J. Electroanal. Chem.* 745 (2015) 80.
49. J. Ma, X. Zhang, G. Zhu, B. Liu, J. Chen. Sensitive electrochemical detection of nitrobenzene based on macro-/meso-porous carbon materials modified glassy carbon electrode, *Talanta* 88 (2012) 696.
50. P. Niu, M. Gich, C. Navarro-Hernandez, P. Fanjul-Bolado, A. Roig. Screen-printed electrodes made of a bismuth nanoparticle porous carbon nanocomposite applied to the determination of heavy metal ions, *Microchim. Acta* 183 (2016) 617.
51. P. Veerakumar, V. Veeramani, S. M. Chen, R. Madhu, S. B. Liu. Palladium nanoparticle incorporated porous activated carbon: electrochemical detection of toxic metal ions, *ACS Appl. Mater. Interfaces* 8 (2016) 1319.
52. F. Liu, X. Chen, F. Luo, D. Jiang, S. Huang, Y. Li, X. Pu. A novel strategy of procalcitonin detection based on multi-nanomaterials of single-walled carbon nanohorns–hollow Pt nanospheres/PAMAM as signal tags, *RSC Adv.* 4 (2014) 13934.
53. Z. Li, Y. Fu, W. Fang, Y. Li. Electrochemical impedance immunosensor based on self-assembled monolayers for rapid detection of Escherichia coli O157: H7 with signal amplification using lectin, *Sensors* 15 (2015) 19212.
54. S. Liu, Z. Zheng, X. Li. Advances in pesticide biosensors: current status, challenges, and future perspectives, *Anal. Bioanal. Chem.* 405 (2013) 63–90.
55. P. Selid, H. Xu, E. M. Collins, M. Striped Face-Collins, J. X. Zhao. Sensing mercury for biomedical and environmental monitoring, *Sensors* 9 (2009) 5446–5459.
56. M. Liu, Z. Wang, S. Zong, H. Chen, D. Zhu, L. Wu, G. Hu, Y. Cui. SERS detectionand removal of mercury(II)/silver(I) using oligonucleotides – functionalized core/shell magnetic silica Sphere @ Au nanoparticles, *ACS Appl. Mater. Interfaces.* 6 (2014) 7371–7379.
57. X. Liu, J. J. Xiang, Y. Tang, X. L. Zhang, Q. Q. Fu, J. H. Zou, Y. Lin. Colloidal gold nanoparticle probe-based immune chromatographic assay for the rapid detection of chromium ions in water and serum samples, *Anal. Chim. Acta.* 745 (2012) 99–105.
58. H. J. Gibb, P. S. Lees, P. F. Pinsky, B. C. Rooney. Lung cancer among workers in chromium chemical production, *Am. J. Ind. Med.* 38 (2000) 115–126.

59. World Health Organization. *Guidelines for Drinking-Water Quality*, vol. 1. Geneva: World Health Organization; 2008.
60. K. D. Beer, J. W. Gargano, V. A. Roberts, V. R. Hill, L. E. Garrison, P. K. Kutty, E. D. Hilborn, T. J. Wade, K. E. Fullerton, J. S. Yoder. Surveillance for waterborne disease outbreaks associated with drinking water-United States, 2011–2012, *MMWR Morb. Mortal. Wkly. Rep.* 64 (2015) 842–848.
61. World Health Organization. *WHO Publishes List of Bacteria for Which New Antibiotics are Urgently Needed.* Geneva: WHO; 2017.
62. M. Ali, A. R. Nelson, A. L. Lopez, D. A. Sack. Updated global burden of cholerain endemic countries, *PLoS Negl. Trop. Dis.* 9 (2015) e0003832.
63. T-Y. Wu, Y-Y. Su, W-H. Shu, A. T. Mercado, S-K. Wang, L-Y. Hsu, Y-F. Tsai, C-Y. Chen. A novel sensitive pathogen detection system based on microbeadquantum dot system, *Biosens. Bioelectron.* 78 (2016) 37–44.
64. M. Martín, P. Salazar, C. Jiménez, M. Lecuona, M. J. Ramos, J. Ode, J. Alcoba, R. Roche, R. Villalonga, S. Campuzano, et al. Rapid Legionella pneumophiladetermination based on a disposable core–shell Fe_3O_4 @ Poly (dopamine) magnetic nanoparticles immune platform, *Anal. Chim. Acta.* 887 (2015) 51–58.
65. J. Hu, K. Fu, P. W. Bohn. Whole-cell Pseudomonas aeruginosa localized surface plasmon resonance aptasensor, *Anal Chem.* 90 (3) (2018) 2326–2332; *Eur J Clin Microbiol Infect Dis.* 30 (2011) 273–278.
66. S. Kumari, N. Dhanda, A. Thakur, V. Gupta, S. Singh, R. Kumar, S. Hameed, P. Thakur. Nano Ca–Mg–Zn ferrites as tuneable photocatalyst for UV light-induced degradation of rhodamine B dye and antimicrobial behavior for water purification, *Ceram. Inter.* 49 (2023) 12469–12480.
67. D. Chahar, D. Kumar, P. Thakur, A. Thakur. Visible light induced photocatalytic degradation of methylene blue dye by using Mg doped Co-Zn nanoferrites, *Mater. Res. Bull.* 162 (2023) 112205.
68. S. Kumar, S. Sachdeva, S. Chaudhary, G.R. Chaudhary. Assessing the potential application of biocompatible tuned Nanosensor of Yb_2O_3 for selective detection of imazapyr in actual samples, *Colloid. Surface. Physicochem. Eng. Aspect.* 593 (2020) 124612.
69. P.J. Vikesland. Nanosensors for water quality monitoring, *Nat. Nanotechnol.* 13 (8) (2018) 651–660.
70. Z.B. Shawon, M.E. Hoque, S.R. Chowdhury. Nanosensors and nano biosensors: agricultural and food technology aspects. In: Kaushik Pal, Fernando Gomes (eds) *Nanofabrication for Smart Nanosensor Applications*, Amsterdam, The Netherlands: Elsevier; (2020) 135–161.
71. A. Dubey, D.R. Mailapalli, Nano fertilisers, nano pesticides, nanosensors of pestand nanotoxicity in agriculture, In: Lichtfouse, E. (eds) *Sustainable Agriculture Reviews*, vol 19. Cham: Springer; (2016) 307–330.
72. G. E. Marchant. What are best practices for ethical use of nanosensors for worker surveillance? *AMA J. Ethics* 21 (4) (2019 Apr 1) 356–362.
73. Y. Su, Z. Zhou. Electromechanical analysis of flexoelectric nanosensors based on nonlocal elasticity theory, *Micromachines* 11 (12) (2020) 1077.
74. Y. Zhou, L. Ding, Y. Wu, X. Huang, W. Lai, Y. Xiong. Emerging strategies to develop sensitive AuNP-based ICTS nanosensors, *Trac. Trends Anal. Chem.* 112 (2019) 147–160.
75. C. Wang, S. Otto, M. Dorn, K. Heinze, U. Resch-Genger, Luminescent TOP nanosensors for simultaneously measuring temperature, oxygen, and pH at a single excitation wavelength, *Anal. Chem.* 91 (3) (2019) 2337–2344.

9 Nanotechnology in Treatment and Bioremediation

Anand Salvi and Manish Shandilya
Amity University Haryana

Fayu Wan
Nanjing University of Information Science & Technology

Preeti Thakur
Amity University Haryana

Atul Thakur
Nanjing University of Information Science & Technology
Amity University Haryana

9.1 INTRODUCTION

In recent years, the field of nanobiotechnology has emerged as a significant contributor to the conversion of soil, water, and air contaminants into environmentally beneficial substances. The quality of the system is contingent upon its composition, which has three essential components: (a) treatment and remediation, (b) sensing and monitoring, and (c) pollution prevention [1]. Most importantly, nano-bioremediation could help protect the environment by getting rid of waste and cleaning up messes. As was already said, combining traditional bioremediation with nanobiotechnology or using direct nano-remediation methods could be a good way to get rid of contaminants in the environment. Nanoparticles (NPs) are used to remove heavy metals, herbicides, pesticides, and insecticides (organic and inorganic toxins) from polluted environments by breaking them down or storing them [2]. New research shows that, in general, NPs could be a good way to control pollution, either directly or by working with other methods. For example, nanoscale zero-valent irons (nZVIs) are used to get rid of organic pollution like herbicides (like atrazine and molinate) and pesticides (like chlorpyrifos). On the contrary, both phytoremediation and the gathering of NPs have been shown to break down organic pollution. Here, the NP-mediated reaction breaks down long-chain hydrocarbons and halocarbons that did not break down with bioremediation. So, it affects the ability of living things to break down, which helps control pollution. Through the change of nZVI/nanoscale Fe0, it may be possible to improve the reactivity of nanomaterials so that contaminants can be broken down and cleaned up [3]. So, using this reaction process to find, extract, and break down environmentally dangerous contaminants like halo and hydrocarbons in general could be a new way to make the environment more sustainable. An extraordinary number of contaminants are released into the environment due to extensive industrialization, urbanization, and modern agricultural practices. These contaminants can pollute soil, air, and water and cause deforestation, biodiversity loss, soil degradation, and damage to human health. Carbon monoxide (CO), chlorofluorocarbons (CFCs), heavy metals (arsenic, chromium, lead, cadmium, mercury, and zinc), hydrocarbons, nitrogen oxides, organic compounds (volatile organic compounds and dioxins), sulphur dioxide, and particles are all examples of these pollutants. Numerous of these pollutants are known or suspected to be carcinogens and mutagens, and they may modify the function of the

DOI: 10.1201/9781003502692-9

ecosystem. Thus, numerous bioremediation-based environmental cleansing techniques have been developed. An extraordinary number of contaminants are released into the environment due to extensive industrialization, urbanization, and modern agricultural practices. These contaminants can pollute soil, air, and water and cause deforestation, biodiversity loss, soil degradation, and damage to human health. CO, CFCs, heavy metals (arsenic, chromium, lead, cadmium, mercury, and zinc), hydrocarbons, nitrogen oxides, organic compounds (volatile organic compounds and dioxins), sulphur dioxide, and particles are all examples of these pollutants. Numerous of these pollutants are known or suspected to be carcinogens and mutagens, and they may modify the function of the ecosystem. Thus, numerous techniques for environmental cleansing have been developed, including bioremediation, phytoremediation, and physical and chemical remediation. The realization that "traditional" methods of treatment (e.g., disposal to landfill, isolation, pump-and-treat) are not sustainable has resulted in an explosion in the development of alternative treatment technologies for environmental remediation [4]. The most important thing is to come up with new, more effective ways to clean up pollution. This can help protect and recover the integrity of natural habitats [5].

9.2 BIOREMEDIATION NANOTECHNOLOGY

Nanomaterials are particles with at least one dimension ranging from 1 to 100 nm. In the realm of remediation technology, certain traits such as high surface-to-volume ratio, improved magnetic and unique catalytic properties, and so on [6,7] make these NPs more valuable than their mass phase counterparts. Distinctive subordinate properties of NPs are frequently observed, which are primarily due to their high surface-to-volume ratio and can lead to very sensitive detection, and their respective bulk material enables them to remediate the contamination at a rapid rate with a low level of hazardous by-products [8]. Other inorganic nanomaterials utilized to remove heavy metal ions in wastewater treatment include NPs generated by carbon nanotubes (CNTs), nanoscale zeolites, dendrimer enzymes, metallic metals, and metal oxides. Nanosized metals or metal oxides, on the contrary, have a large surface area and a high specific affinity. Furthermore, metal oxides have a low environmental impact, limited solubility, and no secondary pollution and have been used as sorbents to remove heavy metals from polluted locations [9,10]. NPs are increasingly being used in enzyme-mediated remediation technology because they offer a biocompatible and inert microenvironment that least interferes with the enzymes' natural properties and aids in maintaining their biological activities [11]. Nevertheless, the selection of nanomaterials for the remediation process is a significant advance because they may be harmful to the microorganisms required for remediation (Figure 9.1).

9.3 BIOTECHNOLOGICAL APPROACH TO FABRICATION OF NPS

In the process of biomanufacturing in most cases, the synthesis of NPs involves the utilization of either a reduction or oxidation technique. The present study reveals that the biomolecules present in bacterial and plant species play a crucial role in the stabilization of metal ions within their respective nanostructures (Figure 9.2).

The way NPs are made is very similar to how they are made in nature. This process, called biomimetics, is much "cleaner" and "greener" [12–14] have all been written about the living systems that are involved in making NPs. In the realm of living organisms, the creation of NPs derived from plants is widely regarded as a more feasible, secure, expeditious, and scalable approach compared to the production of NPs derived from microbes. Due to their reliance on the preservation of microbial cultures and the inclusion of hazardous compounds with detrimental effects on both the environment and human well-being, the utilization of the latter should be avoided [15]. The fact that this way of making NPs is hard makes it less valuable, and the plant-based method is better for making NPs. Since the first report on plants that store metals, which came out in 1855, there have been several other reports about abnormally high levels of metal ions in plant cells [16]. Plant-based

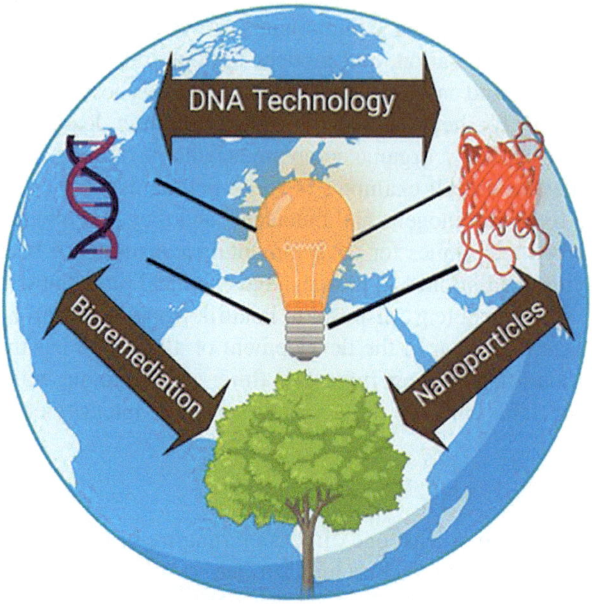

FIGURE 9.1 Nanobiotechnology bioremediation pathway.

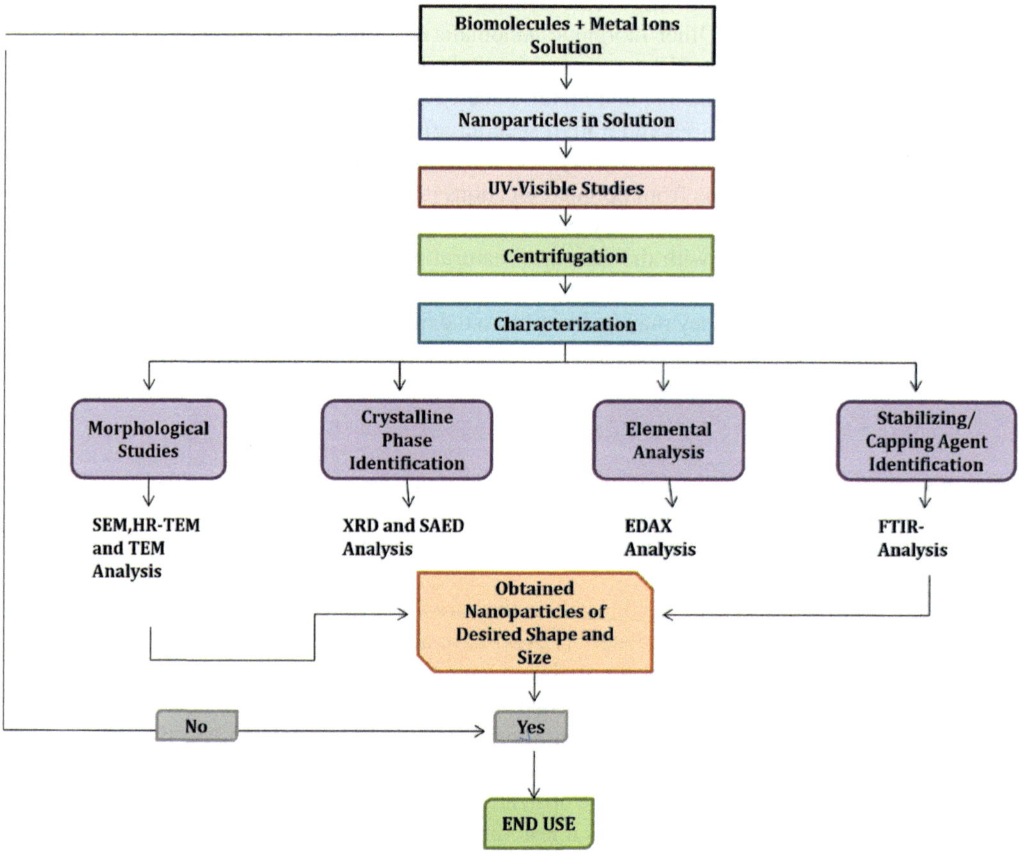

FIGURE 9.2 A general outline showing the steps involved in making nanoparticles outside of cells.

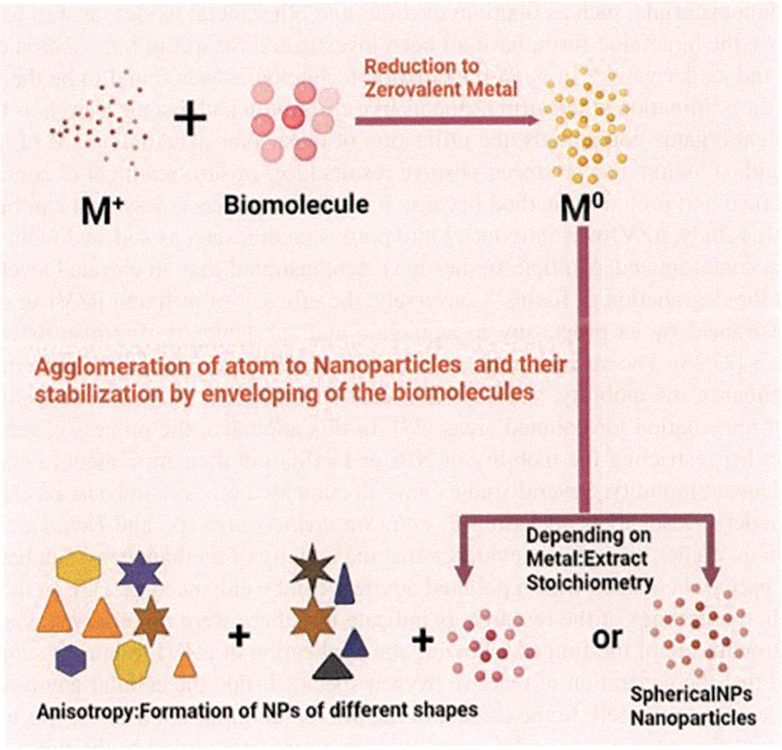

FIGURE 9.3 Nanoparticle production mechanism [13].

NPs can be produced extracellularly or intracellularly. The first study on the creation and characterization of intracellular NPs was published in 2002 by Gardae-Torresday and colleagues. Sastry and colleagues claimed that extracellular synthesis was achieved for the first time by utilizing plant broth [17]. The utilization of weeds and waste materials for the production of NPs has also been documented [12,18–20]. Many techniques employed to produce NPs are complex and require strict control measures. These measures include controlling pH levels, agitation, light exposure, bioagent and metal ion concentration, and other variables to ensure effective management of the shape and size of NPs [21]. By changing the mixing fraction as the process variable, authors have developed novel methods for producing particles with desired morphologies [13,15,18,19,22–25]. Figure 9.3 depicts the general mechanism involved in the production of NPs.

9.4 ADVANCES IN ENVIRONMENTAL APPLICATION OF NANOBIOTECHNOLOGY

Nano-bioremediation is characterized by the utilization of nanobiotechnology, encompassing three notable elements: (a) the employment of environmentally friendly nanomaterials; (b) the provision of remedies for the elimination of harmful substances from polluted areas; and (c) the significance of employing on-site (in situ) or off-site (ex situ) approaches for the treatment of pollutants, which play a crucial role in the pursuit of environmental enhancement. While the latter method dissolves the organic impurities, leaving no toxic by-products like CO_2 and H_2O, the farmer technique involves the removal of heavy metal contaminants through sequestration. As these adaptable remediation technologies continue to expand and improve, it becomes possible to achieve the sustainable reduction of pollutants from the environment.

Biogenic nanomaterials, such as titanium dioxides and other metal oxides, as well as noble metals, particularly the bimetallic form, have all been investigated for use in remediation efforts. The use of nZVI and its derivatives in nano-remediation technologies was found to be the most widespread. With the elimination of harmful elements like chromium and arsenic as well as the destruction of persistent organic compounds, the utilization of these nanostructures made of iron, nickel, gold, silver, and palladium has produced positive results [26]. In situ treatment of contaminants is the most practical and preferred method because it works better, costs less, and can be used on a big scale. In this study, nZVI was introduced into porous media, such as soil, sediments, and aquifers, that were contaminated. Multiple studies have demonstrated that an elevated level of reactivity facilitates the degradation of toxins. Conversely, the efficacy of utilizing nZVI in remediation efforts is constrained by its propensity to aggregate and subsequently diminish its reactivity as time progresses [27,28]. The utilization of organic polymers such as guar gum and lactate has been observed to enhance the mobility, stability, and reactivity of nZVI. This development facilitated the process of remediation for polluted areas [27]. In this approach, the process of remediation is achieved by either restricting the mobility of NPs or facilitating their movement to contaminated areas with adequate mobility. Several studies have documented adverse impacts on the microbial populations under investigation, including *E. coli*, *Staphyloccoides* sp., and *Dehalococcoides* sp. [29–31]. Previous studies have shown evidence that the activity of methanogens, sulphate-reducing bacteria, and bacterial colonies within polluted aquifers can be enhanced [32,33]. In the context of fungal growth, the findings of the researchers indicate that there were no observed changes in the structure or growth rate of the fungus following the application of nZVI to human subjects. It has been reported that the generation of reactive oxygen species inside the cellular environment leads to structural damage to the cell. In the context of the microbial inhibition trials, it was necessary to take into account the inconsistent findings and the constraints pertaining to the types of NPs and microorganisms that were examined. For instance, pumping groundwater or moving sediment above ground level or to other locations for treatment is not part of remediation at an in situ site. However, the treatment efficacy of pollutant concentration at all sites is nearly eliminated when nanomaterials are injected. NPs produced via bioagent binding may have a use in the remediation of soil and groundwater. This is feasible because of the special qualities of nanomaterials, like increased sorption and strong reactivity. Because they are colloidal, they clean up water more successfully than conventional techniques by penetrating deeper into contaminated areas. Recapitulated research on the application of NPs includes the adsorption of several pollutants, including colours and heavy metals. There are several uses for this adsorption and magnetic separation combination in environmental remediation and water purification. To adsorb and decompose organic and inorganic pollutants, three methods of injecting nanoscale zinc dioxide (ZVI) into contaminated sites are as follows: (a) injecting nZVI with high mobility to induce nZVI plumes; and (b) injecting NPs onto topsoil. When it comes to treating organic contaminants, contaminant degradation-influenced nano-remediation holds more significance than adsorption. Biogenic NPs, such as TiO_2, ZnO, nZVI, and noble metal NPs, have the ability to eliminate organic pollutants in ex situ technologies. The study of bimetallic gold-palladium NPs speeds up the breakdown of important water contaminants including mercury and trichloroethylene, acting as an active catalyst. Additionally, gas sensors for measuring pollution levels of nitrogen oxides (NO_x) and CO have been developed using gold NPs. The root and shoot growth of plant species like soybean, wheat, maize, and lucerne has been encouraged by metal oxide NPs like ZnO, TiO_2, and CeO_2, suggesting that nanobiotechnology plays a significant role in phytoremediation. At the nanoscale, TiO_2 facilitated nitrate adsorption, the process by which inorganic nitrogen is converted to organic nitrogen and increases the reactivity of several enzymes. In an independent investigation, silver NPs were utilized on a nanoscale in conjunction with Congo red bouillon that contained *Aspergillus niger.* The results indicated a decolorization rate above 85.8% over a 24-hour timeframe [34]. The utilization of nZVI in the study yielded noteworthy findings, indicating a reduction of 99.9% for lead (Pb) and 95% for hexachlorohexanes in contaminated water. Furthermore, it was observed that these contaminants were effectively adsorbed [35]

TABLE 9.1
A List of the Nanoparticles and Microorganisms Used in On-Site Nano-Bioremediation

Nanoparticles Used	Microorganisms Used	Contaminant	% of NR	% of NBR	Reference
Pd/Fe	*Trametesersicolor*	Triclosan	100	100	[78]
Pd/Fe	*Sphingomonaswittichii*	Dioxin	100	100	[78]
Pd/Fe	*Sphingomonas* sp.	Lindane	99	99	[79]
nZVI	*Burkholderiaxenovorans*	Aroclor1248 (Polychlorinated biphenyls (PCB))	89	90	[80]

Table 9.1 presents a concise overview of the nano-bioremediation technique employed to eliminate pollutants. The initial stage of the nano-bioremediation procedure involves the reduction of pollutant concentrations, followed by the promotion of contaminant biodegradability to levels below the threshold of risk. The advancement of nanotechnology has facilitated research efforts aimed at monitoring and mitigating air pollution, alongside existing applications in the treatment of soil, water, and wastewater. The enhanced surface area exhibited by CNTs in their nanoscale form has been found to result in a significantly higher absorption capacity for dioxin, up to a thousand times greater than that of activated carbon. The current state of technical improvement in the field of nano-biotechnological approach in nano-remediation is still in its early stages due to the suboptimal size, shape, and surface chemistry of the materials involved. The introduction of nanomaterials into the environment is an inevitable consequence of their application in nano-remediation processes. Nevertheless, individuals can be exposed to these NPs by several routes, including ingestion, inhalation, skin contact, and/or injection. A comprehensive examination is necessary to ascertain the ramifications of nanotoxicity on biological systems, to validate the safety of employing engineered nanostructures [36].

9.5 BIOREMEDIATION IMPACT/IMPORTANCE

The long-term impact of heavy metal pollution in wastewater, land, lakes, and streams has adversely affected human health. With the advent of industrialization, conventional therapeutic approaches have become less effective, less targeted, and more costly. Limited research has been conducted on the integration of biological approaches with alternative methodologies for addressing these issues, including biophysical, biochemical, physiochemical, and physiochemical nano-based strategies. Bioremediation, nanotechnology processes and applications, and common therapeutic modalities such as chemical, physical, and biological therapies are the main subjects covered in the following section. New research indicates that metal oxide-based NPs can be used as efficient nano adsorbents to remove organic and heavy metal contaminants from water [37]. The utilization of polymer-templated nanoporous and functional NPs for heavy metal removal has received significant interest, mostly due to its perceived advantages over conventional water treatment techniques.

9.6 DIFFERENT STRATEGIES USED FOR BIOREMEDIATION

9.6.1 Treatment of Physical Methods

9.6.1.1 Precipitation

The method employed for the precipitation of metal salts involves the introduction of a suitable anion. The procedure commonly employs chemicals such as manganese sulphate, copper sulphate, ammonium sulphate, alum, and ferric compounds. The effectiveness of salt (ion) and sediment disposal is influenced by a decrease in pH levels. The aforementioned operation incurs a higher cost. The precipitation method including bisulfide, lime, or ions is insufficient in terms of specificity and efficiency for the removal of metal ions with low concentrations.

9.6.1.2 Ion Exchange

The following is the procedure utilized for the extraction of metals from wastewater generated by industrial processes. This technology relies on a solid-phase ion-exchange material capable of exchanging cations or anions. The most common type of ion-exchange matrix material is synthetic ion-exchange resin. It's not cheap, but it's possible to treat a huge volume and reduce the concentration to the ppb range. One limitation of this technique lies in its inability to effectively manage elevated levels of metal concentrations, mostly attributed to the formation of fouling within the matrix. This fouling phenomenon arises from the coexistence of particulate matter and organic compounds within the effluent. It is also exceedingly sensitive to changes in solution pH and displays no selective behaviour [38].

9.6.1.3 Electrowinning

This technology finds applications in various domains of industrial metallurgy and mining, such as metal processing, acid mine drainage, the electronics and electrical industries, and heap leaching for heavy metal regeneration and removal [38].

9.6.1.4 Electrocoagulation

Similar to electrochemical methods, this one uses electrical current to precipitate metals out of solution. This implies that the utilization of an internal voltage enables the retention of contaminants in wastewater within a dissolved state. When the electrocoagulation system neutralizes the ions and charged particles, the ions with the opposite electric charge destabilize and precipitate them [38].

9.6.1.5 Cementation

Electrochemical precipitation is a process where metals with a higher oxidation ability pour into a solution. Apart from copper, various other metals can be obtained through the process of cementation. These include gallium (Ga), lead (Pb), gold (Au), silver (Ag), antimony (Sb), cadmium (Cd), and tin (Sn). Adsorption is a process wherein metal species adhere to the surface of an adsorbent through mechanisms of physisorption and chemisorption. The efficacy of metal ion removal is influenced by various parameters, such as pH, adsorbent surface area, and surface energy. A variety of adsorbents are available, such as activated alumina, activated carbon, potassium permanganate ($KMnO_4$), granular ferric hydroxide, iron oxide-coated sand, and copper-zinc granules [38].

9.6.1.6 Membrane Filtration

Among these techniques is the use of a semipermeable membrane and a stress gradient to separate metals from water. The primary cause of fouling in this process is the co-precipitation of Fe^{2+} and Mn^{2+} ions in the aqueous solution. The added expense results from factors like monitoring stress variations and water pretreatment [38].

9.6.1.7 Electrodialysis

With the exception of the driving power, this has a resemblance to the reverse osmosis procedure, wherein an electric field is employed across a semipermeable membrane to segregate charged metal ions from polluted water. The method demonstrates enhanced efficacy in the extraction of heavy metals from groundwater, relying on key factors including porosity, pH levels, groundwater flow rate, texture, ionic conductivity, and water content. The integration of this therapeutic technique with additional procedures, including reactive zone therapy, membrane filtration, surfactant flushing, reactive zonal therapy, permeable reactive barriers (PRBs), and bioaugmentation, can contribute to the achievement of effective remediation goals [38,39].

9.6.2 Chemical Treatment Methods

Groundwater is known to exhibit significant dispersion of heavy metal pollutants, posing considerable challenges for conventional treatment methodologies in terms of effective management. This section will provide a discussion of various strategies employed in chemical therapy, along with their respective drawbacks.

9.6.2.1 Reduction

Reagents such as gaseous hydrogen sulphide and dithionites are introduced into contaminated regions characterized by high soil permeability and alkaline pH levels through deep injection. In these polluted locations, contaminants undergo degradation or immobilization. One drawback associated with the reduction process is the generation of hazardous intermediates [40–42]. An example of such an entity is the colloidal ZVI, which can be introduced at considerable depths into the aquifer, undergoing swift degradation and resulting in the production of hazardous residues [43,44].

9.6.2.2 Chemical Washing

The direct approach to heavy metal removal involves the utilization of potent extractants, such as acids. However, it is important to note that this process has the potential to compromise the quality of the soil, hence posing a risk to the surrounding environment. The ex situ remediation of contaminated soil poses inherent risks and presents significant challenges in terms of issue management and hazardous waste handling [36].

9.6.2.3 Chelate Flushing

This methodology entails the extraction of a substantial volume of heavy metals, facilitated by the potential to revitalize and reuse the active agents employed in the aforementioned procedure. Solvent-charged resins exhibit exceptional efficiency and complete regenerability and are extensively employed in PRBs. One limitation associated with this methodology pertains to the financial implications and the potential carcinogenic properties of chelating agents, such as ethylene diamine tetraacetic acid and diethylene triamine pentaacetic acid [45–47].

9.6.3 Biological Treatment Methods

Bioremediation, often known as the application of biotechnology in water treatment, has witnessed significant expansion in its diverse range of environmental applications. This chapter provides a comprehensive discussion of the many biotechnology-based water treatment strategies that are now accessible and widely adopted. Bioremediation is currently recognized as an ecologically sustainable and economically efficient remediation approach for the removal of metal contaminants, particularly in aquatic and terrestrial environments. While bioremediation is generally favoured as a method for addressing environmental contamination, it is important to acknowledge that, in certain cases, the toxins themselves might exhibit toxicity towards the microorganisms involved in the bioremediation process. These challenges prompted scientists to explore an alternate approach by broadening the techniques of bioremediation to enhance resistance in challenging conditions and provide stable remediation features that can sustain a high rate of bioremediation [37].

9.6.3.1 Bioremediation

Bioremediation refers to the effective utilization of a biological system, wherein harmful pollutants are degraded, decayed, transformed, immobilized, or stabilized to achieve a state that is harmless or falls below the concentration limits deemed acceptable by regulatory authorities. Examples of organisms that fall under this category include bacteria, fungi, algae, and certain types of plants. *Escherichia*, *Citrobacteria*, *Klebsiella*, *Rhodococcus*, *Staphylococcus*, *Alcaligenes*, *Bacillus*, and *Pseudomonas* are commonly employed bacterial species in the field of bioremediation.

Bioremediation encompasses a range of remediation strategies, including the utilization of indigenous microorganisms (bioaugmentation) for natural attenuation, the introduction of nutrients to stimulate microbial activity (biostimulation), the application of genetically modified organisms, the use of plants for phytoremediation, and the complete biodegradation of organic substances into inorganic components through biomineralization [48].

9.6.3.2 Biofiltration

The biofilter can be described as a permeable media that facilitates the interaction between water and microorganisms, resulting in the coverage of microorganisms on its surface. The phenomenon under consideration is rooted in the intricate process of complex formation occurring in aqueous environments, involving the interaction between pollutants and organic compounds. During this stage, the porous media undergoes adsorption and subsequently transformation into metabolic by-products, biomass, carbon dioxide, and water. The biofilter undergoes three key processes, including microbe adhesion, growth and degradation, and detachment [49,50].

9.6.3.3 Biosorption

Biological methodologies that utilize biomaterials derived from deceased or dormant microorganisms are frequently employed. The technique is characterized by its passive nature, which necessitates no internal energy input. This approach has numerous advantages, including the very efficient regeneration of biosorbents, the recovery of metals, the minimal production of sludge, and the cost-effective nature of the regeneration process [51]. Biomass functions as a matrix for ion exchange, effectively binding and exhibiting its intrinsic properties to remove heavy metals from very diluted water solutions. The occurrence of some crops, fungi, and bacteria can be attributed to the presence of biosorbent products, such as cell wall composition [52]. Biosorption is considered a more favourable approach for the cleanup of effluents containing metals due to the numerous advantages it offers. Many sorption-based remediation techniques are conducted using a single metal ion. Limited research has been conducted on blended metal alternatives in this context. Bacteria, algae, and yeasts have been identified as possessing several types of biomasses that exhibit absorptive capabilities. The utilization of these economically viable biosorbents renders the approach highly cost-efficient and competitive in the realm of environmental applications [53].

This approach encompasses various attributes, such as the ability to recover specific metals across a broad range of pH and temperature conditions, quick adsorption and desorption kinetics, and cost-effectiveness in terms of resource utilization and operational expenses. The biosorbents under investigation have demonstrated a significant affinity for many types of toxic heavy metals commonly found in industrial waste. Furthermore, the utilization of biosorbents derived from a composite of deceased biomass comprising distinct varieties of microorganisms is also employed. Furthermore, the utilization of biosorbents involves the amalgamation of nonliving biomass comprising distinct types of microorganisms [53]. The utilization of immobilized biomass is suggested as a more suitable option for large-scale deployment in comparison to indigenous biomass. However, it is crucial to thoroughly examine various immobilization techniques to assess their efficacy, user-friendliness, and cost-effectiveness. Multiple strategies were employed to mitigate the employment of cumbersome processes, including complex formation, utilization of chelating agents, electrostatic interactions, and ion exchange with materials derived from agricultural operations and other natural sources. Preliminary treatment with chemical agents is a necessary need for enhancing the efficacy and stability of sorption. The utilization of biosorption as a method is advantageous due to its ability to achieve higher rates of adsorption, easy accessibility to biosorbents, affordability, and absence of hazardous properties [54].

9.6.3.4 Biophysiochemical Method

This approach involves the integration of an adsorption or coagulation technique with a biological process. Due to its numerous advantages in comparison to conventional physiochemical treatment approaches, it is widely recognized as a very effective alternative remediation methodology. Furthermore, biological processes exhibit considerable potential in the implementation of sludge disposal protocols and are a fundamental component of any arsenic therapeutic technology [55]. The chemolithotrophic bacterium *Acidithiobacillus ferrooxidans* BY-3, which is naturally found in mines, has been extensively employed as a bioremediation agent for the removal of both organic and inorganic arsenic compounds from aqueous solutions [56]. Researchers successfully isolated five unique fungal strains and employed them in the remediation of arsenic-contaminated sites. A further method involves the biotic oxidation of iron through the use of microbes, specifically *Gallionella ferruginea* and *Leptothrix ochracea*. The process by which iron oxide is deposited within the filter medium adjacent to the organism is facilitated, hence creating a conducive environment for the sorption and subsequent removal of metal ions from the solution. The approach under consideration offers the advantage of obviating the need for chemical reagents in the oxidation of trivalent arsenic. Furthermore, it eliminates the requirement for meticulous monitoring often associated with sorption methods, as the generation of iron oxides occurs continuously in situ. An additional noteworthy advantage is in its use of a combination of biological oxidation filtration–sorption techniques, enabling the simultaneous removal of many inorganic pollutants from groundwater, including iron, manganese, and arsenic [57].

9.6.3.5 Novel Biosorbents

Novel biosorbents have been developed to enhance the selectivity and accumulation of microorganisms employed in the bioremediation process. The utilization of genetic engineering in a biosorption approach has the potential to enhance the therapeutic activity of microorganisms. Future studies will focus on developing microbes that are specifically engineered to have strong adsorption ability for hazardous metal ions. Several experiments have been conducted to assess the suitability of these biosorbents for the remediation of industrial effluents. In general, the task of changing long-standing conventional practices is a significant challenge. Despite the high cost associated with it, the utilization of biosorbents is widely recognized for its exceptional performance, therefore indicating significant prospects for this technology. Consequently, a comprehensive investigation is necessary about the pilot and full-scale biosorption procedures [54].

9.6.3.6 Bioaugmentation

The present approach entails the utilization of genetic engineering to enhance the metabolic activity of microorganisms, hence facilitating in situ bioremediation. Certain researchers have engaged in the genetic manipulation of a strain of *Escherichia coli*, wherein they have deliberately increased the expression of ArsR genes. This deliberate alteration has led to the notable accumulation of arsenic within the organism. The utilization of selected ligands is widely recognized as an effective approach for enhancing arsenic accumulation and sequestration [58,59]. A collection of metagenomes was generated from the sludge obtained from manufacturing wastewater processing plants. Through this analysis, a previously unidentified gene (arsN) conferring resistance to As(V) was identified. Notably, the protein encoded by this gene was shown to exhibit complete similarity to acetyl transferase. The excessive expression of this protein leads to a higher disparity in response to arsenic in *E. coli*. Novel biological methodologies, including metagenomics research, directed evolution, and genome shuffling, have the potential to be employed in the development of arsenic-resistant mechanisms that are well-suited for arsenic remediation purposes [60] The process of modifying an arsenic resistance operon by DNA shuffling was effectively elucidated in a study [61].

9.6.3.7 Bacterial Sulphate Reduction

In their study, Jong and Parry [62] employed a sulphate-reduction bacteria-based up-flow anaerobic packed-bed reactor to treat mercury and several acidic metals, including copper (Cu), magnesium (Mg), zinc (Zn), aluminium (Al), nickel (Ni), iron (Fe), and sulphate pollutants. The study conducted by [63] showed a significant reduction of over 77.5% in the initial levels of arsenic. Furthermore, their findings provided strong evidence for the effective removal of chromium (Cr) and arsenic from solution, with removal rates ranging from 60% to 80%, employing sulphate-reducing bacteria *Desulfovibrio desulfuricans*. In a comparable study [64], additionally, a sulphate-reducing biological method was devised to address the objective of eliminating heavy metals from acid mine drainage. Furthermore, a study conducted by Fukushi et al. [65] employed an alternative indirect method to eliminate mercury from the system. This method involved the sequestration of the metal into insoluble sulphides through the metabolic activity of sulphate-reducing bacteria. The bacteria utilized a diverse range of organic substrates; in an anaerobic environment, the terminal electron acceptor is represented by SO_4^{2-}. The enhanced mobility and toxicity of As (III) relative to As(V) necessitates microbial-mediated transformations between these two forms, hence improving the mobility of arsenic.

9.6.3.8 Phytoremediation

Numerous plant species are available for soil remediation, capable of extracting toxins from various sources such as land, surface water, groundwater, and sediments. Phytofiltration harnesses the inherent ability of plants to collect heavy metals, such as arsenic, and exhibit effective tolerance. Hydrilla crops are commonly employed for phytofiltration. To further improve this technology, it is possible to explore the cultivation of these crops under actual field conditions in the presence of contaminated water. Phytoextraction refers to the process by which plant roots absorb and transport pollutants within agricultural crops. This methodology is commonly employed for the remediation of soil contaminated with metals, although it is frequently associated with various drawbacks including the removal of plant biomass, metal recovery, disposal of biomass, and the potential phototoxic effects of metals. Phytodegradation, alternatively referred to as phytotransformation, denotes the process by which plants metabolically break down pollutants inside their structures or the ex situ destruction of contaminants facilitated by enzymes generated by plants as depicted in figure 4. The main concern associated with this approach pertains to the deterioration of goods and the generation of hazardous intermediates. Phytovolatilization is an additional technique employed for the sequestration of heavy metals through plant-mediated processes. This method involves the uptake of contaminants by crops, which are then released into the atmosphere by transpiration. This release may involve the original contaminants or modified forms of contaminants [66–68].

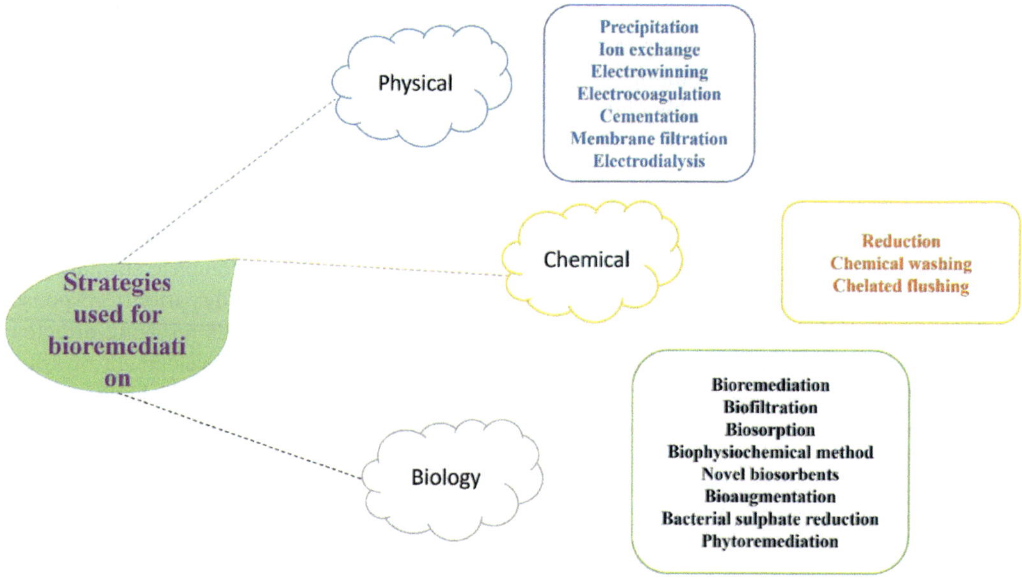

FIGURE 9.4 Three types of bioremediation strategies: chemical, physical, and biological.

9.7 ROLE OF NANOMATERIALS IN BIOREMEDIATION

Nano-remediation has emerged as an innovative and effective approach for environmental remediation due to its significant role in mitigating pollution through prevention, detection, monitoring, and remediation processes [69] (Figure 9.5). The implementation of this solution has the potential to address various challenges encountered during the process of cleaning a certain area, hence enhancing overall efficiency and reducing associated costs. Both chemical reduction and catalysis are viable methods for the remediation of contaminants using NPs. Nanomaterials possess advantageous properties that render them very suitable for various in situ applications. NPs have the potential to traverse minute interstices within the subsurface and maintain suspension within groundwater due to their diminutive size and possession of distinctive surface coatings. This phenomenon implies that NPs possess the capability to traverse greater distances compared to larger macroparticles, hence enabling them to achieve a more extensive dispersion. The utilization of NPs in technologies reliant on solid–water and solid–gas interfaces is advantageous due to their elevated surface area-to-mass ratio. Several technologies are employed for water and waste gas treatment, including adsorption, which is utilized to remove impurities, and photocatalytic processes, which facilitate the degradation of contaminants. The impact of nanoscale dimensions on the chemical reactivity of materials is influenced by the distinct compositions observed in near-surface regions compared to bulk regions. The significance of interfacial free energy in dissolution–precipitation processes is heightened by this observation [70].

NPs offer significant versatility for both in situ and ex situ remediation due to their ability to be easily utilized in ex situ slurry reactors for the treatment of contaminated soils, sediments, and solid wastes. Alternatively, they have the potential to be affixed to a stable substrate, such as carbon, zeolite, or a membrane, to enhance the purification of water, effluent, or gaseous process streams. Nano-remediation encompasses the utilization of reactive nanomaterials, including nanoscale zeolites, metal oxides, CNTs and fibres, and bimetallic NPs, to facilitate the conversion and elimination of contaminants for environmental remediation purposes.

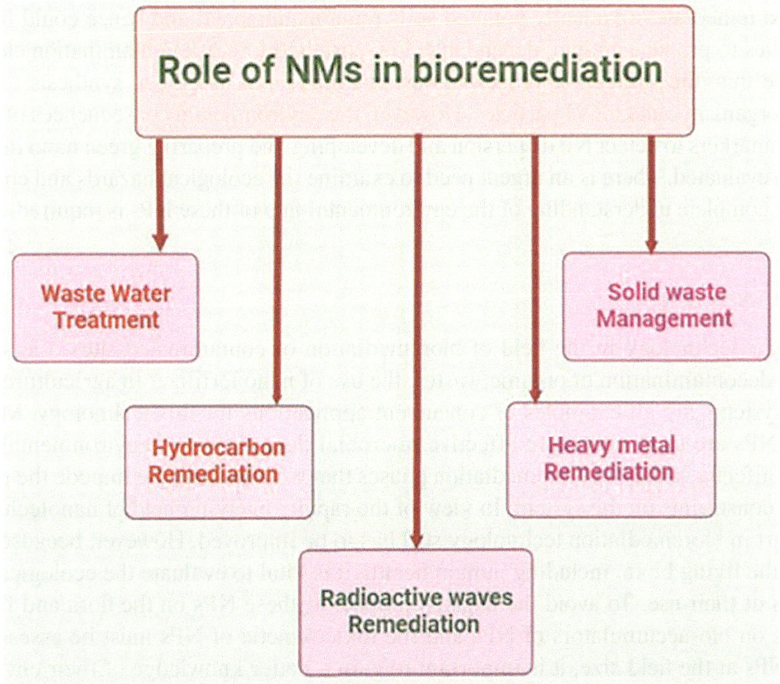

FIGURE 9.5 Nanoparticles in bioremediation: applications.

9.8 FUTURE PERSPECTIVE

Nano-remediation has become a big area of study and development, and it has a lot of potential for cleaning up polluted sites and keeping pollution out of the environment. It makes it possible to clean up contaminated sites with deeper soils, works with other technologies like bioremediation, and is a growing tool for cleaning up contaminated sites. To make the most of the huge importance of nano-remediation in Polycyclic aromatic hydrocarbon (PAH) removal, it could be looked into how to make and use nano fertilizers (biostimulation and bioaugmentation), nano minerals (biostimulation), or green synthesized nano oxidizers (PAH oxidation) [71]. Recently, polluted soils have been treated with electrokinetic (EK) remediation, enzyme-mediated bioremediation, phytobioremediation, and biosensors. EK remediation is a good way to clean up low-permeability soils when other methods, like natural decline, are not enough. In this method, a low-voltage direct current is sent through the dirt between electrodes that are in the right places. Electromigration moves ionic waste to the anode with the opposite charge at the moment; EK treatment of a PRB is the most innovative way to get rid of heavy metals or organic molecules from polluted soils [72]. The development of electrochemical biosensors, coupled with metallic (mono/bi) NPs like nanowires [73], nanorods, and nanospheres, has many benefits, such as high sensitivity and real-time detection of heavy metals. Scientists found that eukaryotic bacteria are better than prokaryotic cells at detecting pesticides and heavy metals [74]. This is an area where microbial biosensors could be used. This is mostly because it is helpful to make whole-cell biosensors that can identify heavy metal and pesticide toxicity selectively and sensitively. In the future, these microbial biosensors will be used in more ways to track metal pollution in the environment and make clean energy [75].

At present, the metagenomic-based molecular method is seen as a new way to study the variety of microbes. It is an accurate way to find out the genetic content in very high concentrations of arsenic and antimony that have contaminated a very complex environment [76]. Nano-remediation, transgenic approaches, and photo-hetero microbial systems are some new ways to clean up pollution that are mostly biological and are expected to lead to an "era of green biotechnology" soon. Several organic and inorganic toxins have been cleaned up with the help of these new technologies [77]. However, their ability to remediate organically polluted soils remains untapped and hence could be the focus of future studies to produce a rapid, dependable, low-cost, and low-risk contamination cleanup technique. Despite this, the strategic approach to sustainability is to use green synthesis of NPs from plants, microorganisms, and nZVI particles. However, the environmental consequences of developing diagnostic biomarkers to detect NP dispersion and developing and preparing green nano initiative particles must be evaluated. There is an urgent need to examine the ecological hazards and consequences of NPs, and a complete understanding of the environmental fate of these NPs is required.

9.9 CONCLUSION

The use of nanotechnology in the field of bioremediation of contaminated sites offers additional benefits. The decontamination of organic wastes, the use of nano fertilizer in agriculture, and water purification systems are all examples of concurrent applications for this technology. Many different kinds of NPs are used to achieve effective microbial degradation of environmental pollutants because they affect a variety of bioremediation phases that would otherwise impede the process and place kinetic constraints on the system. In view of the rapidly evolving field of nanotechnology, the state-of-the-art in bioremediation technology still has to be improved. However, because these NPs might affect the living biota, including human health, it is vital to evaluate the ecological risks and repercussions of their use. To avoid the negative impact of these NPs on the flora and fauna, additional studies on bio-accumulators of NPs and the toxicokinetic of NPs must be assessed. Before using these NPs at the field size, it is important to gain a better knowledge of their environmental destiny. Additionally, choosing a low-cost, ecologically safe nanotechnology that is compatible with a varied range of human existence, such as biomedicine, agriculture, and bioremediation, is essential for the successful and sustainable implementation of nanotechnology.

REFERENCES

1. E. Carata, E. Panzarini, and L. Dini. "Environmental nanoremediation and electron microscopies," In *Nanotechnologies for Environmental Remediation*, Cham: Springer International Publishing, 2017, pp. 115–136. doi: 10.1007/978-3-319-53162-5_4
2. M. K. Bharti, S. Gupta, S. Chalia, I. Garg, P. Thakur, and A. Thakur. Potential of magnetic nanoferrites in removal of heavy metals from contaminated water: mini review. *Journal of Superconductivity and Novel Magnetism*, 33, 2020, 3651–3665. doi: 10.1007/s10948-020-05657-1
3. N. Sakulchaicharoen, D. M. O'Carroll, and J. E. Herrera. Enhanced stability and dechlorination activity of pre-synthesis stabilized nanoscale FePd particles. *Journal of Contaminant Hydrology*, 118(3–4), 2010, 117–127. doi: 10.1016/j.jconhyd.2010.09.004
4. A. B. Cundy, L. Hopkinson, and R. L. D. Whitby. Use of iron-based technologies in contaminated land and groundwater remediation: a review. *Science of the Total Environment*, 400(1–3), 2008, 42–51. doi: 10.1016/j.scitotenv.2008.07.002
5. G. Bhandari. Environmental nanotechnology: applications of nanoparticles for bioremediation. In *Nanotechnology in the Life Sciences*, Cham: Springer Science and Business Media B.V., 2018, pp. 301–315. doi: 10.1007/978-3-030-02369-0_13.
6. N. Dhanda et al. Green-synthesis of Ni-Co nanoferrites using aloe vera extract: structural, optical, magnetic, and antimicrobial studies. *Applied Organometallic Chemistry*, 37, 2023, e7110. https://doi.org/10.1002/aoc.7110
7. A. Thakur, P. Punia, R. Dhar, R. K. Aggarwal, and P. Thakur. Separation of cadmium and chromium heavy metals from industrial wastewater by using Ni-Zn nanoferrites. *Advances in Nano Research*, 12, 2022, 457–465. https://doi.org/10.12989/anr.2022.12.5.457
8. P. Thakur and A. Thakur. Nanomaterials, their types and properties. In: Thakur, A., Thakur, P., Khurana, S. P. (eds) *Synthesis and Applications of Nanoparticles*. Springer, Singapore, 2022, pp. 19–44. https://doi.org/10.1007/978-981-16-6819-7
9. M. T. Amin, A. A. Alazba, and U. Manzoor. A review of removal of pollutants from water/wastewater using different types of nanomaterials. *Advances in Materials Science and Engineering*, 2014, 2014, 1–24. doi: 10.1155/2014/825910
10. M. I. F. Guedes, E. O. P. Tramontina Florean, F. De Lima, and S. R. Benjamin. Current trends in nanotechnology for bioremediation. *International Journal of Environment and Pollution*, 1(1), 2020, 1. doi: 10.1504/ijep.2020.10023170
11. P. Ferreira, P. Alves, P. Coimbra, and M. H. Gil. Improving polymeric surfaces for biomedical applications: a review. *Journal of Coatings Technology and Research*, 12(3), 2015, 463–475. doi: 10.1007/s11998-015-9658-3
12. T. Abbasi, N. Neghi, and S. U. Ganaie. Gainful utilization of four otherwise worthless and problematic weeds for silver nanoparticle synthesis. *Official Journal of the Patent Office*, 28, 11869, 2011.
13. J. Anuradha, T. Abbasi, and S. A. Abbasi. Rapid and green synthesis of gold nanoparticles by the use of an otherwise worthless weed lantana (Lantana camara L.). *Journal of Environmental Science and Engineering*, 57(3), 2015, 203–213.
14. P. Singh, Y.-J. Kim, D. Zhang, and D.-C. Yang. Biological synthesis of nanoparticles from plants and microorganisms. *Trends in Biotechnology*, 34(7), 2016, 588–599. doi: 10.1016/j.tibtech.2016.02.006
15. N. Dhanda, P. Thakur, and A. Thakur. Green synthesis of cobalt ferrite: a study of structural and optical properties. *Materials Today: Proceedings*, 73, 2023, 237–240. doi: 10.1016/j.matpr.2022.07.202
16. R. R.-P. of toxic metals: using plants to clean and undefined 2000, "Metal-accumulating plants," *cir.nii.ac.jp*, Accessed: May 21, 2023. Online.. Available: https://cir.nii.ac.jp/crid/1571980075459755648
17. S. S. Shankar, A. Ahmad, R. Pasricha, and M. Sastry. Bioreduction of chloroaurate ions by geranium leaves and its endophytic fungus yields gold nanoparticles of different shapes. *Journal of Materials Chemistry*, 13, 2003, 1822. doi: 10.1039/b303808b
18. P. Sharma et al. Nanomaterial fungicides: in vitro and in vivo antimycotic activity of cobalt and nickel nanoferrites on phytopathogenic fungi. *Global Challenges*, 1(9), 2017, 1700041. doi: 10.1002/gch2.201700041
19. P. Punia, M. K. Bharti, R. Dhar, P. Thakur, and A. Thakur. Recent advances in detection and removal of heavy metals from contaminated water. *ChemBioEng Reviews*, 9(4), 2022, 351–369. doi: 10.1002/cben.202100053
20. M. K. Bharti, S. Chalia, P. Thakur, S. N. Sridhara, A. Thakur, and P. B. Sharma. Nanoferrites heterogeneous catalysts for biodiesel production from soybean and canola oil: a review. *Environmental Chemistry Letters*, 19(5), 2021, 3727–3746. doi: 10.1007/s10311-021-01247-2
21. A. Singh, F. Wan, K. Yadav, A. Salvi, P. Thakur, and A. Thakur. Synergistic effect of ZnO nanoparticles with Cu^{2+} doping on antibacterial and photocatalytic activity. *Inorganic Chemistry Communications*, 157, 2023, 111425. doi: 10.1016/j.inoche.2023.111425

22. N. Dhanda, P. Thakur, A-C. A. Sun, and A. Thakur. Structural, optical and magnetic properties along with antifungal activity of Ag-doped Ni-Co nanoferrites synthesized by eco-friendly route. *Journal of Magnetism and Magnetic Materials*, 572, 2023, 170598. doi: 10.1016/j.jmmm.2023.170598
23. D. Chahar et al. Photocatalytic activity of Cobalt substituted Zinc ferrite for the degradation of methylene blue dye under visible light irradiation. *Journal of Alloys and Compounds*, 851, 2021, 156878. doi: 10.1016/j.jallcom.2020.156878
24. D. Kala, S. Gupta, R. Nagraik, V. Verma, A. Thakur, and A. Kaushal. Diagnosis of scrub typhus: recent advancements and challenges. *3 Biotech*, 10(9), 2020, 396. doi: 10.1007/s13205-020-02389-w
25. P. Punia et al. Recent advances in synthesis, characterization, and applications of nanoparticles for contaminated water treatment- a review. *Ceramics International*, 47(2), 2021, 1526–1550. doi: 10.1016/j.ceramint.2020.09.050
26. A. Thomé, K. R. Reddy, C. Reginatto, and I. Cecchin. Review of nanotechnology for soil and groundwater remediation: brazilian perspectives. *Water, Air, and Soil Pollution*, 226(6), 2015, 121. doi: 10.1007/s11270-014-2243-z
27. K. R. Reddy, A. P. Khodadoust, and K. Darko-Kagya. Transport and reactivity of lactate-modified nanoscale iron particles for remediation of DNT in subsurface soils. *Journal of Environmental Engineering*, 140(12), 2014, 870. doi: 10.1061/(ASCE)EE.1943-7870.0000870
28. W. Yan, H.-L. Lien, B. E. Koel, and W. Zhang. Iron nanoparticles for environmental clean-up: recent developments and future outlook. *Environmental Science: Processes & Impacts*, 15(1), 2013, 63–77. doi: 10.1039/c2em30691c
29. H. Li, Q. Zhou, Y. Wu, J. Fu, T. Wang, and G. Jiang. Effects of waterborne nano-iron on medaka (Oryzias latipes): antioxidant enzymatic activity, lipid peroxidation and histopathology. *Ecotoxicology and Environmental Safety*, 72(3), 2009, 684–692. doi: 10.1016/j.ecoenv.2008.09.027
30. M. Auffan, J. Rose, M. R. Wiesner, and J.-Y. Bottero. Chemical stability of metallic nanoparticles: A parameter controlling their potential cellular toxicity in vitro. *Environmental Pollution*, 157(4), 2009, 1127–1133. doi: 10.1016/j.envpol.2008.10.002
31. T. Gordon, B. Perlstein, O. Houbara, I. Felner, E. Banin, and S. Margel. Synthesis and characterization of zinc/iron oxide composite nanoparticles and their antibacterial properties. *Colloids and Surfaces A: Physicochemical and Engineering Aspects*, 374(1–3), 2011, 1–8. doi: 10.1016/j.colsurfa.2010.10.015
32. Z.-M. Xiu, Z.-H. Jin, T.-L. Li, S. Mahendra, G. V Lowry, and P. J. J. Alvarez. Effects of nano-scale zero-valent iron particles on a mixed culture dechlorinating trichloroethylene. *Bioresource Technology*, 101(4), 2010, 1141–1146. doi: 10.1016/j.biortech.2009.09.057
33. T. L. Kirschling, K. B. Gregory, Jr., E. G. Minkley, G. V. Lowry, and R. D. Tilton. Impact of nanoscale zero valent iron on geochemistry and microbial populations in trichloroethylene contaminated aquifer materials. *Environmental Science & Technology*, 44(9), 2010, 3474–3480. doi: 10.1021/es903744f
34. R. R. Nithya. Decolorization of the dye Congo red by Aspergillus nigernanoparticle, 2010.
35. Y. Xi, M. Mallavarapu, and R. Naidu. Reduction and adsorption of Pb2+ in aqueous solution by nano-zero-valent iron—a SEM, TEM and XPS study. *Materials Research Bulletin*, 45(10), 2010, 1361–1367. doi: 10.1016/j.materresbull.2010.06.046
36. P. Thakur, Y. Verma, and A. Thakur. Toxicity of nanomaterials: an overview. In *Synthesis and Applications of Nanoparticles*. Springer Nature, Singapore, 2022. pp.535–544. doi: 10.1007/978-981-16-6819-7_25
37. S. Kumari et al. Nano Ca–Mg–Zn ferrites as tuneable photocatalyst for UV light-induced degradation of rhodamine B dye and antimicrobial behavior for water purification. *Ceramics International*, 49, 2023, 12469–12480. doi: 10.1016/j.ceramint.2022.12.107
38. S. S. Ahluwalia and D. Goyal. Microbial and plant derived biomass for removal of heavy metals from wastewater. *Bioresource Technology*, 98(12), 2007, 2243–2257. doi: 10.1016/j.biortech.2005.12.006
39. N. F. Gray. *Water Technology*, Second Edition, May 2005, CRC Press, London. doi: 10.1201/B12844/WATER-TECHNOLOGY-SECOND-EDITION-GRAY
40. J. E. Amonette, J. E. Szecsody, H. T. Schaef, Y. A. Gorby, J. S. Fruchter, and J. C. Templeton. Abiotic reduction of aquifer materials by dithionite: a promising in-situ remediation technology. 1994. Accessed: May 22, 2023. Online. Available: https://inis.iaea.org/Search/search.aspx?orig_q=RN:26059390
41. J. S. Fruchter et al. Creation of a subsurface permeable treatment zone for aqueous chromate contamination using in situ redox manipulation. *Groundwater Monitoring & Remediation*, 20(2), 2000, 66–77. doi: 10.1111/J.1745-6592.2000.TB00267.X
42. "Enhancing the design of in situ chemical barriers with multicomponent reactive transport modeling (Conference) | OSTI.GOV." https://www.osti.gov/biblio/10120043 (Accessed May 22, 2023).

43. K. J. Cantrell, D. I. Kaplan, and T. W. Wietsma. Zero-valent iron for the in situ remediation of selected metals in groundwater. *Journal of Hazardous Materials*, 42(2), 1995, 201–212. doi: 10.1016/0304-3894(95)00016-N
44. B. A. Manning, M. L. Hunt, C. Amrhein, and J. A. Yarmoff. Arsenic(III) and Arsenic(V) reactions with zerovalent iron corrosion products. *Environmental Science & Technology*, 36(24), 2002, 5455–5461. doi: 10.1021/es0206846
45. C. N. Mulligan, R. N. Yong, and B. F. Gibbs. Remediation technologies for metal-contaminated soils and groundwater: an evaluation. *Engineering Geology*, 60(1–4), 2001, 193–207. doi: 10.1016/S0013-7952(00)00101-0
46. S. K. Sikdar, D. Grosse, and I. Rogut. Membrane technologies for remediating contaminated soils: a critical review. *Journal of Membrane Science*, 151(1), 1998, 75–85. doi: 10.1016/S0376-7388(98)00189-6
47. L. G. Torres, R. B. Lopez, and M. Beltran. Removal of As, Cd, Cu, Ni, Pb, and Zn from a highly contaminated industrial soil using surfactant enhanced soil washing. *Physics and Chemistry of the Earth, Parts A/B/C*, 37–39, 2012, 30–36. doi: 10.1016/j.pce.2011.02.003
48. N. Tahri, W. Bahafid, H. Sayel, and N. El Ghachtouli. Biodegradation: involved microorganisms and genetically engineered microorganisms. In Rolando, C., Rosenkranz, F. *Biodegradation - Life of Science*. 2013. doi: 10.5772/56194
49. J. S. Devinny, M. A. Deshusses, and T. S. Webster. *Biofiltration for Air Pollution Control*. CRC Press, London, 2017. doi: 10.1201/9781315138275
50. N. K. Srivastava and C. B. Majumder. Novel biofiltration methods for the treatment of heavy metals from industrial wastewater. *Journal of Hazardous Materials*, 151(1), 2008, 1–8. doi: 10.1016/j.jhazmat.2007.09.101
51. M. A. Hashim, S. Mukhopadhyay, J. N. Sahu, and B. Sengupta. Remediation technologies for heavy metal contaminated groundwater. *Journal of Environmental Management*, 92(10), 2011, 2355–2388. doi: 10.1016/j.jenvman.2011.06.009
52. B. Volesky and Z. R. Holan. Biosorption of heavy metals. *Biotechnology Progress*, 11(3), 1995, 235–250. doi: 10.1021/bp00033a001
53. J. Wang and C. Chen. Biosorbents for heavy metals removal and their future. *Biotechnology Advances*, 27(2), 2009, 195–226. doi: 10.1016/j.biotechadv.2008.11.002
54. K. Vijayaraghavan and Y.-S. Yun. Bacterial biosorbents and biosorption. *Biotechnology Advances*, 26(3), 2008, 266–291. doi: 10.1016/j.biotechadv.2008.02.002
55. C. K. Jain and R. D. Singh. Technological options for the removal of arsenic with special reference to South East Asia. *Journal of Environmental Management*, 107, 2012, 1–18. doi: 10.1016/j.jenvman.2012.04.016
56. L. Yan, H. Yin, S. Zhang, F. Leng, W. Nan, and H. Li. Biosorption of inorganic and organic arsenic from aqueous solution by Acidithiobacillus ferrooxidans BY-3. *Journal of Hazardous Materials*, 176(1–3), 2010, 209–217. doi: 10.1016/j.jhazmat.2010.01.065
57. D. Pokhrel and T. Viraraghavan. Biological filtration for removal of arsenic from drinking water. *Journal of Environmental Management*, 90(5), 2009, 1956–1961. doi: 10.1016/j.jenvman.2009.01.004
58. J. Kostal, R. Yang, C. H. Wu, A. Mulchandani, and W. Chen. Enhanced arsenic accumulation in engineered bacterial cells expressing ArsR. *Applied and Environmental Microbiology*, 70(8), 2004, 4582–4587. doi: 10.1128/AEM.70.8.4582-4587.2004
59. N. S. Chauhan, R. Ranjan, H. J. Purohit, V. C. Kalia, and R. Sharma. Identification of genes conferring arsenic resistance to *Escherichia coli* from an effluent treatment plant sludge metagenomic library. *FEMS Microbiology Ecology*, 67(1), 2009, 130–139. doi: 10.1111/j.1574-6941.2008.00613.x
60. M. Dai and S. D. Copley. Genome Shuffling Improves Degradation of the Anthropogenic Pesticide Pentachlorophenol by *Sphingobium chlorophenolicum* ATCC 39723. *Applied and Environmental Microbiology*, 70(4), 2004, 2391–2397. doi: 10.1128/AEM.70.4.2391-2397.2004
61. A. Crameri, G. Dawes, E. Rodriguez, S. Silver, and W. P. C. Stemmer. Molecular evolution of an arsenate detoxification pathway by DNA shuffling. *Nature Biotechnology*, 15(5), 1997, 436–438. doi: 10.1038/nbt0597-436
62. T. Jong and D. L. Parry. Removal of sulfate and heavy metals by sulfate reducing bacteria in short-term bench scale upflow anaerobic packed bed reactor runs. *Water Research*, 37(14), 2003, 3379–3389. doi: 10.1016/S0043-1354(03)00165-9
63. S. Simonton, M. Dimsha, B. Thomson, L. L. Barton, and G. Cathey. Long-term stability of metals immobilized by microbial reduction 1. https://api.semanticscholar.org/CorpusID:14568237

64. V. S. Steed, M. T. Suidan, M. Gupta, T. Miyahara, C. M. Acheson, and G. D. Sayles. Development of a sulfate-reducing biological process to remove heavy metals from acid mine drainage. *Water Environment Research*, 72(5), 2000, 530–535. doi: 10.2175/106143000X138102
65. K. Fukushi, M. Sasaki, T. Sato, N. Yanase, H. Amano, and H. Ikeda. A natural attenuation of arsenic in drainage from an abandoned arsenic mine dump. *Applied Geochemistry*, 18(8), 2003, 1267–1278. doi: 10.1016/S0883-2927(03)00011-8
66. H. Ali, E. Khan, and M. A. Sajad. Phytoremediation of heavy metals—Concepts and applications. *Chemosphere*, 91(7), 2013, 869–881. doi: 10.1016/j.chemosphere.2013.01.075
67. C. D. Jadia and M. H. Fulekar. Phytoremediation of heavy metals: recent techniques. *African Journal of Biotechnology*, 8(6), 2009, 921–928. Available: https://www.academicjournals.org/AJB
68. S. Srivastava, M. Shrivastava, P. Suprasanna, and S. F. D'Souza. Phytofiltration of arsenic from simulated contaminated water using Hydrilla verticillata in field conditions. *Ecological Engineering*, 37(11), 2011, 1937–1941. doi: 10.1016/j.ecoleng.2011.06.012
69. C.S. Rajan. Nanotechnology in groundwater remediation . *International Journal of Environmental Science and Development*, 2, 2011, 3.
70. M. G. Kanatzidis et al. Report from the third workshop on future directions of solid-state chemistry: the status of solid-state chemistry and its impact in the physical sciences. *Progress in Solid State Chemistry*, 36(1–2), 2008, 1–133. doi: 10.1016/j.progsolidstchem.2007.02.002
71. M. Villen-Guzman et al. Scaling-up the acid-enhanced electrokinetic remediation of a real contaminated soil. *Electrochimica Acta*, 181, 2015, 139–145. doi: 10.1016/j.electacta.2015.02.067
72. E. M. Ramírez, C. S. Jiménez, J. V. Camacho, M. A. R. Rodrigo, and P. Cañizares. Feasibility of coupling permeable bio-barriers and electrokinetics for the treatment of diesel hydrocarbons polluted soils. *Electrochimica Acta*, 181, 2015, 192–199. doi: 10.1016/j.electacta.2015.02.201
73. S. Gunti, A. Kumar, and M. K. Ram. Comparative organics remediation properties of nanostructured graphene doped titanium oxide and graphene doped zinc oxide photocatalysts. *American Journal of Analytical Chemistry*, 06(08), 2015, 708–717. doi: 10.4236/ajac.2015.68068
74. J. C. Gutiérrez, F. Amaro, and A. Martín-González. Heavy metal whole-cell biosensors using eukaryotic microorganisms: an updated critical review. *Frontiers in Microbiology*, 6, 2015, 48. doi: 10.3389/fmicb.2015.00048
75. J.-Z. Sun et al. Microbial fuel cell-based biosensors for environmental monitoring: a review. *Water Science and Technology*, 71(6), 2015, 801–809. doi: 10.2166/wst.2015.035
76. J. Luo, Y. Bai, J. Liang, and J. Qu. Metagenomic approach reveals variation of microbes with arsenic and antimony metabolism genes from highly contaminated soil. *PLoS One*, 9(10), 2014, e108185. doi: 10.1371/journal.pone.0108185
77. S. Kuppusamy, T. Palanisami, M. Megharaj, K. Venkateswarlu, and R. Naidu. In-situ remediation approaches for the management of contaminated sites: a comprehensive overview. *Reviews of Environmental Contamination and Toxicology*, 9, 2016, 1–115. doi: 10.1007/978-3-319-20013-2_1
78. V. Bokare, K. Murugesan, J. Kim, E. Kim, and Y. Chang. Integrated hybrid treatment for the remediation of 2,3,7,8-tetrachlorodibenzo-p-dioxin. *Science of the Total Environment*, 435–436, 2012, 563–436. doi: 10.1016/j.scitotenv.2012.07.079
79. R. Singh, N. Manickam, M. K. R. Mudiam, R. C. Murthy, and V. Misra, An integrated (nano-bio) technique for degradation of γ-HCH contaminated soil. *Journal of Hazardous Materials*, 258–259, 2013, 35–41. doi: 10.1016/j.jhazmat.2013.04.016
80. T. Le, K. Nguyen, J. Jeon, A. J. Francis, and Y. Chang. Nano/bio treatment of polychlorinated biphenyls with evaluation of comparative toxicity. *Journal of Hazardous Materials*, 287, 2015, 335–341. doi: 10.1016/j.jhazmat.2015.02.001

10 Biochar in Environmental Remediation

Deepa Suhag, Raksha Rathore, and Moni Kharb
Amity University Haryana

Hui-Min David Wang
National Chung Hsing University

Preeti Thakur and Atul Thakur
Amity University Haryana

10.1 INTRODUCTION

Often referred to as "biochar," char derived from biomass is a versatile renewable energy source that may generate power, heat, and liquid biofuels. Although the conversion of biomass resources into black carbon has been known since the birth of modern science, and there is archaeological evidence of far earlier usage, the use of biochar for energy generation and agricultural development is not a novel idea [1]. A growing number of people are interested in producing solid biomass products (biochars) for carbon sequestration as their knowledge of renewable energy sources and mitigating climate change has increased [2]. The various xenobiotics found in the environment are depicted in Figure 10.1. Burning biomass at low pressure and temperatures between 300°C and 1,000°C produces biochar. In essence, biochar is a stable, recalcitrant organic carbon (C) molecule. Biochar is one of a class of materials known as black carbons. They are all created by transforming the original biomass material chemically or thermally [3]. The range of pyrolysis products included in this categorization has been described by Spokas [3] and others in terms of increasing carbon content. On this continuum, combustion leftovers with O:C molar ratios less than 0.6 frequently imply biochar [4].

The agronomic uses of biochar have been the subject of recent research studies. Given the cost projections of black carbon generation for direct agricultural use, the use of biochar in engineering applications has received far less attention, even if it would be a more practical use for the material. To create a sizable market for biochar and enhance the profitability of biochar production, smaller volumes of biochar are advised to be used in particular applications [3]. As long as the biochar is not burned to produce energy, the basic goal of carbon sequestration is maintained even if it is not immediately integrated into the soil. Biochars can have very different physical and chemical properties depending on the feedstock and thermochemical conversion (production) processes used [5]. The composition of the source material and the manufacturing methods employed thus have a major influence on the performance of biochar in different field applications. The relationship between the qualities of biochar, the conditions under which it is created, and the composition of the feedstock must be described to better comprehend the range of biochars that are now available and the implications for their usage as engineered materials. To achieve this, we provide an overview of our present knowledge of how source material and manufacturing processes affect the properties of biochar. Biochar offers a great deal of promise for usage in a variety of environmental remediation applications due to its special qualities. We talk about the potential uses, fresh difficulties, and hopes for biochar in environmental remediation in the future.

DOI: 10.1201/9781003502692-10

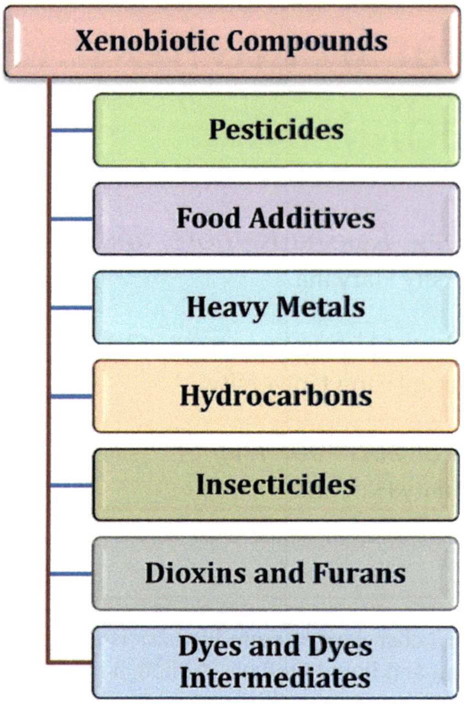

FIGURE 10.1 Different xenobiotics in the environment.

The durability of biochar in the soil is important when using it to clear up pesticide residue. H/C and O/C molar ratios have been found as critical indicators for assessing the potential of biochar for application in the bioremediation of contaminants, in addition to pyrolysis conditions and raw material type [6]. With a van Krevelen diagram, the C, H, and O contents of raw biomass and biochar can be examined [7]. It compares H/C to O/C, which was traditionally used to determine coal and petroleum origin and maturity. It is now commonly used to describe biochar to demonstrate how biomass components changed throughout the pyrolysis process [8]. The van Krevelen diagram can be used to evaluate the degree of aromaticity and carbonization in raw biomass and biochar produced since it is frequently used to compare atomic ratios of H/C and O/C [7]. The molar H/C ratio is therefore seen to be a reliable indicator of unsaturation [8].

10.2 PRODUCTION, PROPERTIES, AND APPLICATIONS OF BIOCHAR

When biomass is thermally transformed in low oxygen settings, black carbon known as biochar is produced (Figure 10.2) [9]. All thermal conversion processes, including gasification, torrefaction, and hydrothermal carbonization (HTC), are less expensive and more successful at producing biochar than pyrolysis. To create biochar, pyrolysis techniques are used, which involve roasting biomass at temperatures between 300°C and 900°C in the absence of oxygen [10]. According to heating rates, pyrolysis can be categorized into two groups: rapid pyrolysis and slow pyrolysis [11]. Slow pyrolysis yields more biochar than other pyrolysis techniques; hence, it is chosen for producing biochar despite having various drawbacks, including (a) inefficient energy consumption and (b) a long production period. Solid, liquid, and gaseous products are produced during the pyrolysis process [12]. Like charcoal, biochar is a solid substance (Hadi). The liquid products are separated into two phases: the bio-oil phase, which is non-aqueous, and the aqueous pyrolysis liquid. Together with hydrogen, methane, carbon monoxide, and carbon dioxide, the gas produced also contains trace amounts of other hydrocarbons, including propane and ethane [13].

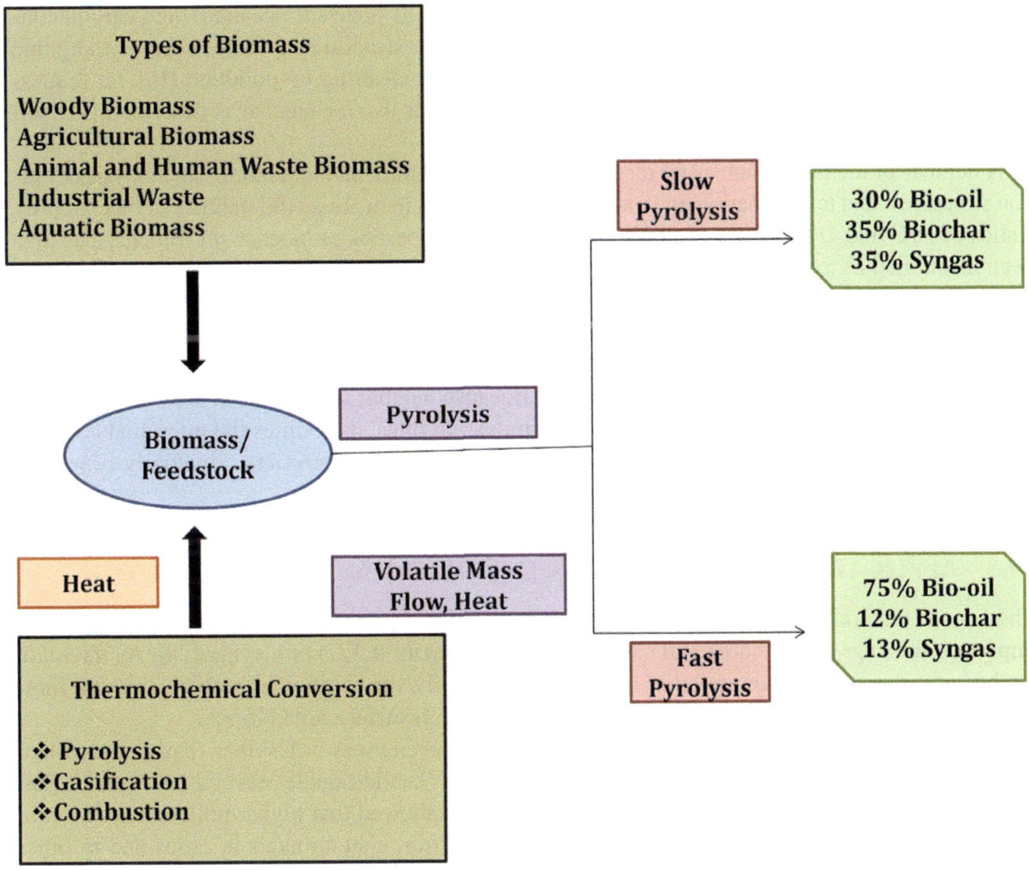

FIGURE 10.2 Biochar production process.

Biochar is commonly made from a variety of materials animal manure, herbaceous wood waste, crop residues, including sewage sludge, agricultural biomass, and municipal solid waste [14]. Petrol is a clean-burning fuel that is employed in energy recovery. The produced bio-oil is caustic to burn since it still contains water, organic acids, and oxygenated organics while having a high energy content [15]. Agriculture and the environment are two potential uses for biochar [16]. The production of liquid products must be reduced to maximize the yield of gas products and biochar; however, this can only be achieved by optimizing the pyrolysis parameters. More charcoal and gas are produced during slow pyrolysis (heating at a rate less than 100°C/min) than during fast pyrolysis (heating at a rate more than 300°C/min), which yields more liquid products.

The surface area and pore size of the resulting biochar are influenced by the temperature of the pyrolysis process and the feedstock. Lehmann [17] found that high temperatures (600–700°C) typically produce biochar with a substantial surface area. He contrasted 400°C biochar with 600°C biochar synthesized from plant debris and chicken litter. The scientists found that the surface area of biochar derived from plants was higher than that of biochar derived from chicken litter. When applied to soil with a high macropore count, a high biochar surface area is beneficial to the soil's capacity to hold water [18]. Water-holding capacity, pore volume, ash, surface area, pH, volatiles, and bulk density are significant biochar attributes that are influenced by feedstock pyrolysis and process factors [10]. High moisture content feedstocks are ideal for composting since the process needs some moisture (60–70%), but because pyrolysis needs more energy to remove water, they are frequently not suitable for biochar synthesis. The characteristics and efficacy of biochar are intimately linked [10].

Biochar has been shown to have a variety of uses due to its huge surface area, high carbon content, enriched surface functional groups, developed porous structures, and presence of inorganic chemicals: (a) reducing greenhouse gas emissions [19], (b) cleaning up pollution [16], (c) sequestering carbon [11], and (d) waste management [20]. Another use for biochar is described as a soil supplement [21].

A number of methods, such as (a) resistance to physicochemical degradation, (b) analysis of carbon structures, and (c) incubation and modeling, can be used to evaluate the stability of biochar [10]. Ratios of H/C and O/C are also helpful markers of the properties of biochar [8]. Biochar has also been mentioned as a possible source of carbon as well as an inorganic fertilizer and trace element source for bacteria. It has the potential to increase microbial activity and metabolism, impacting a number of microbial activities involved in the breakdown of organic matter [22]. Biochar has an effect on the microbial community because it lowers the bulk density of composting heaps while enhancing soil aeration and water retention [23]. Because biochar improves soil aeration and water retention while decreasing the bulk density of composting piles, it modifies the microbial community. Moreover, using biomass to create biochar as a soil supplement offers a mutually beneficial solution for the removal of agricultural waste and soil restoration [23].

10.3 APPLICATIONS OF BIOCHAR IN ENVIRONMENTAL REMEDIATION

The source materials, pyrolysis temperatures, and any post-production processing have a major impact on the huge surface area and cation exchange capacity (CEC) of biochar [24]. As the main method of reducing pollutant mobility in contaminated soils, the sorption of both organic and inorganic pollutants on biochar [25] depends critically on a high surface area [26].

The capacity of biochar-treated soil to sorb a particular species may be less than that of unmodified soil in soils with high levels of dissolved organic matter [27]. Additionally, several field studies [28] on larger sizes and longer times of biochar incorporation showed that higher pollutant sorption in biochar-modified soils may be advantageous in reducing soil contaminants in crops and in other areas (ponds, marshes, and so forth). Compared to activated charcoal, biochar is less expensive; causes less site disturbance during passive, in situ treatment; and may result in cost savings if polluted sediments can be removed without the need for expensive dredging or excavation.

10.3.1 Magnetic Biochar

Because of its extraordinary properties, which include a huge surface area, porosity, and an abundance of functional groups, biochar is becoming more and more recognized as a multifunctional material that may be employed in a range of fields, including water treatment and pollution reduction. However, its application is severely limited due to its low density and small particle size, which make it difficult to separate from water. A method that preserves the high adsorption capacity of biochar throughout wastewater treatment and enables its efficient separation from other sorbents is an interesting one. By pyrolyzing biomass with $FeCl_3$ and $FeCl_2$, Chen [12] created an effective method for the first time to overcome these difficulties by giving biochar magnetic properties that enabled solid–liquid separation.

Orange peel was the initial biomass material, while magnetite was the intended magnetic medium. Iron hydroxides were chemically precipitated on powdered orange peel, and the mixture was pyrolyzed to create magnetic biochar. The special magnetic biochars' ability to adsorb phosphate and organic pollutants increased significantly when compared to untreated biochars, suggesting that magnetic biochars may be a powerful sorbent type for use in environmental applications [29]. This discovery goes beyond magnetic separation. Numerous investigations have since been conducted to create magnetic biochars using a range of methodologies, biomass, and magnetic media.

High-quality magnetic biochars can be made using a variety of well-known procedures, including co-precipitation, thermal breakdown and/or reduction, and hydrothermal synthesis [12,29,30].

For instance, to precipitate the iron hydroxides on biochar, Han et al. [31] used a post-treatment that involved precipitating biochar with $FeSO_4$ and $FeCl_3$ at base conditions. After that, the biochar was rinsed and dried to produce biochar with magnetic iron oxides. This procedure was not the same as pretreatment, which involved pyrolysis after the co-precipitation of iron ions and biomass. Magnetic biochar can also be produced by hydrothermally treating impregnated biomass and heating it with a microwave [32], in addition to pyrolysis, which reduces or transforms ferric salt into magnetic iron oxides [12,33]. When pure metals like Fe and Co are applied to charcoal, magnetic biochars can be created [34]. Particularly unstable in the atmosphere is nanoscale zero-valent iron (nZVI), also referred to as pure nanoscale Fe. Easy separation is the main advantage of magnetic biochar over natural biochar. Because of this property, it would be possible to employ magnetic biochar directly to remove contaminants from aqueous solutions without worrying about the biochar particles dissolving and causing more pollution [35]. Magnetic biochar has the ability to remove a variety of contaminants through sorption, such as heavy metals, organic pollutants, and inorganic anions [29]. Furthermore, the catalytic reduction activity for heavy metals in magnetic biochar is bestowed or enhanced by magnetic components (nZVI, nano-$Fe3O_4$, and nano-FeO) [36].

It is important to remember that hazardous and poisonous materials could be added or produced during the magnetic biochar synthesis process.

10.3.2 2D Membrane Built on Biochar

Modern separation methods like membranes have been effectively applied to wastewater treatment, gas separation, and saltwater desalination. The development of membranes based on biochar is an additional intriguing approach to mitigate the possible dispersion and migration of minute biochar particles while maintaining exceptional efficacy, specifically strong adsorption capacity, in the context of wastewater treatment or soil restoration [38]. In mixed matrix membranes (MMMs), which consist of a dispersed particle phase (filler) and a continuous phase (matrix), biochar is most frequently used as a filler. Graphene and graphene oxide can readily be integrated into two-dimensional membranes through mutual interactions, while biochar is typically used as a filler in MMMs [36].

Since polymers are abundant in species and easy to turn into MMMs, they are the most widely used type of matrix material. At the moment, polysulfone (PSF), polydimethylsiloxane (PDMS), polyvinylidene fluoride (PVDF), and polyacrylonitrile (PAN) are the most often used polymers in biochar-based membranes (Table 10.1). As inorganic and organic fillers for MMMs, metal oxides, activated carbon, graphene, and metal-organic frameworks have all been studied [37].

TABLE 10.1
Biochar-Based 2D Membrane Fabrication and Its Applications

2D Membrane	Biochar Precursor	Polymer Added	Method	Use
	Wood biochar	Polyvinylidene fluoride		Water purification
	Activated pine biochar	Polyacrylonitrile	Electrospinning process	Aqueous chlotretracycline removal
	Sunflower seed hull biochar and biogas biochar	Polysulfone	Phase inversion and spinning	Heavy metal removal
	Wheat straw biochar	Waterborne polyacrylate	Distributed and dried	For the controlled release of fertilizers
	Bark biochar	Polydimethylsiloxane		Alcohol water separation

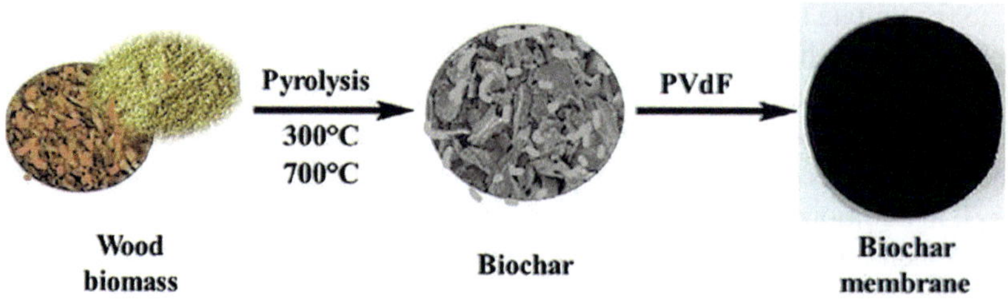

FIGURE 10.3 Two-dimensional membrane-making process from biochar.

Biochar has the ability to provide both distinctive features and cost reductions because it is more sustainable and inexpensive than graphene and other fillers. Based on polymer characteristics and manufacturing procedures, a biochar–polymer combination might be cast and electro-spun to make biochar-based membranes. PVDF and wood biochar were combined to create the casting solution, which was then consistently cast on glass plates with the aid of an automatic film applicator. The biochar composite membranes were produced following additional thermal phase inversion and washing to get rid of any remaining PVP. Because the biochar particles are evenly distributed throughout the membrane's surface and cross section, the resultant biochar composite membranes have a porous structure and great mechanical strength. Due to the synergistic effects of the PVDF and biochar in the composite membrane, as well as its high retention for *Escherichia coli* bacteria and ability to adsorb Rhodamine B, the water flux in biochar composite membranes was significantly bigger than that of pure PVDF membranes. After a straightforward physical cleaning, the biochar composite membranes' adsorption and sieving capacities can likewise be satisfactorily restored [38]. In various studies, researchers have employed casting to fabricate MMMs incorporating charcoal and polymers such as PAA and PDMS. Additionally, the biochar–polymer combination has been utilized to create membranes through electrospinning. For instance, a hollow PS membrane was produced through electrospinning, employing micron-sized biochar particles embedded in a PSF matrix. The results demonstrated significant enhancements in the membrane's structure, including surface area, porosity, and hydrophilicity, leading to improved water flux with the addition of biochar particles. Notably, He et al.'s 2017 study highlighted that the biochar/PSF membrane exhibited a high recovery rate exceeding 93% for both copper and lead, indicating superior filtration regeneration ability [39].

Finally, a biochar-based polymer membrane could be simply created by mixing biochar with polymers as a casting solution; the shape and characteristics could then be adjusted by utilizing other biochar's or polymers. The creation of membranes based on biochar may be aided by the improved adsorption capabilities of biochar particles. A 2D membrane-making technique using biochar is shown in Figure 10.3. Biochar-based 2D membranes were primarily used for water purification by adsorption and filtration. Additionally, the barrier effectively separated a mixture of alcohol and water [40].

10.3.3 3D Macrostructure Made of Biochar

To preserve the remarkable properties of biochar-based nanomaterials and expand their applicability across diverse fields, researchers have introduced macroscopic and recyclable biochar-containing structures. This strategic development not only aims to harness the unique qualities of biochar-based nanomaterials but also seeks to address potential environmental concerns associated with the release of small biochar particles. One safe and innovative approach involves the utilization of 3D biochar-based macrostructures, such as biochar aerogel, complementing the previously discussed 2D biochar-based membranes. Unlike biochar-based membranes, 3D biochar aerogel can be fabricated through various processes, offering more diverse and programmable topologies.

10.4 DIFFERENT USES OF BIOCHAR

The development of biochar-based membranes is illustrated in Figure 10.3 through a process involving casting a biochar–PVDF solution, electrospinning with charcoal and PAN [41], and filtration with a biochar and Reduced graphene oxide (rGO) solution. The conversion of the monolith into a porous 3D biochar macrostructure can be achieved through direct pyrolysis and hydrothermal treatment [42]. Biomass tar production is accomplished by pyrolyzing natural wood biomass in a high-performance, low-tortuosity flow-through reactor. The 3D biochar macrostructure exhibits unique features, including aligned and deformed open channels (5–60 micrometers in diameter) facilitating effective gas flow, mixing, and gas reactions. Ni nanoparticles, sized 20–60 nm, are anchored on channel walls by a thin layer of graphitic carbon, preventing coking or sintering [43].

HTC successfully produces carbonaceous hydrogels and aerogels with sponge-like characteristics using shaped biomass. HTC, applied to various materials [44], demonstrates its potential for separating water and oil, making it a viable technique for 3D biochar macrostructure synthesis from sustainable biomass. An alternative approach, the "bottom-up" process, forms the 3D biochar macrostructure from easily obtained cellulose nanofiber. As cellulose is abundant, sustainable, and ecologically friendly, this method is adaptable across a broad range of biomass sources.

Moreover, the alteration of the form, structure, and characteristics of the 3D biochar macrostructure is achievable through various raw materials and adjustments in synthesis procedures, such as carbonization, freezing dry, and cross-linking reactions. Bacterial cellulose, an abundant biomass material produced through microbial fermentation, consists of interconnected networks of cellulose nanofibers. When biochar is mixed into a composite, 3D biochar macrostructures can be formed similarly to biochar-based membranes. Combining fine biochar particles with cross-linking agents can result in a porous hydrogel, which, when frozen dry, produces a 3D biochar aerogel. The fabricated 3D biochar macrostructure demonstrates significant potential for remediating heavy metals and organic pollutants in water and soil, thanks to its rich porosity, water retention, exceptional hydrophilicity, and ability to maintain the gel shape without collapsing in water or soil. In a separate study, biochar derived from packaging waste underwent grinding, sieving, and fusion with recovered Polyethylene terephthalate (PET). Tensile biochar/PET composite filaments, measuring 1.75 mm, were 3D printed using a melt extruder [45]. These filaments can be utilized to construct a three-dimensional biochar macrostructure.

10.4.1 Biofilters Made of Biochar

The concept of biofilters, or artificial wetlands, involves utilizing substrates such as soil, sand, and gravel, alongside aquatic plants and microorganisms, for the effective removal of pollutants from wastewater. This engineered filtration system is described by Li [35,36]. The significance of the support substrate in biofilters for plant and microbe growth and development, as well as its role in removing impurities through processes like filtration, adsorption, chemical transformation, and biological transformation, is emphasized in studies such as [46]. The careful selection of substrate materials becomes paramount to meet the increasing demand for eco-treatment systems, ensuring biofilters excel in pollutant removal. The strength of interactions within these systems significantly influences their behavior and functionality. A detailed comparison of biochar with other substrates for biofilters is provided in Table 10.2. Biochar emerges as a promising substrate due to its affordability, widespread availability, and ability to efficiently remove a diverse range of pollutants while fostering the growth of microorganisms and plants [47]. These attributes position biochar as a viable and advantageous option for biofilter applications.

One effective strategy to mitigate the limitations of biochar, such as small particle size and low mechanical strength, while simultaneously enhancing the performance of biofilters is through its incorporation into engineered eco-infiltration systems like biofilters or constructed wetlands.

TABLE 10.2
The Classic Modification Ways of Biochar

Feed Stock	Reagent	Treatment Methods	Applications
Herb residue	$FeCl_3.6H_2O$ and $FeSO_4.7H_2O$	Added to mix solutions and then stirred under N_2 atmosphere	Chromium removal
Precursor material	Alkaline	Soaking and stirring	Contaminant removal
Precursor material	Steam activation	Biochar activated by steam	Contaminant removal
Coconut	Ammonia	Mixed and shook	Remove lead
Chestnut shell	$FeCl_2$ and $FeCl_3$	Use $FeCl_3$ and $FeCl_2$ and added to biochar	Remove as V from industry waste
Precursor material	Acidic modification	Soaking and second pyrolysis	Contaminant removal
Coconut	H_2O_2	Mixed and shook	Remove Pb^{2+}
Coconut	HNO_3	Mixed and shook	Remove Pb^{2+}
Cotton straw	NH_4Cl, NH_4Br, NH_4I	Separately stirred	Mercury removal
Cotton straw	Deionized water	Activated by microwave for 4 hours	Mercury removal
Cotton straw	Microwave	Activated by microwave for 4 hours	Mercury removal

This approach leverages biochar's multifunctionality, including adsorption, water and nutrient retention, and serving as an electron shuttle for microbes [48]. In the study conducted by Chen [48], biochar-amended biofilters were tested for the treatment of stormwater containing 200 g/L of bisphenol A. The experimental design included batch sorption, fixed-bed column experiments, and biochar-amended biofilters. Biochar-based biofilters can be categorized into vertical and horizontal flow configurations, where biochar is utilized as a pure layer or module, either independently or mixed with other geomedia. The concept of a "biochar-based functional module" was proposed as a "functional filter" capable of efficiently removing and accumulating nutrients, metal ions, or organic pollutants through electrochemical or bioelectrochemical mechanisms. This is attributed to biochar's strong adsorption to contaminants and its electrical properties, such as electron shuttles and conductivity [49,50]. An example of this application is the ferric-carbon micro-electrolysis biochar-based biofilter, which demonstrated effective removal of nitrogen and phosphorus [51]. The micro-electrolysis process allowed for the addition of extra electrons to denitrification, and the galvanic corrosion of sacrificial anodes enhanced phosphorus removal through flocculation, adsorption, and precipitation [51]. It's important to note that these discoveries were made in laboratory experiments, primarily focusing on nutrient removal. Further field studies, especially concerning metal ions or organic pollutants, are crucial to validate and confirm the broader beneficial effects of biochar.

10.5 ENVIRONMENTAL REMEDIATION AFTER HEAVY METAL CONTAMINATION

Heavy metals pose significant risks to human health, food safety, and ecosystems due to their toxicity and potential accumulation [52]. To mitigate these hazards, extensive studies have focused on extracting and transforming heavy metals into oxidizable and residual components. Biochar has emerged as a valuable tool in altering the availability and solubility of heavy metals, thereby influencing their behavior in the environment [53]. Additionally, biochar demonstrates the potential to remediate ecosystems contaminated with heavy metals by reducing their concentrations and quantities. Research indicates that biochar effectively extracts heavy metals from sewage and immobilizes them in polluted soil, demonstrating its remediation capabilities [53]. Both historical and recent studies have consistently shown that biochar can immobilize heavy metals and mitigate their toxicity. Bogusz et al. [54] highlight various factors influencing heavy metal adsorption and immobilization on biochar, including biochar properties, environmental temperature, pH value, biochar dosage, concentration of heavy metal ions, inorganic ions, and the types of heavy metal ions. The adsorption capacity of biochar for heavy metals, such as Pb^{2+}, Cd^{2+}, and Ni^{2+}, in aqueous solutions is influenced by factors like pH levels and biochar dosage. The presence of light metal ions, such as K+ or Na+, can limit the adsorption capacity of biochar for certain heavy metal ions. Furthermore, specific biochar types exhibit varying adsorption capabilities for different heavy metals. For instance, research on sesame straw biochar revealed different adsorption orders for mono-metal and multi-metal adsorption isotherms [55]. In mono-metal adsorption, the order was Pb > Cd > Cr > Cu > Zn, while in multi-metal adsorption, the order was Pb > Cu > Cr > Zn > Cd [55]. These findings underscore the importance of considering biochar characteristics and environmental conditions in optimizing its effectiveness for heavy metal adsorption and immobilization.

The utilization of biochar has been shown to decrease the phyto availability and extractable fraction of heavy metals. Liang et al. [56] have identified adsorption and precipitation as the two primary mechanisms responsible for stabilizing heavy metals in biochar. The immobilization of heavy metals by biochar involves various processes, including chemical precipitation, surface complexation, electrostatic interaction, and adsorption [57]. Different mechanisms come into play for specific heavy metals. For instance, complexation and electrostatic interaction are predominant mechanisms for as (V), while cation exchange, complexation, and precipitation play crucial roles for Cd and Pb. Hg2+ immobilization involves cation exchange, precipitation, and surface complexation. In the case of Cr (VI), surface complexation, chemical reduction, and electrostatic interaction are mechanisms observed with sugar beet tailing biochar. The interaction between biochar and heavy metals is influenced by several mechanisms. Electrostatic interaction, occurring when biochar surfaces carry negative charges, leads to the electrostatic adsorption of heavy metal ions [58]. Physical adsorption relies on the pore structure and specific surface area of biochar to provide adsorption sites. Cation exchange occurs as Ca2+, Na+, Mg2+, and K+ cations interact with heavy metals on the charcoal surface [59]. The formation of surface complexes happens when heavy metal ions react with functional groups (amide, carboxylic, carbonyl, alcoholic, ether, and hydroxyl) on the biochar surface. Additionally, biochar has the potential to reduce the bioavailability of heavy metals by increasing soil organic matter and enhancing microbial activity, further contributing to the overall remediation process [56].

10.6 REMOVING ORGANIC TOXINS

Biochar has demonstrated remarkable adsorption potential for various organic pollutants, including herbicides, dyes, pesticides, antibiotics, and more. For instance, the application of biochar resulted in impressive elimination efficiencies of 97.1% for metronidazole and 96.4% for dimetridazole, showcasing its potential utility [60]. Different biochar types exhibit specific adsorption capabilities. Hydrothermal biochar, for example, has shown excellent adsorption capacity for pesticides [60]. Korean cabbage and rice straw biochar have exhibited significant adsorption capacity for cationic dyes. The fundamental mechanisms behind biochar's adsorption of organic contaminants differ slightly from those for heavy metals. Two primary mechanisms include (a) partitioning mechanisms, where partially carbonized organic matter on the biochar surface adsorbs nonpolar organic pollutants; and (b) hydrophobic effects, where hydrophobic organic compounds bind with hydrophobic groups on the biochar surface [61]. These mechanisms collectively contribute to the effective adsorption of a diverse range of organic pollutants by biochar.

10.6.1 Phosphorous and H_2S Adsorption

Phosphorus removal with biochar appears to be promising [62]. Phosphorus can be absorbed more effectively when biochar has high levels of ferrum (Fe), calcium (Ca), aluminum (Al), and magnesium (Mg). According to the study, biochar has nearly seven times the activated carbon's ability to adsorb phosphorus (15 mg/g) [63].

Studies on ferric oxide hydrate biochar, for example, have shown a substantial removal capability for phosphorus in the range of 51.71–56.15 mg/g. Using biochar to increase soil fertility makes sense because it has absorbed the phosphorus found in sewage. Applications of biochar fully utilize available resources and deal with the grave problem of eutrophication in water bodies. Phosphorus is physically adsorbed into the pores and surface regions of biochar by the phosphate radical, which forms a coordination compound with the help of functional groups and magnesium oxide particles. This process is known as surface complexation. The findings revealed that biochar performed very well in terms of H_2S removal, with a maximum H_2S removal capacity of 60 mg/g.

10.7 IMPROVING THE SOIL'S STRUCTURE

Soil pH levels can be raised, and CEC can be enhanced by biochar. Because biochar has a high level of organic carbon, it can increase the amount of organic matter and humus in the soil as well as improve soil nutrients. Additionally, because of its huge specific surface area and pore structure, biochar may make use of the nutrients, airflow, and water-holding capacity of the soil, giving soil microorganisms a better home and place to live. It encourages microbial life in the soil to be active. The findings imply that biochar plays a crucial role in the creation and maintenance of soil aggregates.

10.7.1 Process Improvement for Composting

Biochar has shown considerable environmental benefits when used as a bulking ingredient in aerobic composting. This is strongly associated with the characteristics of biochar, including its high porosity, high adsorption capacity, high CEC, and surface functional groups [50]. Compost and charcoal work well together [64]. Aerobic composting requires ecologically friendly biochar with distinct physicochemical features. It enhances the chemical and physical characteristics of the compost mixture, lowers greenhouse gas emissions, reduces ammonia emissions, and reduces total nitrogen loss. It also speeds up the composting maturity. These modifications are advantageous for cleaning up a dirty environment [64].

It is encouraging that the stabilization, safety, and resource efficiency of dredged sediment are benefited by the use of biochar as the aerobic composting bulking agent. Despite the existence of short-term studies, biochar cannot yet be effectively applied to challenging and real environment rehabilitation. Focusing future research on long-term studies in the in situ environment is crucial to maximizing the potential benefits of biochar. To fully comprehend the underlying application mechanisms of biochar, additional research is also necessary.

10.7.2 Altered Biochar

To enhance our living environment, addressing soil and water pollution is crucial, and biochar has emerged as a versatile solution. The efficacy of biochar in pollutant removal is attributed to its large specific surface area, porous structure, and surface functional groups. However, to maximize its effectiveness in polluted areas, it is essential to modify the properties of the original biochar. Researchers have explored various approaches, categorized into feedstock pretreatment and primitive biochar modification, to enhance biochar characteristics. Methods such as stirring, soaking, heating, and second pyrolysis have been employed in this modification process. Tables 10.1 and 10.2

illustrate diverse modification approaches, emphasizing the multitude of processes involved. These modifications result in increased specific surface area, porous texture, oxygen-containing functional groups, minerals, metallic particles [65], hydrophobic and hydrophilic properties, surface charge, and, crucially, adsorption ability. Modified biochar outperforms traditional adsorbents and finds extensive use in adsorbing contaminants like phosphorus, organic pollutants, and heavy metals.

10.8 CONCLUSION AND FUTURE PROSPECTUS

This chapter presented a summary of recent biochar developments in structural characterization, reactivity research, functionalization and device development, and environmental field applications. The structural characteristics of biochar, both microscopic and macroscopic, were examined for the first time. Nano, dissolved, and bulk biochar are among the different-sized particles that make up macroscopic biochar. The multilayer architectures of molecules, phases, elements, and surface chemistry were conjectured from a microscopic perspective. Biochar's reactivity, which encompasses both chemical and biological activities like sorption, redox reactions, and catalysts, as well as interactions with bacteria, is significantly influenced by its structure and characteristics. The structure of biochar and its reactivity are inextricably linked. It is particularly stated how biochar's quaternary structure impacts chemical activity. The functionalization and device of biochar are underlined to address the disadvantages of direct environmental usage of virgin biochar particles (low mechanical strength, difficult separation, simple transportation, and possible toxicity). Magnetic biochar, biochar-based 2D membranes and 3D macrostructures, immobilized microorganisms on biochar, and biochar-based biofilters are discussed, as well as their synthesis methodologies and environmental applications. Innovative biochar-based products and materials provide new perspectives on challenging environmental concerns. It is still necessary to fully comprehend the structure, reactivity, functionalization, and uses of biochar. Despite some sincere efforts, no undisputed molecular structural model for biochar has yet been presented [57]. This makes it difficult to fully comprehend biochar's reactivity and functionalize it for use in environmental applications.

Numerous issues in science and technology still need to be resolved. Further investigation into the domains delineated in Figure 10.4 is imperative to propel forward our comprehension of the interplay of structure, reactivity, functionalization, application, and the development of biochar-based materials and technologies for environmental applications.

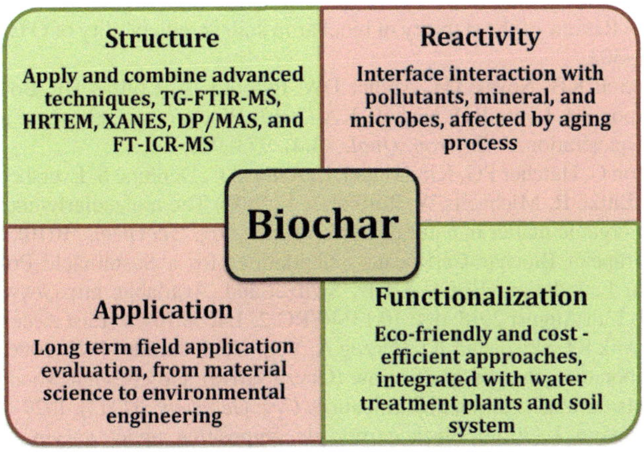

FIGURE 10.4 Applications of biochar, its functionality, and structure.

1. There are two approaches to unraveling the multilevel and multiple-level structure of biochar. The first is to understand the processes that result in the formation of biochar during the pyrolysis of complex biomass, and the second is to integrate and use cutting-edge technologies to analyze the structure of biochar. To appreciate the macroscopic and microscopic aspects of biochar's physical and chemical composition, sophisticated instruments and a suitable combination of techniques should be used.
2. It is critical to thoroughly investigate biochar's sorption and catalytic performance toward a variety of organic compounds and metal ions, as well as the underlying mechanisms. Meanwhile, it is critical to thoroughly understand the redox property and how it affects both direct and indirect reactions with contaminants via interactions between microorganisms and biochar, such as electron transfer between bacteria, contaminants, and biochar (shuttle). Reactivity and the structural characteristics of biochar are strongly related, so it is important to consider how aging affects reactivity.
3. With the intention of protecting the environment, the large-scale production of biochar by design should be enhanced, as has been suggested for years [66]. It's crucial to use eco-friendly and economical methods to transform and functionalize biochar for a particular use. Future research should prioritize incorporating functionalized biochar in water treatment facilities and soil pollution remediation. It should be highlighted that the majority of research on biochar-based products or systems is currently conducted at the lab scale; therefore, attention should be directed to more field investigations. Environmental engineering should be used in biochar research instead of material science or environmental science. More research will make it possible to understand the long-term application performance, underlying mechanisms, and economic analyses of developed applications.

Finally, more attention should be drawn to the interaction and structure of biochar and its reactivity, functionalization, and usage. It necessitates a complete understanding of biochar's reactivity, exact functioning, and long-term field application evaluation. We believe that intelligently functionalized biochar-based goods and equipment will provide the greatest environmental benefits and agricultural.

REFERENCES

1. Boehm HP (1994) Some aspects of the surface-chemistry of carbon-blacks and other carbons. *Carbon*, 32(5), 759–769.
2. Goldberg ED (1985) *Black Carbon in the Environment: Properties and Distribution*. John Wiley & Sons Inc, New York.
3. Spokas KA (2010) Review of the stability of biochar in soils: predictability of O:C molar ratios. *Carbon Manag.* 1(2), 289–303.
4. Spokas KA, Cantrell KB, Novak JM, Archer DW, Ippolito JA, Collins HP, Boateng AA, Lima IM, Lamb MC, Mcaloon AJ, Lentz RD, Nichols KA (2012) Biochar: a synthesis of its agronomic impact beyond carbon sequestration. *J. Environ. Qual.*, 41(4), 973–989.
5. Hedges JI, Eglinton G, Hatcher PG, Kirchman DL, Arnosti C, Derenne S, Evershed RP, Kogel-Knabner I, de Leeuw JW, Littke R, Michaelis W, Rullkotter J (2000) The molecularly-uncharacterized component of nonliving organic matter in natural environments. *Org. Geochem.*, 31(10), 945–958.
6. EBC (2012) European Biochar Certificate - Guidelines for a Sustainable Production of Biochar. European Biochar Foundation (EBC), Arbaz, Switzerland. Available: https://www.european-biochar.org/en/download. 13th August 2018; doi: 10.13140/RG.2. 1.4658.7043) (Last Accessed: 04.11.2018).
7. Hammes K, Smernik RJ, Skjemstad JO, Herzog A, Vogt UF, Schmidt MWI (2006) Synthesis and characterisation of laboratory-charred grass straw (Oryza sativa) and chestnut wood (Castanea sativa) as reference materials for black carbon quantification. *Org. Geochem.* 37 (11), 1629–1633.
8. Nguyen BT, Lehmann J (2009) Black carbon decomposition under varying water regimes. *Org. Geochem.* 40, 846–853.

9. Werle S, Wilk RK (2010) A review of methods for the thermal utilization of sewage sludge: the Polish perspective. *Renew. Energy* 35, 1914–1919.
10. Meyer S, Glaser B, Quicker P (2011) Technical, economical, and climate-related aspects of biochar production technologies: a literature review. *Environ. Sci. Technol.* 45 (22), 9473–9483.
11. Crombie K, Mašek O, Sohi SP, Brownsort P, Cross A (2013) The effect of pyrolysis conditions on biochar stability as determined by three methods. *Global Biol. Bioenergy* 5, 122–131.
12. Hadi P, Xu M, Ning C, Lin CSK, McKay G (2015) A critical review on preparation, characterization and utilization of sludge-derived activated carbons for wastewater treatment. *Chem. Eng. J.* 260, 895–906.
13. Inguanzo M, Dominguez A, Menendez JA, Blanco CG, Pis JJ (2002) On the pyrolysis of sewage sludge: the influence of pyrolysis temperature on biochar, liquid and gas fractions. *J. Anal. Appl. Pyrolysis* 63, 209–222.
14. Ahmad M, Rajapaksha AU, Lim JE, Zhang M, Bolan N, Mohan D, Vithanage M, Lee SS, Ok YS (2014) Biochar as a sorbent for contaminant management in soil and water: a review. *Chemosphere* 99, 19–33.
15. Sohi SP (2012) Carbon storage with benefits. *Science* 338 (6110), 1034–1035.
16. Uchimiya M, Chang S, Klasson KT (2011) Screening biochars for heavy metal retention in soil: role of oxygen functional groups. *J. Hazard. Mater.* 190, 432–441.
17. Lehmann J, Rillig MC, Thies J, Masiello CA, Hockaday WC, Crowley D (2011a) Biochar effects on soil biota: a review. *Soil Biol. Biochem.* 43, 1812–1836.
18. Glaser B, Lehmann J, Zech W (2002) Ameliorating physical and chemical properties of highly weathered soils in the tropics with charcoal - a review. *Biol. Fertil. Soils* 35, 219–230.
19. Agyarko-Mintah E, Cowie A, Singh BP, Joseph S, Van Zwieten L, Cowie A, Harden S, Smillie R (2017) Biochar increases nitrogen retention and lowers greenhouse gas emissions when added to composting poultry litter. *Waste Manag.* 61, 138–149.
20. Inyang M, Dickenson E (2015) The potential role of biochar in the removal of organic and microbial contaminants from potable and reuse water: a review. *Chemosphere* 134, 232–240.
21. Camps M, Tomlinson T (2015) The Use of Biochar in Composting. International Biochar Initiative. https://www.biochar-international.org/sites/default/files/Compost_biochar_IBI_ final.pdf (Last accessed: 27.08.2018).
22. Anderson CR, Condron LM, Clough TJ, Fiers M, Stewart A, Hill RA, Sherlock RR (2011) Biochar induced soil microbial community change: implications for biogeochemical cycling of carbon, nitrogen and phosphorus. *Pedobiologia* 54, 309–320.
23. Diaz LF, De Bertoldi M, Bidlingmaier W (2007) *Compost Science and Technology*. Elsevier, New York.
24. Ahmed A, Pakdel H, Roy C, Kaliaguine S (1989) Characterization of the solid residues of vacuum pyrolysis of populus-tremuloides. *J. Anal. Appl. Pyrol.* 14(4), 281–294.
25. Boehm HP (1994) Some aspects of the surface-chemistry of carbon-blacks and other carbons. *Carbon* 32(5), 759–769.
26. Beesley L, Marmiroli M, Pagano L, Pigoni V, Fellet G, Fresno T, Vamerali T, Bandiera M, Marmiroli N (2013) Biochar addition to an arsenic contaminated soil increases arsenic concentrations in the pore water but reduces uptake to tomato plants (Solanum lycopersicum L.). *Sci. Total Environ.* 2013, 454–455.
27. Cabrera A, Cox L, Spokas KA, Celis R, Hermosin MC, Cornejo J, Koskinen WC (2011) Comparative sorption and leaching study of the herbicides fluometuron and 4-chloro-2-methylphenoxyacetic acid (mcpa) in a soil amended with biochars and other sorbents. *J. Agr. Food Chem.* 59 (23), 12550–12560.
28. Chai YZ, Currie RJ, Davis JW, Wilken M, Martin GD, Fishman VN, Ghosh U (2012) Effectiveness of activated carbon and biochar in reducing the availability of polychlorinated dibenzo-p-dioxins/dibenzofurans in soils. *Environ. Sci. Technol.* 46 (2), 1035–1043.
29. Chen B, Ding J (2012) Biosorption and biodegradation of phenanthrene and pyrene in sterilized and unsterilized soil slurry systems stimulated by Phanerochaete chrysosporium. *J. Hazard. Mater.* 229–230, 159–169.
30. Thakur, P., Thakur, A. (2022). Introduction to nanotechnology. In: Thakur, A., Thakur, P., Khurana, S. P. (eds) *Synthesis and Applications of Nanoparticles*. Springer, Singapore. https://doi.org/10.1007/978-981-16-6819-7
31. Han Z, Sani B, Mrozik W, Obst M, Beckingham B, Karapanagioti HK, Werner D (2015) Magnetite impregnation effects on the sorbent properties of activated carbons and biochars. *Water Res.* 70, 394–403.
32. Cheng S, Liu F, Shen C, Zhu C, Li A (2019) A green and energy-saving microwave-based method to prepare magnetic carbon beads for catalytic wet peroxide oxidation. *J. Clean Prod.* 215, 232–244.

33. Zhang H, Xue G, Chen H, Li X (2018a) Magnetic biochar catalyst derived from biological sludge and ferric sludge using hydrothermal carbonization: preparation, characterization and its circulation in Fenton process for dyeing wastewater treatment. *Chemosphere* 191, 64–71.
34. Wang S, Gao B, Li Y, Creamer AE, He F (2017) Adsorptive removal of arsenate from aqueous solutions by biochar supported zerovalent iron nanocomposite: batch and continuous flow tests. *J. Hazard Mater.* 322, 172–181.
35. Li XP, Wang CB, Zhang JG, Liu JP, Liu B, Chen GY (2020) Preparation and application of magnetic biochar in water treatment: a critical review. *Sci. Total Environ.* 711, 134847.
36. Liu G, Zheng H, Jiang Z, Zhao J, Wang Z, Pan B, Xing B (2018a) Formation and physicochemical characteristics of nano biochar: insight into chemical and colloidal stability. *Environ. Sci. Technol.* 52 (18), 10369–10379.
37. Lewis J, Miller M, Crumb J, Al-Sayaghi M, Buelke C, Tesser A, Alshami A (2019) Biochar as a filler in mixed matrix materials: synthesis, characterization, and applications. *J. Appl. Polym. Sci.* 136 (41), 48027.
38. Ghaffar A, Zhu X, Chen B (2018) Biochar composite membrane for high performance pollutant management: fabrication, structural characteristics and synergistic mechanisms. *Environ. Pollut.* 233, 1013–1023.
39. He J, Song Y, Chen JP (2017) Development of a novel biochar/PSF mixed matrix membrane and study of key parameters in treatment of copper and lead contaminated water. *Chemosphere* 186, 1033–1045.
40. Lan Y, Yan N, Wang W (2016) Application of PDMS pervaporation membranes filled with tree bark biochar for ethanol/water separation. *RSC Adv.* 6 (53):47637–47645.
41. Taheran M, Naghdi M, Brar SK, Knystautas E, Verma M, Surampalli RY, Valero JR (2016) Development of adsorptive membranes by confinement of activated biochar into electrospun nanofibers. *Beilstein J. Nanotechnol.* 7, 1556–1563.
42. Yang H, Yan R, Chen H, Lee DH, Zheng C (2007) Characteristics of hemicellulose, cellulose and lignin pyrolysis. *Fuel* 86 (12–13), 1781–1788.
43. Wang Y, Sun G, Dai J, Chen G, Morgenstern J, Wang Y, Kang S, Zhu M, Das S, Cui L, Hu L (2016a) A high-performance, low-tortuosity wood-carbon monolith reactor. *Adv. Mater.* 29 (2), 1604257.
44. Yin A, Xu F, Zhang X (2016) Fabrication of biomass-derived carbon aerogels with high adsorption of oils and organic solvents: effect of hydrothermal and post-pyrolysis processes. *Materials (Basel)* 9 (9), 758. https://doi.org/10.3390/ma9090758
45. Idrees M, Jeelani S, Rangari V (2018) Three-dimensional-printed sustainable biochar-recycled PET composites. *ACS Sustain Chem. Eng.* 6 (11), 13940–13948.
46. Grebel JE, Mohanty SK, Torkelson AA, Boehm AB, Higgins CP, Maxwell RM, Nelson KL, Sedlak DL (2013) Engineered infiltration systems for urban stormwater reclamation. *Environ. Eng. Sci.* 30 (8), 437–454.
47. Mohanty SK, Valenca R, Berger AW, Yu IKM, Xiong X, Saunders TM, Tsang DCW (2018) Plenty of room for carbon on the ground: potential applications of biochar for stormwater treatment. *Sci. Total Environ.* 625, 1644–1658.
48. Chen T, Zhang Y, Wang H, Lu W, Zhou Z, Zhang Y, Ren L (2014) Influence of pyrolysis temperature on characteristics and heavy metal adsorptive performance of biochar derived from municipal sewage sludge. *Bioresour. Technol.* 164, 47–54.
49. Wu S, Wu H (2019) Incorporating biochar into wastewater eco-treatment systems: popularity, reality, and complexity. *Environ. Sci. Technol.* 53 (7), 3345–3346.
50. Punia P, Bharti MK, Chalia S, Dhar R, Ravelo B, Thakur P, Thakur A (2021) Recent advances in synthesis, characterization, and applications of nanoparticles for contaminated water treatment-A review. *Ceram. Inter.* 47 (2), 1526–1550.
51. Shen Y, Zhuang L, Zhang J, Fan J, Yang T, Sun S (2019) A study of ferric-carbon micro-electrolysis process to enhance nitrogen and phosphorus removal efficiency in subsurface flow constructed wetlands. *Chem. Eng. J.* 359, 706–712.
52. Galal TM, Shehata HS (2015) Impact of nutrients and heavy metals capture by weeds on the growth and production of rice (Oryza sativa L.) irrigated with different water sources. *Ecol. Indic.* 54, 108–115.
53. Lu K, Yang X, Gielen G, Bolan N, Ok YS, Niazi NK, Xu S, Yuan G, Chen X, Zhang X, Liu D, Song Z, Liu X, Wang H (2017) Effect of bamboo and rice straw biochars on themobility and redistribution of heavy metals (Cd, Cu, Pb and Zn) in contaminated soil. *J. Environ. Manag.* 186 (Part 2), 285–292.
54. Bogusz A, Oleszczuk P, Dobrowolski R (2017) Adsorption and desorption of heavy metals by the sewage sludge and biochar-amended soil. *Environ. Geochem. Health* 41 (4), 1663–1674.

55. Park J, Ok YS, Kim S, Cho J, Heo J, Delaune RD, Seo D (2016) Competitive adsorption of heavy metals onto sesame straw biochar in aqueous solutions. *Chemosphere* 142, 77–83.
56. Liang Y, Cao X, Zhao L, Arellano E (2014) Biochar- and phosphateinduced immobilization of heavy metals in contaminated soil and water: implication on simultaneous remediation of contaminated soil and groundwater. *Environ. Sci. Pollut. Res.* 21, 4665–4674.
57. Xiao X, Chen B, Zhu L, Schnoor JL (2017a) Sugar cane-converted graphene-like material for the super-high adsorption of organic pollutants from water via coassembly mechanisms. *Environ. Sci. Technol.* 51 (21), 12644–12652.
58. Shi K, Qiu Y, Ben L, Stenstrom MK (2016) Effectiveness and potential of straw- and wood-based biochars for adsorption of imidazolium-type ionic liquids. *Ecotoxicol. Environ. Saf.* 130. 155–162.
59. Sewu DD, Boakye P, Woo SH (2017) Highly efficient adsorption of cationic dye by biochar produced with Korean cabbage waste. *Bioresour. Technol.* 224, 206–213.
60. Sun K, Gao B, Ro KS, Novak JM, Wang Z, Herbert S, Xing B (2012) Assessment of herbicide sorption by biochars and organic matter associated with soil and sediment. *Environ. Pollut.* 163, 167–173.
61. Gu J, Zhou W, Jiang B, Wang L, Ma Y, Guo H, Schulin R, Ji R, Evangelou MWH (2016) Effects of biochar on the transformation and earthworm bioaccumulation of organic pollutants in soil. *Chemosphere* 145, 431–437.
62. Dai L, Fan L, Liu Y, Ruan R, Wang Y, Zhou Y, Zhao Y, Yu Z (2017a) Production of bio-oil and biochar from soapstock via microwaveassisted co-catalytic fast pyrolysis. *Bioresour. Technol.* 225, 1–8.
63. Antunes E, Schumann J, Brodie G, Jacob MV, Schneider PA (2017) Biochar produced from biosolids using a single-mode microwave: characterisation and its potential for phosphorus removal. *J. Environ. Manag.* 196, 119–126.
64. Wu XL, Wen T, Guo HL, Yang S, Wang X, Xu AW (2013a) Biomassderived sponge-like carbonaceous hydrogels and aerogels for supercapacitors. *ACS Nano* 7 (4), 3589–3597.
65. Sizmur T, Fresno T, Akgul G, Frost H, Moreno-Jimenez E (2017) Biochar modification to enhance sorption of inorganics from water. *Bioresour. Technol.* 246, 34–47.
66. Abiven S, Schmidt MWI, Lehmann J (2014) Biochar by design. *Nat. Geosci.* 7 (5), 326–327.

11 MXenes
Synthesis and Their Biomedical Applications

Nisha Yadav, Ritika Gera, and Chandan Kumar Mandal
Amity University Haryana

Praveen Kumar and Arun Kumar
Doon University

Gyaneshwar Kumar Rao
Amity University Haryana

11.1 INTRODUCTION

Two-dimensional (2-D) nanomaterials with ultrathin layer structures such as hexagonal boron nitrides (h-BN), transition metal dichalcogenides (TMDCs), palladium (Pd) nano-sheets, graphene and its derivatives, transition metal oxides (TMOs), black phosphorus, and transition metal carbides and nitrides also known as MXenes have been widely investigated due to their unique properties [1]. MXenes are characterized by their extreme aspect ratios and exist in the form of a few atomic layers. Typically, the bulk three-dimensional precursors of such materials have a layered structure, where these layers are held together by weak van der Waals forces. The precursors of these materials can be exfoliated physically or chemically, which results in the formation of few-layer MXenes, making them suitable for numerous applications in biomedicine, energy, and the environment [2]. These materials typically consist of transition elements along with nitride, carbonitride, sulfide, and carbide, having a composition of type $M_{n+1}X_nT_x$ [3] where "M" represents the metal atoms such as Nb, Mo, Ti, and V, while T_x represents the surface-ending groups such as OH, F, and O. The "X" in the formula refers to either nitrogen or carbon, with n ranging from 1 to 3 [4]. MXenes have unique properties and consist of three or more atomic layers, which are distinct from their 3-D parent predecessors, that is, MAX phases. More than 60 pure-form MAX phases have already been reported in the literature. The MAX phases are layered ternary nitrides and carbides with a chemical formula of $M_{n+1}AX_n$, where "M" stands for the early transition metal; "A" could be group IIIA or IVA elements such as Al, Ge, Si, or Sn; and "X" is carbon or nitrogen, with n ranging from 1 to 3 [5]. The MAX phases can undergo chemical delamination, resulting in the formation of MXenes after the removal of A. The removal of the A layer from MAX phases is known as the chemical exfoliation approach [3]. The first MXene phase, $Ti_3C_2T_z$, was synthesized in 2011 by dissolving Ti_3AlC_2 in concentrated hydrofluoric acid (HF) for several hours at room temperature [6,7].

MAX phases have covalent, ionic, or metallic bonds between metal M and X, while the bonds between M and A are purely metallic and are weaker. These weaker bonds allow the A elements to undergo selective chemical etching without affecting the M–X bond [8]. The exposure of transition metals in $M_{n+1}X_n$ results in the development of surface functional groups (T_x), leading to the formula $M_{n+1}X_nT_x$, which creates a negatively charged surface, leading to the formation of stable dispersions [7]. The ultrathin structure of MXenes and their enormous surface area allow the creation of a large active site that interacts with molecules. The nature of the "M" transition metals and the

DOI: 10.1201/9781003502692-11

chemistry of the environment during etching and post-treatment methods predict the composition of surface functional groups [8]. The highly organized structure of MXenes makes it possible to anticipate their features theoretically through computational analysis. MXenes have been utilized for energy storage applications since they have high affinity as energy storage devices and in catalysis since they act as catalyst support. These properties in MXenes have emerged due to their unique structure [9]. A single-layer MXene stack typically has a thickness of less than 1 nm compared to the few atomic layers, which can have dimensions ranging from nanometers to micrometers depending on the procedure used for their synthesis [10].

MXenes have been classified into four major categories based on their structures: (a) MXenes having single d-block metal: examples Ti_2C and Nb_4C_3; (b) solid-solution MXenes such as $(Ti,V)_2C$ and $(Ti,V)_3C_2$; (c) ordered double-transition metal MXenes such as $(Cr_2V)C_2$ and $(Mo_2Ti_2)C_3$; and (d) ordered divacancy MXenes such as $Mo_{1.33}C$ and $W_{1.33}C$ [11]. Although many possible hybrids of MXenes with different transition metals and nitrogen/carbon can be formulated, only 14 mono-transition metal MXenes have been reported to date [8]. Among these, only a few nitride-based MXenes such as $Ti_4N_3T_x$ have been reported. However, $Ti_3C_2T_x$ and its related colloidal solutions are the most extensively studied among MXenes. $Ti_3C_2T_x$ has shown outstanding metallic conductivity (approximately 6,500 S/cm; in thin films), great volumetric capacitance as supercapacitors (greater than 900 F/cm^3), and high-rate capability and capacity in Li-ion batteries/capacitors [6]. All three possible forms of $Zr_3C_2T_2$ MXenes, where "T" is OH, O, and F, are metallic, while MXene $Zr_3C_2O_2$ have the strongest mechanical strength (392.9 GPa) [3]. MXenes can delaminate into a mono-layer-enriched MXene colloidal solution, which is possible due to the presence of secondary interactions such as hydrogen bonds and van der Waals forces present among 2-D sheets [7]. MXenes have unique properties due to the presence of hydrophilic fluorine/oxygen/hydroxyl terminal groups in which conductivity originates due to the presence of transition metal nitrides/carbides. Due to the presence of secondary interactions, transition metal nitrides/carbides, and terminal fluorine/oxygen/hydroxyl groups, MXenes possess desirable mechanical, optical, magnetic, and electrical properties, as shown in Figure 11.1.

A wide variety of MXene-based nanomaterials with different geometries or morphologies (such as nano-sheets, quantum dots [QDs], nanoribbons, and nanoflakes) have been produced for multidirectional applications. The nano-scale size of these materials not only enables them to

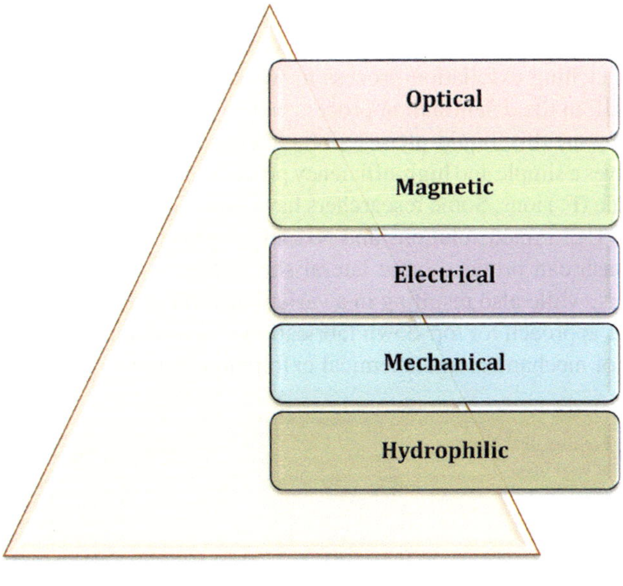

FIGURE 11.1 Few unique properties of MXenes.

circulate for a longer period in biological systems but also gives them novel characteristics like size effect-induced luminescence and closer interactions with nearby molecules [10]. MXenes could offer new approaches to illness treatment and ultrasensitive diagnosis. Despite the many outstanding accomplishments, the synthesis processes, characteristics, modification of surface functional groups, and biocompatibility of these materials still need to be addressed, along with additional work required to investigate their further applications.

MXenes have found suitable candidates for biomedical applications due to their ultrathin and layer-structured topology [12]. The presence of surface functional groups is responsible for their hydrophilicity which makes them biologically compatible. Also, many MXene phases have been found to be non-toxic and biocompatible with living organisms as their main constituents are elements such as carbon and nitrogen which are part of biological systems. More importantly, MXenes showed strong absorption in the near-infrared (NIR) region, which makes them potentially useful for photothermal therapy (PTT) and photo-acoustic imaging in vivo [10]. The different approaches used for MXenes synthesis and their biomedical applications are summarized in the subsequent section.

11.2 SYNTHESIS OF MXENES

Several MXenes phases have been reported due to the rapid development of synthetic techniques. Two primary methods used for synthesizing such 2-D layered nanomaterials are the (a) top-down and (b) bottom-up methods, both of which have been used to produce MXene nanostructures with single, few, or multiple layers. However, the choice of synthesis approach is important as it governs the chemical and physical properties such as size, shape, and functionality [13]. Despite the theoretical prediction of dozens of various possible compositions of MXenes, around 30 MXenes of various compositions have been synthesized to date [14]. The surface modifications are an important approach to tuning the biocompatibility of MXenes and reducing their cytotoxicity for biomedical applications [13]. The production and surface modification of MXenes are summarized below.

11.2.1 Top-Down Synthesis Method

The top-down method has been widely used for the synthesis of MXene materials (Figure 11.2). Multilayer-stacked MXenes can be obtained from the parent MAX-phase ceramics in two stages: (a) delamination etching using HF and (b) disintegration through intercalation of organic base molecules or breaking through sonication to obtain few- or single-layered MXenes [14]. The basic concept is to use the etching exfoliation process to weaken the adhesion between the layers [15]. However, the use of HF in the delamination process can have adverse effects on medicinal efficacy and biosafety. To overcome this, liquid-phase exfoliation has been the focus of the manufacturing of MXenes, which involves a simple and high-efficiency procedure to break the M–A bonds [16] through the presence of fluoride (F^-) ions. Some researchers have explored various fluoride-based compound etchants, such as LiF/HCl mixed solutions and NH_4HF_2, instead of the use of dangerous HF [15]. The top-down approach can produce large lateral sheet sizes of multilayered MXenes with fine nano-scale dimensions, while also resulting in a variety of surface terminations with –OH, –F, and –O groups. A popular approach for top-down fabrication of few or single-layered MXenes involves using a combination of mechanical and chemical exfoliation methods. Fluoride-free etchants have

FIGURE 11.2 Representation of top-down synthesis of MXene.

also gained interest recently due to the harmful liquid waste produced by fluoride-based chemicals. For example, Nb_2CT_x, a fluoride-free MXene, has been synthesized by electrochemical etching exfoliation and acts as a biosensor to detect phosmet, and showed superior performance compared to Nb_2CT_x biosensor synthesized using hydrofluoric acid [17]. Other techniques, such as molten fluoride salt etching, have also been used to overcome challenges associated with traditional synthetic processes, such as difficulties in removing Al layers from nitride-based MXenes and the use of aqueous acidic solutions [18]. Many 2-D-MXenes, such as multilayered Mo_2C, V_2C, Nb_2C, Zr_3C_2, and Ta_4C_3, have been extensively studied for their functionalities and wide-ranging applications. MAX phase such as $Zr_3Al_3C_5$ has been used to synthesize carbide MXene (2-D-Zr_3C_2Tz) by the action of hydrofluoric acid at room temperature. This MXene phase has better 2-D stability under ambient conditions, whereas MXenes based on Ti need to be stored under inert conditions [5]. This ultrathin 2-D nano-agent has small lateral sizes to ensure maximum accumulation and easy transport to lesion locations and easy excretion out of the organism [12]. The top-down approach is generally preferred over the bottom-up method due to the large size of precursor MAX phases.

11.2.2 Bottom-Up Synthesis Method

The bottom-up method is the opposite of the top-down method for synthesizing 2-D nanomaterials. This method allows for controlling the size and composition at the atomic level, which is challenging to achieve through simple bulk exfoliation. It involves the use of atoms or small organic-/inorganic molecules as building blocks, which are assembled to produce ordered layers with two-dimensional morphology. Wet chemical synthesis and chemical vapor deposition (CVD) are the two main bottom-up synthetic methodologies used for producing high-quality 2-D nanomaterials [19]. The CVD growth process developed for transition metal carbides provides a versatile and straightforward method for producing high-quality, ultrathin 2-D transition metal carbide crystals with large lateral sizes. CVD involves the formation of single- or few-layered 2-D nano-sheets by evaporating transition metal precursor under high vacuum and temperature. Sulfurization or selenization of metal and thin films of metal oxide are commonly used in the CVD method [15].

MXenes synthesized using bottom-up approaches produce uniform composition and morphology; however, this method requires more complex and expensive equipment than top-down approaches, and the scaling-up of production can be challenging. However, bottom-up methods can produce the MXenes of the desired quality for a particular application. For instance, the first high-quality MXene (Mo_2C) has been synthesized through the CVD process using methane gas (CH_4) as a precursor for carbon on Cu/Mo metal foil as a growth substrate and at a temperature over 1,085°C (Figure 11.3) [20,21]. The size and thickness were found in the range of 100 μm and a few nanometers, respectively (Figure 11.4). This method has also been used to synthesize various composite materials, such as Mo_2C-graphene (Mo_2C-Gr) hybrid films and heterostructures of graphene/Mo_2C. Furthermore, highly crystalline and defect-free Mo_2C films have been synthesized, which indicates the absence of surface termination functional groups. Such MXenes

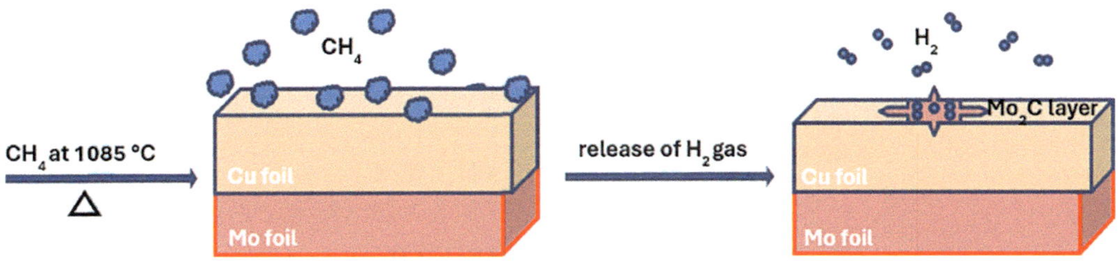

FIGURE 11.3 Synthesis of 2-D Mo_2C by chemical vapor deposition using Cu and Mo foil and CH_4 as the carbon source.

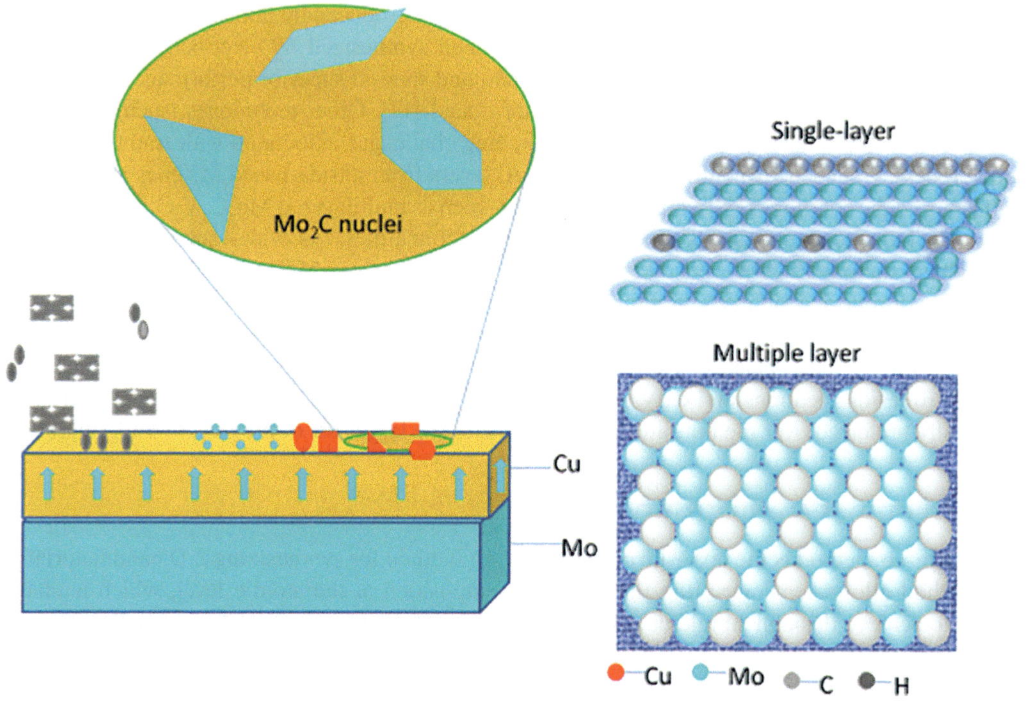

FIGURE 11.4 Atomic model and development of Mo_2C 2-D nanocrystals.

cannot be used for biomedicine applications due to their larger size, which doesn't allow them for effective cell permeation [13].

Moreover, using W or Ta foils in place of Mo during the CVD process, other superior MXene-structural crystals, including TaC and WC, have been successfully synthesized [12]. For the synthesis of MN/MC MXenes, TMOs are commonly used as templates which involve a two-step process: CVD, followed by the insertion of "N" or "C". For example, MoO_3 nano-sheets were first synthesized by researchers elevated temperatures using Mo-precursor-coated on NaCl crystals, which were then used as substrate to synthesize 2-D h-MoN nano-sheets [20]. Similarly, a research group used hot-filament CVD to create 2-D-layer MoO_3 and then converted the resulting material into 2-D-MoN at 800°C using an ammonia atmosphere [22]. This technique has also been used to produce various other MXene compositions. Although the bottom-up approach offers advantages for the synthesis of MXenes such as tunable structure, morphology, composition, and size [23], it has limited success in comparison to the synthesis of other standard members of 2-D materials. The complicated structures of MXenes in layer morphology explain why the ongoing advancement of bottom-up strategies to synthesize MXene has achieved limited success. The complex elemental makeup and structure of MXenes, as well as the mechanisms that govern the interactions between the components, remain unclear. Wet chemical synthesis is the traditional method of the synthesis of MXenes by in situ HF acid etching [24]. To destroy the bond between the "A" element and transition metal, the MAX phase has to be immersed in an acid [25]. The following chemical reaction takes place when Ti_3AlC_2 is soaked in HF:

$$Ti_3AlC_2 + 3HF \rightarrow AlF_3 \frac{3}{2} + H_2 + Ti_3C_2$$

$$Ti_3C_2 + 2H_2O \rightarrow Ti_3C_2(OH)_2 + H_2$$

$$Ti_3C_2 + 2HF = Ti_3C_2F_2 + H_2$$

TABLE 11.1
Different Approaches for the Synthesis of MXenes

Synthesis Method/Reagents Used	Advantages/Disadvantages
HF etching/HF	High yield/unsafe
Alkaline solution etching/TMAOH-NaOH	Anti-acid corrosion, fluorine-free terminal groups/need pre-treatment of HF and TMAOH, acute etching conditions
Fluoride-based acid etching/HCl, NH_4^+, HF_2, KF/LIF/NaF	Safe method/time consuming, impurity of fluoride salts
HCl-based hydrothermal etching/hydrothermal condition, HCl	Simple operation, fluorine-free terminal groups/complex etching condition
CVD/bimetal foil (Mo/Cu) and methane	Highly pure, great thickness/high temperature, and vacuum
Molten F salt-based Lewis acid etching/KF, LiF, NaF, Lewis acids	Production of nitride MXenes, control of surface-ending groups/presence of impurity, complex etching process
Electrochemical etching/TMAOH, NH^{4+}	Safe etching, fluorine-free ending groups/salt impurities in the product
Water-free etching/HF_2NH_4, organic solvents	Organic solvents can be used, delamination by ultrasonication/ complex workup procedure, longer etching time
Halogen etching/ICl, IBr, I_2, Br_2	Safe etching condition, control of surface terminal groups/ complex purification process
Lithiation-microexplosion method/Li^+	Safe method, fluorine-free terminal groups/low productivity, high cost
In situ electrochemical synthesis/Zn $(OTf)_2$, LiTFSI	Eco-friendly and safe method/expensive

Some unconventional wet chemical methods are fluoride-based acid etching, water-free etching, alkaline solution etching, etc. [24]. According to a report, high electronic conductivity is obtained by a wet chemical approach with a fewer number of atomic defects. However, this method is conventional for the synthesis of only C-based MXenes [26]. Several methods used for the synthesis including their advantage and disadvantage are summarized in Table 11.1. MXenes have an advantage over other members of 2-D materials; however, the industrial-scale application of such materials is limited due to the unavailability of large-scale synthetic procedures of monolayer MXene nano-sheets.

11.3 BIOMEDICAL APPLICATIONS

Two-dimensional nanomaterials exhibit outstanding physiochemical, electrical, and biological properties [25]. There has been a huge focus on the biomedical application of 2-D nanomaterials such as graphene, TMOs, hexagonal boron nitrides (h-BN), and MXenes. Among them, MXenes are more compatible with biomedical applications due to the presence of surface terminal groups, tunable compositions, and hydrophilicity [22]. There are various applications where MXenes has been used (Figure 11.5).

11.3.1 MXenes in Biosensing

Biosensing has emerged as a promising approach for detecting specific biomolecules for various applications. Several 2-D-based biosensors are being used for the diagnosis of metabolites like glucose, adenosine, and lactose and the detection of proteins, bacterial cells, and neurotransmitters like serotonin and dopamine [27]. Two-dimensional nanomaterials have enormous potential in biosensing systems to identify biomacromolecules beyond their numerous uses in therapeutic and diagnostic imaging [27,28]. Several methods such as calorimetric biosensors, field-effect transistor (FET) biosensors, fluorescent biosensors, and electro-chemi-luminescence have been developed

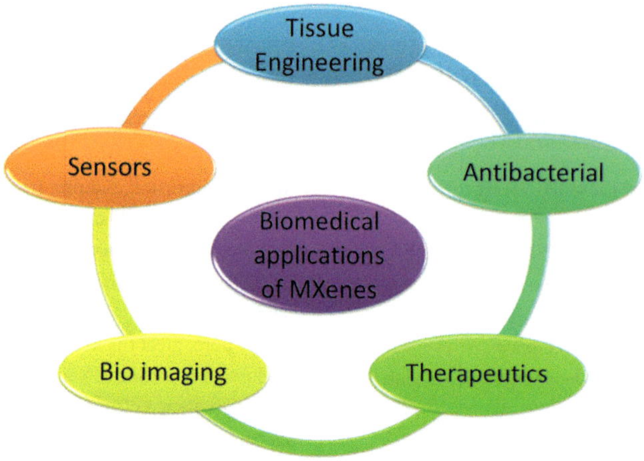

FIGURE 11.5 Biomedical applications of MXenes.

which utilize 2-D materials. The metallic conductivity combined with the hydrophilic surface of MXenes makes them potential candidates for developing biocompatible FETs for rapid, simple, and label-free identification of biological processes. Also, MXenes have been used as electrochemical biosensors. Ti_3C_2-Au has been used for the detection of pesticides, whereas Ti_3C_2-based wearable patches have been used for sensing lactate and glucose from perspiration [29].

Micropattern-based FET devices using MXenes have been used to detect dopamine, a neuromodulatory hormone that plays a critical role in maintaining brain functions and cells. When MXene surface terminations interact with dopamine, it creates a vacancy, resulting in enhanced conductance [30,31]. In addition, Fermto-liter resolution micro-injectors have been used to maintain the quantity of pulse, i.e., the amount of dopamine released in the biological recognition chamber. MXene-based FETs have also been used for continuous tracking of spiking activity in primary hippocampal neurons, highlighting their potential as cutting-edge biosensing systems [31].

MXenes have been used for biosensing purposes due to their high sensitivity to electrochemical reactions, making them highly responsive to signal disturbances caused by action potentials and neurotransmitter release [12]. The complex neurites of neuronal cells on the surface of MXene substrates demonstrate their high level of biocompatibility with neural cells [31]. Moreover, optical signal-based biosensors can be broadly classified into electrochemiluminescence (ECL) biosensors and surface plasmon resonance (SPR) biosensors. ECL biosensors offer low background signals, controllability, rapidity, low cost, and high sensitivity, making them widely used for protein, DNA, and enzyme detection and clinical diagnosis [32]. Zhang et al. developed a novel ECL biosensor using Ti_3C_2 MXene nano-sheets to recognize aptamers and catalyze ECL signal amplification. Materials based on MXene exhibit excellent electrical characteristics and conductivity [33]. The unique 2-D structure of MXene-based materials and their high surface area make them promising materials for biosensing applications and suitable for enzyme conjugation [17].

11.3.2 MXenes in Bioimaging

Bioimaging is an important clinical technique to visualize biological activity with the aid of the optical properties of 2-D materials. New 2-D nanomaterials, such as MXenes and black phosphorous nano-sheets, offer ideal nano-platforms for bioimaging applications (Figure 11.6) due to their unique intrinsic physicochemical properties, such as an adjustable band gap, transport anisotropy, high carrier mobility and excellent biocompatibility and biosafety [34]. They have also been used in the bioimaging of tissues and cells as nano-probes. However, the standard MXene-based materials lack photoluminescence response in aqueous solutions, which limits their biological uses [35].

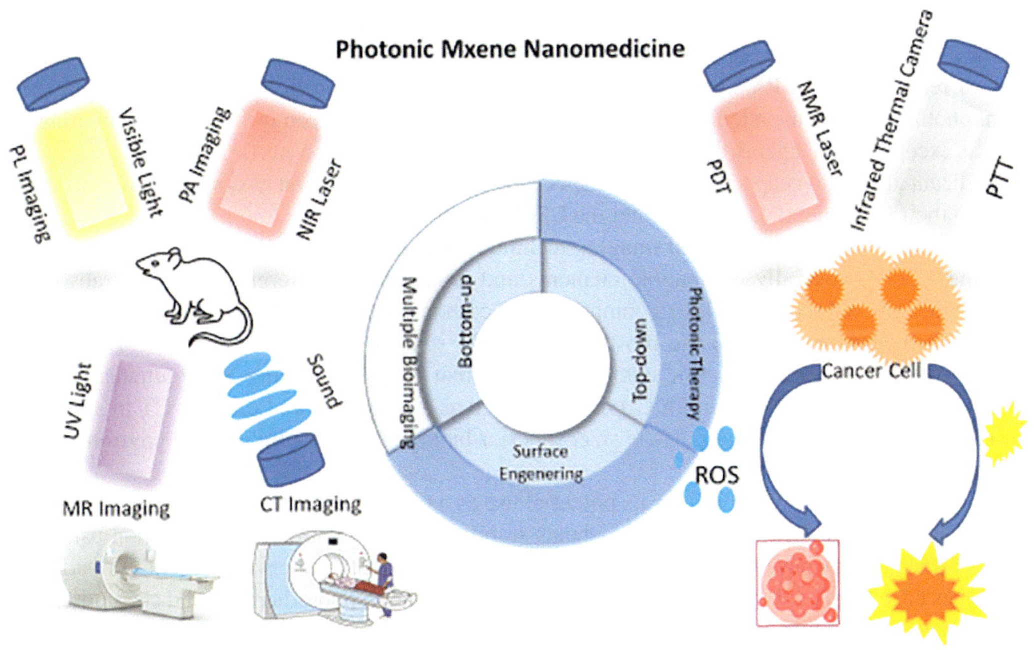

FIGURE 11.6 Bioimaging applications of MXenes.

In contrast, fluorescent nanoparticles can be a valuable tool as biosensors that label specific proteins due to their small size, strong photoluminescence, and optical absorption properties. MXenes also have great potential for diagnostic and oncological applications as they can be incorporated with different elements which can tune their physicochemical properties. Thus, the intrinsic photothermal performance, efficient loading of functional contrast agents, quantum size effects, element-enhanced contrast, and ferromagnetism make them ideal for diagnostic imaging, radiation therapy, photothermal therapy, and in vivo photo-acoustic (PA) imaging [36]. MXene-based materials have an active surface, which makes them useful in chemotherapy to monitor drug absorption efficiency, and their hydrophilic surface allows for use in radiation therapy. Furthermore, the MXenes can be used in phototherapy and in vivo PA imaging as they show strong light absorption in both NIR-I and NIR-II regions. Moreover, MXene like $MnOx/Ta_4C_3$ is non-toxic and showed good biocompatibility without using any surface-coating biocompatible substances [37]. The exceptional properties of MXene-based materials make them highly attractive for a range of bioimaging applications, such as magnetic resonance imaging (MRI), PA imaging, luminescence imaging (LI), and X-ray computed tomography (CT) which can identify the location of tumors and guide PTT [38]. In some cases, multimodal imaging can also be employed in combination with PTT. Fluorescence microscopy is a popular bioimaging technique due to its ease of use and cost-effectiveness [39]. However, approaches based on fluorescence are not ideal for medical diagnosis due to the limited penetration depth of ultraviolet-visible light in tissues. Ti_3C_2 nano-sheets have been used for cancer detection by the covalent immobilization of bio-receptors as the functional groups in Ti_3C_2–MXene having a high density allow for quicker access to the analyte and enhanced bio-molecule adsorption [40]. The researcher demonstrated that MXene-based bioimaging using luminous Ti_3C_2 nanoflakes produced a delamination reaction in an aqueous solution of tetramethylammonium hydroxide, followed by sonication [41]. Some other methods based on the use of ultrasonic or X-ray radiations used for bioimaging are discussed below.

PA imaging is a technique based on the utilization of non-ionizing visible-IR LASER and has limited penetration depth which depends on the LASER source. However, it offers improved spatial resolution for living tissues and can image deeper tissues by converting photo energy to sound energy through the PA effect. Ta_4C_3-SP/MnO_x nano-sheets composite have been synthesized and their PI imaging property as contrast-advancing agents has been studied. These composites showed linear relationships of PA signal with the Ta element concentration in MnO_x/Ta_4C_3-SP and excellent photothermal conversion capabilities [42]. MXenes have also been utilized for bioimaging and act as excellent PTT agents due to their high light-to-heat conversion performance [43]. They have significant absorption capacity for a wide spectrum of light and function as powerful contrast agents due to their semi-metallic qualities and the LSPR effect [44].

X-ray CT is a popular medical imaging technique which better suited to the use of compounds having more biologically compatible elements and relies on the differences in X-ray absorption among lesions and tissues [45]. In a comparison of carbon-based materials, MXenes and Transition metal chalcogenides (TMDs) are most suited for CT scans out of many possible 2-D materials, because they contain high atomic number elements that have outstanding X-ray attenuation capabilities. Tantalum-rich MXene (Ta_4C_3) has been used as a potential candidate for CT imaging contrast agents due to their ideal particle size, superior biological compatibility, and environmentally friendly manufacturing [46].

Ta_4C_3 has also been explored for its potential use as a CT and PA contrast agent for the bioimaging of tumors. The Ta-rich Ta_4C_3 nano-sheets stabilized through surface modification using soybean phospholipids have been used as a bioimaging agent by Lin and co-workers [42]. In addition, MXenes have been explored for their synergistic photodynamic, photothermal, and chemotherapeutic effects, which could lead to non-invasive and effective anticancer therapies. MXenes containing high atomic number elements have also been used as X-ray CT contrast agents for breast cancer treatment. The super-paramagnetic Ta_4C_3 MXene-based therapeutic nano-platform made from Ta_4C_3 MXene and surface-super-paramagnetic iron-oxide composite MXenes synthesized by Liu et al. showed promising potential for the effective treatment of breast cancer [47]. MXene-derived QDs have been proven effective in being used as alternative bioimaging agents among various luminous nanomaterials. These QDs can be modified in terms of size, shape, and surface, which led to their increased use in bioimaging. MXene-derived QDs have been reported to have high dispersibility, photoluminescence capabilities, and chemical and photochemical stability, which make them promising candidates for bioimaging applications [48]. Unlike non-emissive MXene nano-sheets, the newly discovered emissive MXene QDs have shown excellent luminescent characteristics for bioimaging, including high photo-stability, tunable wavelength, and desirable quantum yields (~10%) [49]. In addition, the quantum confinement effect arises due to the size of these materials and vacancy defects further enhance the luminescent properties of MXene QDs, making them ideal for bioimaging applications.

11.3.3 MXenes in Tissue Engineering

MXenes have shown potential for tissue engineering and regenerative medicine, particularly in bone, skin, nerve, and cardiac tissue engineering. Although they have shown much higher applications in biomedical, for example, bioimaging, tumor therapy, and antimicrobial applications compared to tissue engineering, they can play a remarkable role in the development of novel inorganic composite materials for bone tissue regeneration. For instance, MXenes have been used to create nano-composite films of Ti_3C_2 nano-sheets and hydroxy-apatite nanowires, which have intriguing bone regenerative properties [50]. Chen et al. developed a reliable way to create biologically compatible and strong $Ti_3C_2T_z$-enhanced polylactic acid nano-composite membranes involving n-octyltriethoxysilane (OTES) as an interfacial mediator [51]. The ultimate tensile strength of OTES-$Ti_3C_2T_z$ and Poly-Lactic Acid (PLA) nano-composite has been found significantly higher compared to their precursor due to the significant connection between them [52]. Additionally, Ti_3C_2 MXene films possessing excellent conductivity and specific surface area have been used to create flexible microelectrodes for nerve tissue engineering. This electrode records neural activity with accuracy and sensitivity due to its very low resistance [53]. MXenes have also demonstrated remarkable possibilities in stem cell-based tissue therapies and applied sciences. For example, Ti_3C_2 MXenenano-fibers have been used to create biological materials for tissue engineering and cell culture. Such intelligent biomaterials with unique biological properties have great potential in tissue engineering and repair. MXene composite nanofibers have been shown to be biocompatible with Bone Marrow Mesenchymal Stem Cell (BMSCs) and can enhance their capacity for osteogenesis. These nanofibers can also be molded into different shapes and sizes to meet various needs, making them potential biomaterials for bone tissue engineering [54]. In addition to bone tissue engineering, MXene-based materials have been explored for other biomedical applications. For instance, zero-dimensional (0-D) Ti_3C_2 QDs have been used to improve post-injury tissue repair by selectively decreasing the activation of human IFN-γ^+ cytotoxic $CD4^+$ T-cells and promoting the growth of immune-suppressive $CD4^+CD25^+FoxP3^+$ regulatory T-cells [55]. MXene QDs have also been designed with accurate surface modifications for sub-cellular nano-medicine applications. Another surprising application of MXene is its ability to adsorb urea from aqueous solutions and dialysate, with a maximum adsorption capacity of 10.4 mg/g at room temperature. MXenes with excellent biological compatibility and enhanced urea adsorption selectivity can potentially be used to develop miniaturized dialysate regeneration devices for a portable artificial kidney [50]. MXene-based materials continue to show great promise in the field of biomedical engineering and may lead to the development of innovative solutions for various medical conditions.

11.3.4 MXenes as Antimicrobial Agents

Antimicrobial nanoparticles have shown great promise in the field of public health. Many bacterial infections occur due to antibacterial resistance. To overcome this, the ROS-generating capability of 2-D nanomaterials makes them ideal for advances in antimicrobial and antibacterial agents [56]. For example, 2-D material like GO produces O_2 for antibacterial purposes and its antibacterial activity can be increased by decorating the surface of 2-D GO with nanoparticles [57]. Also, materials such as MoS_2 and graphene, among 2-D materials, have attracted considerable attention attributed to their structures and unexpected properties. Graphene nano-sheets, for example, have sharp edges that can cause physical damage to bacterial cell membranes due to oxidative and physical stress that leads to the loss of bacterial membrane integrity, thus showing antimicrobial activity [58]. MXenes have many advantages over other 2-D materials since they have layered structures and surface-ending functionalities such as –OH, –F, or –O. The surface-ending groups are nucleophilic and can show efficient coordination chemistry, whereas metals present in MXenes showed rich redox chemistry. MXenes have proven to show application in many health diagnostics fields including antimicrobial applications and can inhibit the growth of bacteria and fungi [59]. Only a few bacterial strains, such as *E. coli* and *Bacillus subtilis*, have shown resistance against MXenes. The mechanism behind the antibacterial activity of MXenes is thought to be due to their hydrophilicity, high electrical conductivity, semiconductor capabilities, surface-ending electronegative groups, 2-D structure, and optical properties, viz., LSPR phenomenon [56]. Moreover, unlike most conventional antibacterial nanomaterials, MXenes have demonstrated superior antibacterial activity [60]. Researchers suggested a novel approach by combining Ti_3C_2 MXenes with NIR light (808 nm) to increase their antimicrobial window. This method has been tested for 15 microbial strains, including antibiotic-resistant strains like MRSA and VRE, and found to be effective [59,61]. Ti_3C_2 MXenes have been found to exhibit antibacterial activity, and several mechanisms have been suggested for their mode of action. The first mechanism suggests that the pointed edges of the MXenes enable effective adhesion to the surfaces of microorganisms, which may result in membrane damage. A second mechanism of action suggests that the surface functionalities present at MXenes can interact either with the cell wall of microorganisms or with the biomolecules present in the cytoplasm, resulting in damage to the cell microstructure and leading to the death of microorganisms. Furthermore, the negatively charged surface of MXenes such as Ti_3C_2 nano-sheets can transfer electrons through bacterial cells due to the establishment of a conductive bridge with an insulating lipid bilayer and finally cell death. Thus, the antibacterial activities of MXenes largely rely on the interaction of surface groups present at nano-sheet edges and bacterial cell membranes [62]. Furthermore, the bacterial cell wall can also be disrupted due to the oxidative damage caused by MXenes. The Superoxide Dismutase (SOD) analysis can be used to understand the positive/negative effect of MXenes under oxidation stress conditions. A decrease in SOD activity in the presence of MXenes indicates that the use of MXenes can cause harm to the antioxidant system, leading to an increase in intracellular ROS and in turn oxidative damage. The semiconducting characteristic of MXenes allows the electron transfer process to bacterial cells upon excitation to light leading to cell death by exploiting the reactive metal-F couples. This process is analogous to semiconductors which create an e–h pair upon light excitation, where active electrons can cause cell death similar to type-I Photo Dynamic Therapy (PDT) [63]. To assess the capacity of Ti_3C_2 MXenes to prevent bacterial development, the bacteria were grown on a culture plate in the absence/presence of Ti_3C_2 (0–200 μg/mL), and their colonies were counted [64]. The results showed that the antibacterial activity of Ti_3C_2 MXenes was significantly higher compared to surface-oxidized Ti_3C_2 MXenes. The antibacterial activity of MXenes has been found concentration-dependent [65]. The suggested mechanism of destruction of cellular membranes and cell death is the direct interaction of surface terminal/reactive electron-rich groups with bacterial cell walls or cytoplasm. Thus, in conclusion, MXenes have enormous potential as antibacterial agents, but their biocompatibility and cellular absorption need to be studied extensively before they can be safely used in biomedicine.

11.3.5 MXenes in Therapeutics

Out of many PTAs, 2-D nanomaterials have high PT conversion efficiency due to their outstanding in-plane electron mobility and great biocompatibility, which makes them prosperous to show therapeutic effects and can be used for a variety of applications like invasive PTT, catalysis, water evaporation, sensing, and energy storage [66]. The heat released from photothermal conversion has the ability to kill cancer cells. Cancer is the growth of any undesired tissue at any part and remains in the second position worldwide to cause death among humans. To cure cancer by imaging-guided cancer therapy or by PTT, 2-D nanomaterials have the capability to produce high temperatures at tumor sites to surgically remove tumor cells without damaging normal cells. The surgical procedures leave the human body vulnerable to infection and recurrence, while chemotherapy indiscriminately kills both healthy and malignant cells. PTT has emerged as an attractive alternative with fewer side effects, where tumor cells are destroyed by taking advantage of NIR light, which can be used to produce heat [67]. Likewise, 2-D materials, MXenes have gained significant attention in the field of cancer treatment due to their small size and layered structure, resulting in a high surface-to-volume ratio, high photothermal conversion efficiency, hydrophilicity, and functional groups that make them suitable for PTT. Several MXene varieties have been found to exhibit substantial NIR absorption, making them well-suited for deep-tissue PTT. In addition, MXenes can serve as carriers for anticancer drugs and contrast agents for bioimaging, which allow them to be used for drug delivery and treatment progress tracking [68]. By leveraging these synergistic effects, the success rates of cancer treatment can be significantly improved, while also increasing convenience. Several carbides of Ti, Nb, and Ta (Ti_3C_2, Nb_2C, and Ta_4C_3) have been found suitable for in vivo PTT of cancers using 808 nm LASER [69]. In addition, such MXenes have also been found suitable for drug delivery in the chemotherapy process. The field of cancer therapies has been greatly advanced by the use of biological optical imaging and phototherapy. Recently, 2-D MXene derivative (Nb_2C) has been found suitable as a phototherapeutic agent in both NIR-I and NIR-II bio-windows and showed in vivo photothermal ablation of mouse tumors. MXenes have been utilized as transporters for cargo loading to produce a synergistic therapeutic effect and have been found to operate through three mechanisms: (a) generate ROS for effective photodynamic cell killing, (b) show photothermal effect for effective cell killing, and (c) show cargo loading for synergistic therapy [10].

The unique electronic structure and photo-electronic characteristics of MXenes made them ideal to be employed in PDT and PTT/PDT/chemo synergistic therapy [10]. First MXene-based photothermal agent (Ti_3C_2 nano-sheets) has been found to show significant in vivo PTT efficacy. However, their relatively poor photothermal conversion efficiency has limited their commercial applications [70]. The use of MXenes in PTT is advantageous because it allows for the remote triggering of therapy with NIR light, reducing adverse side effects and enabling precise spatiotemporal modulation of local heating effects. Multi-functional ultrathin Ta_4C_3 MXenes (size < 100 nm) having biocompatible "Ta" showed effective PTT of tumors. It exhibits strong absorption bands in their absorption spectra resembling reported 2-D nanomaterials, which makes them ideal for subsequent photothermal transduction processes [66]. Active targeting is a promising approach for enhancing the accumulation of drug carriers or therapeutic substances at the site of the disease. Nano-structured MXenes, such as RGD-targeted Ti_3C_2@mMSNs-RGD, have been found suitable for targeted drug release and delivery and photothermal conversion performance for PTT due to the presence of mesopores [12]. MXenes have significant potential for their applications in energy transfer since they can be combined with dyes. By serving as a carrier for different cargos, MXenes have significant potential for multimodal imaging, therapies, and theragnostic.

11.4 CONCLUSION

MXenes have been developed by the etching of MAX layers, which exhibit remarkable characteristics over other 2-D materials due to their layered structure topology, surface-ending functionalities, hydrophilicity, etc. Top-down and bottom-up techniques have been used to synthesize MXene nanostructures, while the top-down approach is preferred over the bottom-up due to the large size of precursor MAX phases. However, their complex synthetic procedure limits their use for commercial applications. MXenes showed different applications in biological systems from illness treatment to ultrasensitive diagnosis due to their unique physiochemical and electrical properties. Properties like hydrophilic surface and metallic conductivity of MXenes are useful for the development of biocompatible FETs, which have been used for continuous tracking of spiking activity. MXenes nano-probes have also been used in the bioimaging of tissues and cells. MXenes such as Ti_3C_2 nano-sheets have been used for cancer detection, while Ta_4C_3 has been used as CT imaging contrast agents. Such intelligent biomaterials with unique biological properties have great potential in tissue engineering and repair. Encouragingly, these numerous applications confirmed that MXene shows remarkable performance in the biomedical field, which is quite difficult to achieve by other conventional 2-D nanocrystals, but their biocompatibility and cellular absorption need to be studied extensively before they can be safely used in biomedicine.

11.5 ABBREVIATION TABLE

CT	Computed tomography
CVD	Chemical vapor deposition
ECL	Electrochemiluminescence
FETs	Field-effect transistors
GO	Graphene oxide
Gr	Graphene
h-BN	Hexagonal boron nitride
HF	Hydrofluoric acid
LSPR	Localized surface plasmon resonance
MRI	Magnetic resonance imaging
MRSA	Methicillin-resistant *Staphylococcus aureus*

(Continued)

(*Continued*)

NIR	Near infrared
OTES	n-Octyltriethoxysilane
PA	Photo-acoustic
PTAs	Photothermal agents
PTT	Photothermal therapy
QDs	Quantum dots
ROS	Reactive oxygen species
SPR	Surface plasmon resonance
TMDCs	Transition metal dichalcogenides
TMOs	Transition metal oxides
2-D	Two dimensional
VRE	Vancomycin-resistant enterococci

REFERENCES

1. S. Wang, L. Zhou, Y. Zheng, L. Li, C. Wu, H. Yang, M. Huang, and X. An, *Colloids Surf, A: Physicochem. Eng. Asp*, 2019, 583, 124004.
2. S. Z. Butler, S. M. Hollen, L. Cao, Y. Cui, J. A. Gupta, H. R. Gutiérrez, T. F. Heinz, S. S. Hong, J. Huang, A. F. Ismach, E. Johnston-Halperin, M. Kuno, V. V. Plashnitsa, R. D. Robinson, R. S. Ruoff, S. Salahuddin, J. Shan, L. Shi, M. G. Spencer, M. Terrones, W. Windl, and J. E. Goldberger, *ACS Nano*, 2013, 7, 2898–2926.
3. N. M. Abbasi, Y. Xiao, L. Peng, Y. Duo, L. Wang, L. Zhang, B. Wang, and H. Zhang, *Adv. Mater. Technol.*, 2021, 6, 2001197.
4. C. J. Zhang, S. Pinilla, N. McEvoy, C. P. Cullen, B. Anasori, E. Long, S.-H. Park, A. Seral-Ascaso, A. Shmeliov, D. Krishnan, C. Morant, X. Liu, G. S. Duesberg, Y. Gogotsi, and V. Nicolosi, *Chem. Mater.*, 2017, 29, 4848–4856.
5. P. Eklund, M. Beckers, U. Jansson, H. Högberg, and L. Hultman, *Thin Solid Films*, 2010, 518, 1851–1878.
6. Y. Gogotsi, and B. Anasori, *ACS Nano*, 2019, 13, 8491–8494.
7. F. Dixit, K. Zimmermann, R. Dutta, N. J. Prakash, B. Barbeau, M. Mohseni, and B. Kandasubramanian, *J. Hazard. Mater.*, 2022, 423, 127050.
8. W. Hong, B. C. Wyatt, S. K. Nemani, and B. Anasori, *MRS Bull.*, 2020, 45, 850–861.
9. A. VahidMohammadi, J. Rosen, and Y. Gogotsi, *Science,* 2021, 372, 6547.
10. K. Huang, Z. Li, J. Lin, G. Han, and P. Huang, *Chem. Soc. Rev.*, 2018, 47, 5109–5124.
11. M. Ding, C. Han, Y. Yuan, J. Xu, and X. Yang, *Sol. RRL*, 2021, 5, 2100603.
12. H. Lin, Y. Chen, and J. Shi, *Adv. Sci.*, 2018, 5, 1800518.
13. Wan, F., Thakur, A., Thakur, P. (2023). Classification of nanomaterials (carbon, metals, polymers, bio-ceramics). In: Suhag, D., Thakur, A., Thakur, P. (eds) *Integrated Nanomaterials and their Applications*. Springer, Singapore. https://doi.org/10.1007/978-981-99-6105-4_3
14. K. Hantanasirisakul and Y. Gogotsi, *Adv. Mater.,* 2018, 30, 1804779.
15. H. Cui, Y. Guo, W. Ma, and Z. Zhou, *ChemSusChem*, 2020, 13, 1155–1171.
16. J. Wang, W. Zhang, Y. Wang, W. Zhu, D. Zhang, Z. Li, and J. Wang, *Part. Part. Syst. Charact.*, 2016, 33, 825–832.
17. M. Song, S. Pang, F. Guo, M. Wong, and J. Hao, *Adv. Sci.*, 2020, 7, 2001546.
18. B. Bhattacharjee, Md. Ahmaruzzaman, R. Djellabi, E. Elimian, and S. Rtimi, *J. Environ. Manage.,* 2022, 324, 116387.
19. Thakur, P., Thakur, A. (2022). Nanomaterials, their types and properties. In: Thakur, A., Thakur, P., Khurana, S. P. (eds) *Synthesis and Applications of Nanoparticles*. Springer, Singapore. https://doi.org/10.1007/978-981-16-6819-7_2
20. L. Verger, C. Xu, V. Natu, H.M. Cheng, W. Ren, and M. W. Barsoum, *Curr. Opin. Solid State Mater. Sci.*, 2019, 23, 149–163.
21. C. Xu, L. Wang, Z. Liu, L. Chen, J. Guo, N. Kang, X.-L. Ma, H.-M. Cheng, and W. Ren, *Nat. Mater.*, 2015, 14, 1135–1141.
22. J. Huang, Z. Li, Y. Mao, and Z. Li, *Nano Select*, 2021, 2, 1480–1508.

23. Z. Meng, R. M. Stolz, L. Mendecki, and K. A. Mirica, *Chem. Rev.*, 2019, 119, 478–598.
24. J. Chen, Y. Ding, D. Yan, J. Huang, and S. Peng, *Sus. Mat.*, 2022, 2, 293–318.
25. Krishnia, L., Thakur, P., Thakur, A. (2022). Synthesis of nanoparticles by physical route. In: Thakur, A., Thakur, P., Khurana, S. P. (eds) *Synthesis and Applications of Nanoparticles.* Springer, Singapore. https://doi.org/10.1007/978-981-16-6819-7_3
26. K. A. Papadopoulou, A. Chroneos, D. Parfitt, and S. R. G. Christopoulos, *J. Appl. Phys.,* 2020, 128, 170902.
27. B. T. Murti, A. D. Putri, Y. J. Huang, S.-M. Wei, C. W. Peng, and P. K. Yang, *RSC Adv.*, 2021, 11, 20403–20422.
28. A. Murali, G. Lokhande, K. A. Deo, A. Brokesh, and A. K. Gaharwar, *Mater. Today*, 2021, 50, 276–302.
29. N. Rohaizad, C. C. Mayorga-Martinez, M. Fojtu, N. M. Latiff, and M. Pumera, *Chem. Soc. Rev.,* 2021, 50, 619–657.
30. X. Gan, H. Zhao, and X. Quan, *Biosens. Bioelectron.*, 2017, 89, 56–71.
31. B. Xu, M. Zhu, W. Zhang, X. Zhen, Z. Pei, Q. Xue, C. Zhi, and P. Shi, *Adv. Mater.*, 2016, 28, 3333–3339.
32. X. Huang, Y. Liu, B. Yung, Y. Xiong, and X. Chen, *ACS Nano*, 2017, 11, 5238–5292.
33. H. Zhang, Z. Wang, Q. Zhang, F. Wang, and Y. Liu, *Biosens. Bioelectron.*, 2019, 124–125, 184–190.
34. J. C. Ranasinghe, A. Jain, W. Wul, K. Zhang, Z. Wang, and S. Huang, *J. Mater. Res.*, 2022, 37, 1689–1713.
35. Z. Lei and B. Guo, *Adv. Sci.,* 2021, 9, 2102924.
36. J. Yao, P. Li, L. Li, and M. Yang, *Acta Biomater.*, 2018, 74, 36–55.
37. A. Sundaram, J. S. Ponraj, C. Wang, W. K. Peng, R. K. Manavalan, S. C. Dhanabalan, H. Zhang, and J. Gaspar, *J. Mater. Chem. B*, 2020, 8, 4990–5013.
38. J. Zhao, J. Chen, S. Ma, Q. Liu, L. Huang, X. Chen, K. Lou, and W. Wang, *Acta Pharm. Sin. B*, 2018, 8, 320–338.
39. X. Cai, J. Tian, J. Zhu, J. Chen, L. Li, C. Yang, J. Chen, and D. Chen, *Chem. Eng. J.*, 2021, 426, 131919.
40. S. Kumara, Y. Leia, N. H. Alshareefb, M. A. Quevedo-Lopezb, and K. N. Salamaa, *Biosens. Bioelectron.*, 2018, 121, 243–249.
41. L. Liu, M. Orbay, S. Luo, S. Duluard, H. Shao, J. Harmel, P. Rozier, P.-L. Taberna, and P. Simon, *ACS Nano,* 2022, 16, 111–118.
42. H. Li, R. Fan, B. Zou, J. Yan, Q. Shi, and G. Guo, *J. Nanobiotechnol.*, 2023, 21, 73.
43. M. Ashfaq, N. Talreja, D. Chauhan, S. Afreen, A. Sultana, and W. Srituravanich, *J. Drug Deliv. Sci. Technol.*, 2022, 70, 103268.
44. W. Tao, N. Kong, X. Ji, Y. Zhang, A. Sharma, J. Ouyang, B. Qi, J. Wang, N. Xie, C. Kang, H. Zhang, O. C. Farokhzad, and J. S. Kim, *Chem. Soc. Rev.*, 2019, 48, 2891–2912.
45. X. Han, K. Xu, O. Taratula, and K. Farsad, *Nanoscale*, 2019, 11, 799–819.
46. C. Dai, Y. Chen, X. Jing, L. Xiang, D. Yang, H. Lin, Z. Liu, X. Han, and R. Wu, *ACS Nano* 2017, 11, 12, 12696–12712.
47. Z. Liu, H. Lin, M. Zhao, C. Dai, S. Zhang, W. Peng, and Y. Chen, *Theranostics*, 2018, 8, 1648–1664
48. S. Iravani and R. S. Varma, *Nanomaterials*, 2022, 12, 1200.
49. Z. Huang, X. Cui, S. Li, J. Wei, P. Li, Y. Wang, and C.-S. Lee, *Nanophotonics*, 2020, 9, 2233–2249.
50. S. Iravani and R. S. Varma, *Mater. Adv.*, 2021, 2, 2906–2917.
51. K. Chen, Y. Chen, Q. Deng, Seol-Ha Jeong, Tae-Sik Jang, S. Du, Hyoun-Ee Kim, Q. Huang, and Cheol-Min Han, *Mater. Lett.*, 2018, 229, 114–117.
52. S. M. George and B. Kandasubramanian, *Ceram. Int.*, 2020, 46, 8522–8535.
53. N. Driscoll, A. G. Richardson, K. Maleski, B. Anasori, O. Adewole, P. Lelyukh, L. Escobedo, D. K. Cullen, T. H. Lucas, Y. Gogotsi, and F. Vitale, *ACS Nano*, 2018, 12, 10419–10429.
54. A. Murali, G. Lokhande, K. A. Deo, A. Brokesh, and A. K. Gaharwar, *Mater. Today*, 2021, 50, 276–302.
55. E. R. Molina, B. T. Smith, S. R. Shah, H. Shin, and A. G. Mikos, *J. Control. Release*, 2015, 219, 107–118.
56. L. Wang, Y. Li, L. Zhao, Z. Qi, J. Gou, S. Zhang, and J. Z. Zhang, *Nanoscale*, 2020, 12, 19516–19535.
57. Thakur, A., Thakur, P., Baccar, S. (2023). Structural properties of nanoparticles. In: Suhag, D., Thakur, A., Thakur, P. (eds) *Integrated Nanomaterials and their Applications.* Springer, Singapore. https://doi.org/10.1007/978-981-99-6105-4_4
58. W. Sun and F.-G. Wu, *Chem. Asian J.*, 2018, 13, 3378–3410.
59. N. Dwivedi, C. Dhand, P. Kumar, and A. K. Srivastava, *Mater. Adv.*, 2021, 2, 2892–2905.
60. M. S. Salmi, U. Ahmed, N. Aslfattahi, S. Rahman, J. G. Hardy, and A. Anwar, *RSC Adv.*, 2022,12, 33142–33155.
61. Z. Yu, L. Jiang, R. Liu, W. Zhao, Z. Yang, J. Zhang and S. Jin, *Chem. Eng. J.*, 2021, 426, 131914.
62. M. E. One-Sun Lee, Madjet, and K. A. Mahmoud, *Nano Lett.* 2021, 21, 19, 8510–8517.
63. H. Huang, W. Feng, and Y. Chen, *Chem. Soc. Rev.*, 2021, 50, 11381–11485.

64. K. Rasool, M. Helal, A. Ali, C. E. Ren, Y. Gogotsi, and K. A. Mahmoud, *ACS Nano*, 2016, 10, 3674–3684.
65. K. S. Rizi, *J. Mol. Struct*, 2022, 1262, 132958.
66. H. Ma and M. Xue, *J. Mater. Chem. A*, 2021, 9, 17569–17591.
67. H. S. Han and K. Y. Choi, *Biomedicines*, 2021, 9, 305.
68. S. Iravani and R. S. Varma, *Nanomaterials*, 2022, 19, 3360.
69. B. Lu, Z. Zhu, B. Ma, W. Wang, R. Zhu, and J. Zhang, *Small*, 2021, 17, 2100946.
70. H. Lin, X. Wang, L. Yu, Y. Chen, and J. Shi, *Nano Lett.*, 2016, 17, 384–391.

12 Green Synthesized Metal Oxide Nanostructures for Sensing Applications

Taranga Dehury and Chandana Rath
Indian Institute of Technology (BHU)

12.1 BACKGROUND OF SENSORS: BASIC PRINCIPLES AND TYPES

Sensors help us recognize various forms of relevant information in the surrounding environment and convert this information into required forms of information output through specific mechanisms to meet numerous needs in industries, the environment, and healthcare. Gas and liquid sensors are one of the most used types of sensors. The primary purpose of gas sensors is to identify chemicals that endanger human health or whose concentration is below the olfactory threshold for humans (ppm/ppb level). In environmental assessments, gas sensors are used to monitor the concentration of gases such as H_2S, NO_2, and volatile organic compounds to ensure that the immediate environment is suitable for human survival. Gas sensors are being used to regulate the concentration of combustible and explosive gases in industrial units. With the increasing use of natural gas and hydrogen, the use of gas sensors turned out to be more prevalent. Another significant application of gas sensors is breath analysis. Ethanol sensors are employed in road safety, and such sensors can instantly determine whether a person is driving under the influence of alcohol and can quantitatively analyze ethanol intake. In the medical field, as a non-invasive technique, the analysis of breathing by using gas sensors can enable better diagnoses. For instance, the concentration of acetone is a crucial indicator to determine whether a patient has diabetes. In addition, gas sensors play an important role in the detection of automotive exhaust gas, in aquaculture, and in many other fields. Moreover, liquid sensors play a vital role in the detection of toxic compounds in the laboratory and industrial wastes. Liquid sensors are also crucial for detecting toxic compounds in the water supply systems for homes, commercial establishments, industry, and irrigation. It is extremely desirable to develop sensors that have excellent selectivity, sensitivity, reversibility, and durability combined with microelectronic circuitry and software. An adequate set of metrics, such as sensitivity selectivity, response/recovery time, operating temperature, and life cycle, are used to assess the performance of a sensor. The gas sensors may be divided into chemical-type gas sensors operated by gas adsorption and physical-type gas sensors operated by field ionization.

12.2 CHEMICAL SENSORS

Chemical information is converted into an analytical signal by a chemical sensor. Gaseous substances are often adsorbed onto the surface of active materials. The kind and concentration of the gas may be identified by looking at changes in the material's characteristics. These kinds of sensors employ metal oxides, carbon nanotubes (CNTs), polymers, porous silicon, and other materials as their active layer. The active layer typically adsorbs the gas molecules, changing the electrical resistance of the sensor.

DOI: 10.1201/9781003502692-12

$$S(\text{sensitivity}) = (R_g - R_i)\backslash R_i \quad (12.1)$$

where R_i and R_g are the initial resistance and the resistance after exposure, respectively.

Except for a few polymer-based sensors, the working temperature is high (200–500°C) in this type of sensor. Although chemical-type sensors may have low cost, a simple operating system, and good sensitivity, these sensors lack high accuracy in the detection of gases with low adsorption energy. Another disadvantage is that the nature of the detecting element of these sorts of sensors changes as a result of gas adsorption and is typically irreversible.

12.3 PHYSICAL SENSORS

Gaseous-breakdown ionization sensors provide a number of benefits in terms of miniaturization and are selectivity compared to other physical types of gas sensors that are available on the market or described in the literature. The fabrication and assessment of an ionization microsensor that detects various gaseous compounds based on the electrical breakdown of gases are reported by Modi et al. [1]. Huang et al. incorporated CNTs grown on silicon substrate into their gas ionization sensor [2]. However, in an atmosphere comprising oxygen, the CNTs can easily decompose because of the oxidation process. Although ZnO nanowires have breakdown voltages that are greater than those of CNTs, they are far more stable and have superior anti-oxidation properties. Liao et al. [3] show the ZnO nanowires exhibit greater stability and anti-oxidation characteristics than CNTs when used as the field ionization anode. The material surface of metal oxide semiconductors (MOSs) exhibits strong adsorption characteristics due to the abundance of free electrons in the conduction band and oxygen vacancies. As a result, measurements based on electrical properties are possible. These materials also have a large band gap, which enables them to have a wide range of electrical characteristics. The material size frequently has a significant impact on the characteristics of MOSs. A material will exhibit distinct characteristics at the nanoscale because of the nano effect. For instance, major modifications to the electrical characteristics at the nanoscale result in the development of an excellent gas-sensing material. Further, for practical use, these MOSs must have higher stability and rapid responsiveness in challenging situations. Acceptance of this material is also largely due to its affordability and ease of manufacture. The electron depletion layer (EDL) theory, the hole accumulation layer (HAL) theory, the bulk resistance control mechanism, and the gas diffusion control mechanism are a few of the intriguing mechanisms for metal oxide sensors that have been discovered by analyzing the data from various perspectives. There are two types of commonly used gas-detecting methods. One type comprises processes like Fermi-level control theory, grain boundary barrier control theory, and EDL/HAL theory that characterize the changes in electrical characteristics from a microscopic point of view. The focus of the second type, which is comparatively macroscopic, is primarily on the interaction between materials and gases. This category includes the gas diffusion control mechanism, the bulk resistance control mechanism, and the adsorption/desorption model. These theories allow us to analyze the process of gas sensing by the sophisticated material analysis methods based on real physical phenomena.

12.4 CONVENTIONALLY SYNTHESIZED METAL OXIDE SENSORS

Semiconductor gas sensors are the most promising among electrochemical, contact combustion, and semiconductor gas sensors. The use of a variety of oxide semiconductors, such as NiO, ZnO, SnO_2, and CdO, has triggered extensive studies into metal oxides for gas sensing. The p-type semiconductor known as an oxidation semiconductor is one whose conductivity rises with the presence of an oxidizing environment. The n-type semiconductor known as a reduction semiconductor is one whose conductivity rises with the presence of a reduction environment. The term amphoteric semiconducting refers to a semiconductor whose conductivity type produces a p-type or n-type semiconductor when exposed to atmospheric oxygen partial pressure. There are several ways to improve the gas-sensing performance, such as modification of surface with organic molecules, conducting inorganic heterojunction sensitization, making hybrid structures, and modification of oxygen vacancy [4]. Moreover, adding a specific amount of rare earth metals to the material is also an efficient method to enhance the sensitivity and selectivity. In this context, Song et al. synthesize porous NiO nanotubes with core-shell structure using a hydrothermal method to sense 50ppm ethanol [5]. The material exhibits outstanding gas sensitivity because the hollow porous core-shell structure allows ethanol molecules to permeate and transport instantly into the interior of the sensor. Wang et al. synthesize a series of nanocomposites of polyaniline–TiO_2 on the TiO_2 surface using in situ chemical oxidation polymerization of aniline [6]. It was learned that the polyaniline chain made of polyaniline and TiO_2 may boost CO adsorption, which resulted in a significant increase in the transfer of electrons from CO to polyaniline. For hydrogen gas detection, Nulhakim et al. deposit highly Ga-doped thin films of ZnO polycrystalline using radio-frequency magnetron sputtering [7]. The effectiveness of hydrogen gas sensing in relation to the microstructure of preferred c-axis oriented thin films is investigated. Under the operating temperature of 330°C, it is discovered that the sample's sensitivity to hydrogen marginally increases as the microcrystalline size decreases. Additionally, by expanding the preferred orientation distribution, the sensitivity is enhanced significantly. Rai et al. fabricate the Au/NiO core-shell structure for hydrogen sulfide (H_2S) gas detection using a two-step hydrothermal technique [8]. The H_2S gas is adsorbed on the surface of the Au particles by the sulfurization layer, which decreases the reaction potential energy, facilitate the electron transport from the Au particles to the NiO shell, and enhanced the resistance value. Using hydrothermal method, Yang et al. synthesize very long MoO_3 nanobelts with an average length of 200μm and width of 200–400nm [9]. The findings show that the ultralong MoO_3 nanobelt based gas sensor has high trimethylamine gas detection capability at a temperature of 240°C. Amani et al. synthesize a WO_3 based gas sensor using electron beam evaporation technique and study the WO_3 nano film's gas-detecting properties when it is triggered by Pt and Au [10]. The findings show that the activation of the WO_3 nano thin film by the Pt layer greatly reduces the operating temperature of the sensor.

For H_2S gas sensing at ambient temperature, N-type semiconducting metal oxide nanostructures (SMONs) based on ZnO, In_2O_3, CeO_2, and Fe_2O_3 have often been reported (RT) [11–14]. Among them, wide band-gap semiconductors ZnO and In_2O_3 stand out due to their respective band gaps of 3.3 and 3.6eV. Because H_2S molecules may be quickly broken down and can react with the chemisorbed oxygen species on the surface of these sensing materials due to the low bond energy of H-S-H, they are particularly effective for H_2S sensing. The H_2S molecules react with ZnO or In_2O_3 on their surfaces, forming ZnS or In_2S_3 in addition to SO_2 and H_2O from their reactions with the oxide ions of O_2. Resistance is significantly reduced as a result of the interactions with the oxide ions that increase the electron concentrations on the surface of ZnO or In_2O_3. Since ZnS and In_2S_3 are metallic conductors, their formation also reduces sensor resistance, which considerably improves the responses to the gases at RT. The formation of these metal sulfides, which do not react with many other gases, including NH_3, H_2, NO_2, CO, CH_4, C_2H_5OH, and HCHO, results in good H_2S selectivity for RT sensors composed of nanostructured ZnO or In_2O_3. Metal sulfide reactions during the sensing phase and their conversion back to metal oxides during the recovery step can both be very sluggish at RT. As a result, the reaction and recovery times are frequently prolonged, sometimes by several hours, for the RT H_2S gas sensors [15,16]. Dendritic ZnO nanostructures synthesized utilizing a vapor-phase transport technique using Cu as the catalyst at 930°C can increase the response and recovery rates of RT H_2S gas sensors [17]. It has also been observed that other N-type SMONs, including Fe_2O_3 and CeO_2, are suitable sensing materials for H_2S sensing at RT [13,14]. For instance, after annealing a precursor FeOOH nanoparticle, porous α-Fe_2O_3 nanoparticles with a diameter of 34nm and pore sizes from 2 to 10nm are obtained [14]. This porous α-Fe_2O_3 nanoparticle-based sensor has a high sensitivity (~38 for 100ppm H_2S) and a limit of detection (LOD) of 50 ppb.

Moreover, it exhibits strong repeatability and good selectivity to H_2S in comparison to other gases (such as C_2H_5OH, CO, H_2, and NH_3). It shows fast response time (180 seconds) with long recovery period, measuring ~3,750 seconds for 100 ppm H_2S. Another simple hydrothermal method is used to synthesize CeO_2 nanowires, which exhibit quick reaction and recovery times (24 and 15 seconds for 50 ppb H_2S, respectively) [13]. By annealing $Sn(OH)_4$ precursor powders at 550°C in both vacuum and ambient air, Wei et al. synthesize SnO_2 nanocrystals [18]. Compared to air annealed SnO_2, the response value of the vacuum annealed SnO_2 sensor to 5 ppm of NO_2 at RT is higher. This is mostly due to the fact that the surface oxygen vacancies on the vacuum annealed SnO_2 nanocrystals are much higher than those on the air annealed SnO_2 nanocrystals. For NO_2 detection at RT, Yu et al. synthesize ZnO nanowalls by a solution method with evenly distributed and cross-linked nanowalls of 20 nm [19]. The sensor demonstrates strong repeatability, a high response value, and quick reaction and recovery times toward 50 ppm NO_2 at RT. Some often mentioned nanomaterials for NO_2 sensing are TiO_2 and In_2O_3. In order to synthesize porous TiO_2 nanoparticles with a high concentration of oxygen vacancies and the interstitial defect states that are essential for the effective adsorption and desorption of NO_2 gas molecules, Tshabalala et al. use a hydrothermal method [20]. These nanostructures are used to fabricate a sensor that has a high response, fast response/recovery times, and a low LOD of 20 ppb at RT. With its relatively strong responses to several different gases, such as H_2, NH_3, and CH_4, this sensor has poor selectivity. The sol-gel method has also been used to synthesize In_2O_3 octahedra for NO_2 sensing, and the resulting sensor has a response value of 63–200 ppm NO_2 at RT and exhibits strong selectivity to NO_2 against CO, H_2, and NH_3 [21]. One of the most explosive and highly flammable gas is the hydrogen gas. To reduce the risk of an explosion, it is essential to detect any hydrogen gas utilizing RT gas sensors. The sensors for this application must be quick, extremely sensitive, and selective. An instant response is crucial for the timely detection of a potential hydrogen leakage. The interaction of H_2 molecules with chemisorbed O^{2-} ions on the surface of the SMONs serves as the basis for the sensing process. ZnO based sensors have strong sensitivities to H_2 but have sluggish response and recovery times [22]. Lim et al. synthesize vertical ZnO nanorods utilizing atomic layer deposition using anodized aluminum oxides as nano-templates. They have use these vertical ZnO nanorods to fabricate a highly sensitive and quick response/recovery H_2 gas sensor [23]. A quicker response using [001] oriented α-MoO_3 nanoribbons, a hydrogen gas sensor is also reported with a response time of ~14 s for 1,000 ppm and a low LOD of 500 ppb [24]. Against ethanol, CO, and acetone, it is very selective and has remarkable reproducibility.

The performance of metal oxide sensors is significantly influenced by environmental humidity. Nevertheless, the way that water vapor and other toxic gases like CO, NO_2, and H_2S are sensed is different. Ionic-type humidity sensors are the most prevalent designs for metal oxide humidity gas sensors. The dissociation of adsorbed water, which bounces between neighboring hydroxyl groups, is what drives the conduction process, which is dependent on H^+ or H_3O^+ [25]. The metal oxide surface will not provide electrons to the detecting active layers when water adsorbs on it. Moreover, it will reduce the sensitivity of metal oxide sensors due to the following factors. The gas sensor's baseline resistance decreases due to the interaction between surface oxygen and water molecules, which also lowers the sensitivity. [26]. The reduction in surface area caused by the adsorption of water molecules results in reduced chemisorption of oxygen species on the SnO_2 surface, which affects the sensor response [27]. Besides, water molecules also function as a barrier to prevent C_2H_2 adsorption. The sensitivity drops, the response/recovery times rise as the superficial migration of the C_2H_2 on the SnO_2 surface becomes challenging. Metal oxide gas sensors' sensitivity will drop dramatically as a result of water adsorption. Moreover, continual exposure to humid conditions causes the surface to develop persistent chemisorbed OH^-, which gradually reduces the sensitivity of gas sensors [25]. Nevertheless, surface hydroxyls begin to desorb at around 400°C, and the removal of the hydroxyl ions requires heating over 400°C [28]. The sensor resistance does not return to its initial value after a number of humidity pulses; however, heating the sensor to a temperature of 450°C for many minutes causes the signal to fully recover [27]. With the metal oxide gas sensors, temperature is also crucial. The responses of gas sensors increase and reach the peak at a certain temperature, and then decrease rapidly with further increasing in the temperature [29], [30]. This is because sluggish kinetics at low temperatures and accelerated desorption at high temperatures are in competition with one another [30].

12.5 GREEN SYNTHESIS OF METAL OXIDE NANOSTRUCTURES

Metal oxide nanoparticles are inspiring materials that have unique characteristics and a wide range of applications. However, the primary techniques now used in their manufacture are not eco-friendly. Among physical, chemical, and biological synthesis methods of nanoparticles, the chemical approach has received the greatest attention in recent years. As compared to precursor chemicals, the substance produced chemically is either less toxic, as in the case of Au, or similarly harmful, as in the case of Ag [31,32]. The chemically synthesized nanoparticles would, however, still have certain undesirable characteristics. This barrier is the main cause of the present growth of environmentally friendly green nanoparticle synthesis. We can get nontoxic byproducts by the development and usage of efficient alternatives, such as synthesizing nanoparticles using plant extracts. The product is more ecologically friendly since the production processes are considerably more similar to those found in nature. As a result, the fundamental ideas of green synthesis, which call for utilizing harmless materials and reducing waste and pollution, are also adopted. The main aspects of green synthesis techniques are the methodology and their yields. Extracts used to biosynthesize metallic and metal oxide nanoparticles contain a variety of types of secondary metabolites. The bulk of the secondary metabolites found in plant extracts are bioactive chemicals, such as flavonoids, alkaloids, terpenoids, phenolic compounds, and enzymes that aid in reducing metal ions and stabilizing nanoparticles.

Leaf extracts of *C. fistula* and *M. azedarach* undergo physio-chemical alterations in the aqueous solution when zinc acetate dihydrate is added [33]. The most noticeable of them is the color change that may be seen in the reaction mixture within a short period. This is regarded as the first indication of nanoparticle formation. The conversion of Zn ions to ZnO nanoparticles is hypothesized to be caused by flavonoids and phenolic substances. The color of the solution stops changing after a few hours, indicating that the ZnO salt had fully undergone bioreduction to become nanoparticles. The synthesis of the nanoparticles is thought to be greatly influenced by temperature. Furthermore, it is well known that the size of the nanoparticles decreases with increasing reaction temperature during the synthesis. As a result, the reactants are incubated at a considerably higher temperature of 70°C, which results in the formation of extremely small-sized ZnO nanoparticles. Darroudi et al. have used gelatin (GL, type B) for synthesizing ZnO nanoparticles to study the cytotoxicity effects [34]. It is discovered that the production of ZnO nanoparticles in gelatinous medium is equivalent to that of standard reduction techniques utilizing hazardous polymers or surfactants. CuO and ZnO nanoparticles are synthesized by Maruthupandy et al. using *Camellia japonica* leaf extract for optical sensor applications [35]. Different sizes of ZnO nanoparticles are synthesized by *Parthenium hysterophorus* plant extracts to study the antifungal activity with respect to the particles size [36]. Further, Sharma et al. synthesized CuO nanoparticles of 30–40 nm by green synthesis method using *Calotropis gigantean* leaf extract [37].

Controlling particle size and shape, as well as attaining monodispersity in the solution phase, is a common challenge for the bio-production of nanoparticles. Normally, nanoparticles are synthesized using complex procedures, and the outcomes depend on numerous parameters. If the intended particle size is extremely small, precise control of the parameters is required. Although there are many variables that can impact the green synthesis of metal oxide nanoparticles, there have been several attempts to resolve these issues by adjusting growth parameters such as light, pH, reactant concentration, temperature, and incubation period.

12.6 GREEN SYNTHESIZED HFO_2 NANOPARTICLES

There is well-explained literature on metal oxide-based NH_3 gas sensors which work on the chemoresistance principle. In such sensors, the surface oxide layer adsorbs the reducing gas and alters the resistivity of the metal oxide by injecting electrons in the conduction band, thus facilitating the flow of electrons [38]. However, only a few reports are available for sensing liquid ammonia. Moreover, considering environmental sustainability and the impact of the conventional synthesis methods of nanoparticles for sensing applications, increased emphasis is being directed toward developing or adopting green (environment-friendly) methods for the synthesis of relevant nanoparticles. Although there are many reports available on the synthesis of HfO_2 nanoparticles using conventional methods, there is no work reported on the green synthesis of HfO_2 nanoparticles using biomaterials. Thus, the overall objective of this study is to develop a green synthesized HfO_2-based ammonia sensor.

We have used the orange peel extracts as the biomaterials for the synthesis of HfO_2 nanoparticles. To obtain the orange peel extract, oranges were initially washed two times with deionized water. The peeled-off orange peels were then dried in an oven for 9 hours at 65°C. Dried peels were further ground into moderately fine powder using a mortar and pestle. Required amount of orange peel powder (1, 2 and 4 g) was taken in a glass beaker with 100 mL of deionized water. The mixture was stirred for 3 hours at room temperature. Then the mixture was kept for water bath at 60°C for 60 minutes. After cooling down to the room temperature, the mixture was filtered first with a muslin cloth and then with a Whatman grade 2 filter paper. The filtered orange peel extract was kept in an airtight Nalgene bottle at 4°C for further use. HfO_2 nanoparticles were synthesized using a green sol-gel method. The HfO_2-orange nanoparticles were synthesized by mixing 3 g of hafnium (IV) chloride with 45 mL of 1 wt% orange peel extract. The solution was stirred for 60 minutes at room temperature and then placed in a water bath at 60°C for 60 minutes. Subsequently, the solution was dried at 100°C for 2 hours and ground into fine powder using a mortar and pestle. The sample was then calcined at 900°C for 1 hour using a porcelain crucible with lid in a muffle furnace. Following the above method, HfO_2 nanoparticles were also synthesized using 2 and 4 wt% of orange peel extract. The HfO_2 nanoparticles synthesized using 1 wt% orange peel extract (OPE) was denoted as HO-1-OPE. Similarly, the HfO_2 nanoparticles synthesized using 2 and 4 wt% OPE were denoted as HO-2-OPE and HO-4-OPE, respectively.

X-ray diffraction patterns of the green synthesized HfO_2 samples were recorded using a PANalytical Empyrean powder X-ray diffractometer. The bright field transmission electron micrographs of the HfO_2 samples were obtained by a Tecnai 20 TEM operating at an acceleration voltage of 200 keV. The green synthesized HfO_2 nanoparticles were characterized through a Cary 60 UV–Visible spectrometer to record the absorption spectra.

12.6.1 STRUCTURAL, MICROSTRUCTURAL, AND OPTICAL PROPERTIES OF THE HFO_2 NANOPARTICLES

The formation of the HfO_2 nanoparticles by the OPEs is due to the presence of biomolecules (such as flavonoids, limonoids, and carotenoids) in the peel extracts [39]. The biomolecules present in the peel extracts act as the ligation agents. The aromatic hydroxyl groups from the biomolecules form complexing agents with the hafnium chloride and ligate with hafnium ions. This starts the nucleation process that goes into reverse micellization, which in turn causes the reduction and shaping of nanoparticles. Subsequently, when calcined at a suitable temperature HfO_2 samples were formed by direct decomposition. The HfO_2 nanoparticles were synthesized using different concentrations of OPE to examine the effect of OPE concentration on the properties of the HfO_2 samples and to choose the suitable sample for sensing applications. We have calcined aliquots of the sample at 600–900°C for different time periods. We observe that HfO_2 samples calcined at 900°C for 1-hour show sharp X-ray diffraction (XRD) peaks. We fixed the calcination temperature of all HfO_2

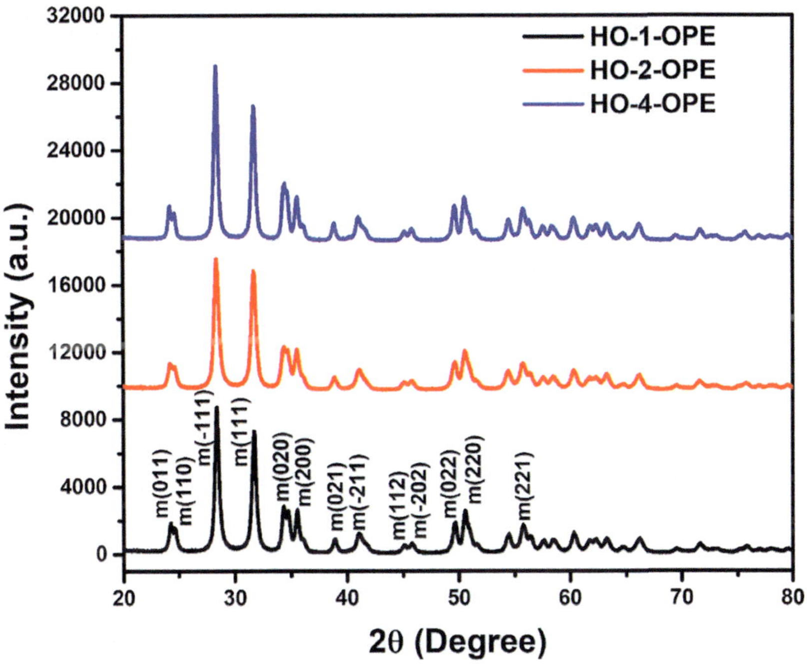

FIGURE 12.1 XRD patterns of HfO_2 samples synthesized using 1, 2, and 4 wt% orange peel extract.

samples synthesized using different concentrations of OPE as 900°C for 1 hour. The time for calcination of green synthesized HfO_2 is smaller than the reported calcination times of HfO_2 synthesized by conventional chemical methods. The yields of HO-1-OPE and HO-2-OPE are 91% and 95%, respectively, whereas the yield of HO-4-OPE is 97%. HO-4-OPE sample shows a higher yield than the other samples. It is to be noted that high yield of HfO_2 nanoparticles using citrus fruit extracts has significance from large scale industrial production perspective.

Figure 12.1 shows the XRD patterns of HfO_2 samples synthesized using 45 mL of 1, 2, and 4 wt.% OPE. The sharp diffraction peaks such as (011), (110), (–111), (111), (020), (200), (021), (–211), and (112) are indexed as the monoclinic phase of HfO_2 ($P2_1/c$) with JCPDS number: 78-0049.

We have done the Le Bail profile fitting for the XRD data of HfO_2 samples synthesized using OPE as the reducing and stabilizing agent. Le Bail profile fitting is carried out by the FullProf_suite software to find out the cell parameters of the HO-4-OPE. We used the pseudo-Voigt analytical function to fit the experimental profiles as it considers both crystallite size and strain broadening of the experimental profiles [40]. A typical Le Bail refinement plot of HO-4-OPE is shown in Figure 12.2a.

Structural cell parameters of the monoclinic HfO_2 phase with $P2_1/c$ space group are obtained from the refined data and are shown in Table 12.1 along with the cell parameters of standard HfO_2 from JCPDS card number 78-0049. The refined cell parameters match well with the standard cell parameters of monoclinic HfO_2 as well as with the reported cell parameters for HfO_2 nanoparticles [41].

The X-ray diffraction analysis was employed to determine the full width at half maximum (β), Bragg angle (θ), strain (ε), average crystallite size (D), and X-ray wavelength (λ) for HO-4-OPE. The Williamson–Hall plot in Figure 12.2b reveals an average crystallite size (D) of 34 nm and lattice strain (ε) of 0.00182 for HO-4-OPE, with k as a constant (0.94 for spherical particles) [42].

TEM micrographs in Figure 12.3a–c illustrate the morphology of HO-1-OPE, HO-2-OPE, and HO-4-OPE, respectively. Notably, HO-1-OPE and HO-2-OPE exhibit agglomeration in large

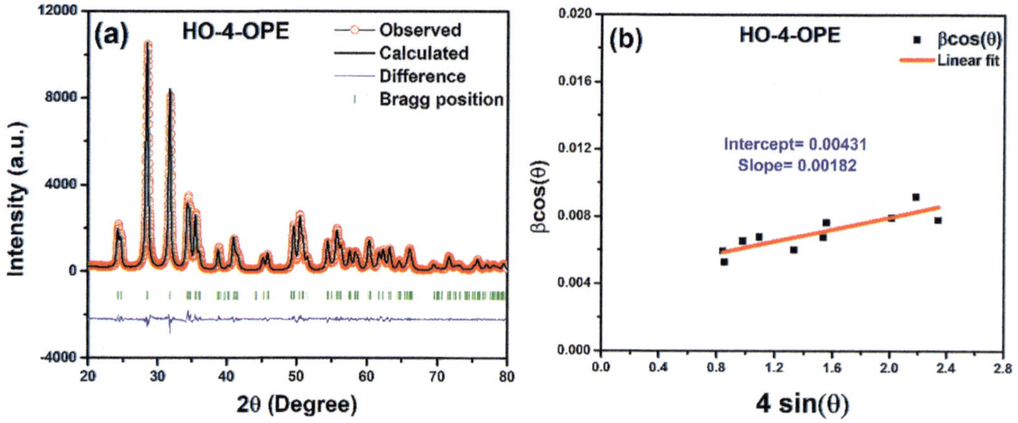

FIGURE 12.2 (a) Le Bail profile fitting and (b) Williamson–Hall plot of HfO_2 sample synthesized using 4 wt% orange peel extract (HO-4-OPE).

TABLE 12.1
Refined Cell Parameters of HO-4-OPE Obtained from Le Bail Profile Fitting

			Cell Parameters				
JCPDS Card No./Sample	Crystal Structure	Space Group	*a* (Å)	*b* (Å)	*c* (Å)	*β* (°)	Volume (Å³)
78-0049	Monoclinic	$P2_1/c$	5.117	5.175	5.291	99.225	138.32
HO-4-OPE	Monoclinic	$P2_1/c$	5.118	5.175	5.290	99.241	138.31

clusters, while HO-4-OPE nanoparticles are well-dispersed, indicating increased surface area. The sensing layer for electrochemical impedance spectroscopy (EIS) measurements employed HfO_2 nanoparticles synthesized with 4 wt% OPE (HO-4-OPE). ImageJ software was utilized to calculate the average particle sizes, as depicted in the histograms in Figure 12.3d. The average particle size of HO-4-OPE, determined as 35 nm, aligns closely with the average crystallite size obtained from the Williamson–Hall plot.

TEM micrographs in Figure 12.3a–c illustrate HfO_2 samples synthesized with varying concentrations of OPE, specifically 1, 2, and 4 wt.%. The average particle size histogram for HO-4-OPE is presented in Figure 12.3d. Furthermore, Figure 12.3e displays UV–Visible absorption spectra for HO-1-OPE, HO-2-OPE, and HO-4-OPE samples.

The UV–Visible spectra reveal distinctive features for HO-4-OPE, displaying an absorption edge and peak indicative of HfO_2, with a peak at 215 nm corresponding to its absorption band [41]. Notably, no absorption peaks are observed in the visible region. In contrast, HO-1-OPE and HO-2-OPE exhibit no absorption edges or peaks within the 200–800 nm wavelength range, suggesting the absence of absorption bands in visible light. This observation, coupled with the absence of peaks in agglomerated particles, underscores the selection of HO-4-OPE as the active material for fabricating the liquid ammonia sensing device. The decision is grounded in the unique spectral characteristics observed in HO-4-OPE [41].

12.6.2 Device Fabrication and Performance for Liquid Ammonia Sensing

In this investigation, the liquid ammonia sensing properties of HfO_2 nanoparticles, synthesized through a green approach, were examined using an electrochemical method. EIS served as the

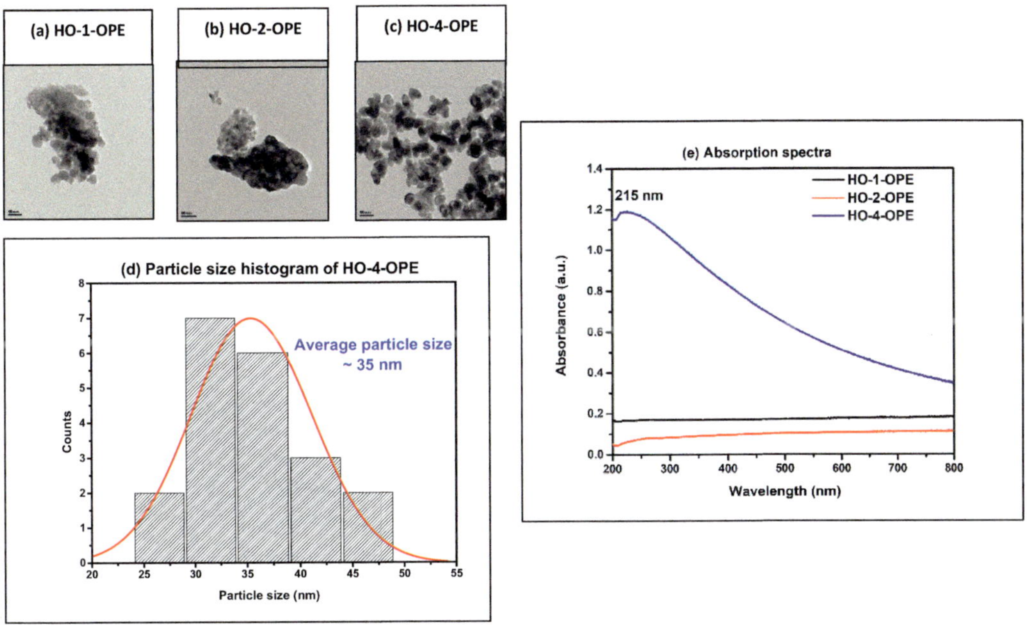

FIGURE 12.3 (a–c) TEM micrographs of HfO_2 samples synthesized using 1,2, 4 wt% orange peel extract, (d) average particle size histogram of HO-4-OPE, (e) UV-Visible spectra of HO-1-OPE, HO-2-OPE and HO-4-OPE samples.

primary technique for studying liquid ammonia sensing, employing zinc oxide-coated interdigitated electrodes [43].

The experimental setup involved three-electrode configuration screen-printed electrodes (SPEs) obtained from BASi, USA. These electrodes featured a 4-mm diameter gold working electrode, a gold counter electrode, and a silver/silver chloride reference electrode. Prior to nanoparticle deposition, electrodes underwent cleaning with acetone, followed by air-drying at room temperature. For the preparation of the HfO_2 slurry, a polyvinylidene fluoride (PVDF) N-methyl pyrrolidone (NMP) binder was employed. The PVDF NMP binder was created by adding 10 mL of NMP to 300 mg of PVDF, resulting in a 3 wt% PVDF solution in NMP. The mixture was stirred for 12 hours to achieve complete homogeneity. The HfO_2 slurry was then prepared by combining 8 mg of HO-4-OPE sample with 600 μL of distilled water (DI) water in a mortar and pestle. Subsequently, 5 μL of the prepared PVDF NMP binder was added, and the mixture was ground until complete homogeneity. To this, 400 μL of 200 proof ethanol was added and thoroughly mixed. A 10 μL volume of the resulting homogeneous mixture was drop-casted onto the working electrode, left for 15 minutes, and then dried in an oven at 60°C for 2 hours, forming the electrode coating [43].

For the liquid ammonia sensing experiments via EIS, we investigated six distinct concentrations of liquid ammonia: 50, 100, 200, 300, 400, and 500 ppm. To prepare these concentrations, a 29 w/w% ammonia solution was appropriately diluted with deionized water. The EIS spectra were recorded using Gamry's Reference 600 potentiostat. The electrochemical measurements were conducted in a three-electrode cell configuration, as illustrated in the schematic diagram in Figure 12.4a. The experimental setup for measuring EIS, depicted in Figure 12.4b, utilized Gamry's Reference 600 potentiostat. The EIS measurements for various ammonia concentrations were performed using the HO-4-OPE-coated SPE in the frequency range of 100 mHz–1 MHz. The applied AC voltage was maintained at a constant 10 mV for each concentration [43].

In this study, the liquid ammonia sensing behavior of HfO_2 nanostructures, a novel exploration, was conducted utilizing the HO-4-OPE-coated electrode for EIS at concentrations of 50, 100,

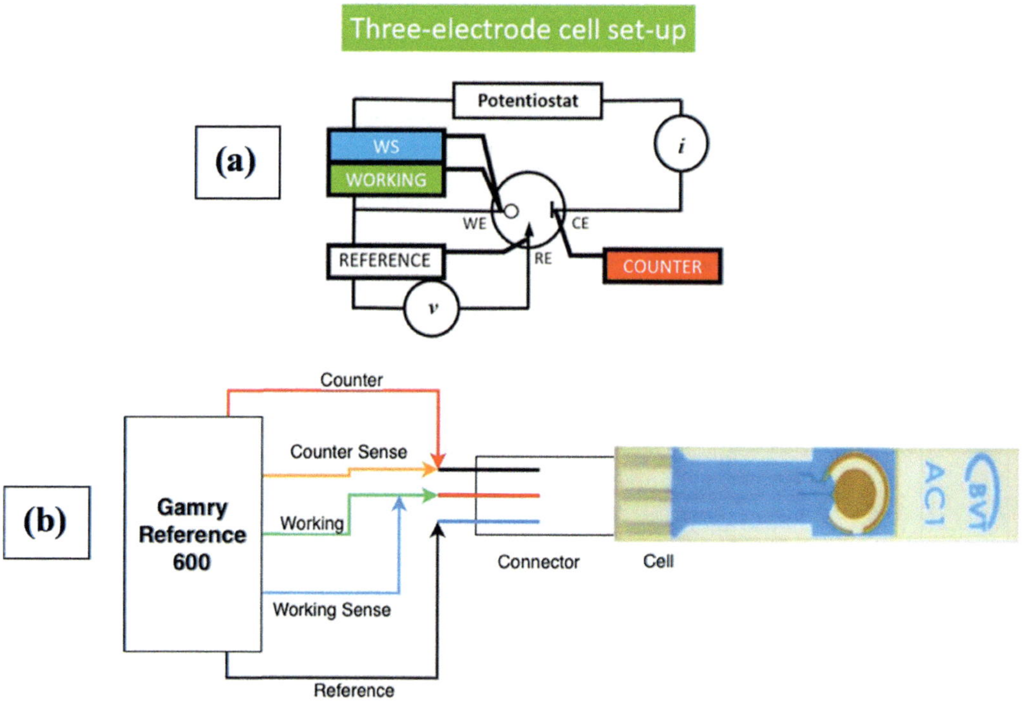

FIGURE 12.4 (a) General setup for three electrodes configuration and (b) schematic diagram of our connection for the screen-printed electrode to Gamry Reference 600 potentiostat (diagram not to scale).

200, 300, 400, and 500 ppm ammonia. EIS measurements were performed with an AC voltage of 10 mV over a frequency range of 100 mHz to 1 MHz, and the resulting Bode and Nyquist plots are presented in Figure 12.5a and b. The Bode plot indicates a phase angle approaching −90 degree at higher frequencies, signifying diffusion phenomena due to the constant phase element. The observed hump in the phase angle at 12.5 kΩ is attributed to a frequency filter change in the potentiostat. For liquid ammonia sensing, the charge transfer resistance was considered as the sensing parameter. The Nyquist plot reveals a decrease in charge transfer resistance with increasing ammonia concentration. The acquired Bode and Nyquist plots were fitted using the equivalent circuit illustrated in Figure 12.5c, featuring two constant phase elements in series. In this circuit, R1 represents the charge transfer resistance through the double layer formed at the electrode–analyte interface. The variations in charge transfer resistances, obtained after fitting, are presented in Figure 12.5d [43].

For liquid NH_3 sensors, the electrochemical reaction occurs at the surface of the electrode. Liquid NH_3 is catalyzed by the superoxides (O_2^-) adsorbed on the surface of the electrode [44]. It is due to the mechanism of nitrification of NH_4^+ to form nitrate ions (NO_3^-). The energy generated during oxidation is enough to cause the electrons to move into the conduction band, thus reducing the materials resistance [45]. The reaction for nitrification is given below.

$$NH_4^+ + 2O_2^- \rightarrow NO_3^- + H_2O + 2H^+ \quad (12.2)$$

We have studied the detailed quantification of the aqueous ammonia sensing through the green synthesized HfO_2 nanoparticle. The variations of charge transfer resistance for HO-4-OPE-coated SPE with respect to ammonia concentration are fitted with the following exponential decay function,

$$y = y_0 + A_1 \exp(-x/t_1) \quad (12.3)$$

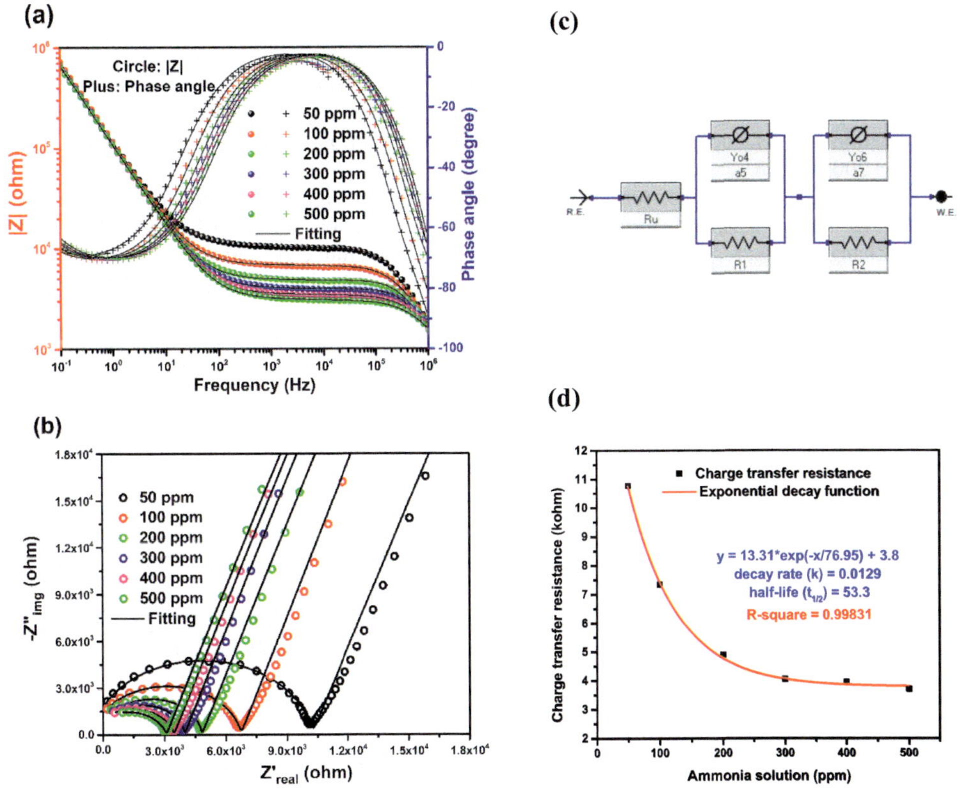

FIGURE 12.5 (a) Bode and (b) Nyquist plots obtained from electrochemical impedance spectroscopy (EIS) at different ammonia concentrations with 10 mV ac voltage in 100 mHz to 1 MHz frequency range, (c) model used for fitting EIS data, and (d) variation of charge transfer resistance with respect to ammonia concentration.

where y_0 is the offset, A_1 is the amplitude, and t_1 is the time constant of the exponential decay function (Figure 12.5d). The pertinent derived parameters such as decay rate ($k = 1/t_1$) and half-life ($t_{1/2} = t_1 * \ln(2)$) are also mentioned in Figure 12.5d. The coefficient of determination (R^2) for the developed model is greater than 0.99.

12.7 CONCLUSIONS

In summary, this study provided an overview of sensor principles, encompassing chemical and physical sensor types. A comprehensive literature survey focused on conventional synthesis methods for metal oxide gas sensors, while highlighting recent developments in room temperature-operating metal oxide-based gas sensors. The impact of environmental factors on sensor poisoning throughout its lifespan was explored. The research delved into the green synthesis of HfO_2 nanostructures, specifically nanoparticles, using OPE. Structural, microstructural, and optical properties were characterized through XRD, TEM, and UV–Visible spectroscopy. Notably, HfO_2 synthesized with 4 wt% orange peel extract and calcined at 900°C for 1 hour (HO-4-OPE) exhibited well-dispersed crystalline nanoparticles. For liquid ammonia sensing, EIS was employed at six concentrations ranging from 50 to 500 ppm. The study reported successful detection of ammonia concentrations as low as 50 ppm in liquid samples. Consequently, HfO_2 nanoparticles emerged as a promising sensing material for ammonia, holding significant implications for healthcare applications [44,45].

REFERENCES

1. A. Modi, N. Koratkar, E. Lass, B. Wei, and P. M. Ajayan, "Miniaturized gas ionization sensors using carbon nanotubes," *Nature*, vol. 424, no. 6945, pp. 171–174, 2003, doi: 10.1038/nature01777.
2. J. Huang, J. Wang, C. Gu, K. Yu, F. Meng, and J. Liu, "A novel highly sensitive gas ionization sensor for ammonia detection," *Sens Actuators A Phys*, vol. 150, no. 2, pp. 218–223, 2009, doi: 10.1016/j.sna.2009.01.008.
3. L. Liao et al., "A novel gas sensor based on field ionization from ZnO nanowires: Moderate working voltage and high stability," *Nanotechnology*, vol. 19, no. 17, 2008, doi: 10.1088/0957-4484/19/17/175501.
4. C. Zhang, Y. Luo, J. Xu, and M. Debliquy, "Room temperature conductive type metal oxide semiconductor gas sensors for NO_2 detection," *Sens Actuators A Phys*, vol. 289, no. 2, pp. 118–133, 2019, doi: 10.1016/j.sna.2019.02.027.
5. X. Song, L. Gao, and S. Mathur, "Synthesis, characterization, and gas sensing properties of porous nickel oxide nanotubes," *Journal of Physical Chemistry C*, vol. 115, no. 44, pp. 21730–21735, 2011, doi: 10.1021/jp208093s.
6. Z. Wang, X. Peng, C. Huang, X. Chen, W. Dai, and X. Fu, "CO gas sensitivity and its oxidation over TiO2 modified by PANI under UV irradiation at room temperature," *Appl Catal B*, vol. 219, no. 2, pp. 379–390, 2017, doi: 10.1016/j.apcatb.2017.07.080.
7. L. Nulhakim, H. Makino, S. Kishimoto, J. Nomoto, and T. Yamamoto, "Enhancement of the hydrogen gas sensitivity by large distribution of c-axis preferred orientation in highly Ga-doped ZnO polycrystalline thin films," *Mater Sci Semicond Process*, vol. 68, no. July, pp. 322–326, 2017, doi: 10.1016/j.mssp.2017.06.045.
8. P. Rai, J. W. Yoon, H. M. Jeong, S. J. Hwang, C. H. Kwak, and J. H. Lee, "Design of highly sensitive and selective Au@NiO yolk-shell nanoreactors for gas sensor applications," *Nanoscale*, vol. 6, no. 14, pp. 8292–8299, 2014, doi: 10.1039/c4nr01906g.
9. S. Yang et al., "High sensitivity and good selectivity of ultralong MoO_3 nanobelts for trimethylamine gas," *Sens Actuators B Chem*, vol. 226, pp. 478–485, 2016, doi: 10.1016/j.snb.2015.12.005.
10. E. Amani, K. Khojier, and S. Zoriasatain, "Improving the hydrogen gas sensitivity of WO3 thin films by modifying the deposition angle and thickness of different promoter layers," *Int J Hydrogen Energy*, vol. 42, no. 49, pp. 29620–29628, 2017, doi: 10.1016/j.ijhydene.2017.10.027.
11. A. D. Faisal, "Synthesis of ZnO comb-like nanostructures for high sensitivity H_2S gas sensor fabrication at room temperature," *Bull Mater Sci*, vol. 40, no. 6, pp. 1061–1068, 2017, doi: 10.1007/s12034-017-1461-6.
12. Z. Li et al., "Ultra-sensitive UV and H_2S dual functional sensors based on porous In_2O_3 nanoparticles operated at room temperature," *J Alloys Compd*, vol. 770, pp. 721–731, 2019, doi: 10.1016/j.jallcom.2018.08.188.
13. Z. Li et al., "Hydrothermally synthesized CeO_2 nanowires for H_2S sensing at room temperature," *J Alloys Compd*, vol. 682, pp. 647–653, 2016, doi: 10.1016/j.jallcom.2016.04.311.
14. A. Thakur, P. Thakur, and S. Baccar, Structural properties of nanoparticles. In: Suhag, D., Thakur, A., Thakur, P. (eds) *Integrated Nanomaterials and their Applications*. Springer, Singapore (2023). https://doi.org/10.1007/978-981-99-6105-4_4.
15. Z. S. Hosseini, A. I. Zad, and A. Mortezaali, "Room temperature H_2S gas sensor based on rather aligned ZnO nanorods with flower-like structures," *Sens Actuators B Chem*, vol. 207, no. Part A, pp. 865–871, 2015, doi: 10.1016/j.snb.2014.10.085.
16. C. Wang, X. Chu, and M. Wu, "Detection of H_2S down to ppb levels at room temperature using sensors based on ZnO nanorods," *Sens Actuators B Chem*, vol. 113, no. 1, pp. 320–323, 2006, doi: 10.1016/j.snb.2005.03.011.
17. N. Zhang, K. Yu, Q. Li, Z. Q. Zhu, and Q. Wan, "Room-temperature high-sensitivity H_2S gas sensor based on dendritic ZnO nanostructures with macroscale in appearance," *J Appl Phys*, vol. 103, no. 10, 2008, doi: 10.1063/1.2924430.
18. Y. Wei, C. Chen, G. Yuan, and S. Gao, "SnO_2 nanocrystals with abundant oxygen vacancies: Preparation and room temperature NO_2 sensing," *J Alloys Compd*, vol. 681, no. 2, pp. 43–49, 2016, doi: 10.1016/j.jallcom.2016.04.220.
19. L. Yu et al., "Both oxygen vacancies defects and porosity facilitated NO_2 gas sensing response in 2D ZnO nanowalls at room temperature," *J Alloys Compd*, vol. 682, no. 2, pp. 352–356, 2016, doi: 10.1016/j.jallcom.2016.05.053.
20. F. Wan, A. Thakur, and P. Thakur, Classification of nanomaterials (carbon, metals, polymers, bioceramics). In: Suhag, D., Thakur, A., Thakur, P. (eds) *Integrated Nanomaterials and their Applications*. Springer, Singapore (2023). https://doi.org/10.1007/978-981-99-6105-4_3

21. J. Gao et al., "Mesoporous In_2O_3 nanocrystals: Synthesis, characterization and NO_x gas sensor at room temperature," *New J Chem*, vol. 40, no. 2, pp. 1306–1311, 2016, doi: 10.1039/c5nj02214b.
22. K. Vijayalakshmi and D. Gopalakrishna, "Influence of pyrolytic temperature on the properties of ZnO films optimized for H_2 sensing application," *J Mater Sci: Mater Electron*, vol. 25, no. 5, pp. 2253–2260, 2014, doi: 10.1007/s10854-014-1868-4.
23. Y. T. Lim, J. Y. Son, and J. S. Rhee, "Vertical ZnO nanorod array as an effective hydrogen gas sensor," *Ceram Int*, vol. 39, no. 1, pp. 887–890, 2013, doi: 10.1016/j.ceramint.2012.06.035.
24. S. Yang et al., "Highly responsive room-temperature hydrogen sensing of α-MoO_3 nanoribbon membranes," *ACS Appl Mater Interfaces*, vol. 7, no. 17, pp. 9247–9253, 2015, doi: 10.1021/acsami.5b01858.
25. E. Traversa, "Ceramic sensors for humidity detection: the state-of-the-art and future developments," *Sens Actuators B Chem*, vol. 23, no. 2–3, pp. 135–156, 1995, doi: 10.1016/0925-4005(94)01268-M.
26. J. Gong, Q. Chen, M. R. Lian, N. C. Liu, R. G. Stevenson, and F. Adami, "Micromachined nanocrystalline silver doped SnO_2 H_2S sensor," *Sens Actuators B Chem*, vol. 114, no. 1, pp. 32–39, 2006, doi: 10.1016/j.snb.2005.04.035.
27. A. Tischner, T. Maier, C. Stepper, and A. Köck, "Ultrathin SnO_2 gas sensors fabricated by spray pyrolysis for the detection of humidity and carbon monoxide," *Sens Actuators B Chem*, vol. 134, no. 2, pp. 796–802, 2008, doi: 10.1016/j.snb.2008.06.032.
28. M. Egashira, S. Kawasumi, S. Kagawa, and T. Seiyama, "Temperature programmed desorption study of water adsorbed on metal oxides. I. Anatase and Rutile," *Bull Chem Soc Japan*, vol. 51, no. 11, pp. 3144–3149, 1978. doi: 10.1246/bcsj.51.3144.
29. Z. Jing and J. Zhan, "Fabrication and gas-sensing properties of porous ZnO nanoplates," *Adv Mater*, vol. 20, no. 23, pp. 4547–4551, 2008, doi: 10.1002/adma.200800243.
30. A. Kolmakov, D. O. Klenov, Y. Lilach, S. Stemmer, and M. Moskovitst, "Enhanced gas sensing by individual SnO_2 nanowires and nanobelts functionalized with Pd catalyst particles," *Nano Lett*, vol. 5, no. 4, pp. 667–673, 2005, doi: 10.1021/nl050082v.
31. M. Ehsan, D. Suhag, R. Rathore, A. Thakur, and P. Thakur, Metal Nanoparticles in the Field of Medicine and Pharmacology. In: Suhag, D., Thakur, A., Thakur, P. (eds) *Integrated Nanomaterials and their Applications*. Springer, Singapore (2023). https://doi.org/10.1007/978-981-99-6105-4_7
32. M. Thwala, N. Musee, L. Sikhwivhilu, and V. Wepener, "The oxidative toxicity of Ag and ZnO nanoparticles towards the aquatic plant Spirodela punctuta and the role of testing media parameters," *Environ Sci: Proc Impacts*, vol. 15, no. 10, pp. 1830–1843, 2013, doi: 10.1039/c3em00235g.
33. M. Naseer, U. Aslam, B. Khalid, and B. Chen, "Green route to synthesize Zinc oxide nanoparticles using leaf extracts of Cassia fistula and Melia azadarach and their antibacterial potential," *Sci Rep*, vol. 10, no. 1, pp. 1–10, 2020, doi: 10.1038/s41598-020-65949-3.
34. M. Darroudi, Z. Sabouri, R. Kazemi Oskuee, A. Khorsand Zak, H. Kargar, and M. H. N. Abd Hamid, "Green chemistry approach for the synthesis of ZnO nanopowders and their cytotoxic effects," *Ceram Int*, vol. 40, no. 3, pp. 4827–4831, 2014, doi: 10.1016/j.ceramint.2013.09.032.
35. M. Maruthupandy et al., "Synthesis of metal oxide nanoparticles (CuO and ZnO NPs) via biological template and their optical sensor applications," *Appl Surf Sci*, vol. 397, pp. 167–174, 2017, doi: 10.1016/j.apsusc.2016.11.118.
36. P. Rajiv, S. Rajeshwari, and R. Venckatesh, "Bio-Fabrication of zinc oxide nanoparticles using leaf extract of Parthenium hysterophorus L. and its size-dependent antifungal activity against plant fungal pathogens," *Spectrochim Acta A Mol Biomol Spectrosc*, vol. 112, pp. 384–387, 2013, doi: 10.1016/j.saa.2013.04.072.
37. J. K. Sharma, P. Srivastava, G. Singh, M. S. Akhtar, and S. Ameen, "Catalytic thermal decomposition of ammonium perchlorate and combustion of composite solid propellants over green synthesized CuO nanoparticles," *Thermochim Acta*, vol. 614, pp. 110–115, 2015, doi: 10.1016/j.tca.2015.06.023.
38. A. Dey, "Semiconductor metal oxide gas sensors: a review," *Mater Sci Eng B*, vol. 229, no. July 2017, pp. 206–217, 2018, doi: 10.1016/j.mseb.2017.12.036.
39. C. K. Tagad, S. R. Dugasani, R. Aiyer, S. Park, A. Kulkarni, and S. Sabharwal, "Green synthesis of silver nanoparticles and their application for the development of optical fiber based hydrogen peroxide sensor," *Sens Actuators B Chem*, vol. 183, pp. 144–149, 2013, doi: 10.1016/j.snb.2013.03.106.
40. H. Dutta and S. K. Pradhan, "Microstructure characterization of high energy ball-milled nanocrystalline V_2O_5 by Rietveld analysis," *Mater Chem Phys*, vol. 77, no. 3, pp. 868–877, 2003, doi: 10.1016/S0254-0584(02)00169-4.
41. T. Dehury, S. Kumar, and C. Rath, "Structural transformation and bandgap engineering by doping Pr in HfO_2 nanoparticles," *Mater Lett*, vol. 302, no. June, pp. 130413, 2021, doi: 10.1016/j.matlet.2021.130413.
42. G. K. Williamson and W. H. Hall, "X-ray line broadening from filed aluminium and wolfram," *Acta Metallurgica*, vol. 1, no. 1, pp. 22–31, 1953, doi: 10.1016/0001-6160(53)90006-6.

43. F. F. Franco, L. Manjakkal, D. Shakthivel, and R. Dahiya, "ZnO based screen printed aqueous ammonia sensor for water quality monitoring," *Proceed IEEE Sens*, vol. 2019, pp. 1–4, 2019, doi: 10.1109/SENSORS43011.2019.8956763.
44. R. Ahmad, N. Tripathy, M. Y. Khan, K. S. Bhat, M. S. Ahn, and Y. B. Hahn, "Ammonium ion detection in solution using vertically grown ZnO nanorod based field-effect transistor," *RSC Adv*, vol. 6, no. 60, pp. 54836–54840, 2016, doi: 10.1039/c6ra09731f.
45. J. X. Wang et al., "Zinc oxide nanocomb biosensor for glucose detection," *Appl Phys Lett*, vol. 88, no. 23, pp. 98–101, 2006, doi: 10.1063/1.2210078.

13 Graphene-Based Materials for Various Green Nanotechnology Applications

Nibedita Mohanty
GIET University

Tapan Dash
International PranaGraf Mintech Research Centre
Centurion University of Technology and Management

Tapan Kumar Patnaik and
Sushree Subhadarshinee Mohapatra
GIET University

Sunita Dhar
Centurion University of Technology and Management

Surendra Kumar Biswal
International PranaGraf Mintech Research Centre

13.1 INTRODUCTION AND SCOPE OF GREEN NANOTECHNOLOGY

Nanotechnology explores various potential approaches in material sciences on a molecular level [1–8]. The technology that is utilized to create clean technologies to reduce threats to human health and the environment is known as "green nanotechnology." It improves the sustainability of processes. Green nanotechnology integrates various principles of green chemistry and green engineering to develop innocuous and eco-friendly nano assemblies to reduce the problems which affect human health and the environment. It is connected to the manufacturing and use of nanotechnology products. To create new nano-products, green nanotechnology promotes the replacement of existing products. The creation of novel nano-products helps to protect the environment. There is a lot of excitement about nanotechnology, which is regarded as a critical technology for the twenty-first century. However, progress has been hampered due to a lack of knowledge about the risks involved with nanotechnology and a lack of regulations to cope with these threats. The hurdles spanning from managing, producing, funding, regulatory, and technical concerns, however, have not stopped researchers from moving further. Green technology is a subset of green technology that incorporates the principles of green chemistry and green engineering. The name "green" alludes to the use of plant-based materials. Green nanotechnology aims to create nanomaterials and products that don't affect the environment or people's health, as well as nano-products that solve environmental issues. It makes use of the most up-to-date concepts in green chemistry and green engineering to produce nano-based materials and products that are free of harmful components, at low temperatures, with less energy and renewable inputs when possible, and by using lifecycle thinking throughout all design and engineering stages.

DOI: 10.1201/9781003502692-13

Green nanotechnology refers to the use of nanotechnology to develop nano-based materials and products with lower environmental impacts in addition to making current production processes for non-nano materials and goods more environmentally friendly. Nanoscale membranes can help for example, by helping to separate plant waste from desired chemical reaction products. Chemical reactions can be improved and used less wastefully with the help of nanoscale catalysts. When used in conjunction with nano-enabled information systems, nanoscale sensors can be a component of process control systems. Another approach to "green" industrial procedures is to use alternative energy sources, which nanotechnology has enabled.

Green nanotechnology's second goal is to produce goods that either directly or indirectly benefit the environment. Directly treating pollutants, desalinating water, cleaning hazardous waste sites, or sensing and monitoring environmental contaminants are all possible with nanomaterials or goods. Self-cleaning nanoscale surface coatings could indirectly minimize or even do away with the need for numerous cleaning agents utilized in standard maintenance procedures; lightweight nanocomposites for cars and other forms of transportation could save fuel and reduce materials used for production; fuel cells and light-emitting diodes powered by nanotechnology could reduce pollution from energy generation and help conserve fossil fuels; and improved battery life could result in less material consumption. By taking a comprehensive systems approach to nanomaterials and products, green nanotechnology makes sure that unanticipated effects are minimized, and impacts are anticipated throughout.

Green nanotechnology is the necessity to meet the needs of an ever-developing society. Nanotechnology should be explored in a sustainable and possibly recycling manner. The impact of nanomaterial and nanotechnology for developing advanced sustainable materials is highly required for society. Green nanotechnology generally uses low-cost and biosafe techniques in various areas of catalysis, solar cells, pharmaceuticals, industrial industries, sensors, water desalination, water purification, and air purification [9–11]. The popularity of green route approaches is due to the fact that they make it possible to produce nanomaterials in a controlled and clean setting, making them ecologically friendly. Utilizing plant extracts and creating nanoparticles with the help of fungi, bacteria, viruses, and algae are only a couple of the accessible green-friendly ways and methodologies. Each of these environmentally friendly practices illustrates particular benefits over environmental preservation and drawbacks. When it comes to environmentally friendly, biocompatible, and secure nanoparticles, plant extracts are the first green route methodology and are regarded as the most trustworthy method [12–14]. The strategy has a number of benefits, including the free, readily available, safe-to-handle, and inexpensive source of material to enable the synthesis of different nanoparticles. Utilizing bacteria as a green synthesis method is an additional strategy that has a number of benefits. Microorganism's ability to decrease metal ions has advantages, especially when used to manipulate bacteria. Graphene-based materials have been widely considered as potential materials for various industrial and biomedical applications due to their good biocompatibility and environmentally friendly nature with high mechanical stiffness and strength, anticorrosion, excellent electrical conductivity, high optical transparency, high surface area [15–30]. They have numerous environmentally sustainable applications in various fields such as energy storage, nanoelectronics devices, biomedical applications, advanced composites, anticorrosion coating, and water purification technology [9–11]. They are used in biomedical applications like biosensors/antibacterial/cell imaging/drug delivery, detectors, actuators, memory, and nano-communication devices. Among the various graphene derivatives, graphene oxide (GO) is the soluble oxidized product mostly used for biomedical applications, as it promptly interacts with cells, proteins, and bacteria in unique ways. But the mass-scale production of high-quality graphene and its derivatives under an environmentally sustainable model is the major focus of research and development [9–12].

13.2 OVERVIEW OF GRAPHENE

One of the elements that is most prevalent in the crust of the planet and useful is carbon [31]. Each of its various allotropes is beneficial to developing science and technology. Among these is graphene, which is considered one of the wonderful materials that has two-dimensional sp^2 hybridization. It shows outstanding thermal, mechanical, and electrical qualities, which make it a supercarbon material for various industrial applications.

13.3 FORMS OF GRAPHENE

13.3.1 Graphene with Different Layers

It is a sheet of hexagonally organized, sp^2-bound carbon atoms that is only one atom thick and is either attached to or suspended freely from a foreign substrate. Its lateral measurements can range from a few nanometers to microscale. For usage in high-frequency electronics, monolayer (single-layer) is the purest form currently understood. Bi-layer and tri-layer graphene display various characteristics in material research. Few-layer or multi-layer graphene has few (between 2 and 10) stacks of extended-lateral-dimension graphene layers (Figure 13.1). They could be flaky, film-like, free-standing, or linked to the substrate. These are used as mechanical reinforcement for composite materials.

13.3.2 GO and Reduced Graphene Oxide

A monolayer material with a high oxygen concentration called GO has a C/O atomic ratio of between 2 and 3. Exfoliation and oxidation are used to prepare it, and the basal plane is heavily modified by oxidation after that. Hazardous gases cannot pass through GO, but water can penetrate it. When GO is reduced in oxygen content through chemical, photochemical, thermal, photothermal, microwave, or microbial/bacterial processes, reduced GO (RGO) is formed.

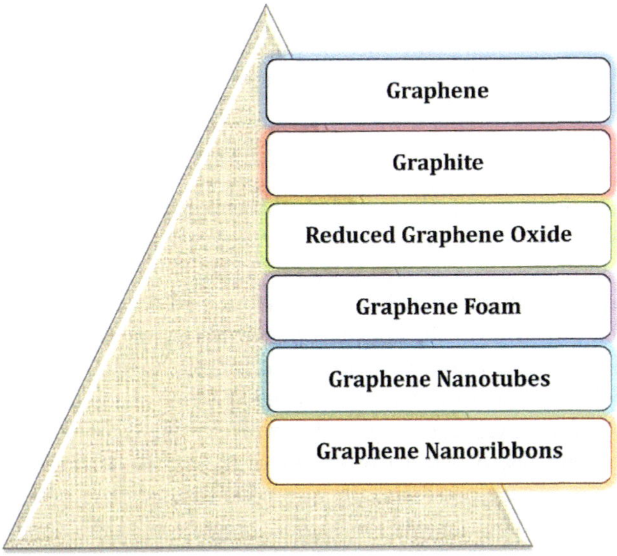

FIGURE 13.1 Various forms of graphene.

13.3.3 Graphene Nanomaterials

It consists of graphene nanosheets, GO, RGO, nanoribbons, nanoflakes, and more. These two-dimensional graphene materials show 100 nm thick and/or in lateral dimension. They are either freely suspended or bonded to a non-carbon substrate. Graphene-based composite materials show extraordinary electrical conductivity.

13.4 GRAPHENE PRODUCTION TECHNIQUES

Numerous methods, from basic to cutting-edge technologies, can be used to create graphene. Studies have been carried out worldwide for the production of graphene derivatives in different ways, broadly separated into "(a) chemical intercalation methods (chemical oxidation), (ii) growth methods using chemical vapor deposition (CVD) mechanical exfoliation methods and (iii) unzipping of carbon nanotubes" [12–30]. Each method has some advantages or disadvantages. However, the composition, morphology, and production process route decide the quality of GO. Some important techniques are discussed below.

13.4.1 Mechanical Exfoliation

This technique, as of 2014, creates graphene with the fewest flaws and greatest electron mobility [32]. R. Ruoff et al. were the ones who first popularized this approach [33]. The graphene layer was separated from the graphite flakes using an adhesive tape. To get single layers, several exfoliation procedures are needed. The "Scotch tape" or "drawing" method is another name for this procedure. The planetary ball milling is one of the most effective mechanical exfoliation techniques to produce graphene. When mechanical force is applied, a layer-by-layer stack of sheets of graphene is exfoliated. The minimal cost and tremendous scalability of this technology are its major advantages.

13.4.2 Chemically Derived Graphene from GO

In this process, graphite is converted into GO, which is then used to create graphene. R. Ruoff and team gave the first demonstration of this method in 2006 [34–37]. Using Hummers' process, graphite is chemically changed into a water-dispersible intermediate GO.

13.4.3 Epitaxial Graphene and Chemical Vapor Deposition

The high-temperature reduction of silicon carbide through an epitaxial technique was pioneered by De Heer and his team at the Georgia Institute of Technology [38–40]. In this process, small islands of graphitized carbon form as silicon desorbs at approximately 1,000°C under extremely high vacuum conditions. This method is alternatively known as the thermal decomposition of graphene. Another substrate-based approach involves the chemical vapor deposition of graphene on transition metal films, typically utilizing nickel films and methane gas [41–43]. During cooling, the solubility of carbon in the transition metal decreases, leading to the precipitation of a thin carbon layer on the surface. These techniques showcase excellent compatibility with existing technology, making them advantageous for graphene synthesis. An attractive feature of these methods is their seamless integration into modern semiconductor devices, especially by producing a single graphene layer across an entire wafer. These advancements contribute to the ongoing progress in graphene synthesis [38–43].

13.5 PROPERTIES OF GRAPHENE

Graphene, characterized as a zero-gap semiconductor, possesses exceptional electrical conductivity due to the intersection of its conduction and valence bands at the Dirac points (Figure 13.2) [44]. With six electrons available for chemical bonding in each carbon atom, the sp^2 hybridization in

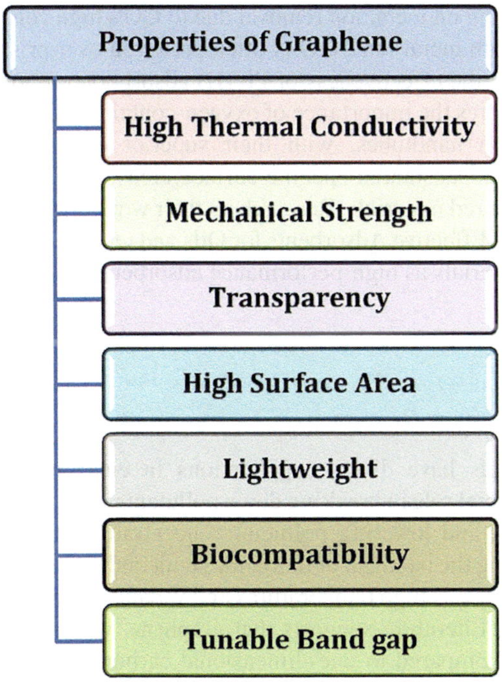

FIGURE 13.2 Properties of graphene.

graphene leaves one electron free for three-dimensional electrical conduction. Remarkably, graphene exhibits high electronic mobility at ambient temperatures, and its electron mobility has been experimentally proven to be nearly temperature independent [45]. The material boasts impressive mechanical properties, with a Young's modulus of 1 TPa, a tensile strength of 130 GPa, and an incredibly low weight of only 0.77 mg/m^2 [46].

Graphene's opacity in a vacuum is notable, absorbing 2.3% of white light for each atomic monolayer. This absorption percentage remains consistent with the addition of more graphene layers [47]. Additionally, graphene demonstrates remarkable thermal conductivity, reaching approximately 4,000 W/m/K. Its high Seebeck coefficient and figure of merit make it particularly suitable for converting electrical current into heat. These intrinsic properties position graphene as an extraordinary material with diverse applications [44–47].

13.6 USES OF GRAPHENE-BASED MATERIALS IN GREEN NANOTECHNOLOGY

13.6.1 Graphene Materials for Contaminant Adsorption

The surge in population and intensified agricultural and industrial activities have led to a significant increase in environmental pollutants. To address this global concern, there is a concerted international effort to develop reliable technologies for efficient toxin removal from air and water. Among these technologies, adsorption emerges as a quick, cost-effective, and efficient method for eliminating pollutants from aquatic environments. Physicochemical interactions during the adsorption process bind the pollutant (adsorbate) to the nanomaterial (adsorbent), making materials based on graphene particularly promising for removing both inorganic and organic pollutants. This discussion highlights various adsorption mechanisms and weighs the advantages and disadvantages of employing graphene materials as decontamination adsorbents. Numerous studies have explored the use of graphene-based materials, particularly GO, as adsorbents to remove inorganic species from aqueous solutions, focusing on metal ion removal due to GO's high concentration of oxygen groups capable of interacting with metal ions. GO is often preferred over pristine graphene for metal ion adsorption. For instance, the effectiveness of Pb (II) adsorption on both unoxidized and oxidized graphene sheets underscores the importance of oxygen-containing functional groups in the process.

Graphene and carbon nanotubes, with their superior electrical, mechanical, and chemical properties, along with substantial specific surface areas, have garnered attention as effective carbon-based nanostructured materials. Researchers their work titled "Graphene and Carbon-Based Nanomaterials as Highly Effective Adsorbents for Oils and Organic Solvents," explore the potential applications of these materials as high-performance adsorbents for the removal of oils and organic solvents [47].

13.6.2 The Role of Graphene-Based Materials in Highly Efficient Pollutant Removal

Graphene-based materials have diverse applications in environmental cleanup and pollutant removal. They play a crucial role in breaking down pollutants into less hazardous molecules, reducing low-valency species, and lowering pollutant concentrations through adsorption. Challenges associated with modifying the bandgap of inorganic semiconductors and the high costs of advanced production techniques, such as high temperatures and vacuum conditions, limit the widespread use of inorganic solar cells. Literature suggests that graphene, with its extensive surface area, provides more active sites compared to one-dimensional carbon nanotubes, and pH plays a crucial role in the adsorption process. In the removal of heavy metal ions, modified graphene surpasses pristine graphene in adsorption capability. Graphene-based materials, characterized by high specific surface areas and low aggregation, exhibit strong adsorption capacities for organic contaminants, particularly molecules like benzene, where π-interactions dominate. To prevent aggregation between layers, magnetic particles are incorporated into adsorbents, creating magnetic graphene composites for easy separation. Graphene–Fe_3O_4 nanocomposites, known for their excellent performance in pollutant removal and ease of separation from aqueous solutions, have gained significant attention in research. Experimental findings and theoretical considerations indicate that pristine graphene has a substantially higher adsorption capacity than GO due to strong interactions between graphene and organic molecules. In the realm of pollutant removal, graphene-based photo degradants stand out. Their unique electrical characteristics enable rapid electron transport from excited semiconductors to the graphene sheet, coupled with adjustable semiconductor sizes and reduced graphene sheet aggregation. The transparency of graphene sheets, resulting from their one- or multi-atom thickness, enhances the utilization of stimulating light in photodegradation processes. This makes graphene-based photo degradants an appealing choice for reducing high-valent metal ions and degrading organic molecules. The distinctive structure and qualities of graphene make it valuable for modification and complexation with other nanomaterials, leading to the production of modified graphene and graphene-based composites with high adsorption capacities and photocatalytic abilities.

13.7 USES OF GRAPHENE IN SOLAR CELLS FOR ENHANCING RENEWAL ENERGY EFFICIENCY

Conventional photovoltaic cells, employing inorganic semiconductors like amorphous silicon, gallium arsenide, and sulfide salts, have been widely used for generating free holes and electrons upon photon absorption [48]. While inorganic multijunction solar cells have demonstrated impressive power conversion efficiency (PCE) exceeding 40% in laboratory settings [49], their general application is hindered by challenges in altering the bandgap of inorganic semiconductors and the high costs associated with sophisticated production methods involving extreme temperatures and high

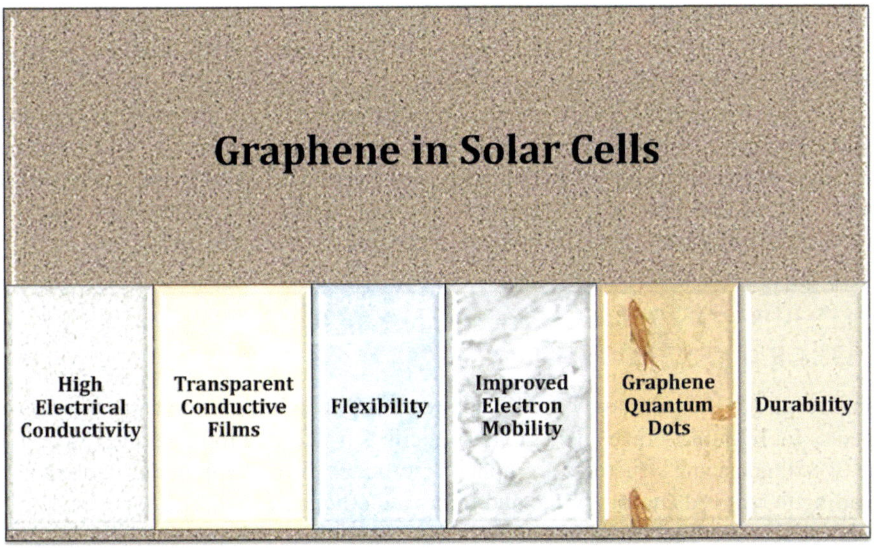

FIGURE 13.3 Advantages of graphene in solar cells.

vacuum [50]. The expense of these inorganic solar cells makes them less competitive with conventional grid power [51]. In contrast, the accessibility, low weight, versatility, and solution processability of organic or polymer materials have attracted significant interest [52,53]. Conjugated polymers, in contrast to inorganic counterparts, form bound electron–hole pairs (excitons) when absorbing photons at ambient temperature. The dissociation of these excitons occurs at the interface of semiconducting materials with different ionization potentials and electron affinities (donors and acceptors), leading to charge generation [54]. Polymer solar cells (PSCs) typically consist of an active layer comprising a combination of donor and acceptor materials, with cathode and anode. One of these electrodes needs to be optically transparent for sunlight passage [55]. PSCs, upon exposure to sunlight, generate electrons and holes that move within the cell following electrical principles. PSCs also incorporate an electron-extraction layer between the cathode and the active layer, along with a hole-extraction layer between the anode and the active layer [56]. Carbon nanoparticles have been explored for various applications in solar cells. Indium tin oxide (ITO) is a popular transparent electrode in PSCs, but it suffers from brittleness and expensive manufacturing, as well as limited indium availability in nature. Graphene, with its high specific surface, excellent optical properties, and high electrical conductivity, presents a promising alternative to ITO as a transparent electrode (Figure 13.3). Graphene sheets exhibit low sheet resistance (a few hundred ohms per square) and high transparency (>80%) [57–59]. Furthermore, GO variants have demonstrated superiority as hole- and electron-extraction layers in PSCs [60–62]. GO, used as a hole-extraction layer, achieved a PCE comparable to the state-of-the-art poly (3,4-ethylene dioxythiophene): poly (styrene sulfonate) (PEDOT: PSS) layer [60]. Conversely, cesium-neutralized GO (GO-Cs) serves as an outstanding electron-extraction layer, leading to a demonstrated PCE of 2.5% in PSC when combined with MoO3 as the hole-extraction layer [61].

13.8 SUPERCAPACITORS WITH GRAPHENE ELECTRODES

The specific surface area of the electrode material is directly related to the electric double-layer capacitance, making graphene an excellent candidate for supercapacitor electrodes. This is attributed to its theoretically high specific surface area of $2.63 \times 10^6 m^2/kg$, exceptional conductivity, tunable microstructure, and remarkable thermal/mechanical stability [62–65].

RGO, prepared through a chemical route by Stoller et al., demonstrated a specific capacitance ranging from 99,000 to 135,000 F/kg [62,63]. Furthermore, when RGOs were produced via straightforward microwave heating, they exhibited capacitance values of 191,000 F/kg in organic electrolytes. In comparison, RGOs created through reduction in propylene carbonate solution showed a capacitance of 120,000 F/kg [63]. Supercapacitors developed by Wang et al., utilizing graphene-based materials, displayed an energy density of 28.5 Wh/kg and a maximum specific capacitance of 205,000 F/kg [65]. Notably, these supercapacitors maintained a remarkable cyclic lifespan, retaining at least 90% of the original capacitance after 1,200 cycles.

13.9 LITHIUM-ION BATTERIES WITH ELECTRODES OF GRAPHENE FOR ENERGY SAVING

Graphene surpasses the commonly used graphite electrode in lithium-ion batteries, offering twice the capacity for lithium storage and a greater specific surface area [66]. Recent research indicates that the two-dimensional structure of graphene facilitates the adsorption and diffusion of lithium ions, leading to reduced charging time and improved efficiency [67]. The nanosized holes in graphene sheets contribute to the high-rate discharge capabilities of lithium-ion battery anodes [68]. Pan et al. utilized electrodes made of graphene sheets in lithium-ion batteries, achieving a capacity of 540 A h/kg, surpassing that of graphite [67]. Furthermore, carbon nanotubes/fullerenes-reinforced graphene electrodes demonstrated capacitance up to 784 A h/kg [69]. Wu et al. explored the use of boron or nitrogen-doped graphene, obtaining high power and energy densities simultaneously due to the outstanding properties of graphene. Graphene-reinforced composites exhibited fast surface Li+ absorption, ultra-fast Li+ diffusion, and improved electrical conductivity [70]. However, challenges such as irreversible reactions with lithium and the development of solid electrolyte interfaces still affect graphene's efficiency as a lithium-ion battery electrode [71]. Hybrid electrodes made of graphene doped with specific metal or metal-oxide nanoparticles exhibit improved reversible storage capacity for lithium ions [72,73]. Nanoporous graphene sheets adorned with loosely packed SnO_2 nanoparticles demonstrated a reversible capacity of 810 A h/kg [72,73–75].

13.10 CONCLUSION

In conclusion, the exploration and utilization of graphene-based materials present a promising avenue for advancing green nanotechnology applications. The exceptional properties of graphene, such as its high electrical conductivity, mechanical strength, and flexibility, make it an ideal candidate for various eco-friendly technological solutions. By incorporating graphene nanomaterials into different platforms, we can enhance the efficiency and sustainability of a wide range of applications, contributing to a greener and more environmentally conscious future. The versatility of graphene-based materials in green nanotechnology is evident in their potential applications, from energy storage devices to water purification systems. The development of graphene nanocomposites and hybrids allows for tailored solutions that address specific environmental challenges. These innovations not only harness the unique properties of graphene but also contribute to the design of cleaner technologies that minimize environmental impact and promote sustainable practices. As we delve deeper into the realm of graphene-based materials for green nanotechnology applications, it becomes evident that ongoing research and technological advancements will continue to unlock new possibilities. The interdisciplinary nature of this field encourages collaboration between materials science, engineering, and environmental science, fostering a holistic approach toward sustainable development. With ongoing efforts and a commitment to responsible innovation, graphene-based materials hold the potential to play a pivotal role in shaping a more environmentally friendly and technologically advanced world.

REFERENCES

1. K. Eric Drexler, *Engines of Creation: The Coming Era of Nanotechnology* (Doubleday, New York, 1986).
2. A. Hubler, *Complexity* 15(5) (2010) 48.
3. P. Thakur, A. Thakur, Nanomaterials, their types and properties. In *Synthesis and Applications of Nanoparticles* (pp. 19–44) (Springer Nature, Singapore, 2022).
4. E. Hannachi, et al., *Journal of Physics and Chemistry of Solids* 170 (2022) 110910.
5. M.A. Albrecht, C.W. Evans, C.L. Raston, *Green Chemistry* 8 (2006) 417.
6. P. Punia, M.K. Bharti, S. Chalia, R. Dhar, B. Ravelo, P. Thakur, A. Thakur, *Ceramics International* 47(2) (2021) 1526–1550.
7. A. Goel, S. Bhatnagar, *Green Nanotechnology, BioEvolution* 2014 (2014) 3–4.
8. D. Chahar, D. Kumar, P. Thakur, A. Thakur, *Materials Research Bulletin* 162 (2023) 112205.
9. D.R. Dreyer, A.D. Todd, C.W. Bielawski, *Chemical Society Reviews* 43 (2014) 5288.
10. V. Palmieri, M. Barba, L. Di Pietro, C. Conti, M. De Spirito, W. Lattanzi, et al., *International Journal of Molecular Sciences* 19 (2018) 3336.
11. N. Dhanda, P. Thakur, A. Thakur, *Materials Today: Proceedings* 73 (2023) 237–240.
12. P. Dash, T. Dash, T.K. Rout, A.K. Sahu, S.K. Biswal, B.K. Mishra, *RSC Advances* 6 (2016) 12657.
13. G.R. Kumar, K. Jayasankar, S.K. Das, T. Dash, A. Dash, B.K. Jena, B.K. Mishra, *RSC Advances* 6 (2016) 20067.
14. N. Dhanda, P. Thakur, R. Kumar, T. Fatima, S. Hameed, Y. Slimani, A-C. A. Sun, A. Thakur, *Applied Organometallic Chemistry* 37 (2023) e7110.
15. T. Dash, T.K. Rout, B.B. Palei, S. Bajpai, S. Kundu, A.N. Bhagat, B.K. Satpathy, S.K. Biswal, A. Rajput, A.K. Sahu, S.K. Biswal, *SN Applied Sciences* 2 (2020) 1.
16. P. Punia, M.K. Bharti, R. Dhar, P. Thakur, A. Thakur, *ChemBioEng Reviews* 9(4) (2022) 351–369.
17. B.B. Nayak, R.K. Sahu, T. Dash, S. Pradhan, *Applied Physics A* 124(220) (2018) 1.
18. S. Dhar, T. Dash, B.B. Palei, T.K. Rout, S.K. Biswal, A. Mitra, A.K. Sahu, S.K. Biswal, *Materials Today: Proceedings* 33 (2020) 5136.
19. T. Dash, A.K. Sahoo, R.K. Mishra, B.B. Palei, *Materials Today: Proceedings* 43 (2021) 216.
20. T. Dash, B.B. Palei, N. Mohanty, S.S. Mohapatra, R.K. Mishra, S.K. Biswal, D. Behera, *Materials Today: Proceedings* 43 (2021) 447.
21. B.B. Palei, T. Dash, S.K. Biswal, *PalArch's Journal of Archaeology of Egypt/Egyptology* 17 (2020) 10119–10132.
22. T. Dash, D. Rout, B.B. Palei, *Materials Today: Proceedings* 46 (2021) 11061.
23. B.B. Palei, T. Dash, S.K. Biswal, *International Journal of Materials and Product Technology* 62 (2021) 49.
24. B.B. Palei, T. Dash, S.K. Biswal, R. Saktivel, *International Journal of Innovative Research Physics* 1(3) (2020) 37.
25. T. Dash, A.K. Katika, B.B. Palei, J.P. Dhal, T.K. Rout, *Materials Today: Proceedings* 44 (2021) 4657–4660.
26. T. Dash, G.K. Sahoo, B.B. Palei, T.K. Rout, *Advances in Materials Processing and Manufacturing Applications* 872 (2020) 441.
27. R.K. Sahu, T. Dash, V. Mukherjee, S.K. Pradhan, B.B. Nayak, Growth of spheroidal silicon carbide by arc plasma treatment, applications of microscopy. In: *Materials and Life Sciences* (pp. 77–85) (Springer Nature, Singapore, 2021).
28. T. Dash, B.B. Palei, *Springer Proceedings in Materials* (pp. 281–287) (Springer Nature, Singapore, 2021).
29. P. Sharma, A. Sharma, M. Sharma, N. Bhalla, P. Estrela, A. Jain, P. Thakur, A. Thakur, *Global Challenges* 1(9) (2017) 1700041.
30. B.B. Palei, T. Dash, S.K. Biswal, *Journal of Materials Science* 57 (18) (2022) 8544.
31. "List of Elements of the Periodic Table - Sorted by Abundance in Earth's crust". Israel Science and Technology Homepage. https://www.science.co.il/PTelements.asp?s=Earth.
32. F.V. Kusmartsev, W.M. Wu, M.P. Pierpoint, K.C. Yung, *Application of Graphene within Optoelectronic Devices and Transistors* (Springer Nature, Singapore, 2014)
33. X.K. Lu, M.F. Yu, H. Huang, R.S. Ruoff, *Nanotechnology* 10 (1999) 269.
34. S. Stankovich, D.A. Dikin, G.H.B. Dommett, K.M. Kohlhaas, E.J. Zimney, E.A. Stach, R.D. Piner, S.T. Nguyen, R.S. Ruoff, *Nature* 442 (2006) 282.
35. S. Stankovich, D.A. Dikin, R.D. Piner, K.M. Kohlhaas, A. Kleinhammes, Y. Jia, Y. Wu, S.T. Nguyen, R.S. Ruoff, *Carbon* 45 (2007) 1558.

36. I. Jung, D.A. Dikin, R.D. Piner, R.S. Ruoff, *Nano Letters* 8 (2008) 4283.
37. D. Yang, A. Velamakanni, G. Bozoklu, S. Park, M. Stoller, R.D. Piner, S. Stankovich, I. Jung, D.A. Field, C.A. Ventrice, R.S. Ruoff, *Carbon* 47 (2009) 145.
38. J. Hass, W.A. de Heer, E.H.J. Conrad, *Journal of Physics: Condensed Matter* 20 (2008) 323202.
39. W.A. de Heer, C. Berger, X.S. Wu, P.N. First, E.H. Conrad, X.B. Li, T.B. Li, M. Sprinkle, J. Hass, M.L. Sadowski, M. Potemski, G. Martinez, *Solid State Communications* 143 (2007) 92.
40. C. Berger, Z.M. Song, X.B. Li, X.S. Wu, N. Brown, C. Naud, D. Mayou, T.B. Li, J. Hass, A.N. Marchenkov, E.H. Conrad, P.N. First, W.A. de Heer, *Science* 312 (2006) 1191.
41. A. Reina, X.T. Jia, J. Ho, D. Nezich, H.B. Son, V. Bulovic, M.S.K. Dresselhaus, *Nano Letters* 9 (2009) 30.
42. K.S. Kim, *Nature* 457 (2009) 706.
43. P.W. Sutter, J.I. Flege, E.A. Sutter, *Nature Materials* 7 (2008) 406.
44. D.R. Cooper, B. D'Anjou, N. Ghattamaneni, B. Harack, M. Hilke, A. Horth, N., M. Massicotte, L. Vandsburger, L. Whiteway, E.Y. Victor, *International Scholarly Research Network* 2012 (2012) 1–56.
45. S.V. Morozov, K. Novoselov, M. Katsnelson, F. Schedin, D. Elias, J. Jaszczak, A. Geim, *Physical Review Letters* 100(1) (2008) 016602.
46. C. Lee, *Science* 321(385) (2008) 385–388.
47. R.R. Nair, P. Blake, A.N. Grigorenko, K.S. Novoselov, T.J. Booth, T. Stauber, N.M.R. Peres, A. K. Geim, *Science* 320(5881) (2008) 1308–1308.
48. R.H. Bube, *Photoelectronic Properties of Semiconductors* (Cambridge University Press, Cambridge, UK, 1992).
49. M.A. Green, K. Emery, D.L. King, S. Igari, W. Warta, *Progress in Photovoltaics: Research and Applications* 12 (2004) 365.
50. R.M. Swanson, *Progress in Photovoltaics: Research and Applications* 14 (2006) 443.
51. J. Johnson, *Chemical & Engineering News* 82 (2004) 13.
52. F.C. Krebs, *Polymer Photovoltaics: A Practical Approach* (SPIE Press, Bellingham, WA, 2008).
53. S. Sun, N.S. Sariciftci, *Organic Photovoltaics: Mechanisms, Materials, and Devices* (CRC Press, Boca Raton, FL, 2005).
54. C.W. Tang, *Applied Physics Letters* 48 (1986) 183.
55. G. Yu, J. Gao, J.C. Hummelen, F. Wudl, A.J. Heeger, *Science* 270 (1995) 1789.
56. R. Steim, F.R. Koglera, C.J. Brabec, *Journal of Materials Chemistry* 20 (2010) 2499.
57. Y. Wang, S.W. Tong, X.F. Xu, B. Özyilmaz, K.P. Loh, *Advanced Materials* 23 (2011) 1514.
58. G. Jo, S.I. Na, S.H. Oh, S. Lee, T.S. Kim, G. Wang, M. Choe, W. Park, J. Yoon, D.Y. Kim, Y.H. Kahng, T. Lee, *Applied Physics Letters* 97(2010) 213301.
59. S. Bae, H. Kim, Y. Lee, X.F. Xu, J.-S. Park, Y. Zheng, J. Balakrishnan, Y. Lei, H.R. Kim, Y.I. Song, Y.J. Kim, B. Özyilmaz, J.H. Ahn, B.H. Hong, S. Lijima, *Nature Nanotechnology* 5 (2010) 574.
60. S.S. Li, K.H. Tu, C.C. Lin, C.W. Chen, M. Chhowalla, *ACS Nano* 4 (2010) 3169.
61. J. Liu, Y.H. Xue, Y.X. Gao, D.S. Yu, M. Durstock, L.M. Dai, *Advanced Materials* 24, 2228 (2012).
62. M.D. Stoller, S. Park, Y. Zhu, J. An, R.S. Ruoff, *Nano Letters* 8 (2008) 3498.
63. Y. Zhu, S. Murali, M.D. Stoller, A. Velamakanni, R.D. Piner, R.S. Ruoff, *Carbon* 48 (2010) 2118.
64. Y. Zhu, M.D. Stoller, W. Cai, A. Velamakanni, R.D. Piner, D. Chen, R.S. Ruoff, *ACS Nano* 4 (2010) 1227.
65. Y. Wang, Z. Shi, Y. Huang, Y. Ma, C. Wang, M. Chen, Y. Chen, *Journal of Physical Chemistry C* 113 (2009) 13103.
66. G. Wang, B. Wang, X. Wang, J. Park, S. Dou, H. Ahn, K. Kim, *Journal of Materials Chemistry* 19 (2009) 8378.
67. D. Pan, S. Wang, B. Zhao, M. Wu, H. Zhang, Y. Wang, Z. Jiao, *Chemistry of Materials* 21 (2009) 3136.
68. T. Takamura, K. Endo, L. Fu, Y. Wu, K.J. Lee, T. Matsumoto, *Electrochimica Acta* 53 (2007) 1055.
69. E. Yoo, J. Kim, E. Hosono, H. Zhou, T. Kudo, I. Honma, *Nano Letters* 8 (2008) 2277.
70. Z.-S. Wu, W. Ren, L. Xu, F. Li, H.-M. Cheng, *ACS Nano* 5 (2011) 5463.
71. D.S. Su, R. Schlogl, *ChemSusChem* 3 (2010) 136.
72. S.M. Paek, E.J. Yoo, I. Honma, *Nano Letters* 9 (2009) 72.
73. Z.S. Wu, W.C. Ren, L. Wen, L.B. Gao, J.P. Zhao, Z.P. Chen, G.M. Zhou, F. Li, H.M. Cheng, *ACS Nano* 4 (2010) 3187.
74. D.H. Wang, D.W. Choi, J. Li, Z.G. Yang, Z.M. Nie, R. Kou, D.H. Hu, C.M. Wang, L.V. Saraf, J.G. Zhang, I.A. Aksay, J. Liu, *ACS Nano* 3 (2009) 907.
75. S.L. Chou, J.Z. Wang, M. Choucair, H.K. Liu, J.A. Stride, S.X. Dou, *Electrochemistry Communications* 12 (2010) 303.

14 Green Synthesized Nanomedicine and Its Applications

Arti and Preeti Thakur
Amity University Haryana

An-Cheng Aidan Sun
Yuan Ze University

Atul Thakur
Amity University Haryana
Yuan Ze University

14.1 INTRODUCTION TO "NANO" MEDICINE

The development of nanomedicine, the scale on which nanomedicine operates (1–100 nm), the size of molecules, and biochemical processes account for a large portion of its rhetorical, technical, and scientific power. In 1999, American scientist Robert A. Freitas Jr. released *Nanomedicine: Basic Capabilities*, the first of two books he devoted to the topic. This is when the phrase "nanomedicine" first appeared. According to the molecular assembler theory of American scientist K. Eric Drexler, nanomedicine would enable the development of nanobot devices, which are nanoscale-sized robots that would explore the human body in search of and treat sickness [1]. The underlying promise of doctors being able to hunt for and eliminate cancerous cells or of nanomachines that replace biological components, which still motivates depictions of the discipline, is underscored by the attractive visuals even if most of it is yet unattainable. Such examples continue to be crucial to the discipline and are frequently utilized by researchers, funding organizations, and the media.

The field of medicine known as nanomedicine uses nanotechnology's knowledge and technologies to prevent and treat disease. To diagnose, deliver, sense, or operate upon a living creature, nanomedicine includes the use of tiny materials, such as biocompatible nanoparticles (NPs) and nanorobots. The delivery of medication to patients at the nanoscale is now possible. One nanometer is equal to one billionth of a meter, to give the term "nano" some context. Although it is still in its early phases, this type of technology is now known as nanomedicine, and awareness and first usage are both fast increasing. Research in this area is expanding because medicine at the nanoscale can do things that conventional medicine cannot. The ability to greatly improve treatment precision and efficacy is what makes nanotechnology so appealing for medical applications.

Nanomedicine offers a great deal of promise to advance therapies, targeted drug delivery, diagnostics, and cell repair. This implies that not only may early diagnosis become more accurate, but that therapy can also become much more thoughtful and tailored to the needs of each patient, leading to better success rates. NPs are used in nanomedicine applications, which is the usual method of implementation. Figure 14.1 shows various kinds of nano-pharmaceuticals. For a given job, NPs can be formed into a variety of forms, such as rods, spheres, tubes, and sheets. Spherical NPs, for instance, are frequently utilized for targeted medication administration because they may function

DOI: 10.1201/9781003502692-14

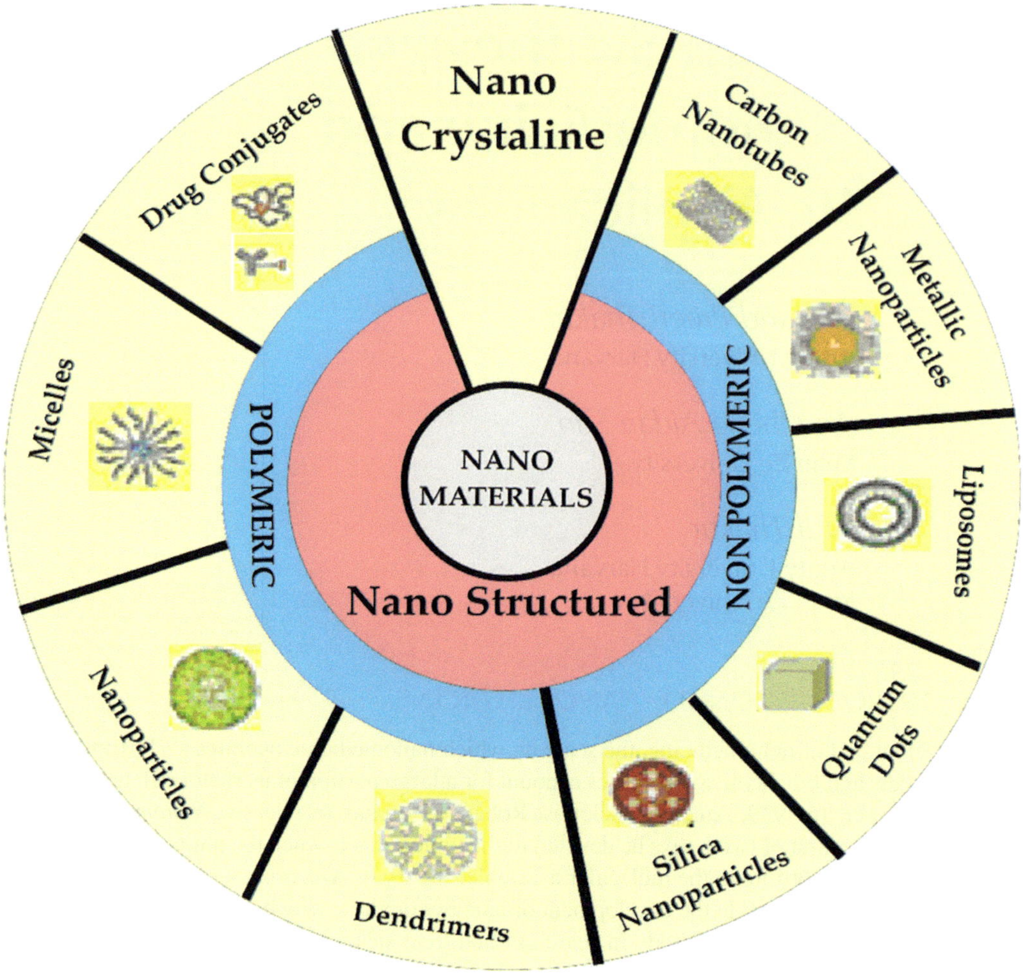

FIGURE 14.1 Illustration showing various kinds of nano-pharmaceuticals [2].

as a carrier to contain pharmaceuticals and convey them to a particular tissue or cell depending on receptor recognition. This is helpful, especially for cancer treatments, as many of the existing medications have dismal success rates.

Working at the molecular level, nanomedicine makes claims about the seamless fusion of biology and technology, the elimination of illness through customized medicine, targeted medication delivery, regenerative treatment, and nanomachinery that can replace small pieces of cells. Although many of these predictions may not come true, some applications of nanomedicine already exist and have the potential to drastically alter medical practice as well as current understandings of biology, health, and disease—aspects that are crucial for modern societies. Due to developments in technology, the field's worldwide market share reached around USD 377.37 million in 2022. The market is anticipated to reach around USD 964.15 billion by the end of 2030.

14.2 GREEN NANOTECHNOLOGY

Green nanotechnology describes a type of nanotechnology that, at the very least, should outperform competing "non-green" technologies in addressing some environmental issues. The discipline of green chemistry and the structure provided by the principles of green chemistry have both been

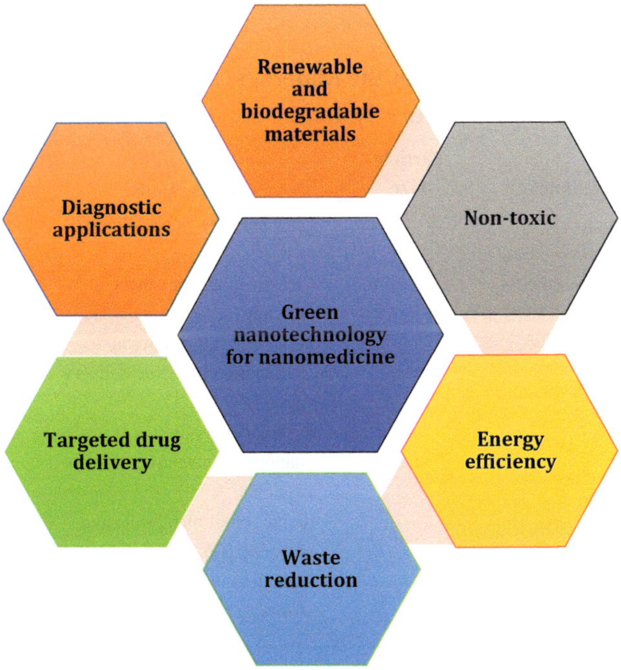

FIGURE 14.2 Features of green nanotechnology for nanomedicine synthesis.

influenced by green nanotechnology. Green chemistry promotes reducing or eliminating the use of hazardous compounds in the development, production, and use of chemical products. Eco-friendly methods are frequently used in the current practices of green nanotechnology to produce nanoparticulate systems using natural and nonhazardous materials and solvents. Figure 14.2 gives some features of green nanotechnology for nanomedicine synthesis.

This review focuses on recent developments in the development of green nanoscale systems for potential biomedical applications, delivery methods for green nanomedicines, potential health risks associated with nanomedicine formulations, current challenges, and potential future developments for the green development of nanomedicines.

14.3 POTENTIAL RISKS OF NANOMEDICINE FORMULATIONS

By resulting in substantial advancements in the field of pharmaceutics, nanotechnology has achieved great progress in the healthcare industry. Potential negative effects on nanomaterials (NMs) and tailored NPs have been noted with the expanding use and manufacture of nanotechnology in medicine delivery. The study of the nature and dangers of nanoscale particles and composites to people and other biological systems is known as nanotoxicology. The toxic substances utilized in nanomedicine formulations have the potential to have negative impacts on the body's biological processes, including irritation and toxicity to organs, tissues, and cells. The most frequently employed materials in the creation of nanoparticulate drug delivery systems are listed in Table 14.1 along with any potential negative health consequences.

14.4 THE STATE-OF-ART OF NANOMEDICINE

The field of "nanomedicine" has emerged through the application of nanotechnology, introducing new possibilities and outcomes. Nanomedicine enables researchers to explore biological systems at the nanoscale, replicating or enhancing mechanisms that mimic healthy natural tissues. It is crucial

TABLE 14.1
Potentially Harmful Health Effects of the Materials Most Frequently Employed to Create Nanoparticulate Medication Delivery Systems

Types of Materials	Examples	Particle Size	Applications	Potential Adverse Health Effects
Nanomaterials	Carbon nanotubes	0.4–100 nm	Drug delivery, gene delivery	Genotoxicity, pulmonary toxicity
	Metal-based NPs	1–100 nm	Cell imaging, anticancer therapy	Cytotoxicity, genotoxicity
Cross linkers	Glutaraldehyde	—	Polymer cross-linking in nanoparticulate drug delivery systems	Irritation and sensitization to skin, eyes, genotoxicity
	Formaldehyde	—	Polymer cross-linking in nanoparticulate drug delivery systems	Skin allergy and contact dermatitis, pulmonary inflammation
Reducing agents	Sodium borohydride	—	Reduction of metal salts and forming NPs	Impact on nervous system, severe irritation to the skin
	Hydrazine	—	Reduction of metal salts and forming NPs	Mild hepatotoxicity, pulmonary edema

to comprehend the intricacies and challenges when integrating human physiology with nanotechnology. In areas like tissue engineering and regenerative medicine, nanoscale biomaterials (such as nano-topographies, NPs, nanotubes, and self-assembled materials) hold the potential to enhance tissue regeneration, minimize immune reactions, and reduce infections. Nanotechnology is increasingly employed in drug delivery, promising improved therapeutic efficiency and efficacy. The nanometer size range, where biologically important components such as amino acids, carbohydrates, nucleotides, proteins, and DNA reside, facilitates interactions between NMs and biological systems. Modifying the surface topography of traditional biomaterials to create nanosized surfaces with unique properties encourages interactions with natural tissues and organs [2].

14.5 GREEN SYNTHESIS AND USE OF SILVER NPS IN NANOMEDICINE

Metal types of NPs with a diameter of less than 100 nm have significantly impacted a variety of biomedical applications during the past few decades, including diagnostic and medicinal devices for individualized healthcare. Since they have distinct physical and optical properties and biochemical functionality tailored by various size- and shape-controlled AgNPs, silver NPs (AgNPs) in particular have great potential in a broad range of applications as antimicrobial agents, biomedical device coatings, drug delivery carriers, imaging probes, and diagnostic and optoelectronic platforms.

Owing to their superior physical, chemical, and biological properties, AgNPs have recently received substantial research attention. AgNPs' superiority over their bulk counterparts is primarily owing to the size, shape, composition, crystallinity, and structure of AgNPs. Numerous initiatives have been made to investigate their appealing features and make use of them in useful applications, including antibacterial and anticancer treatments, diagnostics and optoelectronics, water disinfection, and other clinical and pharmaceutical applications. The application of silver-based NMs has been restricted due to their instability, such as the oxidation in an oxygen-containing fluid, despite the fact that silver has exciting material properties and is a cheap and abundant natural resource [3]. Therefore, compared to relatively stable gold NPs (AuNPs), AgNPs offer an untapped potential. Previous studies have demonstrated that AgNPs' size, distribution, morphological shape, and surface properties—all of which can be altered by a variety of synthetic techniques, reducing agents, and stabilizers—have a significant impact on their physical, optical, and catalytic capabilities. AgNPs can vary in size depending on the application; for example, AgNPs designed for drug delivery are typically larger than 100 nm to account for the amount of drug to be supplied. AgNPs can also be shaped into a variety of shapes with diverse surface properties, including rod, triangular, circular, octahedral, polyhedral, etc. [4]. Additionally, AgNPs are employed in antibacterial applications due to their ability to mimic the antimicrobial properties of Ag^+ ions. AgNPs are used in the disciplines of nanomedicine, pharmacy, biosensing, and biomedical engineering thanks to their outstanding capabilities.

14.6 GREEN CHEMISTRY

As shown in Figure 14.3, the biogenic (green chemistry) metal NP synthesis technique, which uses biological elements such microbes and plant extracts, has recently been proposed as a useful substitute for existing synthesis ways. By decreasing metal ions, microorganisms including bacteria and fungi are recognized to be essential in the cleanup of hazardous compounds. Many bacteria have demonstrated the ability to produce AgNPs inside of their cells, wherein intracellular components function as both reducing and stabilizing agents. Since natural capping agents are used to stabilize AgNPs rather than toxic chemicals and hazardous byproducts, the green synthesis of AgNPs using naturally occurring reducing agents may be a promising alternative to more involved physiochemical synthesis [5].

The biological system of a fungus, *Verticillium* species, was used to investigate a possible method of AgNP generation through green synthesis [6]. The primary theory was that AgNPs do not develop in aqueous solution, but rather underneath the surface of the cell wall. The electrostatic interaction between Ag^+ ions and the negatively charged carboxylate groups of the enzyme causes Ag^+ ions to become trapped on the surface of the fungal cells.

Ag nuclei are then created as a result of the intracellular reduction of Ag^+ ions in the cell wall, which is followed by a further reduction of Ag^+ ions. Figure 14.3 shows plausible green chemistry synthesis process [7]. According to the results of transmission electron microscopy research, AgNPs with a diameter of 25 ± 12 nm were produced in cytoplasmic space as a result of the bioreduction of the Ag^+ ions. It's interesting to note that following the manufacture of AgNPs, the fungal cells kept growing. Nitrate reductase uses the reducing capacity of a reduced form of nicotinamide adenine dinucleotide (NADH) to convert nitrate to nitrite, which is a primary source of nitrogen for bacteria. It is possible to use bacterial nitrate metabolism, namely the conversion of nitrate to nitrile and ammonium, in the bioreduction of Ag^+ ions by an intracellular electron donor. In fact, it

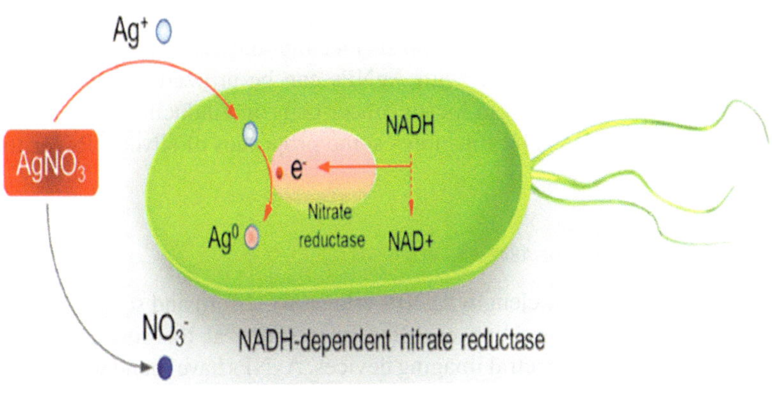

FIGURE 14.3 Plausible green chemistry synthesis process. By using NADH-dependent reductase as an electron transporter to transfer electrons and create NAD^+, the bioreduction is started. By reducing Ag^+ ions to elemental AgNPs, the resultant electrons are obtained [7].

has been discovered that the bioreduction of Ag^+ ions rely heavily on the use of nitrate reductase as a reducing agent. For instance, a justification for an in vitro enzymatic method based on phytochelatin and NADPH-dependent nitrate reductase for the production of AgNPs has been presented by Anil Kumar and colleagues [8]. In vitro, nitrate reductase from the fungus *Fusarium oxysporum* was employed with a co-factor called NADPH. The conversion of NADPH to $NADP^+$ was necessary for the production of AgNPs. To enable the conversion of Ag^{2+} ions to Ag, hydroxyquinoline likely served as an electron shuttle, transferring electrons created during the reduction of nitrate. A stable silver hydrosol (10–25 nm) was created as the Ag^+ ions were reduced in the presence of nitrate reductase, and it was further stabilized by capping peptide. Similar to this, AgNPs have been created in a variety of forms by employing reducing agents that are naturally present, such as supernatants from *Bacillus* species [9]. It was shown that in *Bacillus licheniformis*, the electrons released from NADH were capable of driving the reduction of Ag^+ ions to Ag^0, which produced AgNPs. According to a similar NADH-mediated method, Li et al. previously demonstrated the production of AgNPs by reductase enzymes produced from the fungus *Aspergillus terreus* [10]. Polydispersed nanospheres with a diameter of 1–20 nm, known as synthetic AgNPs, showed antibacterial activity against a variety of harmful bacteria and fungi. Another instance included the manufacture of AgNPs in aqueous $AgNO_3$ using *Pseudomonas stuzeri*, which was isolated from a silver mine [11]. Within the periplasm of the bacterium, the produced AgNPs had a unique shape and well-defined size.

14.7 SILVER NPS BASED NANOMEDICINE

Plasmonic nano-antennas, such as AgNPs, have found extensive applications in various nanomedicine-related fields, including molecular imaging, nanoelectronics, diagnostics, and biomedicine. These applications leverage the enhanced electromagnetic field on and near the surface of AgNPs. While intrinsic Raman scattering of photons from molecules is modest, surface-enhanced Raman scattering (SERS) from molecules near a plasmonic nanoantenna's surface provides significant Raman signal amplification. SERS detection typically requires the adsorption of molecules onto Ag or Au NP aggregates or solid substrates with plasmonic nanostructures, generating strong field amplification in nanogaps or hot spots among interacting plasmonic nanostructures. This effect is crucial for identifying proteins, biomolecules, medication levels in bodily fluids, and early cancer biomarkers. The focus of experimental and theoretical research has been on exploiting hot spots in SERS, which can increase Raman scattering by a factor of 10^8–10^{12}, enabling the detection of even a single molecule.

AgNPs serve as highly sensitive NP probes for in vivo targeting and imaging of small molecules, DNA, proteins, cells, tissues, and tumors. In cellular imaging, AgNPs with stronger and sharper plasmon resonance, achieved through surface modification, are commonly employed. For example, a nanoshell structure coated with AgNPs can be utilized for photothermal treatment and cancer imaging, absorbing light to locate cancer cells and inducing their destruction through the photothermal effect. Figure 14.4 illustrates some properties of silver NPs for the synthesis of nanomedicine [12–15].

14.7.1 Diagnostics with Tunable Wavelength

Diagnostics utilizing tunable wavelength AgNPs efficiently absorb and scatter light, with the nanospheres' broad scattering cross-section enabling observation at the single AgNP level through dark-field microscopy or hyperspectral imaging devices. AgNPs have found widespread applications in cancer cell bioimaging and diagnostics, including the identification of p53 in cancer cells [16]. Nanostructures with silver cores and thick layers of Y_2O_3:Er separated by silica shells, as proposed by Zhang et al., serve as an effective model for studying the interaction of upconversion materials

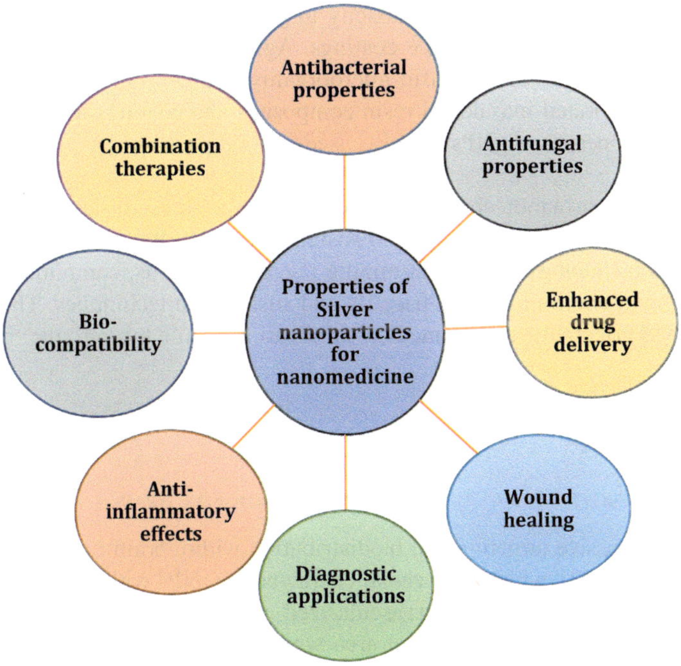

FIGURE 14.4 Properties of silver nanoparticles for the synthesis of nanomedicine.

and metals at the nanoscale, making them suitable as fluorescent markers for single-particle imaging or bioassays requiring low background or tissue-penetrating wavelengths [17]. The size-tunable absorption spectra of AgNPs make them valuable for multiplexed point-of-care diagnostics. Yen et al. presented a multiplexed lateral flow assay based on multicolored AgNPs for detecting various pathogens, allowing for efficient diagnostic investigations with the ability to distinguish between different infections. Triangular plate shaped AgNPs of various diameters, such as 30, 41, and 47 nm, with narrow, controllable absorbance in the visible spectrum, were employed. These AgNPs, conjugated with antibodies, successfully detected the NS protein of the dengue virus, NS1 protein of the yellow fever virus, and the glycoprotein of the Zaire ebola virus, each having a limit of detection of 150 ng/mL in a single channel. Another example is the colorimetric lead detection method developed by Balakumar et al., utilizing AgNPs, which demonstrated remarkable sensitivity for detecting Pb^{2+} as low as 5.2 nM and selective identification even in the presence of competing metal ions. This method provides a rapid and affordable means to identify lead poisoning (saturnism) in a water sample [18,19].

AgNPs have found wide applications as antibacterial and anticancer agents, showcasing inherent cytotoxicity that depends on their surface charges. AgNPs with positive surface charges tend to persist longer on the luminal side of blood vessels or tissue surfaces, a crucial pathway for administering anticancer drugs. These NPs exhibit cytotoxic effects on various cancer cells, including hepatocellular carcinoma, lung, breast, and cervical carcinoma. Smaller-sized AgNPs are more effective in generating reactive oxygen species. Additionally, AgNPs demonstrate anti-angiogenic and anti-proliferative characteristics, damaging DNA, breaking chromosomes, promoting genomic instability, disturbing calcium (Ca^{2+}) homeostasis, leading to apoptosis, and inducing cytoskeletal instability in cancer cells, hindering cell division and cell cycle. In experiments, AgNPs coated with anti-seizure medications showed general anti-amoebic effects against both trophozoite and cyst stages, enhancing fungicidal actions compared to medications alone. The cytotoxic effects of AgNPs have been applied in various sectors, such as medical devices, textiles, food storage, and environmental sensing. Silver, due to its toxicity to germs, is used in products like wound dressings, packaging, and anti-fouling surface coatings. AgNP-coated bandages are particularly interesting for their ability to destroy harmful microorganisms and promote improved healing. Silver ions have been incorporated into dental resin composites and coatings for medical equipment for their antibacterial properties. AgNPs are also used in food packaging to extend shelf life without contamination.

The concept of green nanotechnology in nanomedicine aims to design, synthesize, and use NPs in medicine in ecologically benign and sustainable ways. It seeks to maximize the benefits of traditional nanotechnology while minimizing potential hazards, combining the advantages of nanotechnology with environmentally friendly and sustainable techniques. This ensures the production of safe and effective NMs for medical applications while minimizing their environmental impact [20–26].

14.8 APPLICATIONS

14.8.1 Green Nanotechnology: A Better Solution for Brain Cancer

The success rate of passive targeting and biodistribution within brain tumors is largely dependent on the size of the NPs [27], which is a crucial characteristic. NPs with diameters less than 100 nm take advantage of a brain tumor's hypervascularized and impaired lymphatic drainage system to passively target and enter the intratumoral area while being barred from entering healthy brain tissue. As a result, the enhanced permeability and retention (EPR) effect causes NMs to collect specifically at the location of a brain tumor. However, because to the existence of the NMs, this passive targeting technique has certain built-in restrictions. No matter their composition or structure, all NPs are capable of being opsonized and eliminated by reticuloendothelial system (RES) cells after delivery. Therefore, only a tiny portion of the NPs delivered will really reach the tumor. As a result, for the best EPR effect, NPs should have a diameter of less than 100 nm and biocompatible surfaces to prevent RES removal by cells. This criterion for biocompatibility is crucial, and several researchers are working to improve the biocompatibility of NMs using various methods connected to their functionalization with various coatings [28].

Surface functionality, a critical factor governing the interaction of NMs with biological systems, significantly influences their behavior in both in vitro and in vivo biomedical applications. Researchers have shown that modifying the physiochemical characteristics of coatings can impact the longevity and localization of NMs at the target site. Conventional NP synthesis methods, involving large quantities of toxic compounds like solvents, catalysts, and reducing agents, raise concerns about environmental impact and potential patient reactions. Green nanotechnology presents a solution by promoting the development of clean technologies to minimize risks to human health and the environment. It involves using molecules derived from living organisms as natural starting points for NM synthesis or as catalysts in the conjugation and functionalization of nanostructures. This approach incorporates biological processes and biomolecule-based methodologies, utilizing environmentally friendly platforms that can be integrated with conventional synthesis methods like pulsed laser ablation in liquids. This combination facilitates safer and more sustainable NP production using organic products, extracts, or biological agents obtained or isolated from living organisms [20,21]

Green nanomedicine takes advantage of the possible synergies between green chemistry, drug delivery through nanotechnology, and cellular imaging [29]. In particular, yeast, fungi, bacteria, and extracts from plant tissues were used to produce NMs in a green manner. These green nanostructures have a wide range of uses, including those for magnetically responsive drug delivery, antibacterial and anticancer drugs, photoimaging and photothermal treatment, and nano-detectors of biomolecules and intricate biological structures.

14.9 MICROBIOLOGICALLY DERIVED NANOTECHNOLOGIES

14.9.1 NMs Generated from Bacteria

Due to their widespread availability and viability, bacteria have emerged as primary candidates for the green or biogenic production of NMs. Bacteria, known for their diverse biochemical properties, have historically been utilized for the removal of hazardous metals from the environment by reducing metal ions. Recent interest in NM synthesis by microorganisms has surged, particularly regarding the production of magnetic and noble NMs. Certain bacterial strains, equipped with various enzymes and biochemical processes, possess the ability to capture ions from their environment and reduce them to elemental form. Reduction processes can occur both inside and outside the cells, involving trapping metal ions on the cell surface or transporting them inside microbial cells to generate NPs in the presence of enzymes. Research has focused on understanding and controlling key factors such as particle size, shape, content, and monodispersity in the metallic NPs and nanostructured mineral crystals produced by various microorganisms [30–33].

14.9.2 NMs Generated from Viruses

The design of various nanosized materials, often tailored for delivery purposes, has been inspired by the unique biology and surface attributes of viruses. Viruses exhibit diverse sizes (ranging from 10 to 1,000 nm) and shapes (icosahedrons, spheres, rods, and tubes), featuring a variety of proteins that enhance cellular entry. A notable example is the human immunodeficiency virus, showcasing efficacy in active targeting [34].

14.10 NMS GENERATED BY HUMAN/MAMMALIAN CELLS

Researchers have advanced cell-mediated NP (NM) production as a cutting-edge green nanotechnology method, bypassing the need to internalize NMs within tumors. Leveraging the inherent capability of cells to function as NM biofactories, this approach allows for the direct production of NMs within cell membranes. This cell-mediated NM production proves to be an efficient and environmentally friendly method for entering cells and tumors. Various types of NPs have been derived from diverse cell sources, including cancerous cell lines such as human hepatocarcinoma cells (HepG2), human cervical cancer cells (HeLa, SiHa), and non-cancerous cells like human embryonic kidney cells (HEK-293). These NPs comprise materials such as noble metals (e.g., gold, silver, platinum) and metal oxides (e.g., zinc oxide, magnetite). While research in creating metallic NPs using human cells is ongoing, only a few platforms have been explored for therapeutic applications, and much of the related research is still in its early stages. Nevertheless, NMs hold substantial promise in the field, offering potential anticancer effects and bioimaging capabilities [35,36].

14.11 NMS BASED ON BIOMOLECULES

One of the most versatile green nanotechnology strategies, biomolecule-directed NPs have witnessed a surge in their application, notably in medicine, where they exhibit diverse roles in molecular imaging, drug delivery, theranostics, antimicrobial agents, and the targeting of brain tumors. Recent studies focusing on the development of NMs using biomolecule-based NPs, encompassing carbohydrates, lipids, proteins, and nucleic acids, are outlined in this section. Proteins, nucleic acids, and carbohydrates have garnered special attention due to their potential in crafting and evaluating biomolecule-based NMs. Preliminary proof-of-concept studies have demonstrated enhanced brain drug levels and pharmacodynamic responses following intravenous, nasal, and oral administration. Exemplary biomolecules in this context include chitosan and cellulose polymers, along with carbohydrates, proteins, and nucleic acids, marking these NMs as some of the most investigated and promising carriers [37].

14.12 NANOPLATFORMS BASED ON CARBOHYDRATES

The unique characteristics of carbohydrates, such as their natural origin, biodegradability, and biocompatibility, make them appealing materials for drug carriers. Carbohydrates, also known as sugars, are structurally classified as aldoses and ketoses, composed of carbon (C), hydrogen (H), and oxygen (O) atoms, offering cost-effectiveness and potential for mass production. Chitosan and its derivatives, with remarkable hydrophilicity, biocompatibility, and biodegradability, emerge as promising candidates for nanotechnology-based drug delivery systems. Chitosan magnetic cationic microspheres have been extensively studied for drug delivery to the brain, showcasing controlled release and increased bioavailability. Moreover, chitosan NPs can impact tight junctions, enhance drug permeability across the blood–brain barrier (BBB), and prolong their presence on the nasal mucosa, improving drug delivery from the nasal cavity to the brain. Cellulose, utilized as a feedstock in the synthesis of biomaterials, biofuels, and biochemicals, stands out as an eco-friendly and renewable polymer. In biomedical applications, cellulose and its derivatives have gained attention for tissue engineering, scaffolds, artificial blood vessels, skin grafts, drug transporters, and addressing chronic skin conditions related to cancer therapy. This underscores the versatility of carbohydrates in advancing diverse fields, underlining their significance in contemporary research [34,37].

14.13 NANOPLATFORMS BASED ON PROTEINS

Proteins, abundant in nature, offer favorable attributes such as biocompatibility, biodegradability, easy synthesis, and cost-effectiveness, making them valuable for constructing nano-based drug delivery platforms. Unlike metallic NPs and inorganic/synthetic counterparts, protein-based NMs lack constraints like potential toxicity, excessive size, accumulation, or rapid clearance from the body. Protein-based NMs present opportunities for surface modification through the conjugation of other proteins and carbohydrate ligands. This enables customized delivery to specific tissues and organs, reducing systemic toxicity. Silk protein, a naturally occurring protein fiber, has been extensively employed in the production of nanoconjugate formulations for biomedical applications. Although primarily composed of fibroin, silk is synthesized by certain insect larvae to create cocoons. The affinity of silk protein for various colorants has facilitated its traditional use in the textile industry.

In recent developments, biocompatible silk fibroin (SF) extracted from *Bombyx mori* silkworm cocoons was utilized to produce SF NPs. These SF NPs were employed to dye indocyanine green, resulting in the creation of a therapeutic nanoplatform for the photothermal treatment of glioblastoma. This innovative approach highlights the versatility of protein-based NMs in advancing targeted and effective drug delivery systems [38].

14.14 NANOPLATFORMS BASED ON RNA AND DNA

To diagnose and cure many illnesses, a novel class of medicines known as nucleic acid NPs has emerged [38]. Additionally, the use of nucleic acids as NM coating agents has produced some intriguing applications in the targeting of brain tumors. Using DNA NPs as an example, therapeutic nucleic acids can be delivered at a low cost without running the risk of insertional mutagenesis or the immunogenic responses sometimes seen with virus-based delivery vehicles. The clinical development of gene therapy techniques to treat patients with malignant brain tumors has been sparked by the discovery of potent genetic targets. However, the difficulties of traditional gene vectors to facilitate gene transfer throughout widely dispersed primary brain tumors have been blamed for the lack of success in the clinic.

14.15 ROUTES FOR ADMINISTERING GREEN NANOMEDICINES

Delivering medication molecules to the body's afflicted regions while avoiding any potential negative effects on the organs that are not sick poses the biggest difficulty to drug delivery [39]. By effectively loading drug molecules onto nanoparticulate carriers, safely delivering loaded drug molecules to the body's afflicted organs, and promptly releasing drug molecules at the body's designated places, green nanomedicines should be advantageous for drug delivery. The green nanoparticulate drug formulations can be administered orally, topically, parenterally, and inhaled to the body's specific areas.

14.15.1 Use of Oral Medication

Oral administration is the preferred method of medication administration. When a medicine is administered orally, its gastrointestinal tract solubility and absorption determine its bioavailability and ultimate effectiveness [40]. The medication's low bioavailability is the greatest obstacle to drug delivery via the oral route, though. While peptidase-triggered enzymatic degradation operates as a biochemical barrier, impermeable gastrointestinal epithelium acts as a physical barrier. Proteins and peptides, which are common medications, have a bioavailability of less than 1% [41]. The disadvantages of oral administration might be solved with the help of green nanomedicines. Once the nanomedicine is administered orally, drug molecules begin to leak from the green nanoparticulate system in the gastrointestinal fluids. In the gastrointestinal fluids, the polymeric matrix or wall hydrates and expands, increasing in size. After some time, the matrix or wall breaks or erodes, enabling the loaded medicine to diffuse and release. By shielding the vulnerable medications from the GI tract's digesting enzymes and stomach acid, the green nanocarriers may gradually release them. Due to its huge surface area and the presence of villi and microvilli, the small intestine is where the majority of medication absorption takes place. Due to the high density of villi and microvilli in the duodenum and jejunum, which have the largest surface areas, ileum has the smallest surface area. The majority of the blood supply to the liver—75% of the total hepatic blood flow—comes from the intestines' veins. Prior to hepatic first-pass metabolism, the gut effectively clears the medicines [42]. The retention of the released medicine in the GI tract as well as the duration and intensity of interaction with the mucus layer of the gut are improved by the muco adhesion between the positively charged NPs and the negatively charged mucus and endothelium layer [43]. Nanomedicines' small size also enables them to have a larger surface area to volume ratio, which increases their ability to bind the released medication to surfaces [44].

By lowering metabolic clearance in the liver, nanocarriers safeguard the medication that is entrapped and likely improve drug transport and exposure to the blood circulation system [45].

14.15.2 Topical Delivery

Topical medicine administration is extremely difficult since the skin acts as a barrier against allergies, toxins, germs, fungi, viruses, dust, and big molecules. Three layers typically make up the human skin: The Squamous cell (SC) and the viable epidermis make up the epidermis first. The SC is made up of a 10–20 m thick matrix of corneocytes, or terminally developed keratinocytes, which are dehydrated and dead and are embedded in highly organized lipid layers. The barrier function of the skin and the inadequate absorption of medications are mostly caused by the skin's outermost layer. The first layer of live cells is the viable epidermis, which is the layer directly under the SC and is typically between 0.06 and 0.8 mm thick. There are around four to five layers of keratinocytes and dermal fibroblasts in it. Second, underneath the epidermis lies the dermis, which is typically 0.3–5 mm thick. It has a network of capillaries, lymphatic vessels, nerve endings, sweat glands, hair follicles, and connective tissue.

Third, the subcutaneous tissue, also known as the hypodermis, is the skin's lowest layer. It is made up of loose, white, fibrous connective tissue, and fat acts as a cushion inside it. The disadvantages of oral administration (such as enzymatic degradation and first-pass metabolism of the medication) and injection (such as patient pain, discomfort, and infection at the injection site) are eliminated by topical drug delivery, making it a desirable method of drug administration. Green nanomedicines accumulate and diffuse in hair follicles, where they enter the circulatory system. Lipophilic medications may also diffuse more easily since sebum is present in the pilosebaceous gland's duct. Green nanoparticulate carriers also make it easier for medicine molecules to enter the skin's deeper layers through the SC's outermost layers without triggering skin allergies or irritation. The intimate contact to the SC and increased skin penetration of loaded drug molecules are made possible by nanomedicines with tiny particle size, bigger surface area, and strong affinity for the skin. Additionally, nanocarriers deliver drug molecules to the skin for an extended period of time, improving drug absorption in the skin.

14.16 CONCLUSION

Nanomedicine is a specialized branch of medicine that focuses on using nanotechnology to develop new advancements in healthcare, including the treatment of a variety of illnesses like cancer, infections, and auto-immune disorders. The discipline began to take off in the 1980s, coinciding with the first nanomedical oncological medications to get regulatory agency clearance. Furthermore, nanotechnology was crucial in the creation of the mRNA vaccines used to combat the COVID-19 pandemic, thus demonstrating its pervasive importance in the fields of science and medicinal innovation. Nanotechnology is used in biomedicine for a variety of purposes, although it is often used to transport or protect bioactive chemicals to specific tissues. The overall aim is to develop nanoscopic platforms that may interact with biological systems in many ways, either by altering pharmacokinetics or by preferentially activating a particular biological route. The complex biomolecular interactions that underpin the biological system's functionality occur at the nanoscale. To interact with biological systems in novel ways, NPs can be engineered to be comparable in size to important biological structures like big biomolecules or tiny organelles. This can enhance the effects of medications used to treat certain disorders. In other words, such interactions can lead to extremely distinctive and enhanced biological consequences. This chapter has been set up to highlight the most recent developments in green nanotechnology-based approaches, where all raw materials are derived from living things or natural sources, as well as environmentally friendly approaches, where one or more commonly used raw materials have been replaced with natural or naturally occurring alternatives, or that use one or more biologically derived molecules, to provide a more effective and tailored solution.

REFERENCES

1. Viseu, A. (2020). Nanomedicine. Encyclopedia Britannica. https://www.britannica.com/science/nanomedicine. Accessed 15 August 2024.
2. Sevastre, A.-S., Horescu, C., Carina Baloi, S., Cioc, C. E., Vatu, B. I., Tuta, C., Artene, S. A., et al. (2019). Benefits of nanomedicine for therapeutic intervention in malignant diseases. *Coatings*, *9*(10), 628.
3. Wang, L., Zhang, T., Li, P., Huang, W., Tang, J., Wang, P., & Chen, C. (2015). Use of synchrotron radiation-analytical techniques to reveal chemical origin of silver-nanoparticle cytotoxicity. *ACS Nano*, *9*(6), 6532–6547.
4. Heiligtag, F. J., & Niederberger, M. (2013). The fascinating world of nanoparticle research. *Materials Today*, *16*(7–8), 262–271.
5. Al-Warthan, A., Kholoud, M. M., El-Nour, A., Eftaiha, A., & Ammar, R. A. A. (2010). Synthesis and applications of silver nanoparticles. *Arabian Journal of Chemistry*, *3*, 135–140.
6. Gajbhiye, M., Kesharwani, J., Ingle, A., Gade, A., & Rai, M. (2009). Fungus-mediated synthesis of silver nanoparticles and their activity against pathogenic fungi in combination with fluconazole. *Nanomedicine: Nanotechnology, Biology and Medicine*, *5*(4), 382–386.

7. Lee, S., & Jun, B.-H. (2019). Silver nanoparticles: synthesis and application for nanomedicine. *International Journal of Molecular Sciences, 20*(4), 865.
8. Anil Kumar, S., Abyaneh, M. K., Gosavi, S. W., Kulkarni, S. K., Pasricha, R., Ahmad, A., & Khan, M. I. (2007). Nitrate reductase-mediated synthesis of silver nanoparticles from $AgNO_3$. *Biotechnology Letters, 29*, 439–445.
9. Vaidyanathan, R., Gopalram, S., Kalishwaralal, K., Deepak, V., Pandian, S. R. K., & Gurunathan, S. (2010). Enhanced silver nanoparticle synthesis by optimization of nitrate reductase activity. *Colloids and surfaces B: Biointerfaces, 75*(1), 335–341.
10. Li, G., He, D., Qian, Y., Guan, B., Gao, S., Cui, Y., & Wang, L. (2011). Fungus-mediated green synthesis of silver nanoparticles using Aspergillus terreus. *International Journal of Molecular Sciences, 13*(1), 466–476.
11. Klaus, T., Joerger, R., Olsson, E., & Granqvist, C. G. (1999). Silver-based crystalline nanoparticles, microbially fabricated. *Proceedings of the National Academy of Sciences, 96*(24), 13611–13614.
12. Lee, S. J., Morrill, A. R., & Moskovits, M. (2006). Hot spots in silver nanowire bundles for surface-enhanced Raman spectroscopy. *Journal of the American Chemical Society, 128*(7), 2200–2201.
13. Kneipp, J., Kneipp, H., McLaughlin, M., Brown, D., & Kneipp, K. (2006). In vivo molecular probing of cellular compartments with gold nanoparticles and nanoaggregates. *Nano Letters, 6*(10), 2225–2231.
14. Kleinman, S. L., Frontiera, R. R., Henry, A. I., Dieringer, J. A., & Van Duyne, R. P. (2013). Creating, characterizing, and controlling chemistry with SERS hot spots. *Physical Chemistry Chemical Physics, 15*(1), 21–36.
15. Loo, C., Lowery, A., Halas, N., West, J., & Drezek, R. (2005). Immunotargeted nanoshells for integrated cancer imaging and therapy. *Nano Letters, 5*(4), 709–711.
16. Zhou, W., Ma, Y., Yang, H., Ding, Y., & Luo, X. (2011). A label-free biosensor based on silver nanoparticles array for clinical detection of serum p53 in head and neck squamous cell carcinoma. *International Journal of Nanomedicine, 6*, 381–386.
17. Zhang, F., Braun, G. B., Shi, Y., Zhang, Y., Sun, X., Reich, N. O., & Stucky, G. (2010). Fabrication of Ag@ SiO_2@ Y_2O_3: Er nanostructures for bioimaging: tuning of the upconversion fluorescence with silver nanoparticles. *Journal of the American Chemical Society, 132*(9), 2850–2851.
18. Yen, C. W., de Puig, H., Tam, J. O., Gómez-Márquez, J., Bosch, I., Hamad-Schifferli, K., & Gehrke, L. (2015). Multicolored silver nanoparticles for multiplexed disease diagnostics: distinguishing dengue, yellow fever, and Ebola viruses. *Lab on a Chip*, 15(7), 1638–1641.
19. Balakumar, V., Prakash, P., Muthupandi, K., & Rajan, A. (2017). Nanosilver for selective and sensitive sensing of saturnism. *Sensors and Actuators B: Chemical*, 241, 814–820.
20. Suhag, D., Thakur, P., & Thakur, A. (2023). Introduction to nanotechnology. In: Suhag, D., Thakur, A., Thakur, P. (eds) *Integrated Nanomaterials and their Applications.* Springer, Singapore. https://doi.org/10.1007/978-981-99-6105-4_1
21. Rathore, R., Suhag, D., Wan, F., Thakur, A., & Thakur, P. (2023). Everyday nanotechnology. In: Suhag, D., Thakur, A., Thakur, P. (eds) *Integrated Nanomaterials and their Applications.* Springer, Singapore. https://doi.org/10.1007/978-981-99-6105-4_2
22. Thakur, P., & Thakur, A. (2022). Introduction to nanotechnology. In: Thakur, A., Thakur, P., Khurana, S. P. (eds) *Synthesis and Applications of Nanoparticles.* Springer, Singapore. https://doi.org/10.1007/978-981-16-6819-7_1
23. Jeyaraj, M., Sathishkumar, G., Sivanandhan, G., MubarakAli, D., Rajesh, M., Arun, R., & Ganapathi, A. (2013). Biogenic silver nanoparticles for cancer treatment: an experimental report. *Colloids and surfaces B: Biointerfaces, 106*, 86–92.
24. Gurunathan, S., Han, J. W., Eppakayala, V., Jeyaraj, M., & Kim, J. H. (2013). Cytotoxicity of biologically synthesized silver nanoparticles in MDA-MB-231 human breast cancer cells. *BioMed Research International*, 2013, 535796.
25. Zhang, W. S., Cao, J. T., Dong, Y. X., Wang, H., Ma, S. H., & Liu, Y. M. (2018). Enhanced chemiluminescence by Au-Ag core-shell nanoparticles: a general and practical biosensing platform for tumor marker detection. *Journal of Luminescence, 201*, 163–169.
26. Sondi, I., & Salopek-Sondi, B. (2004). Silver nanoparticles as antimicrobial agent: a case study on E. coli as a model for Gram-negative bacteria. *Journal of Colloid and Interface Science, 275*(1), 177–182.
27. Krishnia, L., Thakur, P., & Thakur, A. (2022). Synthesis of nanoparticles by physical route. In: Thakur, A., Thakur, P., Khurana, S. P. (eds) *Synthesis and Applications of Nanoparticles.* Springer, Singapore. https://doi.org/10.1007/978-981-16-6819-7_3
28. Khaitan, D., Reddy, P. L., Narayana, D. S., & Ningaraj, N. S. (2018). *Recent Advances in Understanding of Blood–Brain Tumor Barrier (BTB) Permeability Mechanisms that Enable Better Detection and Treatment of Brain Tumors.* Elsevier, Amsterdam, The Netherlands, pp. 673–688.

29. Jahangirian, H., Lemraski, E. G., Webster, T. J., Rafiee-Moghaddam, R., & Abdollahi, Y. (2017). A review of drug delivery systems based on nanotechnology and green chemistry: green nanomedicine. *International Journal of Nanomedicine, 12*, 2957–2978.
30. Zhang, X., Yan, S., Tyagi, R. D., & Surampalli, R. Y. (2011). Synthesis of nanoparticles by microorganisms and their application in enhancing microbiological reaction rates. *Chemosphere, 82*(4), 489–494.
31. Varshney, R., Bhadauria, S., Gaur, M. S., & Pasricha, R. (2010). Characterization of copper nanoparticles synthesized by a novel microbiological method. *JOM, 62*(12), 102–104.
32. Thakkar, K. N., Mhatre, S., & Parikh, R. Y. (2010). Biological synthesis of metallic nanoparticles. *Nanomedicine: Nanotechnology, Biology and Medicine, 6*(2), 257–262.
33. Fayaz, A. M., Balaji, K., Morukattu, G., Yadav, R., Kalaichelvan, P. T., & Venketesan, R. (2010). Biogenic synthesis of silver nanoparticles and their synergistic effect with antibiotics: a study against gram-positive and gram-negative bacteria. *Nanomedicine: Nanotechnology, Biology and Medicine, 6*(1), 103–109.
34. Gundogdu, E., & Yurdasiper, A. (2014). Drug transport mechanism of oral antidiabetic nanomedicines. *International Journal of Endocrinology and Metabolism, 12*(1), e8984.
35. Thermo Fisher Scientific. Safety data sheet (2014). www.fishersci.com/us/en/home.html
36. Sciencelab.com, Inc. Material safety data sheet (2005). https://www.sciencelab.com/msds.php?msdsId=9924969
37. Liu, J., Hu, W., Chen, H., Ni, Q., Xu, H., & Yang, X. (2007). Isotretinoin-loaded solid lipid nanoparticles with skin targeting for topical delivery. *International Journal of Pharmaceutics, 328*(2), 191–195.
38. Vaghasiya, H., Kumar, A., & Sawant, K. (2013). Development of solid lipid nanoparticles based controlled release system for topical delivery of terbinafine hydrochloride. *European Journal of Pharmaceutical Sciences: Official Journal of the European Federation for Pharmaceutical Sciences, 49*(2), 311–322.
39. De Jong, W. H., & Borm, P. J. (2008). Drug delivery and nanoparticles: applications and hazards. *International Journal of Nanomedicine, 3*(2), 133–149.
40. Ravichandran, R. (2009). Nanoparticles in drug delivery: potential green nanobiomedicine applications. *International Journal of Green Nanotechnology: Biomedicine, 1*(2), B108–B130.
41. Lam, P. L., & Gambari, R. (2014). Advanced progress of microencapsulation technologies: in vivo and in vitro models for studying oral and transdermal drug deliveries. *Journal of Controlled Release: Official Journal of the Controlled Release Society, 178*, 25–45.
42. Pang, K. S. (2003). Modeling of intestinal drug absorption: roles of transporters and metabolic enzymes (for the Gillette Review Series). *Drug Metabolism and Disposition: The Biological Fate of Chemicals, 31*(12), 1507–1519.
43. Ramesan, R. M., & Sharma, C. P. (2009). Challenges and advances in nanoparticle-based oral insulin delivery. *Expert Review of Medical Devices, 6*(6), 665–676.
44. Kadam, R. S., Bourne, D. W., & Kompella, U. B. (2012). Nano-advantage in enhanced drug delivery with biodegradable nanoparticles: contribution of reduced clearance. *Drug Metabolism and Disposition: The Biological Fate of Chemicals, 40*(7), 1380–1388.
45. Goyal, R., Macri, L. K., Kaplan, H. M., & Kohn, J. (2016). Nanoparticles and nanofibers for topical drug delivery. *Journal of Controlled Release: Official Journal of the Controlled Release Society, 240*, 77–92.

15 Development of Green Nanomaterials for Building and Construction Applications

Prakash Chander Thapliyal
CSIR-Central Building Research Institute

15.1 INTRODUCTION

Through the past few decades, a variety of materials have radically changed the backdrop of science and technology. Nanotechnology serves as a common link for material investigations at the nanoscale. The incessant hunt for stronger as well as lighter materials, that can perform manifold functions, self-repairing materials with sustainable design is prominent nowadays. Designers have added choices for green nanomaterials today. The probable input to sustainability makes nanotechnology one of the key technologies in green building areas aiming to (a) recognize nanotechnology developments that are pertinent to green building design and (b) offer a way for potential nanotechnology developments that could be of use in green building practices. Polymer nanocomposite materials are currently in the making with the integration of nanofillers like nanoclays, nanoparticles, nanotubes, and nanofibers. Furthermore, this integration of nano reinforcements into elastomers noticeably improves their mechanical and thermal barrier properties with clear improvements in adhesion, rheological, and processing behaviors. The superior distribution of nanofillers within the matrix produces high-performance nanocomposites [1]. A variety of nanomaterials have been utilized to transform features of polymeric materials since pure polymer materials do not have strong mechanical properties [2], thermal conductivity, and thermal stability [3]. Adding nanomaterials to polymeric materials provides polymer composites with better mechanical, thermal, and electrical properties and better conductivity. For this reason, polymers (including composites) toughened by way of nanomaterials are used in applications such as medical devices, automobiles, safety, protective clothing, aerospace, electronics, optical devices, military equipment, and constructions and buildings [4]. Nanotechnology is set to change roughly all main technology and industrial sectors: IT, medicine, safety, energy, environmental science, and transportation, among many others [5]. The applications of nanotechnology are shown in Table 15.1. Nanoparticles are competent additives for improving cement commodities, even at minute concentrations (<1%). The main modifications are a decrease in setting time (1–2 hours) and diffusivity (4–75%), enhancement of strength (5–25%), and thermal durability (0–30% boost in residual force) [6].

The growth of well-organized green production utilizing usual capping, reducing, and stabilizing agents with no use of lethal, high-priced chemicals and soaring energy use has fascinated researchers in the direction of biological methods [7]. Bakhoum and his research group [3] studied the utilization of nano-granite waste particles as a substitute for cement, along with fine aggregate, during the making of mortar and found that 5% cement along with 10% sand replacement with nano-granite thrown away in mortar mix improved compression of green mortar by 41%. Environmental and community attributes showed a 10% reduction in resource expenditure, while savings in energy spending along with carbon dioxide emissions reached around 5%. The economic field showed a cutback of 6.5%, signifying hopeful outputs in promoting sustainable construction manufacturing. Cement-based resources such as concrete, mortar, and bricks are extensively utilized structural

DOI: 10.1201/9781003502692-15

TABLE 15.1
Some Nanotechnology Applications

Nanomaterials	Area	Properties Improved	Uses
TiO_2, ZnO	Cosmetics	Antimicrobial, cleansing, antioxidant	Shampoos, sunscreens, creams, lotions
SiO_2, CNT, TiO_2	IT	Self-cleaning, water repellent, antifog, anti-reflective	Displays, eyeglasses
SiO_2, Al_2O_3, ZnO, CNT	Composites	Stiff, lightweight, resilient, durable	Tennis rackets, helmets, bumpers, baseball bats
TiO_2	Environment	Absorption	CNT scrubber, paper towel
SiO_2	Sustainable energy	Stronger, lightweight	Windmills

resources considering the built environment. In addition, energy conservation and reduction in greenhouse emissions in the built atmosphere can be guaranteed because nanomaterials act like insulators along with energy storage materials. Other applications such as antimicrobial surfaces would cut resource spending and extend the life cycle of material usage in business, constraining the direction of sustainable practices as a result [8]. Moreover, soil filling, in addition to the recycling charge of waste resources commencing from industries such as mineral, ceramic, granite, plastic, rubber, textile, etc., results in a waste-dumping catastrophe. Such types of wastes having pozzolanic actions are used more in construction manufacturing like partial substitution by means of cement to lessen carbon dioxide release from cement production along with environmental contamination [9]. In the restoration or pulling down of buildings, engineered nanomaterials restricted in past construction resources are recycled; otherwise, they would accumulate as construction waste [10]. Most nanomaterials' investigations depended on using lofty price tags for unprocessed materials (like silica nanoparticles) in the nano range to augment features of cement-based resources, particularly concrete with mortar. 'Green nanotechnology' is considered the biosynthesis of nanomaterials with natural bioactive agents, for instance, plant materials and microbes, along with a variety of biowastes like farming residues, vegetable residuals, eggshells, and fruit peels, as well as nano products to accomplish sustainability [11], which are low-cost, safe, simple, nontoxic, and low-risk and have environment-friendly approach. The pervasiveness of green course techniques is based on evidence to facilitate the making of nanomaterials in synchronized and hygienic surroundings; hence they are environment friendly [12]. There is an assortment of accessible approaches to the green route comprising plant extracts along with synthesizing nanoparticles using fungi, viruses, bacteria, algae, etc. [13]. Every one of these green methods portrays explicit return above environmental management and disadvantages. There are various chemical, physical, and biological methods to prepare nanomaterials. Nonetheless, physical and chemical techniques have the consequence of environmental contamination as these processes cause toxic by-products as well as use an immense deal of energy. For that reason, green nanotechnology comprising the generation of clean, environmentally friendly, and safe nanoparticles is in demand [14].

Green synthesis has already become an alternative, having unmatched reliability, cost-effectiveness, less usage of chemicals, eco-friendliness [15], simple procedures, and better biocompatibility of nanoparticles [16]. Production of green nanomaterials entails guarded, cleaner, and safer processes promoting environmental easiness and overcoming the consumption of noxious reagents. The synthesis of green nanomaterials is extremely noteworthy since they help in the decline of unnecessary or detrimental by-products via the use of renewable biodegradable plant-based or microorganism-based extractive systems. Plants are bestowed with assets of chemicals, comprising flavonoids, saponins, alkaloids, and tannins [17]. Such 'green nanomaterials' exploiting plants or their extracts are within reach via a straightforward process compared to microorganism-mediated synthesis [18]. In addition, key metabolites akin to fructose, glucose, sucrose, etc. at hand in plant extracts are used in the creation of nanoparticles [19]. The synthesis of 'green nanomaterials' in effect is a one-pot process comprising eco-friendly bio-reduction.

Even though the finest way is using 'zero solvents', if utilization of solvent is necessary, water is the finest option. While the side-effects of green nanomaterials are many-fold lesser than those chemically prepared, durability can be demanding as green nanomaterials tend to shape aggregates. Also, the strength of nanomaterials is frequently plagued by surface complexation processes; all can be synchronized through modifying particle dimensions, using suitable capping reactants, and right functionalization techniques [20–22]. Employing bio-compatible stabilizers like biodegradable green polymers [23,24] has initiated new 'greener' sites on behalf of nanomaterial surface engineering mutually through proper functional groups, which would add not only to the permanence of nanomaterials via lowering ionization of metallic groups but also to lesser toxicity of bio-based nanoparticles [25]. For this reason, greener production not only opens less-toxic options in support of the congregation of nanomaterials but also gives a straightforward, effortless, and expedient way to contact them.

The importance of nanomaterials in construction manufacturing is mostly due to optimistic characteristics including moisture behavior, improved strength, thermal properties, energy efficiency, improvement of air worth, self-cleaning, and antimicrobial effects [26]. At the nano point, magnitude becomes unimportant because electrostatic reactions get control, with quantum special effects being dragged within. In addition, as particles tend to nano dimension, the portion of atoms on the plane increases compared to the interior; furthermore, this concludes the new properties of the material. The broad ex-situ production practice consists of selecting a plant. This biomass is composed, dried, and macerated, and after that particles obtained are placed in aqueous or alcoholic solutions and subjected to heat to produce biomolecules. After the solution is filtered, the remainder is used to carry out the synthesis under ambient conditions [27]. The ability of plants to aid in the production of nanomaterials is linked to their wide-ranging biochemical composition, given that this vegetal matter has extensive diverse compounds, such as lipids, vitamins, proteins, polyols, steroids, pigments, flavonoids, alkaloids, terpenoids, resins, and polyphenols [28–32].

Execution of the 'green chemistry' method in the area of nanomaterials and related technologies can be a complex course since green chemistry is mostly a qualitative scheme, meaning there is a need for a universal smallest description criteria to pertain model "green" to chemical processes [33–36]. This has led to the need to approve and extend experimental parameters to quantitatively establish the extent of 'green' production, or sustainability, in a chemical process. The environmental force is a perceptive notion, yet its function has been hardly implemented for green nanomaterial creation in studies and under non-controlled circumstances; consequently, factors linked with the quantity of material change in concepts like atomic competence, the proportion of residues to product, and energy efficiency focused on LCA [37–41]. In comparison, as soon as these parameters are evaluated in the synthesis of nanomaterials, they are discarded in ideal mode owing to inbuilt polydispersity [42,43]. Thus, it is apparent that there is still a noteworthy mode familiar to convey the concepts of sustainability and green chemistry away from easy intellectual discussion.

The presence of various functional groups like hydroxyl, keto, and/or thiols in these compounds makes probable reduction–oxidation and/or stabilization processes of nanoparticles all through production [43–45]. This explains how the same plant species synthesize nanomaterials of varying types; e.g., *Azadirachta indica* (Neem) manufactures a broad range of metal-based nanoparticles. Neem leaf extracts are used for the synthesis of spherical CeO_2 having an average dimension of less than 15 nm, used to accomplish photocatalytic dilapidation of Rhodamine B colorant with 96% yield in around 100 minutes. In contrast, silver nanoparticles (AgNPs) along with copper nanoparticles (CuNPs) were produced using neem leaf extracts, with the adsorption capability of aromatic hydrocarbon naphthalene being 98.81%.

15.2 APPLICATIONS OF NANOTECHNOLOGY IN BUILDING INDUSTRY

It is recognized that concrete is the principal material in load-bearing applications, with strength, firmness, and outlay being the main attributes. Concrete is a composite material at a large scale, but its properties can be improved mutually on meso and nanoscales [46].

Nanotechnology helps in the understanding of concrete's performance, i.e. progress of its mechanical properties and the lessening of ecological and manufacturing expenditure [47]. The majority of proficient nanoparticles for concrete fabrication are nano silica, titanium dioxide, silica fume, nano clay, ferric oxide, carbon nanotubes (CNTs), calcium carbonate, and graphene oxide. Nano silica or silicon dioxide nanoparticles, with their colloidal-appearing silica fume, are widely used, and advantages include tall compressive force, flexural strength, low permeability, high tensile strength, modulus of elasticity, and enhanced durability. Using nano silica in concrete offers better properties such as strength, fire and abrasion resistance or leaching, workability and improved setting time, heat of hydration, and other behaviors under hostile environments [48]. A comparison of the construction sector with other sectors in their applications of nanomaterials is given in Table 15.2. In the construction sector, nanomaterials can be used for mutually structural and non-structural purposes.

15.2.1 Structural Materials

Structural materials are important for a building for the durability, longevity, and sustainability of the structure. For structural material strength/weight, a relative amount is vital, as stronger and

TABLE 15.2
Comparison of Construction Sector with Others in Their Applications of Nanomaterials

Applications	
Other Sectors	**Construction Sector**
Nano glass: Glass can be made to encompass self-cleaning, anti-fouling, sterilizing, etc. by means of nano TiO_2 as a glass cover.	Nanomaterials in concrete: Nano SiO_2, Al_2O_3, TiO_2 improve concrete features and create tall performance, high-strength self-compacting concrete.
Nano steel: Nano copper particles condensed for the surface roughness of steel and resulted in a boost in steel stress and that's why fatigue cracking. Above and beyond, vanadium and molybdenum nanoparticles improved fracture-related inconvenience resulting from high-strength bolts.	Nano cement: Nanotubes and reactive silica nanoparticles can be blended with cement. Using nano cement in a concrete blend will be more efficient than nano carbon tubes or carbon fibers in concrete.
Nanoparticles for fire resistance: By integrating nano cement into carbon nanotubes, fiber composites having high strength and fire resistance can be constructed.	Nano coat for concrete: Nanometer scale coatings are robust plus boast self-cleaning and self-healing features; used to look after concrete structures on or after scratches, chemicals, etc.
	Nano sensors for concrete structures: Smart nano dust (aggregate) sensors as wireless sensors are rooted in concrete, scattered on the surface of construction, or else integrated into the mix. These sensors can be used in concrete structures for quality management and health monitoring by observing changes in compactness, shrinkage, viscosity, chloride concentration, moisture, etc. of concrete.

lighter materials can bear superior loads for every part of the material. Nanotechnology can offer better structural materials by (a) reinforcing current materials, such as concrete and steel, with nanoparticles; (b) given that all of the structure is constructed for the most part with innovative materials, like CNTs, while technically and cost-effectively realistic. Concrete has the largest yearly production among structural materials and undergoes radical enhancements. Energy utilization, carbon emissions, and waste generation are three main environmental concerns allied with concrete making and use. For each ton of cement produced, 1.3 times carbon dioxide is out into the atmosphere. Global cement-making produces more than 1.6 billion tons of carbon, accounting for an excess of 8% of entire carbon emissions. Waste generated is also sizeable, as concrete makes for more than two-thirds of construction, with demolition waste with merely 5% currently being recycled [49].

Usually, concrete toughened with steel bars is used to defy tension, which is a costly process. On the contrary, nanofiber reinforcement improves the strength of concrete appreciably.

Incorporating nanoparticles in the making of construction materials improved properties such as strength and durability. Toughness, shear, and flexural and tensile strength of construction materials also improved [50]. Table 15.2 shows some of the nanomaterials used in the construction industry.

The growth of well-organized green production utilizing natural reducing, capping, and stabilizing catalysts with zero use of lethal, high-priced chemicals and high energy use has fascinated researchers in the direction of biological methods [51]. Environmental and community attributes showed a reduction of 10% in resource expenditure, whereas savings in energy consumption along with carbon dioxide emissions reached 5%. Economic pasture showed a cutback of 6.5%, demonstrating hope in enhancing sustainable construction production [52].

Cement-based resources such as concrete, mortar, and bricks are far and widely used building resources considering the built environment. In addition, energy conservation, as well as reduction in greenhouse emissions in the building surroundings, can be guaranteed owing to nanomaterials acting as insulators as well as energy storage materials. Other applications such as antimicrobial surfaces would cut resource consumption along with extending the life cycle of the material used in the industry, thereby forcing in the direction of sustainable practices [53,54]. Moreover, soil filling, in addition to recycling expenditure of throw-away materials commencing from industries such as mineral, rubber, granite, plastic, textile, and ceramic, results in waste-throwing-away catastrophe. Such varieties of wastes having pozzolanic performance are used more in construction engineering as selected substitution of cement in a bid to diminish carbon dioxide release generated from cement production along with environmental pollution [55]. In the restoration or pulling down of buildings, engineered nanomaterials limited to former construction resources are recycled or developed into construction waste [56]. Most nanomaterials' investigations depended on tall expenditure unprocessed materials (e.g. silica particles) at the nano level to augment the properties of cement-based materials, particularly concrete with mortar.

The influx of microorganisms in green buildings is a cause of worry because their propagation leads to health issues. Outdoor tests with silver nanoparticles have verified the successful safeguarding of straw against microorganisms. In difference to normally reported results, this revision showed silver nanoparticles prepared using non-hydrolytic sol–gel methods and encompass antifungal properties in alfresco circumstances.

15.2.2 Non-Structural Materials

Glass is one of the non-structural materials inexorably used in buildings. However, glass as a glazing substance principally affected sustainability, given that it acts largely as a barrier between outside and indoor environments, impacting indoor airflow. Mainly heat gain and loss occur on glass surfaces, affecting daylighting. Nanotechnology offers better options for glazing using photochromic, thermochromic, and thin-film coatings, as well as electrochromic technologies. Thin-film coatings sieve out infrared illumination to lessen heat increase in buildings. Thermochromic glass alters lucidity with temperature and controls overheating for passive solar heating applications [57]. Nanotechnology use results in the production of electrochromic glass, composed of a five-layer coating about 1 µm thick. The electrochromic mass consists of slim nickel or tungsten oxide coatings sandwiched among two see-through electrical conductors. Voltage application amid electrical conductors caused the electrical field to move various colored ions reversibly. The entire applications are intended to decrease energy for the cooling of buildings, bringing down energy consumption [58,59].

Plastics are one more example of non-structural materials extensively used in the building industry. Polyvinyl chloride is applied in doors, flooring, windows, etc. but is susceptible to fire [60]. Nano-reinforced polyester has exceptional thermal and electrical insulation while being tough and lightweight, with corrosion resistance, good impact, and fine surface finish. It can be used as a load-bearing structural material in doors, bridges, facades, windows, and structural systems [61]. Nanotechnology innovations make insulation materials extra proficient and less dependent on non-renewable resources, about 30% more efficient than conventional materials. Nanomaterials have the capability to capture air inside them, thereby increasing their surface-to-volume proportion. Nanoscale insulation supplies may be sandwiched amid inflexible panels, applied as slim films, or decorated as coatings. Aerogels are exceptional insulation materials and are highly porous solids with tremendously small density, open pores, and highly specific surface area. Regardless of its lightweight, it can carry over 2,000 times its weight. Since nanoporous aerogels are susceptible to humidity, they are repeatedly sandwiched among wall panels. Architectural applications of aerogels embrace windows, skylights, and translucent wall panels [62].

15.3 CONCLUSION

Nanotechnology presents architects and civil engineers with unprecedented control over the shaping and transformation of the human environment. This cutting-edge technology offers a myriad of applications within the buildings and construction sectors, fundamentally influencing the way they design and construct structures. From internal finishes to city planning, nanotechnology's impact is profound, providing architects and engineers with innovative tools to enhance both aesthetic and functional aspects of the built environment. One notable area where nanotechnology revolutionizes construction is the development of sustainable built environments. By harnessing nanomaterials and processes, architects can create structures with enhanced durability, energy efficiency, and environmental consciousness. Nanotechnology enables the integration of smart materials into building components, facilitating real-time monitoring and adaptive responses to environmental conditions. This not only contributes to the longevity of structures but also aligns with the global push toward sustainable and eco-friendly urban development. In the realm of internal design, nanotechnology applications extend to advanced finishes that offer properties like self-cleaning surfaces, improved insulation, and even responsive color-changing elements. These innovations not only elevate the functionality of indoor spaces but also contribute to resource conservation and reduced maintenance requirements. Additionally, on a city-wide scale, the incorporation of nanotechnology in urban planning allows for intelligent infrastructure systems, optimized traffic flow, and environmentally conscious architectural solutions. In essence, the integration of nanotechnology in construction opens up a realm of possibilities for creating resilient, sustainable, and technologically advanced built environments that can positively shape the future of humankind.

REFERENCES

1. Abbas S, Nasreen S, Haroon A, Ashraf MA (2020) Synthesis of silver and copper nanoparticles from plants and application as adsorbents for naphthalene decontamination. *Saudi J. Biol. Sci.* 27, 1016–1023. https://doi.org/10.1016/j.sjbs.2020.02.011
2. Aoul KAT, Attoye DE, Ghatrif LA (2019) Performance of electrochromic glazing: state of the art review. *IOP Conf. Ser.: Mater. Sci. Eng.* 603(2), 022085, 1–11.

3. Bakhoum ES, Garas GL, Allam ME, Ezz H (2017) The role of nanotechnology in sustainable construction: a case study of using nano granite waste particles in cement Mortar. *Eng. J.* 21(4), 217–227.
4. Behnood A, Ziari H (2008) Effects of silica fume addition and water to cement ratio on the properties of high-strength concrete after exposure to high temperatures. *Cem. Concr. Compos.* 30, 106–112.
5. Cannavale A, Ayr U, Fiorito F, Martellotta F (2020) Smart electrochromic windows to enhance building energy efficiency and visual comfort. *Energies.* 13, 1449, 1–17.
6. Chakrabartty I, Hakeem KR, Mohanta YK, Varma RS (2022) Greener nanomaterials and their diverse applications in the energy sector. *Clean Technol. Environ. Policy.* 24, 3237–3252.
7. Chakrabartty I, Vijayasekhar A, Rangan L (2019) Therapeutic potential of labdane diterpene isolated from Alpinia nigra: detailed hemato-compatibility and antimicrobial studies. *Nat. Prod. Res.* 35, 1000–1004. https://doi.org/10.1080/14786419.2019.1610756
8. Constable DJC, Jimenez-Gonzalez C, Lapkin A (2008) Processs metrics. In: Lapkin, A. (eds) *Green Chemistry Metrics: Measuring and Monitoring Sustainable Processes.* John Wiley and Sons Ltd, New York, pp. 228–257. https://doi.org/10.1002/9781444305432.ch6.
9. Demirdoven JB, Karacar P (2015) Green nano-materials with examples of applications. Green Age Symposium, Mimar Sinan Fine Arts University Faculty of Architecture, 15–17 April 2015, Istanbul, Türkiye.
10. Dikshit PK, Kumar J, Das AK, Sadhu S, Sharma S, Singh S, Gupta PK, Kim BS (2021) Green synthesis of metallic nanoparticles: applications and limitations. *Catalysts.* 11(8), 902.
11. Domun N, Hadavinia H, Zhang T, Sainsbury T, Liaghat GH, Vahid S (2015) Improving the fracture toughness and the strength of epoxy using nanomaterials-a review of the current status. *Nanoscale.* 7(23), 10294–10329.
12. Gomez-lopez P, Puente-santiago A, Castro-beltran A, Nascimento LAS, Balu AM, Luque R, Castro-beltran CG (2020) Nano-materials and catalysis for green chemistry. *Curr. Opin. Green Sustain Chem.* 24, 48–55. https://doi.org/10.1016/j.cogsc.2020.03.001
13. Thakur A, Thakur P, Baccar S (2023) Structural properties of nanoparticles. In: Suhag, D., Thakur, A., Thakur, P. (eds) *Integrated Nanomaterials and their Applications.* Springer, Singapore. https://doi.org/10.1007/978-981-99-6105-4_4
14. Krishnia L., Thakur P., Thakur A (2022) Synthesis of nanoparticles by physical route. In: Thakur, A., Thakur, P., Khurana, S. P. (eds) *Synthesis and Applications of Nanoparticles.* Springer, Singapore. https://doi.org/10.1007/978-981-16-6819-7_3
15. Hu K, Kulkarni DD, Choi I, Tsukruk VV (2014) Graphene-polymer nanocomposites for structural and functional applications. *Prog. Polym. Sci.* 39(11), 1934–1972.
16. Huang C, Qian X, Yang R (2018) Thermal conductivity of polymers and polymer nanocomposites. *Mat. Sci. Eng. R.* 132, 1–22.
17. Hussain I, Singh A, Singh NB, Singh H, Singh SC (2016) Green synthesis of nanoparticles and its potential application. *Biotechnol. Lett.* 38, 545–560.
18. Isah T (2019) Stress and defense responses in plant secondary metabolites production. *Biol. Res.* 52, 39. https://doi.org/10.1186/s40659-019-0246-3
19. Jeevanandam J, Kiew SF, Ansah SB, Lau SY, Barhoum A, Danquah MK, Rodrigues J (2022) Green approaches for the synthesis of metal and metal oxide nanoparticles using microbial and plant extracts. *Nanoscale.* 14(7), 2534–2571.
20. Kolahalam LA, Kasi IV, Diwakar BS, Govindh B, Reddy V, Murthy YLN (2019) Review on nanomaterials. *Mater. Today: Proc.* 18, 2182–2190. https://doi.org/10.1016/j.matpr.2019.07.371
21. Kumar MSS, Raju NMS, Sampath PS, Jayakumari LS (2014) Effects of nanomaterials on polymer composites - an expatiate view. *Rev. Adv. Mater. Sci.* 38, 40–54.
22. Kumar SV, Bafana AP, Pawar P, Rehman A, Dahoumane SA, Jeffryes CS (2018) High conversion synthesis of nanoparticles using microwave technology. *Sci. Rep.*, 8, 5106. https://doi.org/10.1038/s41598-018-23480-6
23. Kuunal S, Wragg D, Kutti S, Rauwel E, Rauwel P, Guha M (2016) Biocidal properties study of silver nanoparticles used for application in green housing. *Int. Nano Lett.* 6, 191–197.
24. Lakhani PK, Jain N (2018) Nanotechnology and green nanotechnology: a road map for sustainable development, cleaner energy and greener world. *Int. J. Innov. Sci. Res. Technol.* 3(1), 580–588.
25. Lateef A, Elegbede JA, Akinola PO, Ajayi VA (2019) Biomedical applications of green synthesized-metallic nanoparticles: a review. *Pan Afr. J. Life Sci.* 22, 157–182.
26. Lateef A, Ojo SA, Elegbede JA (2016) The emerging roles of arthropods and their metabolites in the green synthesis of metallic nanoparticles. *Nanotechnol. Rev.* 5(6), 601–622.

27. Lau D, Jian W, Yu Z, Hui D (2018) Nano-engineering of construction materials using molecular dynamics simulations: prospects and challenges. *Compos. Pt B-Eng.* 143, 282–91.
28. Liu H, Ren M, Qu J, Feng Y, Song X, Zhang Q, Cong Q, Yuan X (2017) A cost-effective method for recycling carbon and metals in plants: synthesizing nanomaterials. *Environ. Sci: Nano.* 4, 461–469. https://doi.org/10.1039/C6EN00287K
29. Liu J, Hui D, Lau D (2022) Two-dimensional nanomaterial-based polymer composites: fundamentals and applications. *Nanotechnol. Rev.* 11, 770–792.
30. Maradini GDS, Oliveira MP, Guanaes GMDS, Passamani GZ, Carreira LG, Boschetti WTN, Monteiro SN, Pereira AC, de Oliveira BF (2020) Characterization of polyester nanocomposites reinforced with conifer fiber cellulose nanocrystals. *Polymers (Basel*) 12(12), 2838.
31. Morales ML, Martinez JO, Reyes-Sanchez LB, Martin O, Arroyo GA, Obaya A, Miranda R (2011) How green an experiment is? *Educacion Quimica.* 22, 240–248. https://doi.org/10.1016/S0187–893X(18)30140-X
32. Mulvihill MJ, Beach ES, Zimmerman JB, Anastas PT (2011) Green chemistry and green engineering: a framework for sustainable technology development. *Annu. Rev.* 36, 271–293. https://doi.org/10.1146/annurev-environ-032009–095500
33. Munoz I (2012) LCA in green chemistry: a new subject area and call for papers. *Int. J. LCA.* 17, 517–519. https://doi.org/10.1007/s11367012-0410-2
34. Nagajyothi PC, Sreekanth TVM (2015) Green synthesis of metallic and metal oxide nanoparticles and their antibacterial activities. In: Shanker, U., Hussain, C.M., Rani, M. (eds) *Green Processes for Nanotechnology.* Springer International Publishing, Chem, pp. 99–117. https://doi.org/10.1007/978-3-319-15461-9_4
35. Nasrollahzadeh M, Atarod M, Sajjadi M, Sajadi SM, Issaabadi Z (2015) Plant-mediated green synthesis of nanostructures. In *An Introduction to Green Nanotechnology.* Elsevier, Amsterdam, The Netherlands, pp. 199–322. https://doi.org/10.1016/B978-0-12–8135860.00006–7
36. Pandey GP (2018) Prospects of nanobioremediation in environmental cleanup. *Orient. J. Chem.* 34(6), 2828–2840. https://doi.org/10.13005/ojc/340622
37. Polshettiwar V, Varma RS (2010) Green chemistry by nano-catalysis. *Green Chem.* 12, 743–754. https://doi.org/10.1039/b921171c
38. Raki L, Beaudoin J, Alizadeh R, Makar J, Sato T (2010) Cement and concrete nanoscience and nanotechnology. *Materials.* 3, 918–942.
39. Rashad AM (2014) A comprehensive overview about the effect of nano-SiO_2 on some properties of traditional cementitious materials and alkali-activated fly ash. *Constr. Build. Mater.* 52, 437–464.
40. Razavi M, Salahinejad E, Fahmy M, Yazdimamaghani M, Vashaee D, Tayebi L (2015) Green chemical and biological synthesis of nanoparticles and their biomedical applications. In: Basiuk, V., Basiuk, E. (eds) *Green Processes for Nanotechnology.* Springer International Publishing, Chem, pp. 207–235. https://doi.org/10.1007/978-3-319–154619_7
41. Reches Y (2018) Nanoparticles as concrete additives: review and perspectives. *Const. Build. Mater.* 175, 483–495.
42. Reid BT, Reed SM (2016) Improved methods for evaluating the environmental impact of nanoparticle synthesis. *Green Chem.* 18(15), 4263–4269. https://doi.org/10.1039/c6gc00383d
43. Salem WM, Haridy M, Sayed WF, Hassan NH (2014) Antibacterial activity of silver nanoparticles synthesized from latex and leaf extract of Ficus sycomorus. *Ind. Crops Prod.* 62, 228–234.
44. Savithramma N, Rao ML, Rukmini K, Devi PS (2011) Antimicrobial activity of silver nanoparticles synthesized by using medicinal plants. *Int. J. Chem. Tech. Res.* 3, 1394–1402.
45. Schmid R, Nielaba P (2019) Stability of nanoparticles in solution: a statistical description of crystallization as a finite particle size effect in a lattice-gas model. *J. Chem. Phys.* 150(5), 054504.
46. Shah M, Fawcett D, Sharma S, Tripathy SK, Ponern GEJ (2015) Green synthesis of metallic nanoparticles via biological entities. *Materials.* 8(11), 7278–7308. https://doi.org/10.3390/ma8115377
47. Thakur, P., Thakur, A. (2022). Introduction to nanotechnology. In: Thakur, A., Thakur, P., Khurana, S. P. (eds) *Synthesis and Applications of Nanoparticles.* Springer, Singapore. https://doi.org/10.1007/978-981-16-6819-7_1
48. Sharma JK, Srivastava P, Ameen S, Akhtar MS, Sengupta SK, Singh G (2017) Phytoconstituents assisted green synthesis of cerium oxide nanoparticles for thermal decomposition and dye remediation. *Mater. Res. Bull.* 91, 98–107. https://doi.org/10.1016/j.materresbull.2017.03.034
49. Sharma VK, Filip J, Zboril R, Varma RS (2015) Natural inorganic nanoparticles – formation, fate, and toxicity in the environment. *Chem. Soc. Rev.* 44, 8410–8423. https://doi.org/10.1039/c5cs00236b

50. Sheldon, RA (2017) Metrics of green chemistry and sustainability: past, present, and future. *ACS Sust. Chem. Eng.* 6(1), 32–48. https://doi.org/10.1021/acssuschemeng.7b03505
51. Silva LP, Garcez I, Bonatto CC (2015) Green synthesis of metal nanoparticles by plants. In: Basiuk, V., Basiuk, E. (eds) *Green Processes for Nanotechnology*. Springer International Publishing, Chem, 259–275. https://doi.org/10.1007/978-3-319-15461-9_9
52. Silvestre J, Silvestre N, de Brito J (2015) Review on concrete nanotechnology. *Eur. J. Environ. Civil Eng.* 20(4), 455–485.
53. Thapliyal PC, Singh K (2014) Aerogels as promising thermal insulating materials: an overview. *J. Mater.* 127049, 10.
54. Tickner JA, Becker M (2016) Mainstreaming green chemistry: the need for metrics. *Curr. Opin. Green. Sustain. Chem.* 1, 1–4. https://doi.org/10.1016/j.cogsc.2016.07.002
55. Torgal FP, Jalali S (2011) Nanotechnology: advantages and drawbacks in the field of construction and building materials. *Constr. Build. Mater.* 25(2), 582–590.
56. Tufvesson LM, Tufvesson P, Woodley JM, Borjesson P (2012) Life cycle assessment in green chemistry: overview of key parameters and methodological concerns. *Int. J. LCA.* 18, 431–444. https://doi.org/10.1007/s11367-012-0500–1
57. Varma RS (2021) Greenness of things. *Clean Technol. Environ. Policy.* 23, 2497–2498. https://doi.org/10.1007/s10098-021-02201-0
58. Rathore R, Suhag D, Wan F, Thakur A, Thakur P (2023) Everyday nanotechnology. In: Suhag, D., Thakur, A., Thakur, P. (eds) *Integrated Nanomaterials and their Applications.* Springer, Singapore. https://doi.org/10.1007/978-981-99-6105-4_2
59. Virkutyte J, Varma RS (2011) Green synthesis of metal nanoparticles: biodegradable polymers and enzymes in stabilization and surface functionalization. *Chem. Sci.* 2, 837–846. https://doi.org/10.1039/c0sc00338g
60. Vishwakarma V, Ramachandran D, Anbarasan N, Rabel AM (2016) Studies of rice husk ash nanoparticles on the mechanical and microstructural properties of the concrete. *Mater. Today Proceed.* 3(6), 1999–2007.
61. Wang X, Narayan S (2021) Thermochromic materials for smart windows: a state-of-art review. *Front. Energy Res.* 9:800382, 1–6.
62. Zhang D, Ma Xl, Gu Y, Huang H, Zhang GW (2020) Green synthesis of metallic nanoparticles and their potential applications to treat cancer. *Front. Chem.* 8, 799.

16 Nanomaterials in Theranostic Applications

Deepa Suhag and Raksha Rathore
Amity University Haryana

Hui-Min David Wang
National Chung Hsing University

Preeti Thakur and Atul Thakur
Amity University Haryana

16.1 INTRODUCTION

It is widely acknowledged that most anti-cancer drugs can induce severe systemic toxicity, which can be dose-limiting in some cases. Whether administered intravenously or orally, the medicine frequently accumulates in and damages healthy, normal tissues. To limit their systemic side effects and increase their therapeutic effectiveness, these drugs must be targeted and released in a controlled manner [1]. Drug delivery systems (DDSs) such as liposomes, polymeric nanoparticles, and nano emulsions have been extensively studied to solve the inadequacies and drawbacks of traditional drugs, such as unregulated release and non-specific biodistribution. Even ordinary DDSs, however, usually struggle to release the cargo in a controlled manner at the given spot. As a result, smart DDSs have been developed to provide drug release at the target region in a spatially and temporally controlled way, to preserve the drug/agent in the target site for a longer period of time, to increase therapeutic efficacy, and to prevent adverse systemic side effects [2].

The invention of thermosensitive liposomes in the late 1970s paved the way for the development of smart DDSs (also known as stimulus-responsive drug delivery platforms). These liposomes may release drugs locally if the tissues are heated from the outside [3]. The basic goal of stimulus-responsive DDSs can be defined as systematic delivery along with local activation. Smart delivery systems can be loaded with a variety of bioactive compounds and only release the cargo when two or more independent stimuli, which can be chemical, biological, or physical in origin, are present.

These advanced or intelligent systems provide several advantages and unique opportunities in biological sensing, tissue engineering, diagnosis, and medicine administration [4]. To produce stimulus-responsive platforms, we must develop materials that can undergo certain structural changes, such as protonation, cleavage, or conformational changes, in response to certain stimuli that promote cargo release [3]. When these systems are subjected to external stimuli such as temperature, pH, enzyme activity, redox potential, ionic strength, or media solvent composition, their physicochemical properties can change. Heat, light, magnetic fields, electric fields, and ultrasound (US) are examples of external stimuli [5–7]. Designing such single-, dual-, or multi-stimulus-responsive smart delivery vehicles can lead to the development of new biomaterials.

The tuning of their reactions to local/environmental stimuli can give better-controlled medication administration and stronger therapeutic results due to the synergistic effects of different environmental stimuli [8,9]. The complex disease environment can make managing endogenous or internal stimuli difficult. Exogenous or external stimulation, on the other hand, may cause tissue

 DOI: 10.1201/9781003502692-16

harm, and the depth of penetration may not be sufficient to elicit drug release deep into tissues and organs. External triggers, on the other hand, would be preferred in general due to their customizable activation properties [2,10].

A multitude of concerns must be considered when designing smart DDSs, including overcoming biological barriers; selecting the best route for administration; lowering toxicity; ensuring biodegradability, biosafety, and efficacy; and protecting against long-term carcinogenesis [2]. Despite the complexity and large number of parameters to be adjusted, in vivo therapeutic evaluation of these smart nanostructures for drug administration is critical, even though animal models cannot accurately mimic every feature of human disease [2]. Nonetheless, clinical trials should be utilized to assess the efficacy and safety of nanomaterials used as drug carriers before they may be used in true clinical settings [2,3,11–13], despite significant breakthroughs and laboratory investigations. The utilization of agents to enhance or functionalize nanoparticle surfaces is highly beneficial in the context of various imaging techniques, including CT, radionuclide imaging (positron emission tomography [PET] and single-photon-emission computed tomography [SPECT]), optical imaging (fluorescence and bioluminescence), and magnetic resonance imaging (MRI) [14]. A wide range of organic and inorganic nanoparticle compositions, such as proteins, polymers, lipids, gold, and iron oxides, can serve as the foundation for theranostic systems [14]. It is crucial for these particles to have a diameter of less than 150 nm, exhibit exceptionally homogeneous distributions, and maintain sufficient in vivo stability. Following functionalization with imaging contrast, therapeutic payload, and targeting properties, these particles must retain their nanoparticle characteristics. Notably, theranostic nanoparticles with surface-targeting ligands enable direct interaction with tumor-expressed receptors [15].

Moreover, these nanoparticles can encapsulate therapeutics designed with either slow-release properties, characteristic of many therapeutic drug combinations, or triggered release capabilities. For instance, triggered release may occur when the nanoparticle undergoes endocytosis, leading to release in response to environmental changes. Overall, this approach facilitates the development of effective theranostic systems for targeted imaging and therapy [15].

16.2 SELECTION OF NANOPARTICLES

16.2.1 Using Nanoparticles to Deliver Drugs

The selected nanoparticle-forming substance can exhibit diverse compositions, shapes, charges, sizes, and hydrophobicity, as illustrated in Figure 16.1, showcasing various nanostructures for medical applications. Overcoming limitations associated with traditional drug delivery methods becomes feasible by employing nanoparticle compositions as drug delivery vehicles. One major challenge is therapeutic solubility, particularly with hydrophobic chemotherapeutics requiring solubilization agents like Cremophor EL, linked to hypersensitivity reactions [15].

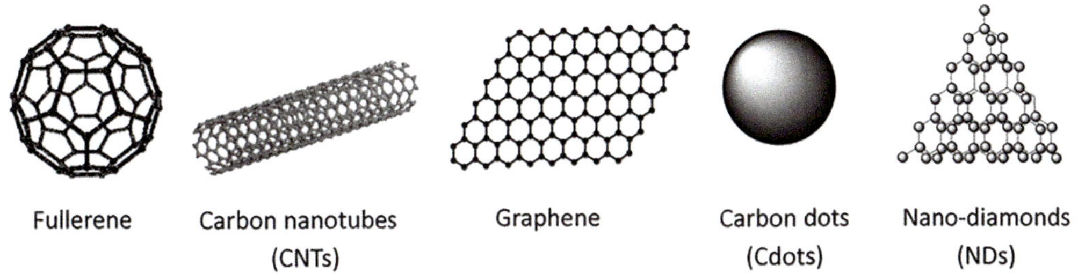

FIGURE 16.1 Different types of nanostructures for medical applications.

The use of nanoparticles, such as polymeric micelles or liposomes, addresses this issue by creating an environment that is both hydrophilic and hydrophobic, thereby enhancing solubility [16]. Another concern is burst release post-consumption, potentially causing harm to adjacent tissues or accelerating in vivo drug breakdown. Successful drug carrier nanoparticles must effectively encapsulate the treatment, delaying its release in the body and protecting it from external impacts in vivo [17]. Traditional DDSs also suffer from unfavorable pharmacokinetic profiles, necessitating frequent high dosages. Nanoparticles offer a solution by enabling delayed drug release and requiring less overall treatment [18].

Biodistribution challenges arise from the lack of specificity in traditional DDSs. To address this, nanoparticles can be equipped with targeting components on their surfaces, reducing systemic toxicity and enhancing therapeutic delivery at the desired site [19]. It is evident that the employment of nanoparticles as medication delivery methods offers numerous advantages and holds significant potential for the future.

16.2.2 Nanoparticle Compositions

The efficiency of cancer cell absorption and interaction with nanoparticles in theranostic applications is heavily influenced by their physicochemical properties [20]. Parameters such as size, shape, charge, hydrophobic/hydrophilic characteristics, and surface chemistry play a crucial role in determining the performance of theranostic systems. Particle size is closely linked to the rate of cellular absorption and the duration of nanoparticle circulation [21]. The uptake mechanism, termed pinocytosis, occurs through passive targeting or adsorptive mechanisms and active targeting with receptor-mediated endocytosis.

For in vivo applications, nanoparticles within the size range of 10–100 nm are optimal, whereas in vitro, particles ranging from 40 to 50 nm exhibit the highest uptake [22,23]. Glomerular filtration, a process in the kidneys, eliminates smaller nanoparticles, typically those less than 10 nm [24]. Conversely, larger particles are cleared by Kupffer cells in the liver and the spleen [25]. These considerations underscore the importance of tailoring nanoparticle characteristics to optimize their therapeutic efficacy and bioavailability in cancer treatment. The size ranges for passive targeting of solid tumors for nanoparticle accumulation are between 70 and 200 nm [26], with anything larger than 400 nm is excessive [19]. Even though these basic size ranges have been defined, there are instances of contradicting results, and it appears that each class of nanoparticle and cell type interaction must be investigated separately. The form of the particles also has a substantial impact on the internalization interaction between nanoparticles and cells. Because hydrodynamic forces play a role in nanoparticle transport, the symmetry of the particles dictates their trajectory [27].

Spheres are more efficiently taken up than rods, and forms that allow for cellular membrane wrapping are more successful in cellular uptake [28]. Because spheres are symmetrical, the forces acting on them are evenly distributed, allowing them to remain in the middle of the blood artery during circulation [29]. Asymmetrical rods are more likely to encounter drag forces and torques as they circulate, altering particle motion and causing them to concentrate near vessel walls [30]. The form of the particles influences both the duration of their circulation and their biodistribution. According to certain reports, spherical particles circulate faster than non-spherical particles [31].

The charge on the surface of the nanoparticles determines whether they bind to the target of interest or undergo non-specific binding elsewhere while in circulation. Greater cellular binding, in turn, will boost cationic nanoparticle absorption by interacting with negatively charged proteins, glycans, and phospholipid head groups on cell surfaces [32]. Anionic particles have been shown to have more uptake than neutral particles [33]. Positively charged particles are internalized via dynamin and F-actin, whereas negatively charged particles are internalized independently of dynamin [34]. The process of absorption is also affected by charge. Positively charged particles are removed by the mononuclear phagocyte system and have a short half-life due to interactions with blood proteins.

These interactions, when they occur, promote the complement pathway [35]. If the hydrophobicity of the designed nanoparticles is larger than that of the cell surface to which they are adhering, there is an increase in absorption as well as protein adsorption [32,36]. On the contrary, hydrophilic nanoparticles will draw fewer molecules which attracts these proteins. This is why PEGylation, the technique of coating the surface of nanoparticles, is frequently used to modify nanoparticles. Particles have a coating of the extremely hydrophilic polymer polyethylene glycol (PEG) [37], which lengthens half-lives and boosts in vivo circulation. It is crucial to research the ideal level of PEGylation in each nanoparticle cell pairing. This level depends on factors including coating thickness, coverage arrangement, and chain length. A variety of parameters interact while designing nanoparticles for theranostic systems. Despite the abundance of data that has been collected and reported, it is impossible to forecast the interactions that might occur in each system. Therefore, it is crucial to establish each of these parameters by experiment for a specific system [38].

16.3 NANO-CARBONS AS THERANOSTICS

According to their bonding structures, nano-carbons are classified as sp^2-carbon nanomaterials or sp^3-carbon nanomaterials. Common sp^2-carbon nanomaterials with well-defined structures include zero-dimensional (0D) fullerene, one-dimensional carbon nanotubes (CNTs), and two-dimensional graphene. Surprisingly, fullerene and graphene were awarded Nobel Prizes in physics in 2010 and chemistry in 1996, respectively. Carbon nanoparticles, also referred to as carbon dots (CDots), are amorphous carbon nanoclusters with sizes less than 10 nm. They can also be considered a type of 0D sp^2-carbon nanomaterial. On the contrary, sp^3-carbon nanomaterials are frequently nanodiamonds (NDs) with crystal sizes in the nano-range.

There are numerous drivers and arguments for using nano-carbons in biological applications such as theranostics. Many nano-carbons, including CNTs, graphene derivatives, Cdots, and NDs, have intriguing inherent optical properties, including fluorescence, making them useful contrast agents in optical imaging and sensing [39–41]. CNTs and graphene, on the other hand, have exceptional electrical properties and can be widely employed in a range of biosensing systems [42–44]. Sp^3 carbon nanomaterials with an extraordinarily large surface area available for efficient drug loading and bioconjugation include single-walled carbon nanotubes (SWCNTs) and graphene, which have all of their carbon atoms exposed on their surfaces [16,31,42]. The interior space of fullerene and CNTs with hollow structures may be used to load additional functional species for theranostic purposes [45,46]. For photothermal ablation of cancer, CNTs and derivatives of graphene having a high optical absorbance in the near-infrared band are also advantageous [47–49].

Furthermore, compared to many other inorganic nanomaterials like quantum dots (QDs), which frequently contain heavy metals, nano-carbons built entirely of carbon are relatively benign, at least in terms of elementary composition. Functionalized fullerenes have not only shown promise in a variety of cancer treatment modalities such as photodynamic therapy, photothermal therapy, radiation, and chemotherapy, but they have also been used as cutting-edge contrast agents in MRI. The biodistribution, metabolism, and toxicity of such functionalized fullerenes are also investigated in this study [50]. Table 16.1 shows some carbon-based nanosystems used for cancer imaging.

The following section contains three articles that highlight the potential of CNTs as a disease diagnostic tool. Authors [51] present a label-free real-time electrical detection of whole viruses using CNT through film field effect devices that are five orders of magnitude more sensitive than conventional electrical impedance sensors without CNTs. Researchers [52] investigated the effect of oxidized multi-walled carbon nanotubes (o-MWNTs) on macrophages. It is unknown how o-MWNTs, which compete with the tumor mass for the recruitment of macrophages from circulating monocytes, could diminish the number of macrophages and blood vessel density in the tumor. This would stop the tumor from growing and spreading. This discovery opens up a whole new avenue for using CNTs in cancer treatment. In addition to these two investigations, Yang et al. [53] discussed the pharmacokinetics, metabolism, and toxicity of CNTs in greater detail. CNT surface

TABLE 16.1
Carbon-Based Nanosystems for Cancer Imaging

System	Tumor Cell/Cancer-Type Disease	Imaging Approach
Graphene oxide	Murine lung metastasis model of breast cancer	Positron emission tomography and fluorescence imaging
Graphene oxide	B16F0 melanoma tumors using the mouse model	Fluorescence imaging
Graphene oxide	Bel-7402, SMMC-77.21, HepG2 cell line hepatocellular carcinoma	Fluorescence imaging
Graphene oxide	MCF-7 cells	Fluorescence imaging
Reduced graphene oxide	KB cell-bearing nude mice	Fluorescence imaging
Reduced graphene oxide	PC-3 cells prostate cancer	Fluorescence imaging
Reduced graphene oxide	4T1 murine breast cancer, MCF-7 human breast cancer, and human umbilical vein endothelial cells (HUVECs)	Positron emission tomography and fluorescence imaging
Reduced graphene oxide	4T1 breast cancer cell-bearing mice	Radioactive nuclear
Graphene	HeLa human cervical cancer cells using zebrafish as the model	Fluorescence imaging
Graphene quantum dots	MCF IOA, SKBR.3, MCF 7, MOA-MB-436, MOA-MB-468, MOA-MB-231, MOA-MB-157, MDA-MB-175V1, HCC1806, and Hs578T cells cancerous and metastatic human breast cells	Fluorescence imaging
Graphene quantum dots	HeLa cell human cervical carcinoma cells, A549 adenocarcinomic human alveolar basal epithelial cells, and HEK.293A normal human embryonic kidney cells	Fluorescence imaging
Tri-malonate derivative of fullerene (G70)	A549 cells lines and 4T1-luc cells using female Balb/c mouse model	Photoluminescence
Nanodiamond	C6-Luc glioblastoma cells, U.251MG-Luc human glioma-bearing NIH nu/nu nude rats	Fluorescence imaging
Nanodiamond	Human hepatoma (HepG2) cancer cell line	Fluorescence imaging

chemistry, according to prior studies and the authors' own research, has a crucial influence in determining their in vivo behaviors. Well-functionalized CNTs (for example, via PEGylation) are much less hazardous in vitro and in vivo than non-functionalized CNTs and hence may be preferred for uses in biomedicine. The experts do, however, state that more research is needed to fully assess the long-term toxicity of CNT exposure. It is also critical to develop quick and dependable analytical tools for studying the in vivo behaviors of CNTs and other freshly developed nanomaterials [54–56].

Graphene has been a rising star in the world of materials science since 2004. Its applications in biomedicine have also piqued people's interest. With few changes, small carbon nanoparticles with a surface function known as Cdots could emit strong light useful for biological imaging. Researchers [57] created and tested highly fluorescent Cdots for optical imaging in mice with and without ZnS doping. Significantly, Cdots perform in vivo at a level comparable to well-known CdSe/ZnS QDs, implying that Cdots may constitute a novel class of optical probes that do not contain heavy metals and could be useful in bioimaging [53,58–60]. In addition to sp^2-carbon nanomaterials, the use of NDs with sp^3-bonded structures in nanomedicine has garnered considerable attention. The in vitro and in vivo toxicity of NDs, as well as their utility in DDSs, is discussed in Zhu et al.'s [61] publication. The properties and behaviors of NDs and ND clusters differ significantly from those of sp^2 carbon nanomaterials such as CNTs and graphene, making them potential candidates for drug administration by fluorescent imaging guided imaging as well as reasonably safe in biological systems. To summarize, the development of novel theranostic procedures based on nano-carbons has sparked tremendous interest and will most likely continue to do so.

16.3.1 US-Responsive Nanomaterials

Radioembolization, employing an innovative cancer therapy technique, has proven to be viable, safe, and successful in treating patients, particularly those with normal liver function, exhibiting a higher rate of tumor response (Table 16.2) [62]. The application of SonoVue has shown promise in lowering US ablation energy by reducing sonication time, thereby achieving a significant change in grayscale during treatment. No significant adverse effects were observed, and SonoVue was found to enhance the ablative effects of high-intensity focused ultrasound (HIFU) treatment for uterine fibroids, with cavitation proposed as a process enhancer [63–67]. A clinical trial involving 102 patients investigated SonoVue's efficacy and safety in treating adenomyosis.

Potential safety concerns, including early massive grayscale changes, lower total energy, and lower mean power, were considered, and SonoVue was suggested to safely enhance HIFU ablation [68]. The only completed clinical trial with published data on US-responsive liposomes focused on lyso-thermosensitive liposomal doxorubicin (NCT02181075). Ten patients, following a single intravenous infusion of Low tolerance long duration (LTLD), underwent extracorporeal targeted US exposure of a single target liver tumor, resulting in a 3.7 times increase in doxorubicin concentration at the cancer site. While some patients experienced serious side effects like neutropenia and anemia, no fatalities occurred. This study demonstrated the safety and feasibility of enhanced intratumoral medication delivery and chemo-ablative therapy for liver cancers resistant to standard chemotherapy.

Despite the development of various US-responsive nanomaterials, only microbubbles (MBs) and liposomes have been studied in humans, likely due to their relative biocompatibility and low toxicity. However, it is crucial to address the fate and potential toxicity of other nanomaterials, such as nonorganic and polymeric nanoparticles, which offer diverse modes of action to improve therapy and diagnostic precision. Further research into the behavior of nanomaterials in biological systems and efforts to enhance their biocompatibility are essential [69,70].

16.3.2 Carbon-Based Nanosystems for Cancer Imaging

Early detection is essential to increase cancer patients' chances of survival. Many cancer imaging approaches rely on the development of contrast chemicals and imaging probes. Because of

TABLE 16.2
Ultrasound-Responsive Nanomaterials (in Trial and in Use)

Nanomaterial	Application	Status
SonoVue+recombinant tissue plasminogen activator	Sonothrombolysis	Completed
Microbubble	Enhanced ultrasonography of blood flow in kidney masses	Completed
Microbubble	Enhanced ultrasonography	Completed
SonoVue	Enhanced ultrasonography	Completed
SonoVue+FOLFIRINOX	Enhanced chemotherapy	Recruiting
SonoVue	Breast and colorectal cancer treatment	Recruiting
Microbubble	Breast cancer treatment	Recruiting
Microbubble	Head and neck cancer treatment	Recruiting
Microbubble	Enhanced ultrasonography in patients with shock	Recruiting
Microbubble	Targeted chemotherapy	Recruiting
Lyso-thermosensitive liposomal (LTSL) doxorubicin	Targeted chemotherapy of liver tumors	Completed
LTSL doxorubicin	Pediatric cancer treatment	Recruiting
LTSL doxorubicin+cyclophosphamide	Primary breast tumor treatment	Recruiting

their specific physical/chemical properties, carbon nanostructure (CNT, graphene, fullerene, and ND)-based platforms for cancer imaging have been the subject of numerous investigations. The physicochemical/biological features of these carbon nanostructures can be designed by functionalizing their surfaces [71]. Imaging modalities such as SPECT, PET, computed tomography (CT), MRI, US imaging, and fluorescence/photoacoustic imaging might all benefit greatly as a result [72]. The majority of commonly used imaging agents have been demonstrated to be incapable of passing through cell membranes. Carbon-based nanostructures (such as CNTs), on the contrary, have significant potential for cancer diagnosis applications since they can be useful for delivering such contrast agents intracellularly for cell tracking with excellent selectivity. Vascular imaging was used to image 1–3 mm deep in the hind limbs in both spatial (30 m) and temporal (200 ms per frame) aspects. The potential benefits of SWCNTs as fluorophores in near infrared (NIR)-II imaging were well demonstrated, exceeding standard imaging modalities [73].

16.4 NANOPARTICLE-BASED THERAPEUTIC AGENTS

16.4.1 Anti-Cancer Drug

In the development of multifunctional nanoparticle systems, the careful selection of therapeutic agents is paramount, considering the solubility of the medicinal compounds for encapsulation, which may require protection or enhancement in solubility. Commonly utilized cancer drugs in theranostic applications include doxorubicin and paclitaxel. Doxorubicin, a widely used chemotherapeutic drug available in PEGylated liposomal form, exhibits potent anti-cancer effects through DNA interaction via intercalation, halting topoisomerase enzymes and stabilizing the compound hindering replication [74,75]. An example is the utilization of doxorubicin in a targeted theranostic system, where PEG-conjugated nanoparticles with hyaluronic acid coating and Cy5.5 NIR fluorescent dye demonstrated enhanced absorption, robust fluorescent signal, increased circulation time, and anti-tumor action when applied to SCC7 cells [76].

Paclitaxel, belonging to the taxane family, is derived from the Pacific yew tree and is widely used in treating various cancers. The FDA-approved Abraxane formulation addresses solubility issues by binding paclitaxel to albumin nanoparticles. Paclitaxel acts on microtubules in cells, stabilizing and protecting them, preventing normal destruction during cell division, and hindering mitosis progression, leading cells back to the G phase of the cell cycle [77]. Indirect side effects of paclitaxel, associated with solubilization agents like Cremophor EL, include rashes, nausea, vomiting, and female infertility [78]. Perfluorocarbon nanoparticles loaded with paclitaxel, targeted at smooth muscle cells, serve as a theranostic device, providing contrast to MRI scans with potent T1-weighted signals, a prolonged release time, and the ability to deliver therapeutics to smooth muscle cells [77].

Moreover, various types of cancer are now treated with nucleic acid-based medicines, including those utilizing small interfering RNA (siRNA) and microRNA (miRNA). These can be exploited in theranostic systems and are referred to as non-coding RNAs [77]. The RNA polymerase II synthesizes the conserved single-stranded RNAs known as miRNAs, which are 19–25 nucleotides long [79]. Abnormal miRNA expression has been associated with a variety of tumors, including leukemia [80] and lung cancer [81]. They can be employed as therapeutic targets because it has been demonstrated that they contribute to the development of chemoresistance in both prostate cancer cells and breast cancer cells [82,83].

Two approaches have been used to use miRNA as a therapeutic agent: the first is to disrupt tumor suppressor miRNA, and the second is to target the genes responsible for their transcription.

Targeted nanoparticle systems can be used to deliver the miRNA to target cells for use in theranostic systems. For example, miR-34a can be delivered into metastatic melanoma using targeted liposomes [55], and miR-29b can be delivered into HeLa cells using gold nanoparticles loaded with the miRNA [83].

Small interfering RNA, a type of double-stranded RNA, is also utilized to treat cancer. These RNAs are typically 2025 nucleotides long and have been shown to block the expression of a gene whose double-stranded DNA complementary sequence they share [84]. The delivery of siRNA specific to cyclin CD-1 in leukocytes utilizing liposomes covered with monoclonal antibodies was successful [85].

16.4.2 Licensed Nanoparticle Medication Delivery Systems

Doxil, developed by Janssen Biotech and clinically licensed in 1995, stands out as a prominent FDA-approved nanoparticle medication formulation widely utilized in the treatment of Sarcoma and Ovarian Cancer. This formulation features a doxorubicin-loaded PEGylated Liposome Drug Carrier. Notably, Doxil demonstrates a distinct distribution pattern, primarily confined to the blood pool, except in areas with heightened vascular permeability such as the liver, spleen, and tumors. With a circulation half-life 100 times longer than free doxorubicin, Doxil also presents a sevenfold lower risk of cardiotoxicity, showcasing its enhanced safety profile [86,87]. Another noteworthy FDA-approved formulation is DaunoXome, a liposomal danorubicin formulation approved for sarcoma treatment in 1996 [88]. Additionally, Abraxane, an albumin-bound paclitaxel formulation, has achieved FDA approval for treating metastatic breast cancer, lung cancer, and pancreatic cancer, further emphasizing the success of nanoparticle-based DDSs in clinical applications [89].

In current clinical trials, numerous micelle-based system solutions are under investigation, showcasing the evolution of nanoparticle technology. Genexol-PM, a micelle-based nanoparticle falling within the size range of 20–50 nm, received FDA approval in 2007 for treating breast and lung cancer [90]. Developed as an alternative to Taxol, Genexol-PM aims to overcome the primary disadvantage associated with paclitaxel treatment, namely, the use of Cremophor EL, a surfactant-solubilization agent necessary for dissolving paclitaxel. This surfactant has been linked to hypersensitivity responses, necessitating premedication before administering the treatment [91].

Genexol-PM encapsulates and protects paclitaxel within the hydrophobic core of the micelle, preserving its solubility through the hydrophilic surface of the micelle. Notably, the formulation demonstrated linear pharmacokinetics, and no hypersensitivity events were recorded in a phase I study, highlighting its improved safety profile [92]. Table 16.3 provides a list of some approved compositions, including nanoparticles, showcasing the growing repertoire of clinically approved nanomedicines.

16.5 IMAGING METHODS

For effective diagnostic applications of theranostic nanoparticles, it is crucial to characterize biological processes at both cellular and subcellular levels using molecular imaging techniques. Utilizing disease state-specific targeting strategies, particularly when combined with contrast agents or genetic probes, enables the identification of various disease states at different stages of progression and facilitates the assessment of therapy efficacy. It is imperative that the chosen imaging modalities

TABLE 16.3
List of Approved Compositions Including Nanoparticles

Drug Name Formulation	Usage	Year Approved
Abraxane albumin-bound paclitaxel	Breast cancer	2005
Cimzia PEGylated fragment of anti-TNF-O antibody	Crohn's disease and arthritis	2008
Doxil PEGylated liposomal doxorubicin	Ovarian cancer and sarcoma	1995
DaunoXome Liposomal danorubicin	Sarcoma	1996
Eligard PLGH-polymer and leuoprolide formulation	Late-stage prostate cancer	2002
Oncaspar PEG-L-asparagine	Acute lymphoblastic leukemia	2006
Genexol-PM PLA-PEG micelle with paclitaxel	Metastatic breast cancer	2007

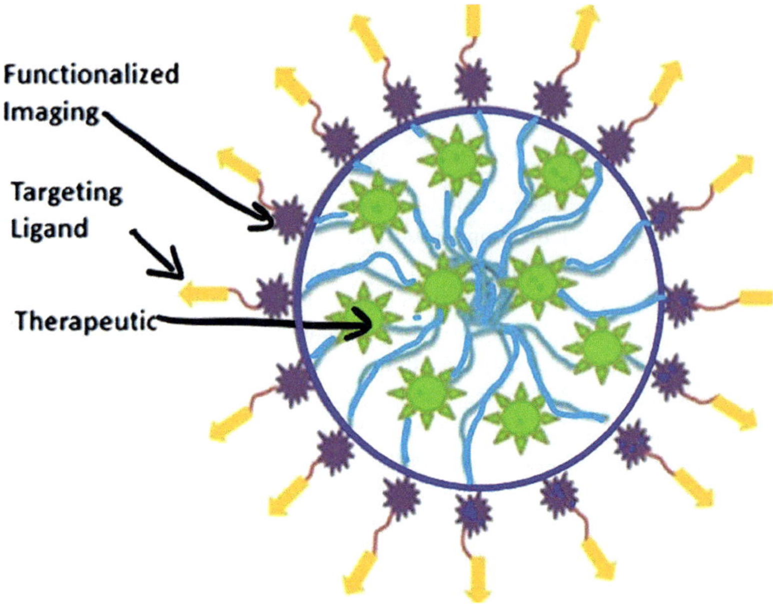

FIGURE 16.2 Nanoparticle-mediated imaging and targeting.

for theranostic systems do not interfere with nanoparticle self-assembly and do not impact the scale of existing particle systems. Both nanoparticle size and disease state-specific targeting, as depicted in Figure 16.2, play pivotal roles in ensuring sufficient accumulation of contrast agents for these modalities to function effectively [93].

16.5.1 Laser Imaging

Optical imaging relies on photons emitted by fluorescent or bioluminescent probes to identify disease conditions, presenting itself as a favorable imaging option due to its high sensitivity, absence of non-ionizing radiation, capability to cover the spectrum from visible to near-infrared, real-time imaging ability, and cost-effectiveness in detecting photons [94].

However, optical imaging techniques come with limitations, including poor tissue penetration, sensitivity to scattering-related noise, and potential autofluorescence [95]. Various probes used in optical imaging include synthetic fluorophores, semiconductor fluorescent crystals, and lanthanide-based probes. In a study by research group, PEGylated hyaluronic acid nanoparticles loaded with a near-infrared fluorescent dye (Cy 5.5) carrying irinotecan were employed as a theranostic agent, enabling precise targeting, treatment, and imaging of colon cancer tumors in vivo [96].

QDs, known for their modifiable fluorescence characteristics, are common probes in optical imaging [97]. A research utilized QDs as a contrast agent encapsulated in chitosan nanoparticles containing HER2/neu siRNA to specifically target SKBR3 cancer cell lines overexpressing HER2. The study demonstrated the detection of gene-silencing behavior using QDs, although concerns about cytotoxicity may limit their utility [98,99].

16.5.2 Computed Tomography

CT is primarily employed for anatomical information due to the varying X-ray attenuation rates of different biological components like bone, muscle, fat, water, and air. CT scans excel in providing detailed three-dimensional images of specific areas with high spatial resolution and comparatively lower radiation exposure compared to other techniques [100]. When utilized in conjunction with a nanoparticle platform, CT scans can prolong clearance rates, allowing the system to remain in circulation without the rapid release of large amounts of the contrast agent before clearance [101]. Nanoparticle platforms, including core-shell, liposome, gold, dendrimer, and bismuth components, have been employed in creating effective CT contrast agents. For instance, some researchers developed a theranostic agent using CT as the contrast agent, incorporating alpha-tocopheryl succinate, folic acid, and multifunctional dendrimer-entrapped gold nanoparticles. This approach demonstrated targeted CT imaging of cancer cells both in vivo and in vitro, along with therapeutic efficacy [102]. In another study, Scientists successfully utilized long-circulating bismuth nanoparticles as an injectable CT contrast agent, demonstrating in vivo imaging of mice, remarkable stability at large doses, and prolonged circulation [103].

16.5.3 PET/SPECT Radionuclide Imaging

Radionuclide-based molecular imaging techniques, specifically SPECT and PET, play a crucial role in diagnostic applications [104]. In SPECT, a dose of radiation is delivered to the tissue of interest, and subsequently, the emitted radiation is detected using a camera. Commonly administered radionuclides such as Tc-99m, I-131, and Ga-67 circulate in the blood [105]. These radionuclides can be attached to a nanoparticle platform, containing both medication and targeting ligands, for theranostic applications. This allows precise steering of radionuclides to the desired tissue locations. Gamma detectors are then utilized and rotated around the body after the nanoparticle platform accumulates, aiding in determining the location of the highest concentration of rays. This process facilitates the creation of 360° images, enabling the generation of three-dimensional representations of the surroundings [106].

The advantages of SPECT include its quantitative nature for precise measurements, low background noise, absence of signal amplification, lower cost compared to PET, flexibility in detecting multiple radionuclides simultaneously, and high sensitivity [107]. However, drawbacks include lower spatial resolution compared to other techniques, higher pricing, radiation use, and relatively large equipment size for detection [108].

16.5.4 Magnetic Resonance Imaging

MRI is a widely used diagnostic modality, offering high-resolution anatomical images with submillimeter resolution and strong contrast. Functional MRI provides detailed neurological and physiochemical information. Advantages of MRI over other imaging modalities include non-invasiveness and the absence of radiation. Tissues with different relaxation times, either longitudinal (T1) or transverse (T2), contribute to three-dimensional images, creating visual contrast. However, disadvantages include the need for lanthanide-based contrast agents, high equipment and maintenance costs, and challenges in imaging patients with medical implants. Wang et al. developed chitosan magnetic nanoparticles as a T2 contrast agent, incorporating doxorubicin and plasmid DNA to target lung and prostate cancer cells. This method exhibited a significant increase in T2 weighting, enhancing MRI of tumors and enabling effective therapy delivery with real-time tracking of therapeutic effects [105–107].

As an example of pH-responsive theranostic systems for MRI contrast agents, Researchers designed a responsive beta cyclodextrin star copolymer linked with doxorubicin, folic acid, and DOTA-Gd moieties. This system is self-assembled into micellar nanoparticles, exhibiting a slow rate of doxorubicin release at neutral pH and rapid release in acidic tumor conditions. In vivo tests demonstrated contrast agent particle accumulation in the rat liver and kidney, strong contrast on T1-weighted images, and higher relaxivity values compared to clinical contrast agents [109].

16.6 CONCLUSION AND FUTURE CHALLENGES

Better and more efficient methods for cancer diagnosis and treatment are provided by these nanomaterials. Although this sector has seen many exciting advancements, it is still in its infancy. Before these carbon-based nanostructures can be used as active therapeutic treatments, a number of problems related to their utilization must be resolved. For instance:

a. It is typically noted that a tumor that is too tiny cannot be accurately diagnosed.
   Therefore, it is essential to create new targeted nanomaterials that can overcome diagnostic limits in the case of extremely small tumors by being coupled with highly effective imaging moieties.
b. It is anticipated that a perfect medication delivery system will enter, eradicate malignant cells, and leave healthy cells alone. To achieve the best loading efficacy and highly active targeting, a medication delivery system should be created. Most significantly, a thorough grasp of the targeting mechanism is required.
c. Because it is difficult to deliver local heat to cancer cells for them to be killed by hyperthermia, efforts should be made to build nanosystems that can efficiently and selectively target cancer cells. These devices should be powerful enough to generate the heat required to destroy cancer cells. The majority of in vivo hyperthermia treatments employ NIR radiation, which has poor penetration into tumors. Longer wavelength electromagnetic or radiofrequency radiation, on the other hand, may be required for the treatment of deep-seated tumor cells, hence the development of additional alternative approaches should be prioritized.
d. One of the other significant obstacles and Among their drawbacks, these carbon-based nanostructures' long-term toxicity needs to be rigorously assessed using in vivo animal models that are more pertinent.
e. A variety of factors influence the pharmacokinetics and toxicity of carbon-based nanomaterials, including their physiochemical and structural features, the dose and duration of exposure, the kind of cell, the mechanism, the residual catalyst, and the manufacturing method. In light of this, it is critical to avoid broad generalizations about the toxicity of carbon nanomaterials. Numerous studies have revealed that the generation of reactive oxygen species in target cells is the most important cytotoxicity mechanism of carbon-based nanomaterials. However, it has been discovered that the toxicity of carbon-based nanostructures can be greatly reduced by functionalizing them with biocompatible polymers or surfactants. Before reaching solid results, further focused and organized research on the toxicity and pharmacokinetics of these carbon-based nanomaterials is required. In this context, a strategy must be devised to provide a valid and acknowledged approach for comparing the toxicity data of various types of nanostructures.
f. Although various in vivo studies in animal models have shown that these nanostructures are safe to some extent, these models are vastly different from people.
g. A comprehensive study of how these nanostructures affect the nervous system, immune system, and reproductive system is necessary. Overcoming each of these challenges will open up new avenues for the application of these nanomaterials in the development of cutting-edge, novel cancer theranostics approaches. Conclusion: The usage of carbon-based nanosystems in cancer theranostics may benefit substantially in the near future.

REFERENCES

1. Hossen S, Hossain MK, Basher MK, Mia MNH, Rahman MT, Uddin MJ. Smart nanocarrier-based drug delivery systems for cancer therapy and toxicity studies: a review. *J Adv Res.* 2019;15:1–18. doi:10.1016/j.jare.2018.06.005
2. Liu D, Yang F, Xiong F, Gu N. The smart drug delivery system and its clinical potential. *Theranostics* 2016;6:1306–23. doi:10.7150/thno.14858

3. Wan F, Thakur A, Thakur P. (2023). Classification of nanomaterials (carbon, metals, polymers, bio-ceramics). In: Suhag, D., Thakur, A., Thakur, P. (eds) *Integrated Nanomaterials and their Applications*. Springer, Singapore. https://doi.org/10.1007/978-981-99-6105-4_3
4. Morey M, Pandit A. Responsive triggering systems for delivery in chronic wound healing. *Adv. Drug Delivery Rev.* 2018;129:169–93. doi:10.1016/j.addr.2018.02.008
5. Xu S, Lu H, Zheng X, Chen L. Stimuli-responsive molecularly imprinted polymers: versatile functional materials. *J Mater Chem C* 2013;1:4406–22. doi:10.1039/c3tc30496e
6. Cao ZQ, Wang GJ. Multi-stimuli-responsive polymer materials: particles, films, and bulk gels. *Chem Rec.* 2016;16:1398–435. doi:10.1002/tcr.201500281
7. Lu Y, Aimetti AA, Langer R, Gu Z. Bioresponsive materials. *Nat Rev Mater.* 2017;2:16075. doi:10.1038/natrevmats.2016.75
8. Wang Y, Zhang XY, Luo YL, Xu F, Chen YS, Su YY. Dual stimuli-responsive Fe_3O_4 graft poly (acrylic acid)-block-poly (2-methacryloyloxyethyl ferrocenecarboxylate) copolymer micromicelles: surface RAFT synthesis, self-assembly and drug release applications. *J. Nanobiotechnol.* 2017;15:76. doi:10.1186/s12951-017-0309-y
9. Yan Y, Sun N, Li F, Jia X, Wang C, Chao D. Multiple stimuli-responsive fluorescence behavior of novel polyamic acid bearing oligoaniline, triphenylamine, and fluorene groups. *ACS Appl Mater Interfaces* 2017;9:6497–503. doi:10.1021/acsami.6b16402
10. Wang Y, Kohane DS. External triggering and triggered targeting strategies for drug delivery. *Nat Rev Mater.* 2017;2:17020. doi:10.1038/natrevmats.2017.20
11. Salvi A, Dhanda N, Kharbanda S, Pathania A, Thakur P, Thakur A. (2023). MXene nanomaterial for medical application. In: Suhag, D., Thakur, A., Thakur, P. (eds) *Integrated Nanomaterials and their Applications*. Springer, Singapore. https://doi.org/10.1007/978-981-99-6105-4_6
12. Crommelin DJA., Florence AT. Towards more effective advanced drug delivery systems. *Int J Pharm.* 2013;454:496–511. doi:10.1016/j.ijpharm.2013.02.020
13. Holzapfel BM, Reichert JC, Schantz JT, Gbureck U, Rackwitz L, Nöth U, Jakob F, Rudert M, Groll J, Hutmacher DW. How smart do biomaterials need to be? A translational science and clinical point of view. *Adv. Drug Delivery Rev.* 2013;65:581–603. doi:10.1016/j.addr.2012.07.009
14. Gelderblom H, Verweij J, Nooter K, Sparreboom A, Cremophor EL. The drawbacks and advantages of vehicle selection for drug formulation. *Eur J Cancer.* 2001;37(13):1590–8.
15. Thakur A, Thakur P, Baccar S. (2023). Structural properties of nanoparticles. In: Suhag, D., Thakur, A., Thakur, P. (eds) *Integrated Nanomaterials and their Applications*. Springer, Singapore. https://doi.org/10.1007/978-981-99-6105-4_4
16. Ehsan M, Suhag D, Rathore R, Thakur A, Thakur P. (2023). Metal nanoparticles in the field of medicine and pharmacology. In: Suhag, D., Thakur, A., Thakur, P. (eds) *Integrated Nanomaterials and their Applications*. Springer, Singapore. https://doi.org/10.1007/978-981-99-6105-4_7
17. Davis ME, Chen Z, Shin DM. Nanoparticle therapeutics: an emerging treatment modality for cancer. *Nat Rev Drug Discov.* 2008;7(9):771–82.
18. Brannon-Peppas L, Blanchette J. Nanoparticle and targeted systems for cancer therapy. *Adv Drug Deliv Rev.* 2004;56(11):1649–59.
19. Kamaly N, Xiao Z, Valencia PM, Radovic-Moreno AF, Farokhzad OC. Targeted polymeric therapeutic nanoparticles: design, development and clinical translation. *Chem Soc Rev.* 2012;41(7):2971–3010.
20. Zhang S, Li J, Lykotrafitis G, Bao G, Suresh S. Size-dependent endocytosis of nanoparticles. *Adv Mater.* 2009;21(4):419.
21. Sahay G, Alakhova DY, Kabanov AV. Endocytosis of nanomedicines. *J Control Release.* 2010;145(3):182–95.
22. Gratton SEA, Ropp PA, Pohlhaus PD, et al. The effect of particle design on cellular internalization pathways. *Proc Natl Acad Sci U S A.* 2008;105(33):11613–8.
23. Wang J, Byrne JD, Napier ME, DeSimone JM. More effective nanomedicines through particle design. *Small.* 2011;7(14):1919–31.
24. Choi HS, Liu W, Misra P, et al. Renal clearance of quantum dots. *Nat Biotechnol.* 2007;25(10):1165–70.
25. Moghimi SM, Szebeni J. Stealth liposomes and long circulating nanoparticles: critical issues in pharmacokinetics, opsonization and protein-binding properties. *Prog Lipid Res.* 2003;42(6):463–78.
26. Torchilin VP. Micellar nanocarriers: pharmaceutical perspectives. *Pharm Res.* 2007;24(1):1–16.
27. Geng Y, Dalhaimer P, Cai S, et al. Shape effects of filaments versus spherical particles in flow and drug delivery. *Nat Nanotechnol.* 2007;2(4):249–55.
28. Verma A, Stellacci F. Effect of surface properties on nanoparticlecell interactions. *Small.* 2010;6(1):12–21.
29. Toy R, Bauer L, Hoimes C, Ghaghada KB, Karathanasis E. Targeted nanotechnology for cancer imaging. *Adv Drug Deliv Rev.* 2014;76:79–97.

30. Doshi N, Prabhakarpandian B, Rea-Ramsey A, Pant K, Sundaram S, Mitragotri S. Flow and adhesion of drug carriers in blood vessels depend on their shape: a study using model synthetic microvascular networks. *J Control Release*. 2010;146(2):196–200.
31. Decuzzi P, Godin B, Tanaka T, et al. Size and shape effects in the biodistribution of intravascularly injected particles. *J Control Release*. 2010;141(3):320–7.
32. Zhao F, Zhao Y, Liu Y, Chang X, Chen C, Zhao Y. Cellular uptake, intracellular trafficking, and cytotoxicity of nanomaterials. *Small*. 2011;7(10):1322–37.
33. Wilhelm C, Billotey C, Roger J, Pons JN, Bacri JC, Gazeau F. Intracellular uptake of anionic superparamagnetic nanoparticles as a function of their surface coating. *Biomaterials*. 2003;24(6):1001–11.
34. Dausend J, Musyanovych A, Dass M, et al. Uptake mechanism of oppositely charged fluorescent nanoparticles in HeLa cells. *Macromol Biosci*. 2008;8(12):1135–43.
35. Karmali PP, Simberg D. Interactions of nanoparticles with plasma proteins: implication on clearance and toxicity of drug delivery systems. *Expert Opin Drug Deliv*. 2011;8(3):343–57.
36. Vonarbourg A, Passirani C, Saulnier P, Benoit JP. Parameters influencing the stealthiness of colloidal drug delivery systems. *Biomaterials*. 2006;27(24):4356–73.
37. Veronese FM, Pasut G. PEGylation, successful approach to drug delivery. *Drug Discov Today*. 2005;10(21):1451–8.
38. Welsher K, Liu Z, Sherlock SP, et al. A route to brightly fluo-rescent carbon nanotubes for near-infrared imaging in mice. *Nat Nanotech*. 2009;4:773–80.
39. Liu Z, Yang K, Lee ST. Single-walled carbon nanotubes in bio-medical imaging. *J Mater Chem*. 2011;21:586–98.
40. Sun YP, Zhou B, Lin Y, et al. Quantum-sized carbon dots for bright and colorful photoluminescence. *J Am Chem Soc*. 2006;128:7756–7.
41. Yu SJ, Kang MW, Chang HC, et al. Bright fluorescent nanodi-amonds: no photobleaching and low cytotoxicity. *J Am Chem Soc*. 2005;127:17604–5.
42. Sun X, Liu Z, Welsher K, et al. Nano-graphene oxide for cellular imaging and drug delivery. *Nano Res*. 2008;1:203–12.
43. Loh KP, Bao Q, Eda G, et al. Graphene oxide as a chemically tunable platform for optical applications. *Nat Chem*. 2010;2:1015–24.
44. Chen RJ, Bangsaruntip S, Drouvalakis KA, et al. Noncovalent functionalization of carbon nanotubes for highly specific electronic biosensors. *Proc Natl Acad Sci U S A*. 2003;100:4984–9.
45. Liu Z, Sun X, Nakayama N, et al. Supramolecular chemistry on water-soluble carbon nanotubes for drug loading and delivery. *ACS Nano*. 2007;1:50–6.
46. Liu Z, Robinson JT, Sun XM, et al. PEGylated nanographene oxide for delivery of water-insoluble cancer drugs. *J Am Chem Soc*. 2008;130:10876–7.
47. Chen C, Xing G, Wang J, et al. Multihydroxylated [Gd@C82(OH)22]n nanoparticles: antineoplastic activity of high efficiency and low toxicity. *Nano Lett*. 2005;5:2050–7.
48. Liang XJ, Meng H, Wang Y, et al. Metallofullerene nanoparticles circumvent tumor resistance to cisplatin by reactivating endocytosis. *Proc Natl Acad Sci U S A*. 2010;107:7449–54.
49. Meng H, Xing G, Sun B, et al. Potent angiogenesis inhibition by the particulate form of fullerene derivatives. *ACS Nano*. 2010;4:2773–83.
50. Hong SY, Tobias G, Al-Jamal KT, et al. Filled and glycosylated carbon nanotubes for in vivo radioemitter localization and im-aging. *Nat Mater*. 2010;9:485–90.
51. Yang K, Wan J, Zhang S, et al. The influence of surface chemistry and size of nanoscale graphene oxide on photothermal therapy of cancer using ultra-low laser power. *Biomaterials*. 2012;33:2206–14.
52. Yang K, Zhang S, Zhang G, et al. Graphene in mice: ultra-high in vivo tumor uptake and photothermal therapy. *Nano Lett*. 2010;10:3318–23.
53. Yang M, Meng J, Cheng X, et al. Multiwalled carbon nanotubes interact with macrophages and influence tumor progression and metastasis. *Theranostics*. 2012;2:258–70.
54. Robinson JT, Welsher K, Tabakman SM, et al. High performance in vivo near-IR (<1 μm) imaging and photothermal cancer therapy with carbon nanotubes. *Nano Res*. 2010;3:779–93.
55. Moon HK, Lee SH, Choi HC. In vivo near-infrared mediated tumor destruction by photothermal effect of carbon nanotubes. *Acs Nano*. 2009;3:3707–13.
56. Chen Z, Ma L, Liu Y, et al. Applications of functionalized full-erenes in tumor theranostics. *Theranostics*. 2012;2:238–50.
57. Mandal HS, Su Z, Ward A, et al. Carbon nanotube thin film biosensors for sensitive and reproducible whole virus detection. *Theranostics*. 2012;2:251–7.
58. Yang S-T, Luo J, Zhou Q, et al. Pharmacokinetics, metabolism and toxicity of carbon nanotubes for biomedical purposes. *Theranostics*. 2012;2:271–82.

59. Shen H, Zhang L, Liu M, et al. Biomedical applications of gra-phene. *Theranostics*. 2012;2:283–94.
60. Cao L, Yang S-T, Wang X, et al. Competitive performance of carbon "quantum" dots in optical bioimaging. *Theranostics*. 2012;2:295–301.
61. Zhu Y, Li J, Li W, et al. The biocompatibility of nanodiamonds and their application in drug delivery systems. *Theranostics*. 2012;2:302–12.
62. Eisenbrey JR, Forsberg F, Wessner CE, Delaney LJ, Bradigan K, Gummadi S, Tantawi M, Lyshchik A, O'Kane P, Liu J.-B, Intenzo C, Civan J, Maley W, Keith SW, Anton K, Tan A, Smolock A, Shamimi-Noori S, Shaw CM. US-triggered microbubble destruction for augmenting hepatocellular carcinoma response to transarterial radioembolization: a randomized pilot clinical trial. *Radiology* 2020;298:450–7. doi:10.1148/radiol.2020202321
63. Peng S, Xiong Y, Li K, He M, Deng Y, Chen L, Zou M, Chen W, Wang Z, He J, Zhang L. Clinical utility of a microbubble-enhancing contrast ("SonoVue") in treatment of uterine fibroids with high intensity focused ultrasound: a retrospective study. *Eur. J. Radiol.* 2012;81:3832–8. doi:10.1016/j.ejrad.2012.04.030
64. Jiang N, Xie B, Zhang X, He M, Li K, Bai J, Wang Z, He J, Zhang L. Enhancing ablation effects of a microbubble-enhancing contrast agent ("SonoVue") in the treatment of uterine fibroids with high-intensity focused ultrasound: a randomized controlled trial. *Cardiovasc. Intervent. Radiol.* 2014;37:1321–8. doi:10.1007/s00270-013-0803-z
65. Jiang H, Luo S, He M, Zhang L, Li K, He J. *J Kunming Med Univ*. 2013;88:58–62.
66. Cheng C, Xiao Z, Huang G, Zhang L, Bai J. Enhancing ablation effects of a microbubble contrast agent on high-intensity focused ultrasound: an experimental and clinical study. *BJOG* 2017;124:78–86. doi:10.1111/1471–0528.14744
67. Isern J, Pessarrodona A, Rodriguez J, Vallejo E, Gimenez N, Cassadó J, De Marco JA, Pedrerol A. Using microbubble sonographic contrast agent to enhance the effect of high intensity focused ultrasound for the treatment of uterine fibroids. *Ultrason. Sonochem.* 2015;27:688–93. doi:10.1016/j.ultsonch.2015.05.027
68. Jingqi W, Lu Z, Jun Z, Yuhong M, Wei Y, Lifeng R, Chengbing J, Dobromir DD, Hui Z, Kun Z. Clinical usefulness of the microbubble contrast agent sonovue in enhancing the effects of high-intensity focused ultrasound for the treatment of adenomyosis. *J. Ultrasound Med.* 2018;37:2811–9. doi:10.1002/jum.14638
69. Lyon PC, Gray MD, Mannaris C, Folkes LK, Stratford M, Campo L, Chung DYF, Scott S, Anderson M, Goldin R, Carlisle R, Wu F, Middleton MR, Gleeson FV, Coussios CC. Large-volume hyperthermia for safe and cost-effective targeted drug delivery using a clinical ultrasound-guided focused ultrasound device. *Lancet Oncol.* 2018;19:1027–39. doi:10.1016/s1470–2045(18)30332-2
70. Welsher K, Liu Z, Sherlock SP, Robinson JT, Chen Z, Daranciang D, Dai H. A route to brightly fluorescent carbon nanotubes for near-infrared imaging in mice. *Nat. Nanotechnol.* 2009;4:773–80.
71. Hong G, Lee JC, Robinson JT, Raaz U, Xie L, Huang NF, Cooke JP, Dai, H. Multifunctional in vivo vascular imaging using near-infrared II fluorescence. *Nat. Med.* 2012;18:1841–6.
72. https://www.doxil.com/.
73. Thorn CF, Oshiro C, Marsh S, et al. Doxorubicin pathways: pharmacodynamics and adverse effects. *Pharmacogenet Genom*. 2011;21(7):440–6.
74. Cho H, Yoon I, Yoon HY, et al. Polyethylene glycol-conjugated hyaluronic acid-ceramide self-assembled nanoparticles for targeted delivery of doxorubicin. *Biomaterials*. 2012;33(4):1190–200.
75. Rowinsky E, Donehower R. Drug-therapy—paclitaxel (taxol). *N Engl J Med*. 1995;332(15):1004–14.
76. Gelderblom H, Verweij J, Nooter K, Sparreboom A, Cremophor EL. The drawbacks and advantages of vehicle selection for drug formulation. *Eur J Cancer*. 2001;37(13):1590–8.
77. Lanza G, Yu X, Winter P, et al. Targeted antiproliferative drug delivery to vascular smooth muscle cells with a magnetic resonance imaging nanoparticle contrast agent implications for rational therapy of restenosis. *Circulation*. 2002;106(22):2842–7.
78. Redis RS, Berindan-Neagoe I, Pop VI, Calin GA. Non-coding RNAs as theranostics in human cancers. *J Cell Biochem*. 2012;113(5):1451–9.
79. Bartel DP. MicroRNAs: genomics, biogenesis, mechanism, and function. *Cell*. 2004;116(2):281–97.
80. Calin GA, Dumitru CD, Shimizu M, et al. Frequent deletions and down-regulation of micro-RNA genes miR15 and miR16 at 13q14 in chronic lymphocytic leukemia. *Proc Natl Acad Sci U S A*. 2002;99(24):15524–9.
81. Hatley ME, Patrick DM, Garcia MR, et al. Modulation of K-rasdependent lung tumorigenesis by microRNA-21. *Cancer Cell*. 2010;18(3):282–93.
82. Zhao J, Lin J, Yang H, et al. MicroRNA-221/222 negatively regulates estrogen receptor alpha and is associated with tamoxifen resistance in breast cancer. *J Biol Chem*. 2008;283(45):31079–86.
83. Fujita Y, Kojima K, Hamada N, et al. Effects of miR-34a on cell growth and chemoresistance in prostate cancer PC3 cells. *Biochem Biophys Res Commun*. 2008;377(1):114–9.

84. Chen Y, Zhu X, Zhang X, Liu B, Huang L. Nanoparticles modified with tumor-targeting scFv deliver siRNA and miRNA for cancer therapy. *Mol Ther.* 2010;18(9):1650–6.
85. Kim J, Yeom J, Ko J, et al. Effective delivery of anti-miRNA DNA oligonucleotides by functionalized gold nanoparticles. *J Biotechnol.* 2011;155(3):287–92.
86. Oh Y, Park TG. siRNA delivery systems for cancer treatment. *Adv Drug Deliv Rev.* 2009;61(10):850–62.
87. Peer D, Park EJ, Morishita Y, Carman CV, Shimaoka M. Systemic leukocyte-directed siRNA delivery revealing cyclin D1 as an antiinflammatory target. *Science.* 2008;319(5863):627–30.
88. Barenholz Y. Doxil (R)—the first FDA-approved nano-drug: lessons learned. *J Control Release.* 2012;160(2):117–34.
89. O'Brien M, Wigler N, Inbar M, et al. Reduced cardiotoxicity and comparable efficacy in a phase III trial of pegylated liposomal doxorubicin HCl (CAELYX (TM)/doxil (R)) versus conventional doxorubicin for first-line treatment of metastatic breast cancer. *Ann Oncol.* 2004;15(3):440–9.
90. Forssen EA, Ross ME. Daunoxome treatment of solid tumors: preclinical and clinical investigations. *J Liposome Res.* 1994;4(1):481– 512.
91. Green MR, Manikhas GM, Orlov S, et al. Abraxane((R)), a novel Cremophor((R))-free, albumin-bound particle form of paclitaxel for the treatment of advanced non-small-cell lung cancer. *Ann Oncol.* 2006;17(8):1263–8.
92. Melmed GY, Targan SR, Yasothan U, Hanicq D, Kirkpatrick P. Certolizumab pegol. *Nat Rev Drug Discov.* 2008;7(8):641–2.
93. Berges R. Eligard (R): pharmacokinetics, effect on testosterone and PSA levels and tolerability. *Eur Urol Suppl.* 2005;4(5):20–5.
94. Dinndorf PA, Gootenberg J, Cohen MH, Keegan P, Pazdur R. FDA drug approval summary: pegaspargase (Oncaspar (R)) for the firstline treatment of children with acute lymphoblastic leukemia (ALL). *Oncologist.* 2007;12(8):991–8.
95. Kim T, Kim D, Chung J, et al. Phase I and pharmacokinetic study of genexol-PM, a cremophor-free, polymeric micelle-formulated paclitaxel, in patients with advanced malignancies. *Clin Cancer Res.* 2004;10(11):3708–16.
96. Janib SM, Moses AS, MacKay JA. Imaging and drug delivery using theranostic nanoparticles. *Adv Drug Deliv Rev.* 2010;62(11):1052–63.
97. Debbage P, Jaschke W. Molecular imaging with nanoparticles: giant roles for dwarf actors. *Histochem Cell Biol.* 2008;130(5):845–75.
98. Choi KY, Jeon EJ, Yoon HY, et al. Theranostic nanoparticles based on PEGylated hyaluronic acid for the diagnosis, therapy and monitoring of colon cancer. *Biomaterials.* 2012;33(26):6186–93.
99. Medintz IL, Uyeda HT, Goldman ER, Mattoussi H. Quantum dot bioconjugates for imaging, labelling and sensing. *Nat Mater.* 2005;4(6):435–46.
100. Tan WB, Jiang S, Zhang Y. Quantum-dot based nanoparticles for targeted silencing of HER2/neu gene via RNA interference. *Biomaterials.* 2007;28(8):1565–71.
101. Derfus A, Chan W, Bhatia S. Probing the cytotoxicity of semiconductor quantum dots. *Nano Lett.* 2004;4(1):11–8.
102. Zhu J, Zheng L, Wen S, et al. Targeted cancer theranostics using alpha-tocopheryl succinate-conjugated multifunctional dendrimerentrapped gold nanoparticles. *Biomaterials.* 2014;35(26):7635–46.
103. Shilo M, Reuveni T, Motiei M, Popovtzer R. Nanoparticles as computed tomography contrast agents: current status and future perspectives. *Nanomedicine.* 2012;7(2):257–69.
104. Rabin O, Perez J, Grimm J, Wojtkiewicz G, Weissleder R. An X-ray computed tomography imaging agent based on long-circulating bismuth sulphide nanoparticles. *Nat Mater.* 2006;5(2):118–22.
105. Zanzonico P. Principles of nuclear medicine imaging: planar, SPEC T, PET, multi-modality, and autoradiography systems. *Radiat Res.* 2012;177(4):349–64.
106. Pysz MA, Gambhir SS, Willmann JK. Molecular imaging: current status and emerging strategies. *Clin Radiol.* 2010;65(7):500–16.
107. Jennings LE, Long NJ. 'Two is better than one'-probes for dualmodality molecular imaging. *Chem Commun.* 2009;24:3511–24.
108. Wang C, Ravi S, Garapati US, et al. Multifunctional chitosan magnetic-graphene (CMG) nanoparticles: a theranostic platform for tumor-targeted co-delivery of drugs, genes and MRI contrast agents. *J Mat Chem B.* 2013;1(35):4396–405.
109. Liu T, Li X, Qian Y, Hu X, Liu S. Multifunctional pH-disintegrable micellar nanoparticles of asymmetrically functionalized betacyclodextrin- based star copolymer covalently conjugated with doxorubicin and DOTA-gd moieties. *Biomaterials.* 2012;33(8):2521–31.

17 Modification and Development of Advanced Nano Polymer Composites

Nancy and Preeti Thakur
Amity University Haryana

Irina Edelman and Sergey Ovchinnikov
Kirensky Institute of Physics

Atul Thakur
Amity University Haryana

17.1 INTRODUCTION

Structural materials that are made of two or more mixed elements are called composites. They are not soluble in one another and are mixed at a macroscopic level. The matrix and the reinforcing phase are two embedded components, respectively. Fibers, particles, or flakes make up the material that makes up the reinforcing phase. Materials used in the matrix phase are typically continuous. It is more valuable to combine two or more materials to create a composite than to simply use classic monolithic metals like steel and aluminum. The requirements of today's cutting-edge technology are not always addressed by monolithic metals and their alloys. Therefore, it can only achieve the necessary performances by mixing several components. Fiber-reinforced composites are frequently more effective. For instance, the very competitive aviation industry is constantly seeking ways to reduce the total bulk of an aircraft without compromising the stiffness and strength of its parts. By using composite materials in place of traditional metal alloys, this is made possible. Despite the increased cost of the composite material, the assembly requires fewer parts, and the fuel savings increase its profitability. About 1,360L of fuel can be saved annually by reducing the mass of a commercial aircraft by 0.453kg; fuel expenses make up about 25% of an airline's overall operating costs. Over traditional materials, composites have several additional benefits. The strength, stiffness, fatigue, impact resistance, thermal conductivity, corrosion resistance, etc. of these materials are all improved. Researchers have recently become interested in nanocomposites made by reinforcing polymers with nanoscale particles for use in real estate construction. Since the previous decade, carbon nanotubes (CNTs), nanofiber, and nano clay with extremely high stiffness and strength have been particularly popular for use as nano reinforcements in polymers. The polymer resin systems greatly benefit rheologically from their nanoscale size. In polymer resins, nanoparticles are dispersed at relatively low concentrations (5wt.%) and frequently add excellent thermal, mechanical, obstacle, and electromagnetic properties. Because of this, a whole new realm of potential structural applications and property advancements has been made possible by polymer nano reinforcement. Nano polymer composites (NPCs) are a novel type of material composites that have yet to be fully explored, despite their extensive potential uses.

DOI: 10.1201/9781003502692-17

17.2 NANOTECHNOLOGY AND NANOSCIENCE

The field of advanced technology known as nanotechnology is concerned with and addresses the creation, development, and use of nanoscale-based materials [1,2].

When Taniguchi first used the term "nanotechnology" in 1974, he described it as "the science of nanoparticles (1–100 nm)" and systems that have parts and structures that exhibit improved biological, chemical, physical, and thermal properties. We use nanofillers to significantly improve mechanical, thermal, optical, and barrier properties [3,4]. Due to their high surface area and exceptional capacity to alter the properties of materials, nanoscience and nanotechnology have flourished over the past 20 years [5,6] and gained importance in fields like biosensors, nanomedicine, computing, aeronautics, polymer modification, and many others. Figure 17.1 [7] illustrates the current applications of nanotechnology across many industrial domains.

When compared to conventional composites [8], recently nanocomposites have attracted a lot of interest in both business and research because of their special characteristics and multifunctionality [9]. Furthermore, the majority of the processing methods used in nanocomposites are essentially the same as those used in micro composites, despite their nanoscale diameters [10–12].

The term nano is used to describe things that are 10^{-9} m in size. A nanometer, or billionth of a meter, is 80,000 times finer than a strand of human hair [13]. The doping agent, also known as a nanoparticle or nanofiller, is dispersed throughout the polymer matrix of a nanocomposite with a minimum of one dimension in the nanoscale domain [14].

Due to their distinct characteristics in comparison to their bulk counterparts, nanofillers derived from either natural or synthetic sources are currently of significant interest and are thought to be the most promising materials of the future. CNTs, laminated aluminosilicates (clays), graphene, ultra-dispersed diamonds (nanodiamonds), fullerenes, nano-metal oxides, $CaCO_3$, metallic nanoparticles, POSS, and nanofibers are some of the most popular nanofillers [15]. In other cases, nanofillers work in concert with conventional or traditional fillers to considerably improve or modify the materials' variable properties, including their physical, mechanical, electrical, thermal, and optical properties [16]. There are numerous ways to create nanomaterials, including chemical, biological, and mechanical processes. In Figure 17.2, the methods used in the preparation of nanomaterials are shown.

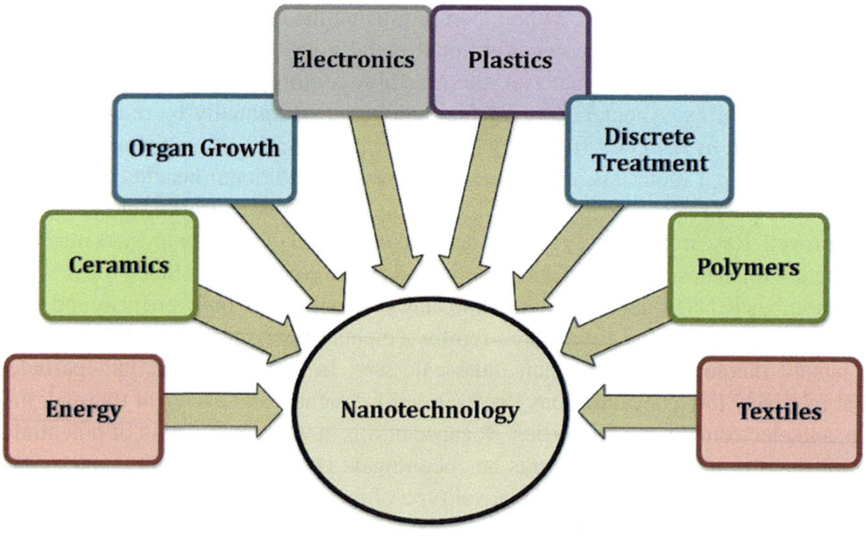

FIGURE 17.1 Various uses of nanotechnology.

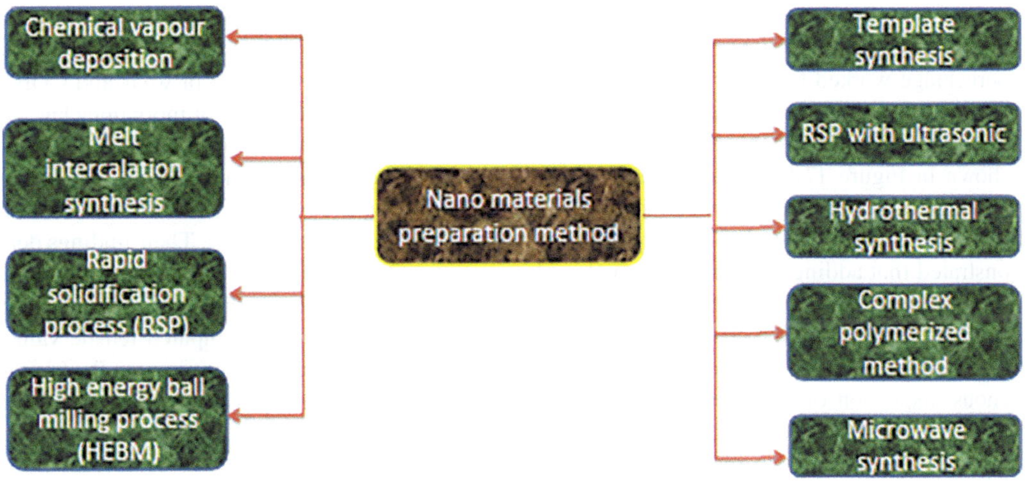

FIGURE 17.2 Various preparation methods of nanomaterials.

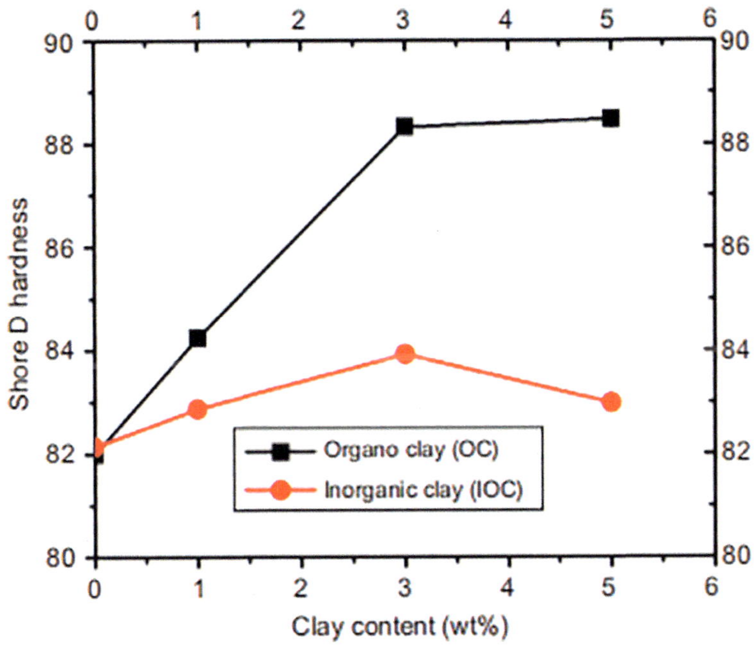

FIGURE 17.3 This graph shows the hardness of nanocomposites [17].

17.2.1 Nano Polymer Composites

Using inorganic clay (montmorillonite clay, MMT) and organo-modified montmorillonite clay (OMMT), respectively, the wear behaviors of conventional composites filled with clay and Nano Composites (NCs) were studied by Jawahar et al. [17]. This indicates that the hardness of the nanocomposites is slightly greater compared to the hardness of pure polyester. The hardness of nanocomposites and standard composites filled with clay is contrasted in Figure 17.3.

Additionally, the use of organoclay reduces the specific wear rate of composites with a coefficient of friction (COF). Furthermore, they found that for a specific wear rate and COF, a 3% clay percentage worked best. Furthermore, they have found that the specific rate of wear and COF of conventional clay-filled composites increase with an increase in the quantity of inorganic clay. The specific wear rates of conventional clay-filled composites and NCs with respect to clay content are shown in Figure 17.4. By using a nanoindentation technique, Dhakal et al. [18] investigated the impact of different nano-clay reinforcement loading levels on the nanomechanical characteristics of layered silicate nano-clay-reinforced unsaturated polyester nanocomposites. Their findings demonstrated that adding 1% w/w of nano clay increased the nano hardness of unreinforced polyester samples from 301 to 387 MPa and adding 5% w/w of nano clay increased the elastic modulus from 5,393 MPa for unreinforced polyester to 6,646 MPa. Table 17.1 lists the computed tensile values. Furthermore, they discovered that obtaining better mechanical characteristics requires the homogenous dispersion of nano-clay particles inside the matrix of unsaturated polyester resin (UPE). Additionally, they discovered that the ideal nano-clay loading for nano hardness and elastic modulus were 1% and 5% w/w, respectively.

Chan et al. [19] have suggested an experimental setting to create homogenous nano-clay/epoxy composites. They discovered throughout their experiment that increasing the nano-clay content of five wt.% as the ideal quantity of nano-clay filling increased the composites' Young's modulus and

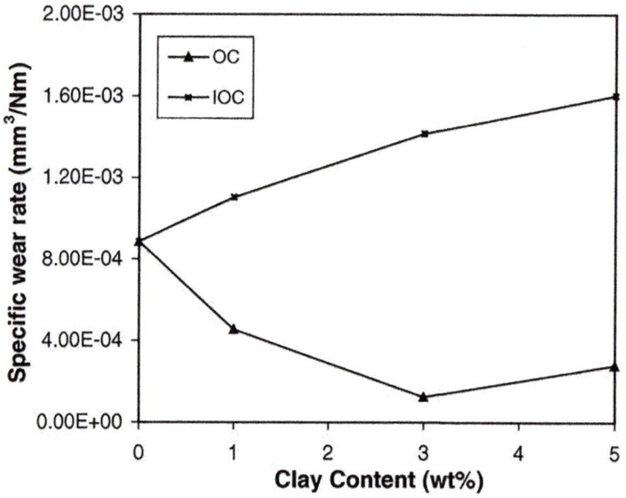

FIGURE 17.4 Specific wear rate of nanocomposites with different clay contents [17].

TABLE 17.1
Calculated Values of Elastic Modulus [18]

Specimens	Elastic Modulus, E Indent (MPa)
Unreinforced unsaturated polyester resin (UPE) only	5,393
1% w/w UPE/LK-EP-C	6,192 (15% increase)
33% w/w UPE/LK-EP-C	6,074 (13% increase)
5% w/w UPE/LK-EP-C	6,646 (23% increase)

tensile strength. The mechanical properties of the resulting composites decreased as the amount of nano-clay was increased further.

The thermal stability of MMT and polyester resin (UP) clays has been researched by Chieruzzi et al. [20]. They claim that the addition of clay particles lowers the coefficient of thermal expansion (CTE) of both pure resin and NCs with different MMT concentrations. At a clay loading of 5%, CTE is observed to be reduced. In their study, the various nanocomposites are evaluated and given in Tables 17.2 and 17.3, which display the CTE values that were found for several composites. Additionally, they assessed the tensile and flexural characteristics of the manufactured nanocomposites shown in Table 17.4.

TABLE 17.2
Analysis of UP/MMT Nanocomposites [20]

Sample	Sample Formulation
A1	Neat resin
A2	5wt%D43B
A3	10wt%D43B+1%VR
A4	5wt%D43B+5%D72T
A5	5wt%D43B+5%D72T+1%VR

TABLE 17.3
UP/MMT Nanocomposites Linear Thermal Expansion Coefficients [20]

	T Range			
	(25–40°C)		(70–100°C)	
Samples	CTE (10^{-6}/°C)	Change (%)	CTE (10^{-6}/°C)	Change (%)
A	117.4±10.9	—	199.8±12.9	—
A1	49.3±18.9	−57.9	131.0±27.5	−34.4
A2	81.3±2.4	−30.8	137.7±5.9	−31.5
A3	126.6±25.2	7.8	192.7±29.4	−3.5
A4	76.6±0.5	−34.8	143.7±4.3	−28.0
A5	137.0±5.3	16.7	165.9±5.2	−16.9

TABLE 17.4
Flexural and Tensile Properties of UP/MMT Nanoparticles [20]

	Tensile			Flexural		
Samples	Stress to Failure (MPa)	Strain to Failure (mm/mm)	Tensile Modulus (MPa)	Stress to Failure (MPa)	Strain to Failure (mm/mm)	Tensile Modulus (MPa)
A	52.68±8.01	0.070±0.010	1,062.2±29.14	75.79±4.05	0.083±0.014	1,562.4±215.4
A1	40.50±6.49	0.032±0.004	1,332.7±218.3	29.07±9.44	0.049±0.008	842.9±215.4
A2	19.30±2.45	0.015±0.002	2,085.9±337.1	40.75±6.43	0.015±0.002	2,674.0±228.0
A3	9.98±2.22	0.040±0.012	576.3±189.4	22.27±7.99	0.085±0.016	527.7±188.9
A4	20.04±2.77	0.015±0.002	2,322.1±231.6	44.31±7.89	0.020±0.003	2,678.5±256.3
A5	18.87±1.44	0.056±0.008	566.3±217.9	33.21±5.45	0.061±0.009	897.2±153.0

By homogeneously dispersing nanofillers, Meguid and Sun [21] have enhanced adhesive epoxy's interfacial strength and properties. The two forms of nanofillers that are utilized are alumina nano powder and CNTs. They discovered that the strength of the interface is significantly influenced by the presence of nanoparticles. Additionally demonstrated that there is a limit to the number of scattered particles, beyond which there is a decline in the attributes.

17.2.2 NPFRCs, or Nano Polymer Fiber-Reinforced Composites

Researchers [22] have looked into the arrangement of traditional fibers (such as glass, carbon, and aramid fibers) and extra nanophase reinforcement (like CNTs). They found that adding very small amounts of CNTs (0.3 weight percent DWCNT-NH_2) increased the matrix-dominated interlaminar shear strength (ILSS) by 20% but had no effect on the fiber-dominated tensile characteristics. In Figures 17.5 and 17.6, the experimentally determined tensile and ILSS of nano-reinforced composites are depicted. Additionally, it demonstrates that in the z-direction, the electrical conductivity in a plane is greater than an order of higher magnitude.

Subramaniyan et al. [23] used scanning electron microscopy (SEM) and transmission electron microscopy (TEM) to study the morphology of nano-clay suspended in acetone and distributed in resin. Additionally, based on the matrix properties, they have created an elastic-plastic model that predicts the compressive strength of fiber-reinforced composites. The anticipated values closely

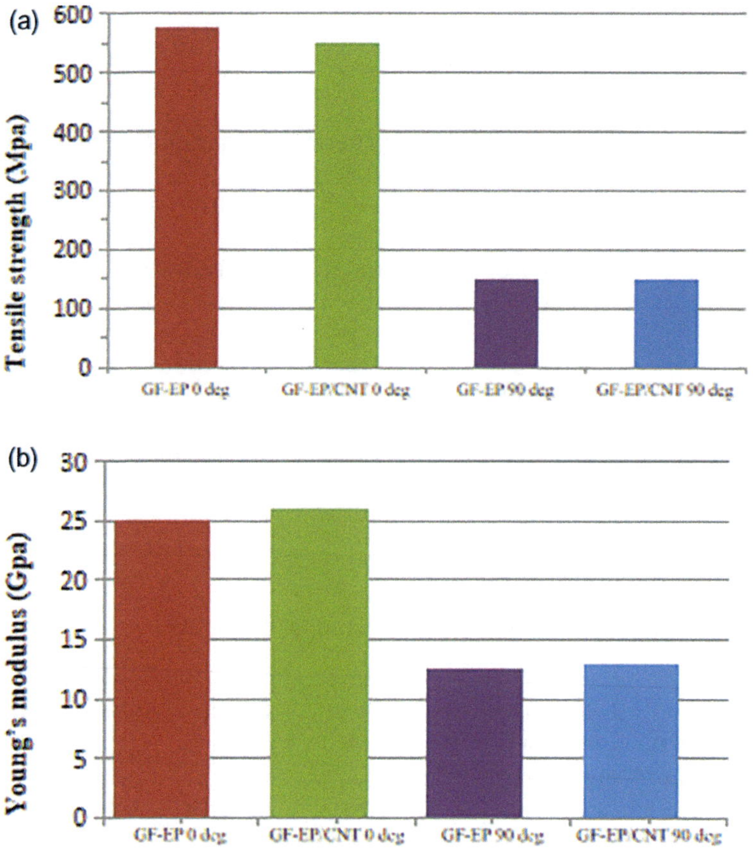

FIGURE 17.5 Tensile strengths (a) and Young's modulus (b) of the Glass Fiber Reinforced Polymer (GFRPs) in 0 and 90 directions, as determined by experiment [22].

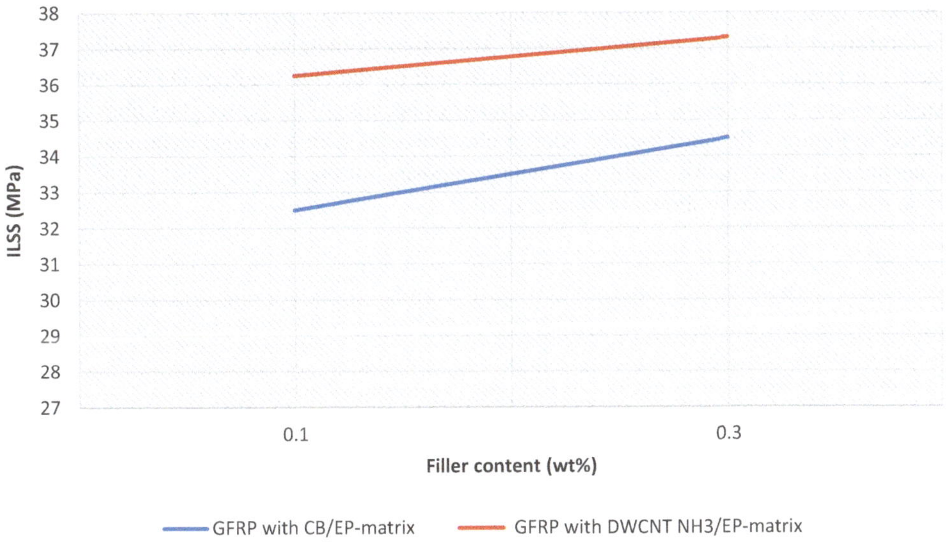

FIGURE 17.6 Interlaminar shear strength (ILSS) of the nano-reinforced GFRPs [22].

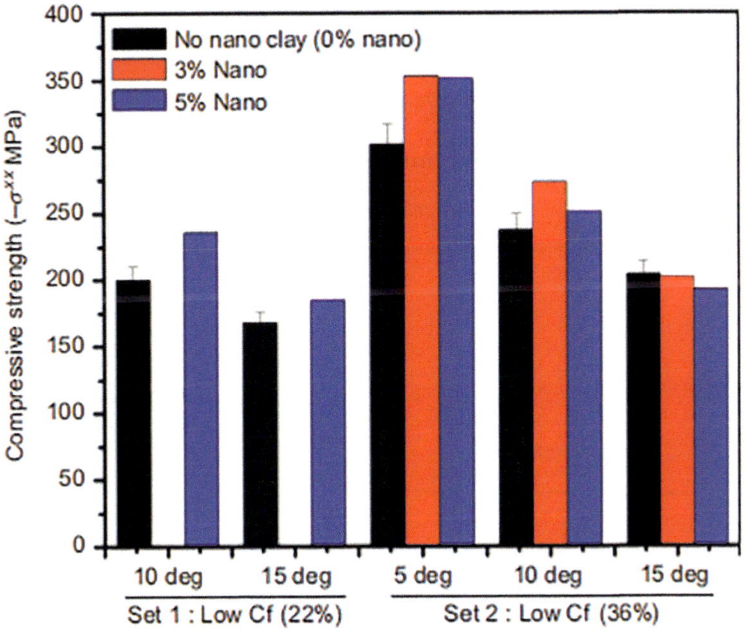

FIGURE 17.7 Results for compressive strength—VAWL specimen (%) values show the increase (+) or decrease (–) in strength as a result of adding nano clay [23].

match the outcomes of the experiments. Nano clay has been added to traditional fiber-reinforced composites using the vacuum-assisted wet layup (VAWL) method, and this specimen demonstrates an increase in the compression strength of the nano-clay-enhanced fiber composites (Figure 17.7). They have also noticed an interesting phenomenon, which is that the increase in compressive strength decreases with increasing off-axis angles.

Bozkurt et al. [24] looked into the mechanical and thermal characteristics of non-crimp glass fiber-reinforced clay/epoxy nanocomposites. According to their findings, the tensile strengths as depicted in Figure 17.8 are not significantly affected by the clay loading. On the other hand, the inclusion of clay enhances the flexural characteristics because of the better glass fiber/epoxy contact depicted in Figure 17.9. Additionally, adding clay particles with a surface treatment has improved the mechanical and dynamic qualities of nanocomposite laminates. The effect of clay loading with storage modulus values is shown in Figure 17.10.

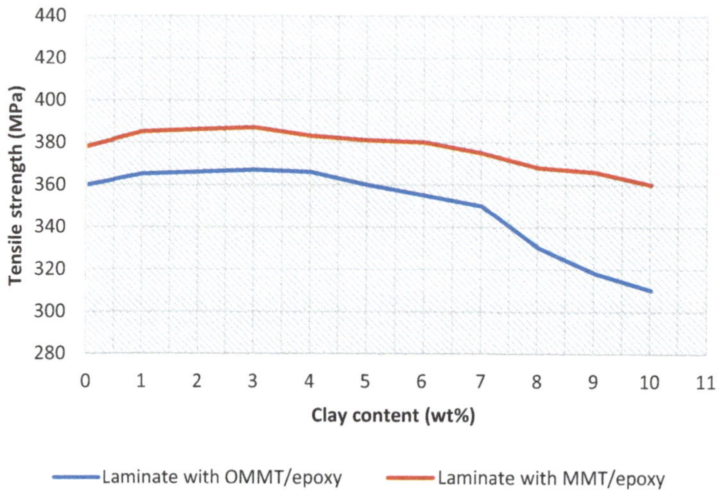

FIGURE 17.8 Tensile strength of clay/epoxy nanocomposites reinforced with no crimp glass fabric [24].

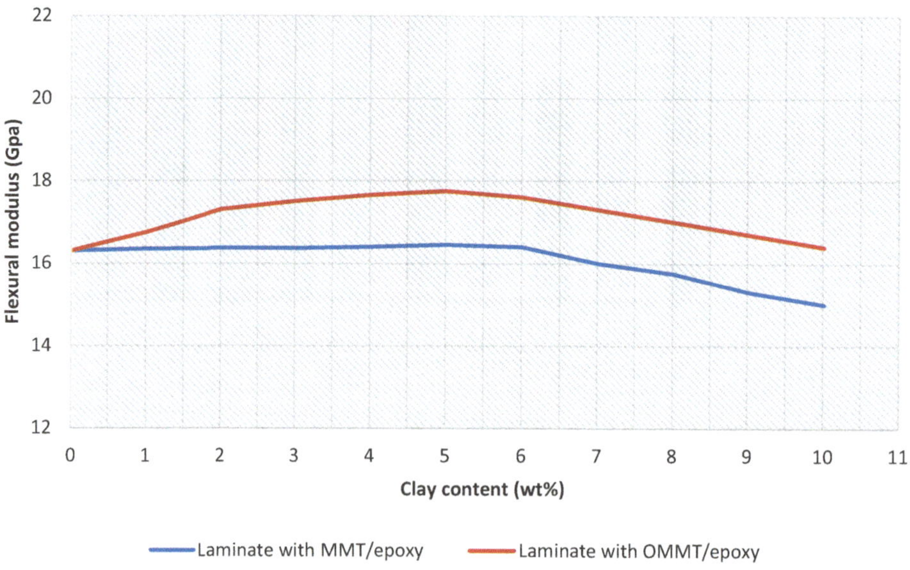

FIGURE 17.9 Flexural strength of clay/epoxy nanocomposites reinforced with non-crimp glass fabric [24].

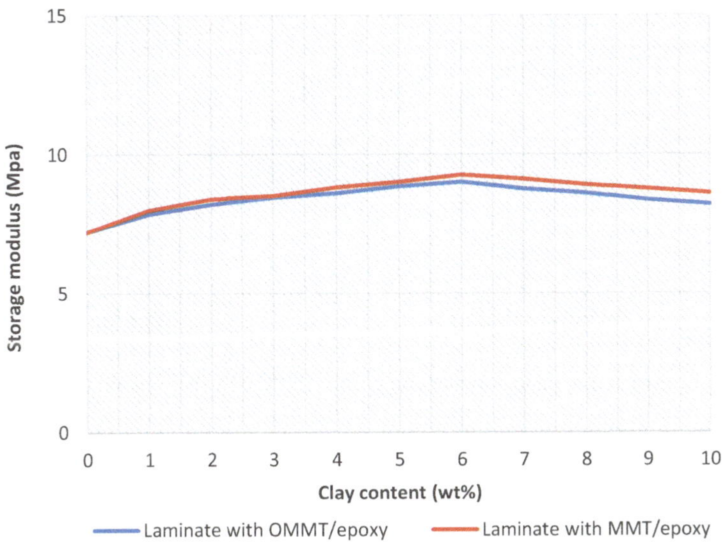

FIGURE 17.10 Storage modulus of non-crimp fabric-reinforced clay/epoxy nanocomposites [24].

17.3 PREPARATION METHODS OF NPCS

17.3.1 Situ Polymerization Method

Since the solvents required dissolving the polymers are highly toxic, this method is often appropriate for polymers that cannot be manufactured economically or safely by using solution methods. The nanoparticles in the polymer matrix are effectively dispersed and distributed by this approach [25]. It's necessary to highlight a few key components of this approach. The first has to do with the process's expense, which may necessitate certain modifications from the usual polymer synthesis. The best catalyst must be chosen with care as well. The equipment used for polymerization without nanoparticles can be the same.

17.3.2 Solution Method

This method is effective when the solvent (water, acetone, chloroform, or alcohol) is less toxic. Due to the effective dispersion of nanoparticles in this approach, various nanoparticle amounts can interact between the polymer and the solvent. To prepare high-quality NCs, this is the easiest method. Since the solvent must be entirely removed after manipulation [26–28], some caution must be exercised. The required equipment (Figure 17.11) is extremely straightforward. Combining spray drying with solution dispersion is one intriguing technique. After a period of homogenization, the polymer solution's dispersed nanoparticles are added to the spray dryer, which turns the material into a fine powder. This method's primary shortcoming is the material's ultimate yield. However, once all circumstances have been altered and the end product exhibits good dispersion and dispersal, this process takes place in a single step. For regulated or targeted delivery, medications can be injected alongside nano systems in this method. Figure 17.12 depicts the tools utilized in this procedure.

17.3.3 Melt Extrusion Method

As opposed to the other methods, this one has a significant benefit because no solvent is required. However, it's crucial to consider how many nanoparticles will be disseminated. Due to their

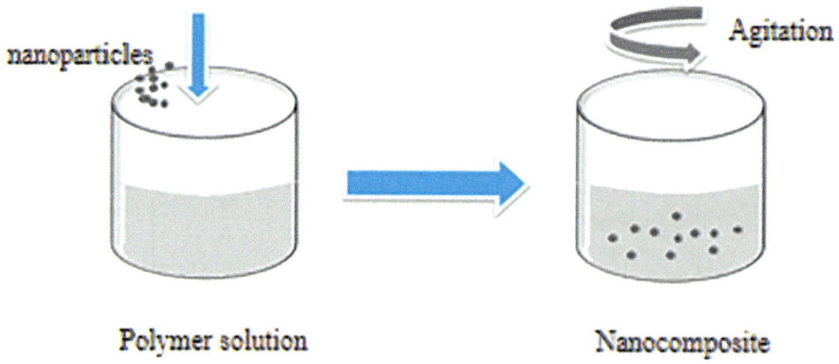

FIGURE 17.11 Preparation of nano polymer composites (NPCs) by the solution method.

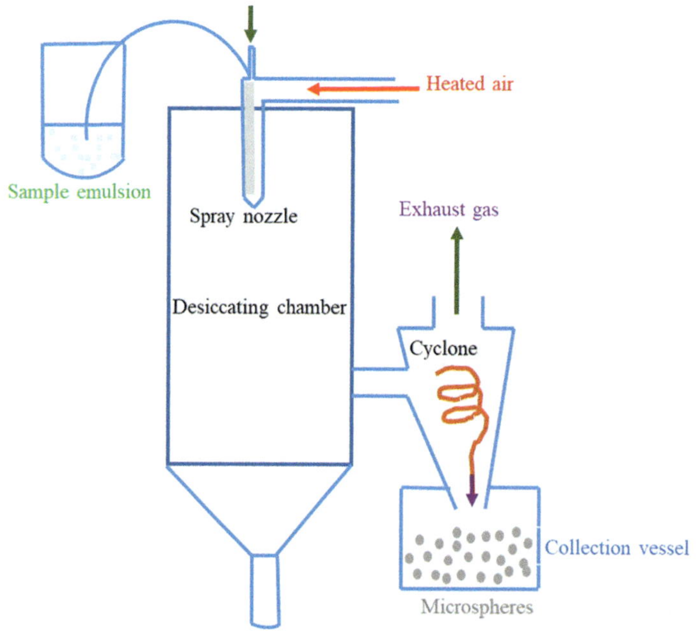

FIGURE 17.12 Preparation of nano polymer composites (NPCs) by spray drying.

propensity to aggregate more readily than in other approaches, this method necessitates careful monitoring of the dispersion of the nanoparticles. The equipment is the same as that used to process polymers without nanoparticles. Therefore, the researcher must be mindful of the temperatures employed to prevent polymer degradation during extrusion, as well as the amount of time required for optimum nanoparticle dispersion. The melting and degrading temperatures are extremely near for most natural polymers and a few biopolymers [29,30]. Figure 17.13 depicts a usual extruder.

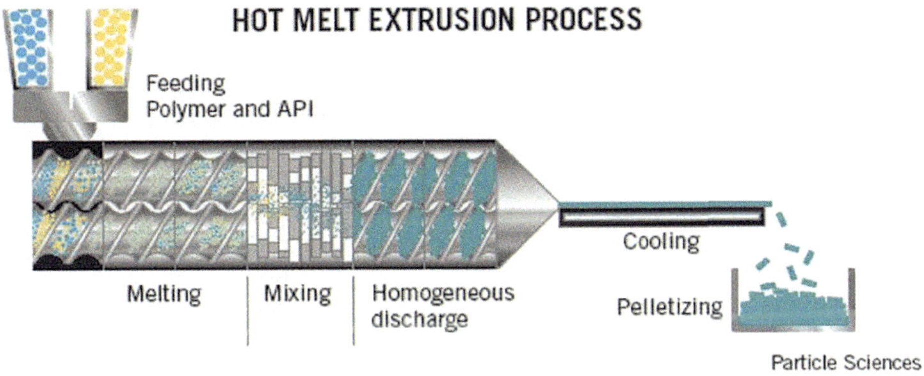

FIGURE 17.13 Technique for the preparation of nano polymer composite (NPC) by melt extrusion.

17.4 PREPARATION OF NPCS USING VARIOUS TECHNIQUES AND EVALUATION OF THEIR PROPERTIES

This part discusses a novel method for characterizing the three strategies and provides examples of each: proton spin-lattice nuclear magnetic resonance relaxometry relaxation period with a T_1H time constant.

The spin-lattice relaxation time is the time it takes for the longitudinal magnetization to return to its initial value of roughly 63% following a perpendicular radio-frequency pulse that flips it into the magnetic transverse plane. This enthalpy process transfers the spin energy to the entire chain that the material adheres to. T_1H can therefore reveal intermolecular interactions' effects on molecular mobility as well as the distribution of the material's constituents. Shorter periods suggest greater mobility, while longer times denote more rigid domains [31–35].

17.4.1 In Situ Polymerization

Getting clay systems and polyvinyl alcohol (PVA) and using relaxometry. The nanomaterials were created by in situ polymerization using different clay ratios (0.75–10% w/w). Vinyl acetate solution was added to Erlenmeyer flasks that held clay dispersions. Before adding 5 mL of a benzoyl peroxide solution in methanol, the acetate vinyl solution was heated for 2 hours at 60°C–65°C. The polymerization reaction took place for 24 hours.

The main characterization was performed with T_1H measurements in the following conditions.

The 40 points that made up the Free Induction Decay (FID) amplitude ranged from 0.1 to 5,000 ms, with each point having four observations and a recycle duration of 5 seconds. Table 17.5 displays the T_1H values for PVA and its nanocomposites made of clay using in situ polymerization.

The relaxation parameter significantly decreased when the T_1 H data for the PVA-clay systems of the material containing 3% clay were analyzed. This suggests that there was good interaction between the clay lamellae and the polymer matrix, which in turn aided in the good dispersion and distribution of the nanoparticles within the polymer matrix. Due to the main chains of the polymer being free to move about the clay lamella, the nanocomposite that had developed was exfoliated. The presence of a paramagnetic metal in the clay composition that acts as a tranquilizer may also help to explain this.

TABLE 17.5
T_1H Values for in situ Polymerization Derived PVA and Its Nanocomposites

Sample/PVA-Clay	T_1H (ms)
PVA	115
PVA-0.25	110
PVA-0.75	120
PVA-1.5	105
PVA-3.0	100
PVA-5.0	105
PVA-7.0	115

TABLE 17.6
T_1H Values for Solution-Cast PVA and Its Nanocomposites

Sample/PVA-Clay	T_1H (ms)
PVA	120
PVA-0.25	115
PVA-0.75	105
PVA-1.5	105
PVA-3.0	95
PVA-5.0	70
PVA-7.0	90

Because these proportions are not free to combine, they form a predominance of intercalated chains, as evidenced by the propensity to raise the relaxation value after this proportion. Because the clay lamellae limit them, they are immobile.

17.4.2 Solution Method

17.4.2.1 PVA-Clay Film Production Using Solution Casting

The films made via solution casting were made by mixing a clay dispersion and a polymer solution in water. Once the solution and dispersion were homogenized, the mixture was heated to 80°C in an oven to remove the solvent. The films were also examined using the same methods as for the in situ polymerization by measuring the proton spin-lattice relaxation time (Table 17.6).

The samples produced by in situ polymerization exhibited similar behavior, as indicated by the relaxation data in Table 17.6. Since 3% and 5% of the clay in the PVA matrix were produced via solution intercalation, the samples showed better exfoliation and the effect was more noticeable. A majority of exfoliated nanocomposite was obtained for both amounts of clay, however, at 5%, this was more pronounced. The clay lamellae were dispersed and distributed in the polymer matrix extremely well for 5% clay.

A majority of exfoliated nanocomposite was obtained for both amounts of clay; however, at 5%, this was more pronounced. The clay lamellae were dispersed and distributed in the polymer matrix extremely well for 5% clay.

A majority of exfoliated nanocomposite was obtained for both amounts of clay, however, at 5% this was more pronounced. The clay lamellae were dispersed and distributed in the polymer matrix extremely well for 5% clay.

TABLE 17.7
T1H Values of Solution-Cast PC Films and PC–Organoclay

Sample/PC–Organoclay (%)	T_1H (ms)
PC	100
PC-1	105
PC-3	90
PC-5	95

TABLE 17.8
T_1H Evaluations of Films Made of PC and PC-TiO_2

Sample/PC-TiO_2 (%)	T_1H (ms)
PC	100
PC-1	95
PC-3	98
PC-5	90

17.4.2.2 Using Solution Casting to Obtain PC–Organoclay Films

PC films and PC–organoclay were made by mixing the polymer solution with the organoclay dispersion and using chloroform as a solvent. The films were air-dried at room temperature, and the T1H values were used to evaluate the final products. The results are listed in Table 17.7.

The introduction of the organoclay in the polycarbonate through intercalation by solution encouraged the formation of mixed nanomaterials, that is, with an exfoliated part and an intercalated part, since there was no appreciable decrease in the spin-lattice relaxation time of the hydrogen nucleus, except for the proportion of 5%, which presented a higher degree of exfoliation. This can be explained by the fact that when the polymer chains are encircling the clay lamellae, the metals in the clay composition—such as iron, calcium, and magnesium—act as relaxing agents, reducing the relaxation times.

17.4.2.3 Using Solution Casting to Obtain PC–TiO_2 Films

The polymer solution was mixed with the TiO_2 dispersion to make PC and PC–TiO_2 films, with chloroform serving as the solvent. T_1H values were used to evaluate the materials that were produced after the films were allowed to air dry at ambient temperature. Table 17.8 is a list of the outcomes. Due to the titanium dioxide's effective dispersion inside the polymer matrix, the inclusion of this compound led to a significant reduction in the hydrogen nucleus's spin-lattice relaxation period, particularly at the 5% ratio.

17.4.3 Melt Extrusion Method

17.4.3.1 Acquisition of PC–Organoclay Systems Through Extrusion

An extruder with twin screws was used to process the materials. The polymer was mixed with the organoclay (PC–organoclay) at 270°C. T_1H was measured to characterize the relaxometry (Table 17.9).

The hybrid nanomaterials with some exfoliation capabilities were generated via melt intercalation at 270°C in the systems containing 3% and 5% organoclay. This system displayed higher intercalation when compared to the same system generated through the solution.

TABLE 17.9
T_1H Measurements of Melt-Intercalation-Produced PC–Organoclay Nanocomposites

Sample/PC–Organoclay (%)	T_1H (ms)
PC	95
PC-1	95
PC-3	90
PC-5	90

TABLE 17.10
T_1H Measurements of PC-TiO_2-Formed Nanocomposites at 270°C Were Acquired by Melt Intercalation

Sample/PC-TiO_2 (%)	T_1H (ms)
PC	95
PC-1	93
PC-3	95
PC-5	93

17.4.3.2 PC–TiO_2 Systems Obtained by Extrusion

In a twin-screw extruder, the PC–TiO_2 systems were processed at 270°C. The results of the relaxation meter T_1H measurement, which was used to describe the circumstances, are displayed in Table 17.10. The numbers are identical to those for PC since the titanium dioxide addition did not affect how long the polymer matrix took to relax. Because the processing temperature was probably too low to promote uniform dispersion of the particles in the polymer matrix, this behavior could be a sign that they were not.

17.5 METHODS USED IN THE CHARACTERIZATION OF POLYMER NANOCOMPOSITES

In this section, the fewest current and widely accepted techniques for assessing the dispersion of nanoparticles in polymeric materials and their range of properties will be covered.

17.5.1 X-Ray Diffraction

Using Bragg's law, which allows for identifying the crystalline periods of nanoparticles and measuring the space between layers of crystalline structure, the analysis of the diffraction design in broad-angle X-beam diffraction is completed. By varying the layer thickness and these grants to reach the class of nanoparticles foundation, it is possible to adjust the viewpoint, width, and power of the distinctive peak in X-ray diffraction (XRD) spectra.

Bragg's law states that "change in d-spacing causes the variances of related peak continuing to lower or higher point of diffraction." Analyzing the peak situation in the nanocomposites may help determine the level of shedding or intercalation. The estimation of d-space and peaks in intercalation designs with a relative reduction in relative point, climb in d-dispersing, and peeling is unreliable due to scattered structures. The spreading of fillers and non-crystalline particles like silica might be evaluated using this technique [36]. The X-beam optical phenomena were applied to epoxy/MMT composites, and it was assumed that variations in the top to bring down points indicated that clay was gently intercalated throughout the lattice but that the filler wasn't completely peeled as indicated by the pinnacle. This technique is used to evaluate the intercalation of clay into frameworks, but it is now also used to peel, shed, and scatter graphene, graphene oxide nanoplatelets, and cellulose nanocrystals individually [37,38]. Nanoplatelets, cellulose nanocrystals, and graphene and its oxides. It was discovered that there was constant dispersion of particles in the network when XRD was used to examine polypropylene (PP)-fused nanoparticles formed of iron and natural clay (OMMT) [39]. There was intercalation between clay layers in another epoxy sample with nano-clay and unidirectional strands, as evidenced by the results [40]. The XRD results showed that platelet peeling increased with increasing interplanar distance, and the nano clay is dispersed with monomer I-28E along with E-glass strands.

17.5.2 Electron Microscopy

TEM and SEM are two modern procedures utilized for achieving enhanced high resolution of scattered particles in the system. The key difference between the two methods is that the amplified filtering in TEM is obtained by focusing a laser beam on the sample, whereas the amplified filtering in the other method is supplied by electronic collaboration by an electrical signal through an ultra-thin component. Since measurements are essentially indistinguishable from the mean free way of electrons that move across specimens, sample components for TEM should be kept in thin storage. A cryo-microtome is mentioned with thermoplastics as it enables the prevention of plastic degradation in the examples while it is coated. The static charges are added together in SEM to assume that the specimens should be semi-conductive. Ultra-thin coating made by sputtering or dissipation is used for polymers. When TEM and SEM were conducted on the ZnO composite, it was found that the particles were widely dispersed throughout the epoxy network [41]. SEM analysis of the organic component ether of bisphenol, which was not treated with natural MMT nano, revealed that it frames a non-miscible stage in the epoxy lattice [42]. Epoxy reinforced with E-glass filaments provided unequivocal evidence that interfacial attraction had improved. Another investigation focused on polystyrene (PS) gum about copper oxide, and to investigate the smaller-scale basic conduct, SEM was used. The results showed that with an increase in mass percentage, the microstructure improved [43].

17.5.3 Thermogravimetric Analysis (TGA) Analysis

This evaluation uses a trial methodology where the percentage of improvement is added up across tests to help distinguish between concoction-based and post-heating corporeal changes [44]. The TGA's basic design combines a systematic offset with high precision, to which a test pan is attached. A test skillet is typically available in an electronically controlled warmer. There are a few different types of test and balance climates, but they all share a controlled environment that was mostly constructed with nitrogen gas. There are a few techniques that provide various settings, control stress, and test or warmer zones. This approach is very beneficial when examining the processes that control mass catastrophes, such as dynamic processes involving fluids and solids.

Investigations showed that the preparation of epoxy nanocomposites containing graphite particles improved the thermal stability of the composites with increasing amounts of filler, primarily due to the high coefficient of conductivity and generally finer interfacial similarity of the nanocomposites. By using TGA analysis, it was determined that the nanocomposites had a high degree of soundness and changed the temperature in a manner consistent with pure epoxy. Glass fiber-reinforced epoxy and CNT tubes exhibit improved performance when exposed to large temperature changes [44]. A further nanocomposite was created using titanium and alumina oxide particles, and Tg showed considerable improvement [45].

17.5.4 Dynamic Mechanical Analysis

Dynamic mechanical analysis (DMA) discovers the property that controls the viscoelastic behavior of polymers. This process investigates the property of material deformed under intense pressure and discovers modulus and damping. These steps can be used to obtain the presentational data of the material. Using this technology, evaluation of a wide range of materials, such as coatings, various types of polymers, and strands, might be conducted. This system is erogenous for measuring the glass change range, and auxiliary relaxations are immediately observed and troublesome in other ones. DMA is carried out on epoxy-containing nanoscale graphite platelets, and the findings showed developed glass transformation temperature with expanding filler measure. The DMA was presented with MMT and Multiwalled Carbon nanotubes (MWCNT) composites, and it concluded that the use of half and half nanoparticles in epoxy demonstrated an increase in glass transition temperature. Epoxy with TiO_2 or Al_2O_3 revealed a rise in stockpiling modulus and a top-down shift in damping at higher temperatures [44].

17.6 DIFFERENT TYPES OF POLYMER NANOCOMPOSITES

17.6.1 CNT–Polymer Composite

With their unique cylindrical nanostructure and hexagon-like appearance, carbon atoms roll into tubes called CNTs. CNTs, with a length-to-diameter ratio of 132,000,000:1, are among the biggest nanocomposites. The CNT's chemical formula is shown in Figure 17.14, along with information about its spherical properties.

Because of the CNT's special chemical makeup and advantageous mechanical, thermal, and electrical characteristics, it can be used to create reinforcing materials, including multifunctional polymer nanocomposites. Their structural properties allow them to be lighter than aluminum yet stronger than steel, and they have higher electrical conductivity than copper [46].

17.6.2 Magnetic–Polymer

Magnetic nanocomposites are made up of magnetic elements dispersed within a solid matrix. The fabrication of polymers smaller than 30 nm often involves a variety of techniques, including thermal

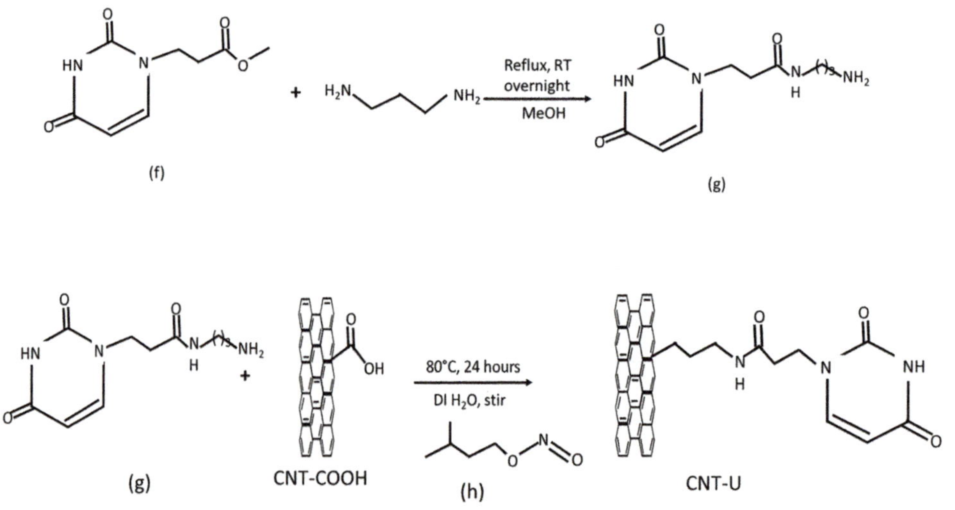

FIGURE 17.14 Chemical formula for the characterization property (tube) and carbon nanotube (CNT).

breakdown, co-precipitation, emulsion processes, liquid laser ablation, hydrothermal, co-evaporation, and co-sputtering.

The CNTs from the instances, which can be created with a wide variety of ferromagnetic nanoparticles, have various structural characteristics. Further significantly, carbon shells protect the ferromagnetic components of the CNT from oxidation and give antiferromagnetic contributions. Furthermore, it has greater thermal stability [47].

17.6.3 Quantum Dot–Polymer Nanocomposites

There are several highly desirable applications for polymers, one of which is wave guiding in conjunction with optical systems. These are especially good for conducting operational wavelengths because it's easy to produce, have low propagation losses, and are nearly 400 nm transparent. The material utilized in the application is nanocrystalline and chemically colloidally bonded to a semiconductor. In semiconductors, the nanocrystals are called quantum dots (QDs) because they are subjected to three-dimensional charge carrier confinement, meaning that their energy levels are discrete and quantized in all spatial directions.

Therefore, using QDs in polymers as the core of wave guiding in photonic devices is a suitable strategy. An appropriate technique for directing wavelengths of various colors is provided by using the proper QD concentration in the matrix (Figure 17.15). The polymers are connected with various kinds of QDs [48].

17.7 PROPERTIES OF NANOCOMPOSITE POLYMERS

In medicine, biology, and pure technology applications, nanoparticles are becoming increasingly important instruments in science and technology. Most molecular or atomic structure materials have a distinctive structure that helps them achieve the characteristics of nanocomposite materials. The polymer matrix's characteristics, however, are very different from those of nanoscale fillers. Nanocomposite materials are more resilient, and they are often reinforced with biodegradable polymeric nanocomposites using particles such as tungsten disulfide, graphene, molybdenum disulfide, and CNTs in medical applications such as the reengineering of bone tissue. Materials made of polymer nanocomposites are more resistant to compressive and flexural characteristics.

Due to their excellent electrical conductivity, nanocomposite materials are used in many different technological applications. A better electrical conductivity polymer nanocomposite has been used to create a multi-walled CNT. Additionally, polymer nanocomposite structural characteristics

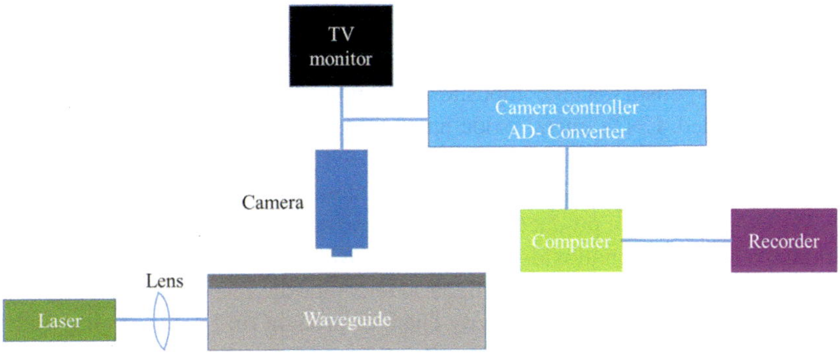

FIGURE 17.15 Setup for the experimental waveguide characterization. The waveguide's structure is displayed in the inset.

are utilized as a material for flame obstruction resistance. When compared to conventional composites, the thermal, mechanical, barrier, and electrical properties of the nanocomposite show observable differences.

17.7.1 Thermal Properties

Depending on the conformation, chemical structures, and bond strength of the polymers, the thermal expansion property of a polymeric material is its primary physical characteristic when it is exposed to a temperature change. Differential scanning calorimetry (DSC) can be used to examine a nanocomposite's thermal characteristics. The thermal stability can be determined from the weight loss that occurs when heating the nanocomposites. The Heat deflection temperature (HDT) can be used to gauge a nanocomposite's heat resistance under external loading. Several academics have looked into how HDT is affected by the amount of clay. Numerous products, including printed circuit boards, thermal interface materials, heat sinks, connectors, and high-performance thermal management systems, use nanocomposite materials with good thermal conductivity [49].

17.7.2 Mechanical Properties

The surface morphology of nanocomposites and the material used to make them have an impact on their mechanical characteristics, such as tensile strength, elongation, and modulus. Because of the high stiffness and high aspect ratio of the nanofillers, as well as their good affinity with the polymer, the mechanical properties of polymer nanocomposite have improved. By reinforcing a polymer matrix, nanofillers are employed to increase mechanical qualities like strength and stiffness [49,50]. Various reinforcement effects were seen in the case of fumed silica, contingent upon the dispersion state, surface area, polydispersity, and organo-modification of the nanoparticles, potentially resulting in their grafting to the matrix [50]. It has been demonstrated that evenly distributed and aligned clay platelets work wonders for increasing the stiffness of polymer matrix materials. When the filler is an organoclay vs. glass fibers, the increase in Young's modulus, E, of injection-molded composites based on nylon6 is compared to the modulus of the neat polyamide matrix, Em [49]. To achieve a two-fold increase in modulus compared to neat nylon6, around three times more mass of glass fibers is needed in this case than with MMT platelets. Consequently, the nanocomposite outperforms the traditional glass fiber composite in terms of being lightweight.

17.7.3 Electrical Properties

Electrical and electronic applications can benefit from the conductive qualities of polymer nanocomposite. Not to be overlooked is the interest in the electrical conductivity of CNTs in insulating polymers. Electromechanical actuators, supercapacitors, transparent conductive coatings, and a variety of electrode applications are among the possible uses [51,52]. For sensor applications such as chemical, bio, and gas sensors, nanocomposite-based polymers with different nanoscale filler inclusions have been studied. CNTs, metal oxide nanowires, and tiny particles of palladium, platinum, copper, nickel, and gold are among the nanofillers used. Still, their electrical characteristics and potential range of use remain the same. For instance, most metal-polymer nanocomposites utilizing silver or gold nanoparticles are produced through a deposition technique. Exposure to certain gases, such as ammonia and nitrogen dioxide, has been discovered to drastically alter the electrical resistance of CNTs. Like single-wall CNT alone, a nanocomposite of polypyrrole and CNT produced gas sensor sensitivity. Based on conductivity changes brought on by gas or chemical interactions between the conjugated polymer and the nanofiller, these nanocomposites can be sensing agents [53]. The electronic transport properties of individual polymer tubes/fibers must be understood to realize the potential applications of conducting polymer (CP) nanotubes and fibers. The simplest and most common method for measuring bulk resistance across the filled membrane using a two-probe method for template-prepared nanofibers is to leave the synthesized polymer fibers inside the template membrane's pores. One can estimate a single fiber's conductivity by measuring its membrane resistance. The conductivity of 30-nm polypyrrole and poly (3-methyl thiophene) fibers was found to be significantly higher at room temperature than in traditional forms of the equivalent polymer.

17.7.4 Property of Absorbing Microwaves

Owing to their facile processing and decreased density, CPs have been investigated as novel microwave-absorbing materials. Traditional films or pellets of doped polypyrrole and polyaniline often show an electrical loss in the microwave frequency range of $f=1$–18 GHz [54]. This could result from a stronger chain ordering that the tubular form causes. Research revealed that doped polyaniline with a fiber-like morphology possesses superior electromagnetic wave absorption properties compared to polyaniline with a particle-like morphology [55].

17.7.5 Barrier Properties

Due to their high aspect ratio and ability to create a winding channel that slows the passage of gas molecules through the matrix resin, nanocomposites have excellent gas barrier properties. The filler's presence inside the nanocomposite structure creates a difficult path for diffusing penetrants to travel. Due to the longer diffusive path that the penetrants must take in the presence of filler, the permeability is decreased. The polyimide nanocomposite exhibits barrier properties against tiny gases such as oxygen, carbon dioxide, helium, nitrogen, and ethyl acetate vapors. This polyimide nanocomposite contains a small fraction of layered silicate. Conventional packaging materials include polymers, paper and cardboard, glass, ceramics, and metallic elements. These days, organic polymers account for around 40% of the market share in the food packaging industry and are the most widely used materials because of their light weight, low cost, ease of processing, and variety of chemical and physical qualities [56]. The most common polymers used in food packaging include PS, polyvinyl chloride (PVC), PP, polyethylene terephthalate, and several kinds of polyethylene (LDPE, HDPE, etc.). The primary drawback of polymeric materials, despite their numerous advantages over conventional materials and their revolutionary impact on the packaging industry, is their intrinsic susceptibility to gases and other tiny molecules.

17.8 POLYMER MODIFICATION TECHNIQUES

A polymer's properties can be changed using the procedures of blending, grafting, curing, and functionalization. To achieve the necessary properties, two (or more) polymers are physically mixed in a process known as blending. The monomers are covalently coupled and crosslinked onto the polymeric chains during graft copolymerization. In contrast, during curing, an oligomer mixture is dispersed to produce a coating that forms a mechanically stressed connection with the substrate. Grafting can take minutes or even days to accomplish, but curing happens very quickly and only needs a small amount of time. In derivatization, simple molecules take the place of the polymer's aggregates.

Modifying polymers to create polymer nanoparticles is another method for improving the characteristics of polymers. Top-down and bottom-up approaches are frequently employed in the production of polymer nanoparticles. The top-down strategy of dispersing prefabricated polymers results in polymeric nanoparticles, whereas the bottom-up strategy of polymerizing subunits results in polymeric nanoparticles. Soft and simple to work with are polymer-modified nanoparticles. Surface roughness can be produced by internalizing metal and other inorganic nanoparticles into the modified polymer. Biological outcomes can be enhanced by particle surface roughness and permeability. The use of polymeric nanoparticles as biomaterials has significantly increased in recent years due to their advantageous traits, including ease of fabrication and configuration, cytocompatibility, a wide structural variation, and impressive bio-imitative capabilities.

The application of organic coatings is a common method for surface modification of inorganic fillers. This can be achieved by taking advantage of the physical and chemical interactions that exist between the fillers and the modifiers. Often, a combination of secondary and chemical linkages results in the ensuing bonding. For micro composites, a great deal of knowledge has been gained regarding processing methods, interfacial characterization, and the impact of interfacial properties on composites. It is well known that, aside from the fact that the modifiers have a very difficult time penetrating the tightly bound nanoparticle agglomerates, there is no appreciable difference between the processes of microparticles and nanoparticles, especially when it comes to physical processes like surfactant treatment and polymer encapsulation.

17.8.1 Physical Methods

Secondary forces such as van der Waals, hydrogen, and electrostatic forces are usually produced between the particles and the modifier through physical treatment. A surfactant has a long aliphatic chain and one or more polar groups. The preferred electrostatic adsorption of the surfactant's polar group onto the fillers' high-energy surfaces is the foundation of surfactant treatment. Certain conditions can result in the formation of ionic bonding. One such illustration is $CaCO_3$ which has been treated with stearic acid [57]. It has been demonstrated that the molecules of the surfactant and the filler surface formed ionic connections. Its decreased surface energy was the result of the shield made of C18 alkyl chains located on the filler surface.

Another type of efficient surface modifier for calcium carbonate fillers is alkyl dihydrogen phosphate, which contains functional groups including olefin, chloromethacryloxy, and mercapto. It has been shown that the reaction of alkyl dihydrogen phosphate molecules with $CaCO_3$ results in the formation of a dibasic calcium salt of phosphate. The mineral oil's outstanding dispersibility indicates that the products get settled on $CaCO_3$ and provide it with a degree of hydrophobicity [58]. Additionally, the surface area of the treated particles always increases as a result of treatment with alkyl dihydrogen phosphates. This can be attributed to calcium salt production or surface erosion. Another physical treatment technique involves encasing inorganic small particles in polymers that are prepared or created in situ. A macromolecular dispersant called a hyperdispersant has been employed successfully in the first instance. Polymeric dispersants have two main components, like surfactants. One type of functional group that helps anchor the dispersants to the particle surface through hydroxyl and electrostatic bonding includes alcohol, amine group, carboxylic group, NR_3z, COO^{2-}, SO_3H, PO_4, and SO_3^{2-}. The other is a solvable macromolecular chain, which can be distributed in a variety of low- to high-polarity media and includes polyolefin, polyester, polyacrylate, and polyether. The following are advantages that hyper dispersants offer over conventional surfactants.

i. They are more securely fixed to the surface of the particles than surfactants, making it difficult for them to remove.
ii. Extended polymeric chains have a greater ability to stop the particles from re-agglomerating.

It is worth noting, nevertheless, that hyperdispersants can only encapsulate nanoparticle agglomerates and hardly ever diffuse into the agglomerates because of their long molecule structures. In comparison, encapsulation using polymers synthesized in situ might result in more equal coverage. When nanoparticles are present, the polymer can be produced using emulsion polymerization or solution polymerization. Vinyl monomers were radical polymerized to create core-shell SiO_2 particles Owing to the negative charges on the silica colloid particle surfaces on an electric double layer, the silica surface can become concentrated for the initiator 2, 29-azobis (2-diamino propane) dihydrochloride, which can then form free radicals. Using a solvent that is less soluble in the polymer facilitated the adhesion of the macromolecular chains to the particles. Applying an emulsion polymerization technique is another method for encasing tiny inorganic particles in a polymer film. Due to the monomer's surface adsorption and subsequent polymerization in the adsorbed layer, polymerization happens largely at the surface of unaltered particles. Using amphiphilic block copolymers, non-ionic surfactants, etc. to pretreat the particles, might cause the production of an initial hydrophobic layer, which is essential for the formation of the polymer shell.

17.8.2 Chemical Method

The covalent attachment modifier has a greater impact on surface modification based on chemical contact than on surface modification based on physical interaction because it keeps the latter from desorbing from the particle surface. Furthermore, the interfacial contact in composites containing chemically pre-treated nanoparticles can be adjusted by adding modifiers with reactive groups. This facilitates the fillers' chemical bonding with the matrix polymer. The most popular and simple approach among the available techniques is coupling agent treatment. A variety of coupling agents, such as silane, titanate, and zirconate, have been employed to improve the binding among organic matrices and inorganic fillers. Fewer details are known about nanoparticles than concerning reaction mechanisms and coupling agent effects in the case of microfillers. The characteristics of fumed silica covered with various silane types were explored by Gunko et al. [59] from both experimental and theoretical perspectives. The connected silanes can reduce silica's hydrophilic characteristics to various degrees, depending on the pre-treatment circumstances. However, it should be noted that using straightforward physical and mechanical methods, it is difficult to achieve the uniform surface covering of the coupling agent on nanoparticles. There are certain benefits of grafting macromolecules onto inorganic particles as opposed to linking or low molecular surfactant modifications. The graft conditions and graft monomer species can be carefully selected to yield polymer-grafted particles with the desired properties. This will make it possible to precisely adjust the interface characteristics between the matrix and the treated nanoparticles to create nanocomposites. A nanocomposite microstructure made up of the nanoparticles, the grafted and ungrafted (homopolymerized) polymer, and the nanoparticles themselves make the previously weak nanoparticle agglomerates stronger. In this area of expanding interest, numerous studies have been conducted to enhance the nanoparticles' compatibility with polymers and their ability to disperse in solvents.

The two main methods for graft polymerization are as follows:

i. "Grafting from" is the process by which active substances (initiators or co-monomers) are polymerized from monomers.
ii. Interaction between the functional groups of the particles and pre-made polymers with reactive terminal groups results in grafting.

The latter is advantageous for managing the grafting polymer's molecular weight (MW). However, in general, the former method is superior to the latter since polymers have a harder time penetrating particle agglomeration than monomers do. Additionally, the "grafting to" technique would leave the particles coated in polymer, which might inhibit the attachment of additional polymers. Two methods for the "grafting from" approach start the polymerization from the particles. One is a surface modification using silane treatment to add polymerizable groups, for example (Figure 17.16). Surface modification using starting groups is the alternative (Figure 17.17). Comparatively, the former is simpler, whereas the latter favors grafting efficiency and a larger grafting percentage. First, methacryloxypropyltrimethoxysilane, a silane coupling agent with a double bond, was used by Espiard and Guyot [60] to modify the surface of silica nanoparticles. After that, polymethyl acrylate was grafted onto the modified silica in an emulsion solution. After immobilizing double bonds on the silica surface with t-butyl lithium, PS and poly(styrene block-isoprene) were grafted upon nano silica (12 nm) by adding monomers to the double bond suspension of the reaction. To graft nanosized alumina and silicon carbide (SC) with PS and polyacrylamide, Rong et al. [61,62] used a similar technique in solution. Polyglycidylmethacrylate (PGMA) was used to graft the identical SiC nanoparticles into a surfactant-free emulsion. Because SiC nanoparticles are more numerous

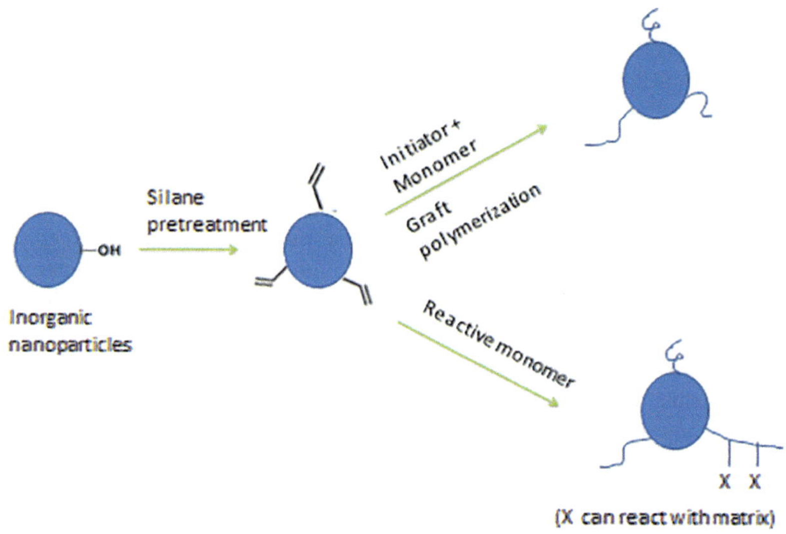

FIGURE 17.16 Diagram showing the polymer that has been grafted from nanoparticles using double bonds on the particles.

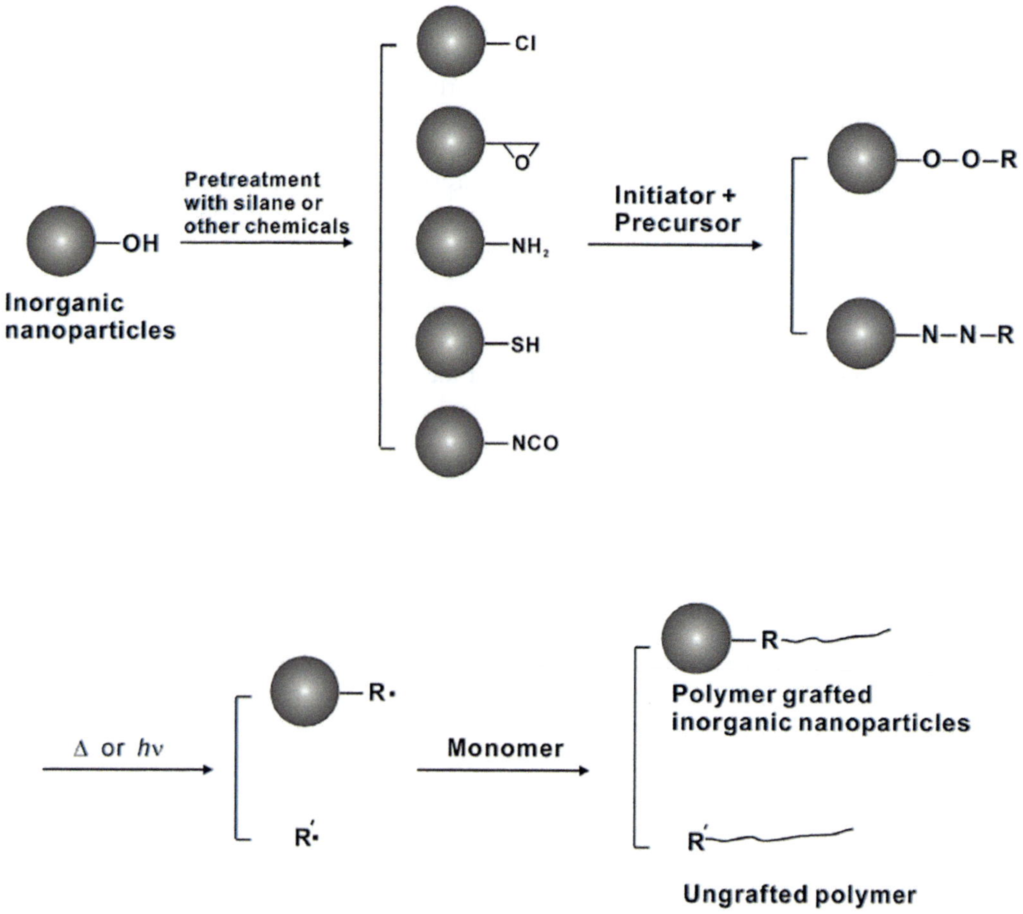

FIGURE 17.17 Utilizing initiators on the particles, the polymer was grafted from nanoparticles.

and have a hydrophobic surface, it has been determined that the loci for emulsion polymerization are on their surface. This reaction process is critical for effective PGMA encapsulation and graft surrounding the nanoparticles. Surface-initiated polymerization can be caused by several mechanisms, such as cationic, free radical, ring-opening metathesis polymerization (the ROMP method), atom transfer radical polymerization (the ATRP method), and anionic polymerization. Using the initiating groups that were previously applied to the surfaces of carbon black and ultrafine silica, Tsubokawa and associates [63,64] methodically investigated the polymerization. Using radical, cationic, and anionic grafts, as well as acylium or benzylic perchlorate, peroxy ester, or azo compounds that already had been added to ultrafine silica, they have carried out polymerizations of a number of monomers. For instance, silica particles were treated with thionyl chloride to produce surface Si-Cl groups, which were then used to react with test-butyl hydroperoxide or di-isopropyl benzene hydroperoxide to create surface-anchored peroxide initiators. There are typically two techniques to introduce the azo groups onto the surface of an ultrafine particle. One occurs when 4,4′- azobis (4-cyano pentanoic acid) (ACPA) reacts with surface epoxide groups that are added when the particle is treated with c-glycidoxypropyltrimethoxysilane. The second involves ACPA's reactivity with surface isocyanate groups that are created when the particle is treated with toluene di-isocyanate (TDI) [65].

Azole initiator may be further immobilized on silica substrates by means of an amide bond formed on the silica between ACPA and a silane coupling agent that contains amine. In addition to chemical grafting, irradiation can also be used to achieve graft polymerization onto nanoparticles. This method is preferable to chemical grafting in terms of graft uniformity, pollution control, and procedure simplification. The high-energy radiation stimulates the nanoparticle surfaces within and outside of the agglomerates in the same way. The relatively low MW monomers are thereby allowed to interact with the active sites of the nanoparticles throughout the agglomerates [66].

Recent research has shown that using this technique, different polymers may be progressively grafted onto the surface of silica nanoparticles in relatively benign conditions (in the air). The nanoparticles' highly active surface may be the essential factor. Additionally, the reactivity of various silica varies. Performance-wise, precipitated silica outperforms fumed silica. The fact that some metal nanoparticles, such as those of iron, nickel, and cobalt, can graft polymers with radiation despite having few hydroxyl groups is another intriguing discovery. This suggests that the reactive site could be either the hydroxyl group or a surface structural flaw. The higher grafting % is facilitated by the silane's insertion of a double bond to the particle surface [67].

17.9 INTERFACIAL INTERACTION'S IMPACT ON POLYMER NANOCOMPOSITES' PERFORMANCE

Composition, structure, interfacial interactions, and component characteristics all have a role in determining the properties of composite materials. The volume and qualities of interphase determine its impact. In other words, the ultimate characteristics of the composite are determined by the quantity of material bound in the interphase and the strength of the contact. Progress in the creation of NPCs would arise from realizing the importance of the fundamental factors governing interfacial adhesion and appropriate surface modification. Creating a wide contact between the polymer matrix and the nanoscale constituents is the fundamental idea behind nanocomposites. However, a large surface cannot always be formed because homogenous dispersion of the nanoparticles is frequently problematic. Additional issues arise from adjusting the components' coupling or interfacial interaction. The issues are crucial to this field's future advancement and the concerns that need to be resolved in the future are described in the paragraphs that follow by analyzing a few carefully chosen references.

Chan et al. [68] introduced nano-$CaCO_3$ (44 nm) to PP via melt mixing. They studied the effects of coupling agents, surfactants, and dispersants. The PP matrix's elastic modulus and impact strength increased because of the nanoparticles. It was demonstrated that the composites' tensile strength and yield stress had the opposite effect. These two amounts declined as the volume fraction of $CaCO_3$ rose and were maximum for the neat PP. But according to a recent study conducted by a different team, adding 0–15 phr nano-$CaCO_3$ particles to PVC simultaneously increased the material's yield strength, elongation at break, impact capacity, flexural modulus, and Vicat softening temperature. Because of its non-polar nature, it is assumed that PP will interact with nanoparticles extremely poorly, whereas PVC composites should exhibit substantially stronger interfacial interactions. A significant factor in the effectiveness of composites is the interaction between the particles and the polymer matrix. Enhancing interfacial contact in nanocomposites is probably achieved by surface-modifying fillers with coupling agents or surfactants, a technique that is frequently employed in the production of micro composites. However, there aren't any universal fixes. Each set of nanocomposites requires its unique approach, which must be created. PP was combined with $CaCO_3$ (50nm) using a twin-screw extruder by Zhang and Hua [69]. A titanate coupling agent that is water soluble was used to pretreat the nanoparticles. In comparison to nano-$CaCO_3$ treated with fatty acid salts, the particles treated with the coupling agent exhibit better dispersion in PP. The impact strength of PP improved by 70% with adding 5–7 phr of coupling agent-pretreated nano-$CaCO_3$; however, the fatty acid salt-treated nano-$CaCO_3$ composites only showed a measured 50% increase. Additionally, nano-$CaCO_3$ particles were altered using homemade phosphonate surfactants. Like this, phosphonate-treated nano-$CaCO_3$/PP composites showed an improvement in Charpy impact strength. A phosphonate coupling agent's effect was assessed in a study by Ren and colleagues [70]. While untreated nano-$CaCO_3$ had essentially no reinforcing effect, treated nano-$CaCO_3$ could boost tensile and impact strength.

The discussion suggests that several coupling agents, including titanate, phosphate, and aluminate, enhance the mechanical characteristics and nanoparticle dispersion of polymer composites. Silane is frequently employed in the treatment of silica but is typically not appropriate for $CaCO_3$ surface modification. Studies on the impact of commercial dispersant (Efka) and silane coupling agent (thiocyanatopropyl-triethoxysilane) pre-treatment of nano silica on tensile characteristics of PP were reported by Kruenate et al. [71]. When PP and untreated nano silica were combined, the tensile strength and modulus increased while the elongation to break dropped.

In contrast, silane treatment is more successful in enhancing the modulus and tensile strength of PP/SiO_2 composites (Table 17.11). Every commercial nano-$CaCO_3$ particle has been pre-treated with stearic acid. Stearic acid treatment was used to solidify $CaCO_3$/high-density polyethylene (HDPE) composites made using a twin-screw mixer-single screw extruder in a work by Lazzeria et al. [72]. The impact strength of the composites was dramatically reduced when HDPE and 10% nano-$CaCO_3$ were mixed, while their yield stress and Young's modulus increased. While the impact strength gradually rose, the use of stearic acid resulted in a modest reduction in the composites' Young's modulus and yield stress when compared to the uncoated $CaCO_3$ composites. For the previously described surface pre-treatment of nanoparticles, a higher-speed mixer is often used to simply mix the particles with a dispersant or a coupling agent, except for the wet treatment approach

TABLE 17.11

Mechanical Characteristics of Nano Silica-Treated and Untreated Polypropylene Composites [71]

Material	Elongation to Break (%)	Tensile Strength (MPa)	Modulus (MPs)
PP	21.12±2.43	36.96±0.40	1,178.29±42.18
PP/2%SiO_2	13.19±1.11	40.21±2.16	1,376.46±63.79
PP/2%SiO_2/Efka	9.14±2.38	38.19±1.44	1,414.67±10.40
PP/10%SiO_2/Efka	10.12±1.05	33.60±0.48	1,603.31±58.43
PP/2%SiO_2/Silane	11.80±1.31	43.54±1.85	1,524.20±45.57

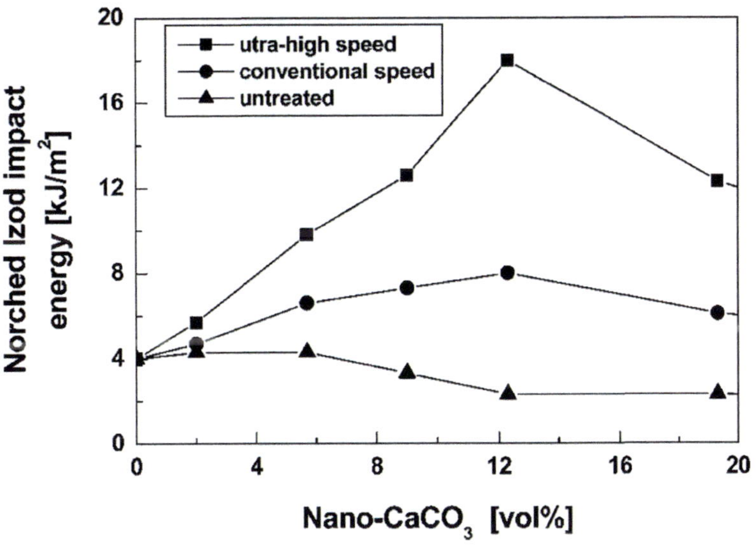

FIGURE 17.18 Effect of filler content on the energy of nano-$CaCO_3$/PP composites treated with stearic acid [70].

(in water). However, achieving a consistent surface coverage of the particles is challenging due to the incredibly low weight ratio of the coupling agent to the nanoparticle agglomerates.

An ultrahigh-speed mixer has been created to address this issue. The mixer features a high speed (6,000 revs per minute), specific rotors made to handle nanoparticles, and an atomizer to add the coupling agent. It was demonstrated that the mechanical attributes of PP, especially its ductility, could be greatly improved by adding nano-$CaCO_3$ enhanced by stearic acid using an ultrahigh-speed mixer (Figure 17.18).

Before being combined with a polymer, nanoparticles were typically treated with coupling agents or dispersants. Since only the external nanoparticles of the agglomerates can engage in the reaction, splitting the treated nanoparticle agglomerates during the following compounding with polymers would reveal and re-agglomerate the unreacted interior nanoparticles. Thus, it is crucial to stabilize the divided nanoparticles. HDPE was introduced to a two-roll mill that had been heated to 150°C initially. Both $CaCO_3$ (40 nm) and stearic acid were added to the fully melted polymer at the required composition. Stearic acid was anticipated to be able to situ activate the dispersed nanoparticles in the polymer melt, preventing the formation of bigger nanoparticle agglomerates. The viability of their concept was demonstrated by microstructural analysis of the composites' broken surfaces. In other words, when the appropriate quantity of the dispersion agent was used, no nanoparticle agglomerates appeared. In this way, composites containing 20% by weight of nano-$CaCO_3$ showed 100% improvements in both impact strength and extension to break HDPE.

17.10 POLYMER MODIFICATION ON NANOMATERIALS

Nanomaterials have been modified using polymers using a variety of techniques. Polymeric grafting and surfactant-assisted modification are the most popular and effective techniques for functionalizing nanomaterials. There are two techniques to graft polymers for nanomaterial modification: either end-functionalized polymers actively react with the appropriate nanoparticles or polymer chains are developed from an initiator-terminated, self-assembling monolayer. By employing polymer-activated domains to induce graft polymerization, it is possible to generate polymer-modified nanomaterials with high grafting percentage yields. A research group created and altered zinc oxide nanoparticles using methyl acrylate acetate and then used polymer grafting on ZnO nanoparticles to create ZnO/MMA nanocomposite films. In situ, mass copolymerization was used to modify undecenyl phosphonic acid-modified TiO_2 nanoparticles and create TiO_2/Poly(methyl methacrylate) (PMMA) nanocomposite.

Nanomaterials are frequently modified using surfactants or macromolecules. The polar molecules of surfactants are drawn to the surface of nanoparticles by electrostatic attraction. Because the macromolecules reduce the intermolecular interactions between the nanomaterials and polymers, which lessen inter-particle contact and prevent aggregation, surfactant-modified nanoparticles can be used as a nanofiller for polymer matrices. SiO_2 nanoparticles were altered by oleic acid and used as a nanofiller for the polyamide matrix.

17.11 ADVANTAGES OF ADVANCED POLYMER NANOCOMPOSITES

Polymer nanocomposite materials provide a variety of advantages and features that make them particularly useful in various applications. Because of their high strength-to-length ratio and high resistance, these materials are used in the reinforcing of structural components. Due to their structural qualities, they are extremely useful in medical applications where they are utilized to reinforce biodegradable polymeric nanocomposites during the reengineering of bone tissue. Materials made of NPCs are more resistant to compressive and flexural characteristics [73]. High thermal resistance and electrical conductivity of polymer nanocomposite materials are further benefits. Unlike other materials, the nanoparticles can sustain higher heat conductivity without melting. Its thermal conductivity also offers resistance against flame or fire due to its ability to withstand high temperatures without breaking the bonds separating their particles [73].

- Polymeric nanomaterials, a novel kind of material that performs better than its microparticle equivalents, are composed of inorganic nanoparticles and organic polymers. Good mechanical strength, strong electrical conductivity, and good photo physical properties are only a few of the distinctive traits that polymer-modified nanomaterials display.
- The mechanical properties of polymer-modified nanoparticles are crucial for complex industrial and technical uses. Researchers created polyurethane nanostructures with elastomeric copolymer grafts that were strengthened by SiO_2 nanoparticles, and they then looked at how their mechanical characteristics varied with graft depth and copolymer molar mass in the SiO_2 nanoparticles. The elasticity, fracture toughness, and crack propagation resistance of the polyurethane matrix were all enhanced by the incorporation of copolymer-grafted SiO_2 nanoparticles.

17.12 APPLICATIONS OF POLYMERIC NANOCOMPOSITES

Due to the advantages of polymer nanoparticles over traditional materials, their use in food packaging has increased dramatically in recent years. The primary purposes of polymer-modified nanomaterials are to enhance gas and ultraviolet barrier capabilities, as well as resistance, toughness, durability, and stiffness, among other attributes.

Polymer-modified nanoparticles are used in the electronics sector as dielectrics, insulators, conducting components, and covering materials (Figure 17.19). Using polymer-grafted TiO_2 nanoparticles, Scientists. examined the electrical performance of Polylelemental (PE) nanomaterials for power line insulation. The researchers found that polymer-modified TiO_2 improved the PMMA matrix's space charge density and stability to dielectric breakdown.

Hydrogels, coatings, and synthetic inorganic or organic polymeric nanoparticles have all shown promise for use in the environmental and public health domains. Additionally, polymeric nanoparticles

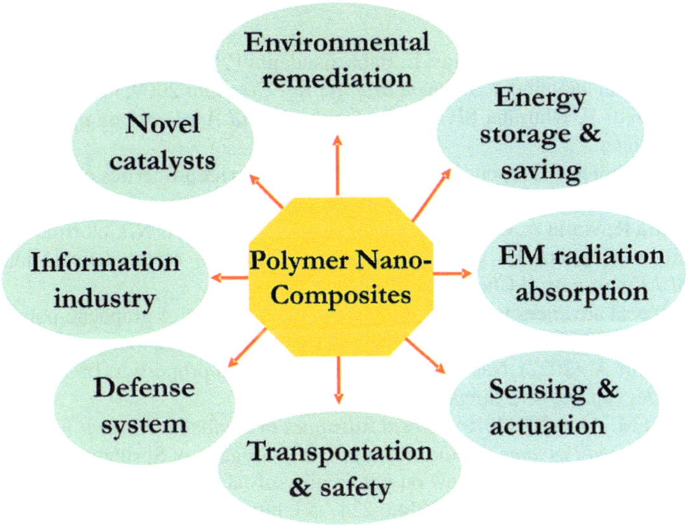

FIGURE 17.19 Applications of NPCs in various fields.

can be viewed as attractive options for vaccine administration, cancer treatment, and targeted antibiotic delivery depending on the polymer utilized and their capacity to control drug release [74].

17.13 CONCLUSION

In the past two decades, polymer nanocomposites have gained importance in the field of material science research. Although this innovative method is still in its early stages of research, it has a bright future in a variety of applications. NPCs offer a lot of potential, which has made them a popular research topic in recent years. How well we can overcome the obstacles will determine how nanocomposites develop and are used. This chapter describes the polymer nanocomposites' ultimate characteristics, filler treatment, and component species. It is clear that coupling agents, such as surfactants and dispersants, frequently work to enhance the mechanical performance of NPCs but rarely make a significant difference in their overall properties. This makes sense given the difficulty in achieving uniform coupling agent coverage around nanoparticles. Due to their stronger interface entanglements, macromolecules that encase nanoparticles perform better than tiny molecules in this comparison. Weak nanoparticle agglomerates can be strengthened by chemically initiating graft polymerization from the outermost layer of the nanoparticles. Interfacial adhesion would be further aided by reactive compatibility that exists between the grafted nanoparticles and the matrix polymers, especially if reactive groups were added to the grafting chains. Even with a range of treatments and modification techniques, agglomeration of nanoparticles can usually be avoided when forming nanocomposites using traditional compounding facilities and procedures. Currently, every effort must be made to create a maximum reduction in quantity. It is significant to highlight that, in addition to interfacial qualities, the mechanical properties of polymer nanocomposites depend on a number of other parameters, such as crystallinity, the amorphousness of the polymer matrix, and multiple morphologies. When all options are carefully considered, the beneficial effects of nanoparticles in the polymer matrix can be fully realized (in reality, many of the previously extensively discussed methods for changing nanoparticles have not yet been used in the production of nanocomposites). As a result, the properties would be far higher than previously reported.

REFERENCES

1. Shabani I, Haddadi-Asl V, Seyedjafari E, Babaeijandaghi F, Soleimani M. Improved infiltration of stem cells on electrospun nanofibers. *Biochem Biophys Res Commun* 2009;382:129–33.
2. Thakur A, Thakur P, Khurana SP. (2022). *Synthesis and Applications of Nanoparticles* (pp. 1–17). Singapore: Springer.
3. Frounchi M, Dadbin S, Salehpour Z, Noferesti M. Gas barrier properties of PP/EPDM blend nanocomposites. *J Membr Sci* 2006;282:142–8.
4. Singh R, Mehra R, Walia A, Gupta S, Chawla P, Kumar H, Kumar N. Colorimetric sensing approaches based on silver nanoparticles aggregation for determination of toxic metal ions in water sample: a review. *Int J Environ Anal Chem* 2023;103(6):1361–1376.
5. Galpaya D. Recent advances in fabrication and characterization of graphene-polymer nanocomposites. *Graphene* 2012;01:30–49. https://doi.org/10.4236/graphene.2012.12005.
6. Thakur P, Thakur A. (2022). Nanomaterials, their types and properties. In *Synthesis and Applications of Nanoparticles* (pp. 19–44). Singapore: Springer Nature.
7. Saba N, Jawaid M, Asim M. (2016). *Recent Advances in Nanoclay/Natural Fibers Hybrid Composites, Nanoclay Reinforced Polymer Composites* (pp. 1–28). Singapore: Springer Nature.
8. Saba N, Tahir P, Jawaid M. A review on potentiality of nano filler/natural fiber filled polymer hybrid composites. *Polymers (Basel, Switz)* 2014;6:2247–73. https://doi.org/10.3390/polym6082247.
9. Babaei I, Madanipour M, Farsi M, Farajpoor A. Physical and mechanical properties of foamed HDPE/wheat straw flour/nanoclay hybrid composite. *Compos Part B Eng* 2014;56:163–70. https://doi.org/10.1016/j.compositesb.2013.08.039.
10. Kumari S, Dhanda N, Thakur A, Gupta V, Singh S, Kumar R, Thakur P. Nano Ca–Mg–Zn ferrites as tuneable photocatalyst for UV light-induced degradation of rhodamine B dye and antimicrobial behavior for water purification. *Ceram Intern* 2023;49(8):12469–80.
11. Saba N, Jawaid M, Hakeem KR, et al. Potential of bioenergy production from industrial kenaf (Hibiscus cannabinus L.) based on Malaysian perspective. *Renew Sust Energ Rev* 2015;42:446–59.
12. Saba N, Paridah MT, Jawaid M. Mechanical properties of Kenaf fibre reinforced polymer composite: a review. *Constr Build Mater* 2015;76:87–96.
13. Raghavan P, Lim D-H, Ahn J-H, Nah C, Sherrington DC, Ryu H-S, et al. Electrospun polymer nanofibers: the booming cutting edge technology. *React Funct Polym* 2012;621(72):915–30. https://doi.org/10.1016/j.reactfunctpolym.2012.08.018.
14. Dastjerdi R, Montazer M. A review on the application of inorganic nano-structured materials in the modification of textiles: focus on anti-microbial properties. *Colloids Surf B: Biointerfaces* 2010;79:5–18. https://doi.org/10.1016/j.colsurfb.2010.03.029.
15. Dey A, Bajpai OP, Sikder AK, Chattopadhyay S, Khan MAS. Recent advances in CNT/graphene based thermoelectric polymer nanocomposite: a proficient move towards waste energy harvesting. *Renew Sust Energ Rev* 2016;53:653–71. https://doi.org/10.1016/j.rser.2015.45909.004.
16. Shalwan A, Yousif BF. Influence of date palm fibre and graphite filler on mechanical and wear characteristics of epoxy composites. *Mater Des* 2014;59:264–73. https://doi.org/10.1016/j.matdes.2014.02.066.
17. Jawahar P, Gnanamoorthy R, Balasubramanian M. Tribological behaviour of clay—thermoset polyester nanocomposites. *Wear* 2006;261:835–40.
18. Dhakal HN, Zhang ZY, Richardson MOW. Nanoindentation behaviour of layered silicate reinforced unsaturated polyester nanocomposites. *Polym Test* 2006;25:846–52.
19. Chan M-L, Lau K-T, Wong T-T, Ho M-P, Hui D. Mechanism of reinforcement in a nanoclay/polymer composite. *Compos Part B* 2011;42:1708–12.
20. Chieruzzi M, Miliozzi A, Kenny JM. Effects of the nanoparticles on the thermal expansion and mechanical properties of unsaturated polyester/clay nanocomposites. *Compos Part A* 2013;45:44–8.
21. Meguid SA, Sun Y. On the tensile and shear strength of nano-reinforced composite interfaces. *Mater Des* 2004;25:289–96.
22. Chahar D, Kumar D, Thakur P, Thakur A. Visible light induced photocatalytic degradation of methylene blue dye by using Mg doped Co-Zn nanoferrites. *Mater Res Bull* 2023;162:112205.
23. Subramaniyan AK, Sun CT. Enhancing compressive strength of unidirectional polymeric composites using nanoclay. *Compos Part A* 2006;37:2257–68.
24. Bozkurt E, Kaya M, Tanoğlu M. Mechanical and thermal behavior of non-crimp glass fiber reinforced layered clay/epoxy nanocomposites. *Compos Sci Technol* 2007;67:3394–403.
25. Yahya K, Hamdi W, Hamdi N. (2022). *Organoclay Nano-Adsorbent: Preparation, Characterization and Applications*. London: Intechopen. https://doi.org/10.5772/intechopen.105903.

26. Thakur P, Chahar D, Thakur A. Visible light assisted photocatalytic degradation of methylene blue dye using Ni doped Co-Zn nanoferrites. *Adv Nano Res* 2022;12(4):415.
27. Valentim ACS, Tavares MIB, Da Silva EO. The effect of the Nb_2O_5 dispersion on ethylene vinyl acetate to obtain ethylene vinyl acetate/Nb_2O_5 nanostructured materials. *J Nanosci Nanotechnol* 2013;13(6):4427–32. https://doi.org/10.1166/jnn.2013.7162
28. Soares IL, Chimanowsky JP, Luetkmeyer L, da Silva EO, Souza, DHS, Tavares, MIB. Evaluation of the influence of modified TiO_2 particles on polypropylene composites. *J Nanosci Nanotechnol* 2015;15(8):5723–32. https://doi.org/10.1166/jnn.2015.10041
29. Antonio PCBC, Maria Inês Bruno T, Emerson Oliveira S, Soraia Z. The effect of montmorillonite clay on the crystallinity of poly(vinyl alcohol) nanocomposites obtained by solution intercalation and in situ polymerization. *J Nanosci Nanotechnol* 2015;15(4):2814–20. https://doi.org/10.1166/jnn.2015.9233
30. Tavares MR, de Menezes LR, do Nascimento DF, Souza, DHS, Reynaud, F, Marques MFV, Tavares MIB. Polymeric nanoparticles assembled with microfluidics for drug delivery across the blood-brain barrier. *Eur Phys J Spec Top* 2016;225(4):779–95. https://doi.org/10.1140/epjst/e2015-50266-2
31. Chahar D, Taneja S, Bisht S, Kesarwani S, Thakur P, Thakur A, Sharma PB. Photocatalytic activity of cobalt substituted zinc ferrite for the degradation of methylene blue dye under visible light irradiation. *J Alloys Compd* 2021;851:156878.
32. da Silva Paulo SRC, Tavares, MIB. *Intercalaçäo por Solugäo de Poliestireno de.* 2013;23(5):644–8. https://doi.org/10.4322/polimeros.2013.047.
33. dos Santos FA, Tavares, MIB. Development of biopolymer/cellulose/silica nanostructured hybrid materials and their characterization by NMR relaxometry. *Polymer Testing* 2015;47:92–100. https://doi.org/10.1016/j.polymertesting.2015.08.008
34. da Silva PSRC, Tavares, MIB. Solvent effect on the morphology of lamellar nanocomposites based on HIPS. *Mater Res* 2015;18(1):191–5. https://doi.org/10.1590/1516-1439.307314
35. Junior JPC, Soares IL, Luetkmeyer L, Tavares, MIB. Preparation of high-impact polystyrene nanocomposites with organoclay by melt intercalation and characterization by low-field nuclear magnetic resonance. *Chem Eng Process* 2014;77:66–76. https://doi.org/10.1016/j.cep.2013.11.012
36. Chaker JA, Dahmouche K, Craievich AF, Santilli CV, Pulcinelli SH. Structure of weakly bonded PPG-silica nanocomposites. *J Appl Cryst* 2000;33:700–3.
37. Kim H, Abdala AA, Macosko CW. Graphene/polymernanocomposites. *Macromolecules* 2010a;43:6515–30. Kim H, Miura Y, Macosko CW. Graphene/polyurethane nanocomposites for improved gas barrier and electrical conductivity. *Chem Mater* 2010b;22:3441–3450.
38. Fortunati E, Peltzer M, Armentano I, Torre L, Jime´nez A, Kenny JM. Effects of modified cellulose nanocrystals on the barrier and migration properties of PLA nano-biocomposites. *Carbohydr Polym* 2012c;90:948–56.
39. Sharma P, Sharma A, Sharma M, Bhalla N, Estrela P, Jain A, Thakur A. Nanomaterial fungicides: in vitro and in vivo antimycotic activity of cobalt and nickel nanoferrites on phytopathogenic fungi. *Global Challenges* 2017;1(9):1700041.
40. Sharma B, Mahajan S, Chhibber R, Mehta R. Glass fiber reinforced polymer-clay nanocomposites: processing, structure and hydrothermal effects on mechanical properties. *Procedia Chem* 2012;4:39–46.
41. Pandey, P. Preparation and characterization of polymer nanocomposites. *Soft Nanosci Lett* 2020;10:1–15. https://doi.org/10.4236/snl.2020.101001
42. Isil I, Ulku Y, Goknur B. Impact modified epoxy/montmorillonite nanocomposites: synthesis and characterization. *Polymer* 2003;44:6371–7.
43. Sinha RS, Okamoto M. Polymer/layered silicate nanocomposites: a review from preparation to processing. *Prog Polym Sci* 2003;28:1539–641.
44. Gupta S, Rahaman A. Effect of carbon nanotube on thermo-mechanical properties of glass fiber. *Epoxy Laminated Nanocompos* 2016;5(2):1–5.
45. Wetzel B, Rosso P, Haupert F, Friedrich K. Epoxy nanocomposite-fracture and toughening mechanisms. *Eng Fracture Mech* 2006;73:2375–98.
46. Bhalla N, Taneja S, Thakur P, Sharma PK, Mariotti D, Maddi C, Thakur A. Doping independent work function and stable band gap of spinel ferrites with tunable plasmonic and magnetic properties. *Nano Lett* 2021;21(22):9780–88.
47. Kim DH, Fasulo PD, Rodgers WR, Paul DR. Structure and properties of polypropylene-based nanocomposites: effect of PP-g-MA to organoclay ratio. *Polymer* 2007;48:5308–23.
48. Shen L. Biocompatible polymer/quantum dots hybrid materials: current status and future developments. *J Functional Biomater* 2011;2:355–72.

49. Fornes T, Paul D. Modeling properties of nylon 6/clay nanocomposites using composite theories. *Polymer* 2003;44:4993–5013.
50. Bugnicourt E, Galy J, Gérard J-F, Barthel H. Effect of sub-micron silica fifillers on the mechanical performances of epoxy-based composites. *Polymer* 2007;48:1596–605
51. Ndoro TV, Böhm MC, Müller-Plathe F. Interface and interphase dynamics of polystyrene chains near grafted and ungrafted silica nanoparticles. *Macromolecules* 2011;45:171–9.
52. Gundupalli SP, Hait S, Thakur A. A review on automated sorting of source-separated municipal solid waste for recycling. *Waste Manag* 2017;60:56–74.
53. Raj A, Thakur A. Fish-inspired robots: design, sensing, actuation, and autonomy—a review of research. *Bioinspir Biomim* 2016;11(3):031001.
54. Ayatollahi M, Shadlou S, Shokrieh M. Multiscale modeling for mechanical properties of carbon nanotube reinforced nanocomposites subjected to different types of loading. *Compos Struct* 2011;93:2250–9.
55. Mortezaei M, Famili MHN, Kokabi M. The role of interfacial interactions on the glass-transition and viscoelastic properties of silica/polystyrene nanocomposite. *Compos Sci Technol* 2011;71:1039–45.
56. Taneja S, Chahar D, Thakur P, Thakur A. Influence of bismuth doping on structural, electrical and dielectric properties of Ni–Zn nanoferrites. *J Alloys Comp* 2021;859:157760.
57. Fekete E, Pukánszky B, Tóth A, Bertóti I. Surface modification and characterization of particulate mineral fillers. *J Colloid Interf Sci* 1990;135:200–8.
58. Gundupalli SP, Hait S, Thakur A. Multi-material classification of dry recyclables from municipal solid waste based on thermal imaging. *Waste Manage* 2017;70:13–21.
59. Gunko VM, Voronin EF, Pakhlov EM, Zarko VI, Turov VV, Guzenko NV, Leboda R, Chibowski E. Features of fumed silica coverage with silanes having three or two groups reacting with the surface. *Colloid Surf A*, 2000;166:187–201.
60. Espiard P, & Guyot A. Poly (ethyl acrylate) latexes encapsulating nanoparticles of silica: 2. Grafting process onto silica. *Polymer* 1995;36:4391–5.
61. Rong MZ, Ji QL, Zhang MQ, Friedrich K. Analysis of the interfacial interactions in polypropylene/silica nanocomposites. *Eur Polym J* 2002;38:1573–82.
62. Ji QL, Rong MZ, Zhang MQ, Friedrich K. Graft polymerization of vinyl monomers onto nanosized silicon carbide particles. *Polym Polym Compos* 2002;10:531–9.
63. Tsubokawa N, Kogure A, Sone Y. *J Polym Sci Polym Chem* 1990;28:1923–33.
64. Tsubokawa N, Ishida H, Hashimoto K. Effect of initiating groups introduced onto ultrafine silica on the molecular weight polystyrene grafted onto the surface. *Polym Bull* 1993;31:457–64.
65. Tsubokawa N, Ishida H. Graft polymerization of methyl methacrylate from silica initiated by peroxide groups introduced onto the surface. *J Polymer Sci Part A: Polymer Chem* 1992;30:2241–46.
66. Bharti MK, Chalia S, Thakur P, Sridhara SN, Sharma PB, Thakur A. Microstructural analysis and effect of spin-canting on magnetic attributes of single-phase polycrystalline cobalt ferrite nanoparticles. *J Supercond Nov Magn* 2022;35:1–9.
67. Rong M, Zhang M, Ruan W. Surface modification of nanoscale fillers for improving properties of polymer nanocomposites: a review. *Mater Sci Technol* 2006;22:787–96. https://doi.org/10.1179/174328406X101247
68. Chan CM, Wu J, Li JX, Cheung YK. Polypropylene/calcium carbonate nanocomposites. *Polymer* 2002;43(10):2981–92.
69. Zhang Z, Hua Y. *Chin Plast (Chin)* 2003;17(6):31–5.
70. Ren X, Bai L, Wang G. *Chem World (Chin)* 2000;41(2):83–7.
71. Kruenate J, Tongpool R, Panyathanmaporn T, Kongrat P. Optical and mechanical properties of polypropylene modified by metal oxides. *Surf Interface Anal* 2004;36:1044–7.
72. Lazzeri A, Zebarjad SM, Pracella M, Cavalier K, Rosa R. Filler toughening of plastics. Part I—The effect of surface interactions on physico-mechanical properties and rheological behavior of ultrafine $CaCO_3$/HDPE nanocomposites. *Polymer* 2005;46:827–44.
73. Dhanda N, Thakur P, Kumar R, Fatima T, Hameed S, Slimani Y, Thakur, A. Green-synthesis of Ni-Co nano ferrites using aloe vera extract: structural, optical, magnetic, and antimicrobial studies. *Appl Organomet Chem* 2023;37:e7110.
74. Müller K, Bugnicourt E, Latorre M, Jorda M, Echegoyen Sanz Y, Lagaron JM, Miesbauer O, Bianchin A, Hankin S, Bölz U, et al. Review on the processing and properties of polymer nanocomposites and nanocoatings and their applications in the packaging, automotive and solar energy fields. *Nanomaterials* 2017;7(4):74. https://doi.org/10.3390/nano7040074

18 Fate of Nanomaterials in the Atmosphere

Preeti Thakur
Amity University Haryana

Shilpa Taneja
NBGSM College Sohna

Ritesh Verma
Amity University Haryana

Yassine Slimani
Imam Abdulrahman Bin Faisal University

Blaise Ravelo
Nanjing University of Information Science & Technology

Atul Thakur
Amity University Haryana
Nanjing University of Information Science & Technology

18.1 INTRODUCTION

Even though the term "nanomaterials" (NMs) became popular only toward the end of the 19th century, ancient civilizations had long employed them. Materials of size range up to 100 nm can be created, processed, imaged, measured, and used with the aid of nanotechnology. Particles and materials are thereby disseminated within porous matrices with nanometer pores in metallic clusters in NMs [1,2]. When compared to non-NMs, NMs exhibit a considerable shift in physical and chemical properties because of their small size or nanoscale effect, such as distinct physical, electrical, optical, thermal, chemical, magnetic, and catalytic capabilities. The surface-to-mass ratio, surface area, and feature ratio of NMs are all quite high. These outstanding features can dramatically change the properties of NMs [3,4]. A vast area of multidisciplinary study, invention, development, and economic activity is made possible by nanotechnology, and this activity has been expanding quickly in recent years. The science of nanotechnology is expanding quickly, and as engineered NMs are used in more and more commercial items, they are becoming more prevalent in the biosphere. Engineered materials have been created with at least one nanoscale (1–100 nm) dimension. Natural processes and human activities (such as acid mine drainage's flocculation of metal oxides with a nanometer-scale) both naturally produce NMs, which are pervasive in the current environment. When compared to a bigger size material with fewer surface atoms, the small sizes of both naturally and artificially created NMs concludes unique characteristics and reactivity. Examples of these designed NMs include quantum dots, which vary in optical and electrical properties due to their size, and gold particles which are generally innocuous but turn catalytic as they go smaller to a few nanometers [5,6]. To maximize their utility, NMs are being built from several NMs, such

DOI: 10.1201/9781003502692-18

as carbon nanotubes (CNTs) doped with quantum dots that have cells in the body. These recently developed, xenobiotic materials will soon make their way into manufacturing supply chains. The prediction of these new materials in environmental systems is made more difficult by the absence of a natural analog for them. There has been an increase in environmental health and safety research designed at determining the likelihood that NMs will damage the environment or human health as a result of effects caused by the novel properties of NMs. Correlating the characteristics of NMs with their behavior in the current environment and also their effects on living things is a major objective of these research initiatives [7–9].

Due to the different physicochemical properties of NMs compared to micro-scale and bulk materials, nanotechnology plays a significant role and has numerous applications in different fields such as medicine, agriculture, water treatment, and electronics. These special qualities are a result of the higher surface energy and atom fraction of tiny particles when compared to bulk materials. The first nanoceramics was created in the 1980s. Since then, there has been a great deal of enthusiasm for its use in numerous technologies. Ultrafine particles with a diameter of <100 nm are used to create nanoceramics. Ceramic nanoparticles (NPs) have proven to be far more stable than metallic ones, and it is generally more affordable to produce them. Compared to conventional biosensors, NM-based biosensors have promising futures [10–12]. It offers several potential benefits, including improved detection sensitivity in a variety of industries, including agriculture and agro-products, environment control, food quality, biodefense, and especially the medical field [13,14].

Understanding the different types of NMs and their exposure routes is essential to characterizing the hazard factors of NMs and their destiny in the environment. The type of NM may be unknown due to the production process, which may expose people to it by consumption, skin absorption, or breathing it in. It is more challenging to predict how these materials will interact with living things and the environment since attributes including melting temperature, mechanical strength, and electrical conductivity are transformed at the nanoscale. Additionally, some naturally occurring NPs have the potential to be harmful and come from a variety of geographical and biological processes [15–17]. It has been noted that due to possible negative exposure to natural NMs, both humans and other creatures have developed, changed, and multiplied. Nanotechnology has been used in environmental management to enhance the quality of soil, water, and air. To track and find the pollutant released in environment, a remediation procedure based on NPs has been created. In the remediation process, waste is also produced that contains NP residues. This wastewater eventually makes its way to various abiotic environmental sources such soil, water, and sediments [16,18,19]. However, scientific academics and regulatory organizations have been interested in the usage and issue of NMs because of their toxicity and probable environmental effects.

18.2 TRANSFORMATIONS IN NANOMATERIALS

We must try to increase our understanding of the changes that NMs go through to accurately predict the pressures on the environment and the effects on human health that are connected to them. A fresh framework for nano Environment, Health and Safety (EHS) research has recently been put forth by the US National Research Council. This committee concentrates on comprehending "critical elements of nanomaterial interactions", which are required for determining exposure, risks, and consequently hazards posed by designed NMs as depicted in Figure 18.1. These elements include physical, chemical, and biological transformations, which influence bioavailability, toxicity, and reactivity [20]. In the environment and in biological systems dissolution, sulfidation, aggregation, and adsorption of macromolecules are all common occurrences. These changes have an important effect on NP behavior. In some circumstances, these changes may increase the probability of toxicity. Other times, it has been demonstrated that these changes have the opposite impact such the breakdown of ZnO NPs, which have the potential to decrease NM determination in the environment. The impact of modifications on biological effects across the board is still very much in question [21–23].

A stabilizer is a tiny anion as shown in Table 18.1, which is frequently used in the production of NMs. Therefore, changes to the material can have an impact on the capping agent. For instance,

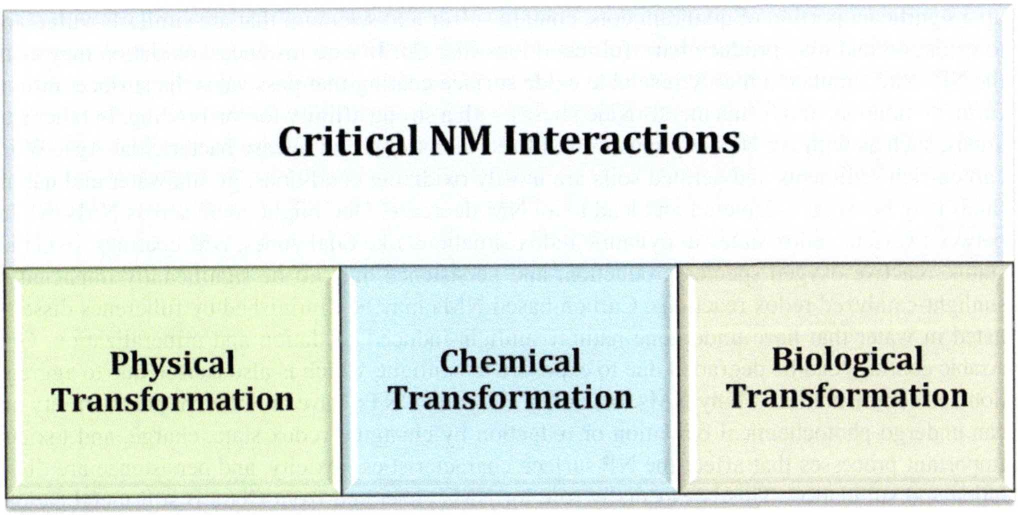

FIGURE 18.1 Nanomaterial transformations affecting their interactions.

TABLE 18.1
Nanomaterials and Capping Agents

	Capping Agents	
Nanomaterials	**Inorganic and Organic Molecules**	**Synthetic and Organic Molecules**
ZnO	2-mercaptoethanol, acetate, polyvinyl pyrrolidone (PVP), polysaccharides	
Silver	Citrate	PVP
Gold	Tannic acid	PVP, PEG
Cerium oxide	Oleic acid	Polypeptides
Titanium dioxide	Oleic acid	PVP
Quantum dots	Silica, zinc sulfide, citrate	Acrylic acid
Iron oxide	Oleic acid	PEG
Zerovalent iron	Au, Pd, Pt	Gum, PEG

the crystalline phase and characteristics of ZnS NPs can be changed by simply coordinating them with water molecules. It is possible for bacteria to digest the capping ingredient and remove it. In fact, the characteristics of the initial NM, its coating, and its surrounding chemical and biological milieu will determine the rate and type of transformations that are conceivable. Importantly, thermodynamics cannot always be used to anticipate many transitions since they are sluggish or effectively irreversible. Here, we discuss what is known about NMs' natural changes mediated by chemistry, physics, and biology [24–26]. We also talk about the gaps in current scientific understanding and instrumentation that prohibit us from biological, chemical, and environmental media.

18.2.1 Chemical Transformations

In natural systems, reduction and oxidation occur in tandem processes that entail the movement of electrons from chemical moieties. Many NMs could be reduced or oxidized in aquatic environments. These include NMs made of basic metals like iron and silver. Ce (III) and Ce (IV) can both be found in ceria NPs, and the ratio of Ce (III) to Ce (IV) on the NP surface can change as a result of subsequent sorption by macromolecules. Some metal sulfides and metal selenides, which make up a significant portion of quantum dots, contain sulfur and selenium that are similarly vulnerable to oxidation and may produce harmful metal ions like Cd. In rare instances, oxidation may cause the NP to accumulate a highly insoluble oxide surface coating that passivates the surface, inhibits further oxidation, and forms metal-oxide phases with a strong affinity for ion binding. In other situations, such as with Ag NPs, Ag must be oxidized to dissolve and release bactericidal Ag+. While carbon-rich sediments and aerated soils are mostly oxidizing conditions, groundwater and natural fluids may be oxygen-depleted and lead to an NM decrease. One might come across NMs cycling between various redox states in dynamic redox situations like tidal zones. NM coatings, oxidation status, reactive oxygen species production, and persistence may all be significantly impacted by sunlight-catalyzed redox reactions. Carbon-based NMs may be diminished by fullerenes disseminated in water that have undergone natural sunlight-induced oxidation and mineralization. Gum Arabic coatings can be degraded due to exposure to sunlight, which is also caused due to aggregation and sedimentation. Many NMs, including TiO_2 and CNTs, have an inherent photoactivity and can undergo photochemical oxidation or reduction by changing redox state, charge, and toxicity. Important processes that affect the NP surface characteristics, toxicity, and persistence are dissolution and sulfidation. This is especially true for NMs generated from Class B soft metal cations, which produce partially soluble metal oxides and have a strong affinity for both inorganic and organic sulfide ligands. Examples of these metals are Ag, Zn, and Cu. Less persistent but more toxic Class B metal NMs often demonstrate toxicity through the dissolution and release of harmful cations. Complete dissolution might allow for the prediction of metal speciation and effects using present models. The high reactivity of Class B metals with biomacromolecules and inorganic sulfur is due to their affinity for electron-dense sulfur molecules. On the surface of the particle, a metal-sulfide changes the surface charge and causes aggregation. To determine a particle's potential in the environment, it is crucial to identify the particle characteristics (such as particle size and capping agent), as well as the environmental factors [27–29]. The surface chemistry of NM surfaces and the consequent behavior in biological and environmental systems can be dramatically changed by the adsorption of macromolecules. For instance, the general decrease in NPs' adhesion to silica surfaces after polymer coating adsorption suggests that NPs are more mobile in the environment and may be more difficult to remove from drinking water. Biomacromolecule adsorption is a crucial transition that is covered separately below. Both the surface of the NM and its coating are capable of adsorbing organic metal cations, which have an impact on the NM's charge, stability, and dissolution.

18.2.2 Physical Transformations

The property of surface area to volume shows effects on NM reactivity are diminished by NP aggregation. Their movement in porous media, sedimentation, and reactivity are all impacted by this rise in aggregate size. Without intentional or accidental coatings to prevent it, the aggregation of particles into clusters over time is unavoidable. There are two types of aggregation: homo-aggregation that arises between the same NMs and hetero-aggregation that arises between NMs, and an environmental particle. In detail, hetero-aggregation will occur as the ambient particles' higher concentration relative to NMs. For instance, Fe NPs with a diameter of 30–70 nm rapidly form micrometer-sized aggregates in water, considerably reducing the value of mobility on the surface [30–32]. Aggregation also decreases the materials' "available" surface area parameter, which lowers their reactivity. The aggregate's fractal dimensions and particle number distribution of sizes will all affect how much specific surface area decreases. Although at distinct locations from the dispersed NPs, aggregation may help to raise the persistence of NM if it slows down the rate of degradation or dissolution. An NP's size may also impact how bioavailable it is to living things. Uptake may be impeded for direct passage through the cell wall [33–35]. Phagocytosis and related mechanisms might be impacted as well. On the contrary, hetero-aggregation of soft biogenic particles may boost the bioavailability of NM. Because uptake is a dynamic process that heavily depends on the species under study, its aqueous chemical environment, and its metabolic state, and because there are currently no tools for tracking NMs in situ or in vivo, it will be challenging to distinguish the effects of aggregation on uptake and any subsequent toxicity.

18.2.3 Biologically Transformations

In living tissues, as well as in environmental media (such as soils), biological modifications of NMs are unavoidable. In all biological systems, redox processes are essential for growth. It is widely known how bacteria respond to redox with naturally occurring, nanoscale iron oxide. Additionally, it has recently been shown that bacteria like *Geobacter* and *Shewanella* spp. may reduce solution-based Ag^+ to create nanoscale silver particles [36–38]. NM core and coatings may undergo biologically mediated changes, and these changes may have an impact on surface charge, aggregation term state, and reactivity. These changes may also have an impact on transport, bioavailability, and toxicity. It has been proven that OH radicals generated by the horseradish peroxidase enzyme can oxidize and carboxylate CNTs. This oxidation decreases hydrophobicity while increasing charge of the CNTs and stability. Additionally, this biological surface functionalization and oxidation may have an impact on CNTs' propensity for toxicity. It is also possible to bio transform the polymer coatings utilized on many NMs for biomedical purposes. Microorganisms obtained from an urban stream, for example, were demonstrated to be accessible to coatings of covalently bonded polyethylene glycol on designed NMs. Additionally, the NMs aggregated as a result of the Poly ethylene glycol (PEG) coating's biotransformation. Biological alterations of NMs, especially carbon-based ones, and their organic coatings may eventually operate to lower their concentrations in the environment or to change mobility, while it is uncertain if these activities occur at rates high enough to be important. The adsorption of biomacromolecules on the surfaces of NMs is possibly the most important biotransformation of NMs [39].

18.3 CONSEQUENCES OF NM TRANSFORMATIONS

NMs are easily transformed in the environment and living organisms, which has a significant impact on their characteristics, actions, and outcomes. However, most of the research on toxicity and transport was conducted to use materials that were largely undisturbed, which will behave differently from the changed materials. Because environmental exposures to aquatic species and humans (such as by drinking water or food ingestion) will modify NMs, information about the destiny and impacts of relatively pristine NMs may not be very helpful. Although direct exposure to relatively clean NMs can occur in production or processing facilities, the effects measured for relatively pure materials may be typical of human exposures occurring there [31,40]. Environmental exposures of NMs are probably long term and low intensity. As a result, it will be necessary to evaluate delayed NM alterations to comprehend their final fate, environmental distribution, and potential for negative impacts. Some transformations are essential processes that have an impact on NM behavior in both the environment and in vivo, such as the adsorption of molecules or dissolution of metal NPs. The rates of these changes under pertinent conditions are now mostly unknown [41,42]. Since progress in this area will be helpful to several nano basis research communities, including fate and transport, toxicity, and sustainable/green nanotechnology, figuring out the possibility for these fundamental processes to occur, as well as the rates of these events, should be a study priority.

18.3.1 Fate of Nanoparticles in the Environment

The aggregation, disaggregation, and chemical aging of NP in the environment are all related. The interaction between these systems and the movement of NP ultimately determines the fate of NP and its potential ecotoxicity (Figure 18.2). Because particle properties and environmental factors control these aging and transport processes, a direct transfer of data to even slightly different

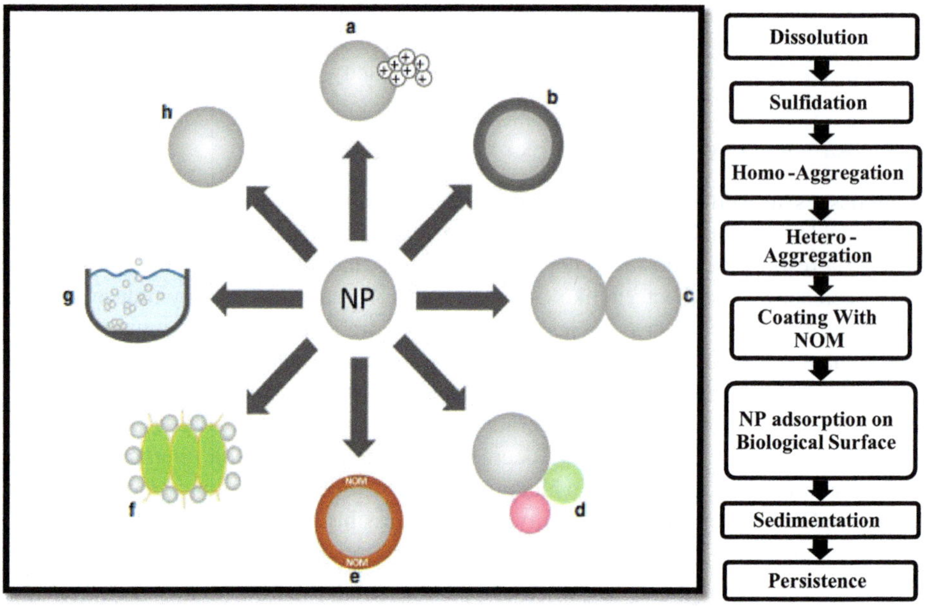

FIGURE 18.2 Interactions and the fate of nanoparticles involving some methods.

settings is difficult and might even be misleading. Here, we review the state-of-the-art in NP destiny research and decide as to whether the data is adequate to extrapolate from well-defined and controlled laboratory situations to difficult real-world scenarios. Changes in chemical speciation, dissolution, and degradation, as well as modifications in surface properties brought on by precipitation and adsorption, are significant. Both aquatic and terrestrial ecosystems have been the subject of in-depth research on these processes. In this context, it is frequently noted that processes of environmental transformation would be significantly influenced by functionalization of NP surfaces, which enhances the properties of the particles for industrial applications. A thorough characterization of NP surface chemistry seems important, yet it is still very challenging to comprehend NP destiny. However, the current concept mainly concentrates on a number of the most significant processes that occur in field settings, such as dissolution, passivation, aggregation, adsorption, sedimentation, and deposition.

Dissolution of NP: The particle chemistry is what directly drives it. For instance, aerobic conditions are necessary for AgNP to dissolve. Ag^+ can be released when an oxide layer (Ag_2O) forms surrounding the particle in these conditions. AgNP dissolution rates are influenced by environmental variables like pH and temperature factors like surface coating, size, shape, and state of aggregation of NM particles. Authors accurately predicted Ag dissolution in a variety of artificial mediums using calculations based on a large body of literature survey. This shows that the scientific community was really able to advance the knowledge to a stage that enables the creation of predictions and models that can be reasonably accurate.

- **Sulfidation:** The sulfidation of NP is a passivation process that commonly takes place under varied environmental conditions. For instance, Ag NP sulfidation can result in the creation of hollow Ag_2S NP or core-shell Ag_0–Ag_2S structures. The sulfidation mechanisms are described elsewhere, but one of the major elements influencing the fate and consequences of NP is their colloidal stability. The NP environment interacts with a variety of inorganic or organic, liquid, or particulate chemicals, changing the dynamics of NP aggregation and consequently colloidal stability. In the end, NP aggregation regulates the exposure circumstances. We also discuss the processes governing NP fate by concentrating on factors governing homo-aggregation and hetero-aggregation.

- **Homo-aggregation of NP:** Their focus on the media is favorably associated with it. Because of the low likelihood of collisions and the projected low ambient concentrations, homo-aggregation is less likely to occur. However, for laboratory studies that frequently use high NP concentrations in comparison to project ambient concentrations, this aspect is pertinent—but largely disregarded. The ionic strength of the medium appears to be more important under field circumstances since aggregation rates rise with ionic strength. In most circumstances, classical theory can be used to describe aggregation dynamics. For instance, Ca^{2+} was able to aggregate citrate-coated Ag NP more effectively than Mg^{2+}, which may be related to Ca^{2+}'s greater capacity to generate citrate complexes. Additionally, the efficiency within both categories of cations is calculated using their respective identities determined by critical coagulation concentrations, with multivalent cations being more effective than monovalent cations. The effects of cations and anions can both vary depending on their concentration. Low concentrations of Cl ions may stabilize NP by the development of negatively charged surface precipitates, but large concentrations of Cl ions increase the aggregation of Ag NPs. Soils and soil extracts also control the aggregation of NP. Numerous environmental factors affect homo-aggregation in addition to ionic strength. The iso-electric point reflects how the surface charge of NP varies with pH.
- **Hetero-aggregation:** Given that natural colloids have a concentration higher than that of NP, this topic has a greater environmental impact. For instance, NP's showed that the primary method for removing hazardous NMs from the aqueous phase by sedimentation is hetero-aggregation. The aggregation kinetics is a very quick process if they have differing charges.
- **Coating with Natural organic matter (NOM):** Due to steric and electrostatic stabilization, the coating decreases hetero-aggregation. Other than electrostatic forces, polymer-based bridging, hydrogen, and chemical bonding have all been identified as processes for hetero-aggregation. As a result, this process is extremely complicated and is influenced by a variety of factors, including the characteristics of natural inorganic, organic, and biological colloids, the age of the NPs, their surface properties, and their interaction with other particle phases. However, just a few articles evaluated aggregation in intricate field-relevant media, making it difficult to draw any generalizations regarding the applicability of the methods described above. The dynamics of these aggregation processes in actual systems cannot be adequately addressed by studies in simpler artificial systems because disaggregation is primarily induced by changing environmental variables. This necessitates the use of experimental layouts that can simulate such erratic situations. Future research should use low and environmentally relevant NP concentrations to avoid any potential confounding effects of NP homo-aggregation, which is, as noted above, less plausible under currently projected ambient NP concentrations than hetero-aggregation.

18.4 CONCLUSION

Our knowledge of the causes, course, and impacts of NP on the environment has significantly improved during the past ten years. In addition to the request to consider NP concentrations that are relevant to the environment and to keep track of what happens to NP by biological testing, there are several unanswered issues that require more thought. Finding out how soil characteristics affect NP destiny and the risk of groundwater pollution by NP requires a more methodical approach immediately. For instance, we now know that, in some cases, the ions released from NP can completely account for the effects seen in living things. However, it is currently unable to adequately characterize the circumstances in which this simplified assumption should be rejected, necessitating the use of other processes. We strongly advise including those coatings in next projects since they are probably of high field relevance. Even though the majority of studies concentrate on effects at extremely high NP concentrations, more recent techniques show sublethal effects at levels that are important for the field, particularly over several generations. Therefore, under existing and future exposure situations, particular attention should be paid to how NPs may alter communities, ecosystems, ecosystem processes, and interactions beyond ecosystem boundaries. Investigations spanning several years of (repeated) exposure and assessment are advised to accurately assess their potential long-term effects in aquatic and terrestrial ecosystems, especially for sparingly soluble or insoluble NPs that may accumulate in specific environmental compartments (such as sediments) over time. The ability to recognize NP-induced alterations in the food web's horizontal and vertical relationships is directly correlated with this component.

REFERENCES

1. T. Dippong, I.G. Deac, O. Cadar, E.A. Levei, L. Diamandescu, G. Borodi, Effect of Zn content on structural, morphological and magnetic behavior of $Zn_xCo_{1-x}Fe_2O_4/SiO_2$ nanocomposites, *J. Alloys Compd.* 792 (2019) 432–443. doi:10.1016/j.jallcom.2019.04.059.
2. M.M. Baig, S. Zulfiqar, M.A. Yousuf, M. Touqeer, S. Ullah, P.O. Agboola, M.F. Warsi, I. Shakir, Structural and photocatalytic properties of new rare earth La^{3+} substituted $MnFe_2O_4$ ferrite nanoparticles, *Ceram. Int.* 46 (2020) 23208–23217. doi:10.1016/j.ceramint.2020.06.103.
3. S. Küchler, N.B. Wolf, S. Heilmann, G. Weindl, J. Helfmann, M.M. Yahya, C. Stein, M. Schäfer-Korting, 3D-Wound healing model: influence of morphine and solid lipid nanoparticles, *J. Biotechnol.* 148(1) (2010) 24–30. doi:10.1016/j.jbiotec.2010.01.001.
4. A.R. Chavan, P.P. Khirade, S.B. Somvanshi, S. V Mukhamale, Eco - friendly green synthesis and characterizations of $CoFe_{2-x}$ Al_xO_4 nanocrystals: analysis of structural, magnetic, electrical, and dielectric properties, *J. Nanostructure Chem.* 11(3) (2021) 469–481. doi:10.1007/s40097-020-00381-7.
5. C. Vinuthna, D. Ravinder, R.M. Raju, Characterization of $Co_{1-x}Zn_xFe_2o_4$ nano spinal ferrites prepared by citrate precursor method, *J. Eng. Res. Appl.* 3(96) (2013) 654–660.
6. D. Chahar, S. Taneja, S. Bisht, S. Kesarwani, P. Thakur, A. Thakur, P.B. Sharma, Photocatalytic activity of cobalt substituted zinc ferrite for the degradation of methylene blue dye under visible light irradiation, *J. Alloys Compd.* 851 (2021) 156878. doi:10.1016/j.jallcom.2020.156878.
7. S.S. Thakur, A. Pathania, P. Thakur, A. Thakur, J. Hsu, Improved structural, electrical and magnetic properties of Mn–Zn–Cd nanoferrites, *Ceram. Int.* 41 (2015) 5072–5078. doi:10.1016/j.ceramint.2014.12.077.
8. D. Maiti, A. Saha, P.S. Devi, Surface modified multifunctional $ZnFe_2O_4$ nanoparticles for hydrophobic and hydrophilic anti-cancer drug molecule loading, *Phys. Chem. Chem. Phys.* 18 (2016) 1439–1450. doi:10.1039/c5cp05840f.
9. T.T. Le, Z. Valdez-Nava, T. Lebey, F. Mazaleyrat, Influence of cold isostatic pressing on the magnetic properties of Ni-Zn-Cu ferrite, *AIP Adv.* 8 (2018), 4994210. doi:10.1063/1.4994210.
10. S. Mallesh, V. Srinivas, A comprehensive study on thermal stability and magnetic properties of MnZn-ferrite nanoparticles, *J. Magn. Magn. Mater.* 475 (2019) 290–303. doi:10.1016/j.jmmm.2018.11.052.
11. O. Id, A.L. Review, wastewater treatment, reuse and disposal associated effects on environment and health, *Water Environ.* 92(10), 1595–1602, doi:10.1002/wer.1406.
12. M.A. Rehman, I. Yusoff, Y. Alias, Fluoride adsorption by doped and un-doped magnetic ferrites $CuCe_xFe_{2-x}O_4$: preparation, characterization, optimization and modeling for effectual remediation technologies, *J. Hazard. Mater.* 299 (2015) 316–324. doi:10.1016/j.jhazmat.2015.06.030.
13. S. Amiri, H. Shokrollahi, The role of cobalt ferrite magnetic nanoparticles in medical science, *Mater. Sci. Eng. C.* 33 (2013) 1–8. doi:10.1016/j.msec.2012.09.003.
14. M. Amiri, M. Salavati-niasari, A. Akbari, Magnetic nanocarriers : Evolution of spinel ferrites for medical applications, *Adv. Colloid Interface Sci.* 265 (2019) 29–44. doi:10.1016/j.cis.2019.01.003.
15. D.H. Kim, S.H. Lee, K.N. Kim, K.M. Kim, I.B. Shim, Y.K. Lee, Cytotoxicity of ferrite particles by MTT and agar diffusion methods for hyperthermic application, *J. Magn. Magn. Mater.* 293 (2005) 287–292. doi:10.1016/j.jmmm.2005.02.078.
16. G. Basina, V. Tzitzios, D. Niarchos, W. Li, H. Khurshid, H. Mao, C. Hadjipanayis, G. Hadjipanayis, Water-soluble spinel ferrites by a modified polyol process as contrast agents in MRI, *AIP Conf. Proc.* 1311 (2010) 441–446. doi:10.1063/1.3530053.
17. M.M. El-kady, I. Ansari, C. Arora, N. Rai, S. Soni, D. Kumar, P. Singh, A. El, D. Mahmoud, Nanomaterials: a comprehensive review of applications, toxicity, impact, and fate to environment, *J. Mol. Liq.* 370 (2023) 121046. doi:10.1016/j.molliq.2022.121046.

18. Q. Yu, Y. Su, Synthesis and characterization of low density porous nickel zinc ferrites, *RSC Adv.* 9 (2019) 13173–13181. doi:10.1039/c9ra01076a.
19. T. Cele, Preparation of nanoparticles, *Eng. Nanomater. - Heal. Saf.* 5 (2020) 1–14. doi:10.5772/intechopen.90771.
20. H.-Y. Park, M.J. Schadt, L. Wang, I.-I.S. Lim, P.N. Njoki, S.H. Kim, M.-Y. Jang, J. Luo, C.-J. Zhong, Fabrication of magnetic core@shell Fe oxide@Au nanoparticles for interfacial bioactivity and bio-separation, *Langmuir.* 23 (2007) 9050–9056. doi:10.1021/la701305f.
21. S.A. Blaser, M. Scheringer, M. Macleod, K. Hungerbühler, Estimation of cumulative aquatic exposure and risk due to silver: contribution of nano-functionalized plastics and textiles, *Sci. Total Environ.* 390 (2007) 396–409. doi:10.1016/j.scitotenv.2007.10.010.
22. E.K. Afshinnia, I. Gibson, R. Merrifield, M. Baalousha, C. Health, The concentration-dependent aggregation of Ag NPs induced by cysteine. *Sci. Total Environ.* 557–558 (2016) 395–403.
23. C. Vannini, G. Domingo, E. Onelli, F. De Mattia, I. Bruni, M. Marsoni, M. Bracale, Phytotoxic and genotoxic effects of silver nanoparticles exposure on germinating wheat seedlings, *J. Plant Physiol.* 171 (2014) 1142–1148. doi:10.1016/j.jplph.2014.05.002.
24. A. Thakur, P. Thakur, J.H. Hsu, Magnetic behaviour of $Ni_{0.4}Zn_{0.6}Co_{0.1}Fe_{1.9}O_4$ spinel nano-ferrite, *J. Appl. Phys.* 111 (2012) 2012–2015. doi:10.1063/1.3670606.
25. K. Rana, P. Thakur, P. Sharma, M. Tomar, V. Gupta, A. Thakur, Improved structural and magnetic properties of cobalt nanoferrites: influence of sintering temperature, *Ceram. Int.* 41(3) (2014) 4492–4497. doi:10.1016/j.ceramint.2014.11.143.
26. A. Thakur, P. Thakur, J.H. Hsu, Smart magnetodielectric nano-materials for the very high frequency applications, *J. Alloys Compd.* 509 (2011) 5315–5319. doi:10.1016/j.jallcom.2011.02.021.
27. R. Landsiedel, M. Diana, M. Schulz, K. Wiench, F. Oesch, Genotoxicity investigations on nanomaterials: methods, preparation and characterization of test material, potential artifacts and limitations—many questions, some answers, *Mutat. Res. Rev. Mutat. Res.* 681 (2009) 241–258. doi:10.1016/j.mrrev.2008.10.002.
28. G. Tortella, O. Rubilar, J.C. Pieretti, P. Fincheira, B.D.M. Santana, A. Fern, A. Benavides-mendoza, A.B. Seabra, Nanoparticles as a promising strategy to mitigate biotic stress in agriculture, *Antibiotics* 12(2) (2023) 338.
29. P. Punia, M. Kumar, S. Chalia, R. Dhar, B. Ravelo, P. Thakur, A. Thakur, Recent advances in synthesis, characterization, and applications of nanoparticles for contaminated water treatment - a review, *Ceram. Int.* 47(2) (2020) 1526–1550. doi:10.1016/j.ceramint.2020.09.050.
30. A. Kermanizadeh, B.K. Gaiser, H. Johnston, D.M. Brown, V. Stone, Toxicological effect of engineered nanomaterials on the liver, *Br. J. Pharmacol.* 171(17) (2013) 3980–3987. doi:10.1111/bph.12421.
31. G. V Lowry, K.B. Gregory, S.C. Apte, J.R. Lead, Transformations of nanomaterials in the environment, *Environ. Sci. Technol.* 46(13) (2012) 6893–6899.
32. B. Rezaei, A. Kermanpur, S. Labbaf, Effect of Mn addition on the structural and magnetic properties of Zn-ferrite nanoparticles, *J. Magn. Magn. Mater.* 481 (2019) 16–24. doi:10.1016/j.jmmm.2019.02.085.
33. S. Arif, S. Shah, K. Zhang, A.R. Park, S. Kim, N. Park, H. Park, P.J. Yoo, Single-step solvothermal synthesis of mesoporous Ag–TiO_2–reduced graphene oxide ternary composites with enhanced photocatalytic activity, *Nanoscale* 5 (2013) 5093–5101. doi:10.1039/c3nr00579h.
34. S. Heo, D. Tahir, J.G. Chung, J.C. Lee, K. Kim, J. Lee, H. Lee, G. Su, S. Heo, D. Tahir, J.G. Chung, J.C. Lee, K. Kim, J. Lee, H. Lee, G.S. Park, S.K. Oh, H.J. Kang, P. Choi, B. Choi, Band alignment of atomic layer deposited $(HfZrO_4)_{1-x}(SiO_2)_x$ gate dielectrics on Si (100). *Appl. Phys. Lett.* 107(18) (2015) 182101. doi:10.1063/1.4934567.
35. K. Ratnaih, V.K. Prasad, X-ray diffraction and magnetic properties of Nd substituted $NiZnFe_2O_4$ characterized by rietveld refinement, *Biointerface Res. Appl.Chem.* 11 (2021) 9062–9070.
36. S. Chandel, P. Thakur, S.S. Thakur, V. Kanwar, M. Tomar, V. Gupta, A. Thakur, Effect of non-magnetic Al3+ doping on structural, optical, electrical, dielectric and magnetic properties of $BiFeO_3$ ceramics, *Ceram. Int.* 44 (2018) 4711–4718. doi:10.1016/j.ceramint.2017.12.053.
37. J. Van Biesen, M. Wagner, Maxwell-Wagner effect in silver bromide emulsions, *J. Appl. Phys.* 41 (2009) 1910–1970. doi:10.1063/1.1659140.
38. M. Bundschuh, J. Filser, S. Lüderwald, M.S. Mckee, G. Metreveli, G.E. Schaumann, R. Schulz, S. Wagner, Nanoparticles in the environment: where do we come from, where do we go to?, *Environ. Sci. Eur.* 30 (2018) 6. doi:10.1186/s12302-018-0132-6.
39. M. Madhukara Naik, H.S. Bhojya Naik, G. Nagaraju, M. Vinuth, H. Raja Naika, K. Vinu, Green synthesis of zinc ferrite nanoparticles in Limonia acidissima juice: characterization and their application as photocatalytic and antibacterial activities, *Microchem. J.* 146 (2019) 1227–1235. doi:10.1016/j.microc.2019.02.059.

40. A. Abu El-Fadl, A.M. Hassan, M.H. Mahmoud, T. Tatarchuk, I.P. Yaremiy, A.M. Gismelssed, M.A. Ahmed, Synthesis and magnetic properties of spinel $Zn_{1-x}Ni_xFe_2O_4$ ($0.0 \leq x \leq 1.0$) nanoparticles synthesized by microwave combustion method, *J. Magn. Magn. Mater.* 471 (2019) 192–199. doi:10.1016/j.jmmm.2018.09.074.
41. S.T. Fardood, K. Atrak, A. Ramazani, Green synthesis using tragacanth gum and characterization of Ni–Cu–Zn ferrite nanoparticles as a magnetically separable photocatalyst for organic dyes degradation from aqueous solution under visible light, *J. Mater. Sci. Mater. Electron.* 28 (2017) 10739–10746. doi:10.1007/s10854-017-6850-5.
42. B. Bhushan, Introduction to nanotechnology: history, status, and importance of nanoscience and nanotechnology education, In: Winkelmann, K., Bhushan, B. (eds) *Global Perspectives of Nanoscience and Engineering Education*, Springer, Cham (2016) 1–31. doi:10.1007/978-3-319-31833-2_1.

19 Fate of Engineered Nanomaterials in Soil and Aquatic Systems

Monika Sohlot, S M Paul Khurana, Sumistha Das, and Nitai Debnath
Amity University Haryana

19.1 INTRODUCTION

Engineered nanomaterials (ENMs) have garnered significant attention and found widespread application in diverse industries, ranging from medical and energy to electronics and environmental sectors. As the commercial use and production volume of ENMs continue to grow, concerns about their potential environmental impact have emerged [1]. The introduction of nanomaterials into aquatic and terrestrial ecosystems and their interactions with organisms have become focal points of investigation. Regulatory bodies face the crucial question of whether nanomaterials necessitate distinct regulations compared to their equivalent micrometer-sized counterparts [2]. The primary concern revolves around determining if nanoscale dimensions result in increased bioavailability and toxicity. Understanding the fate of nanomaterials upon their entry into the natural environment, including changes in particle size, surface charge, and chemical form, is essential for controlling their bioavailability and potential ecological consequences.

One notable example is the environmental impact of silver nanoparticles (NPs) or AgNPs, which can enter aquatic environments through occupational or residential exposure via consumer and hospital products. These AgNPs may undergo oxidative dissolution and sulfidation, potentially affecting the health of the surrounding ecosystem [3]. Balancing the benefits and risks associated with the release of ENMs into the environment becomes a critical consideration. Silica NPs (SNPs) are widely employed in diverse sectors including drug delivery systems for their ease of functionalization and biocompatibility. Their utilization offers the potential to reduce side effects and lower medication dosages, leading to more targeted and efficient treatment approaches [4]. Titanium dioxide (TiO_2), a highly productive and extensively used nanomaterial, is renowned for its photoactive properties. However, metallic NPs such as TiO_2 can generate free radicals and peroxides that might be harmful to microorganisms [5]. Evaluating the potential ecological consequences of their presence is crucial for effectively balancing the benefits and risks associated with their application. Copper NPs (CuNPs), valued for their high catalytic efficiency, ease of isolation, and antimicrobial properties, find preference in specific industries, particularly electronics, the agriculture sector, etc. The interactions between environmental matrices and ENMs are complex, influenced not only by soil and water quality parameters but also by the presence of biomolecules like plant biomass and microorganisms [6].

This chapter aims to thoroughly explore the intricate interplay between ENMs and their surrounding environment. By examining the fate, transformation processes, and potential ecological impacts of popular ENMs, particularly AgNPs, nano TiO_2, mesoporous SNPs, and CuNPs, we aim to contribute to the expanding knowledge base concerning the environmental implications of nanotechnology. Through an investigation of the behavior and interactions of these NPs with their environmental matrices, valuable insights can be gained into their bioavailability, potential toxicity to organisms, and impacts on ecological niches. Understanding these intricate mechanisms is

DOI: 10.1201/9781003502692-19

crucial for informing risk assessments and developing sustainable strategies for the responsible use of ENMs. In the subsequent sections of this chapter, we will delve into the properties and potential environmental impacts of these popular ENMs, with a specific emphasis on their behavior in soil and water. By casting light on these critical aspects, we aim to provide a comprehensive understanding of the intricate dynamics between ENMs and the environment. Ultimately, this knowledge will facilitate informed decision-making for the sustainable utilization of nanomaterials.

19.2 POPULAR ENMS: PROPERTIES AND POTENTIAL IMPACTS ON THE ENVIRONMENT AND LIVING ORGANISMS

ENMs are widely used across diverse industries and various daily-use products starting from outdoor UV-protective paints, floor coatings, textiles, electronic devices, medicines, etc. [7]. Commonly employed ENMs include nano forms of TiO_2, silica, copper, clay, aluminum oxide, zinc, ferric oxide (Fe_2O_3), cerium oxide (CeO_2), carbon nanotube (CNT), and other carbon-based NPs like graphene [8–12]. Each of these ENMs possesses distinctive characteristics and behaviors, significantly influenced by factors such as aggregate size, aquifer chemistry, porosity, and soil particle size [13]. In Table 19.1, the unique properties of a few NPs and their applications are tabulated.

The range of popular ENMs available in commercial applications is extensive, with each having its own distinct properties and potential impacts. TiO_2, one of the most employed nanomaterials, is widely found in sunscreens, skincare products, colorants, and coatings due to their extraordinary UV-blocking characteristics. CNTs have drawn interest and found use in composite materials,

TABLE 19.1
Size, Shape Properties, and Applications of Different Nanoparticles

Nanomaterial	Size Range	Common Shapes	Unique Properties	Applications	References
Silver	10–100 nm	Spherical, nanowire, nanobar, etc.	Sparkling with optical, electrical, and thermal prowess, localized surface plasmon resonance (LSPR)	Used in household utensils to healthcare, food storage, and environmental and biomedical marvels	[13,14]
Gold	10–100 nm	Spherical, nanostar, nanosphere	Exhibit unique optical properties, low toxicity, excellent biocompatibility, LSPR, size-dependent quantum effects	Diverse applications in conductors, therapeutic agent delivery, sensors, and biological imaging	[15]
Palladium	18–100 nm	Cubic, octahedral, rhombic, dodecahedral	Increased surface area over bulk material; exhibits enhanced catalytic, mechanical, and unique electro-analytical properties	Applications in organic reactions, fuel cells, biomedical therapies, and as potent agents against microbes and cancer	[16]
Silica	5–450 nm	Spherical, rod-shaped	Amorphous nature of SNPs allows biocompatibility; SNPs possess abrasive and absorbent properties and well dispersibility	Versatile applications in rubber manufacturing, drug delivery, imaging, contrast agents, and beyond	[17]

(Continued)

TABLE 19.1 (*Continued*)
Size, Shape Properties, and Applications of Different Nanoparticles

Nanomaterial	Size Range	Common Shapes	Unique Properties	Applications	References
Titanium dioxide	5–250 nm	Spherical, polygon, rhomboid, rod	Photo catalytic activity, ability to absorb and scatter light, antimicrobial properties	Applications in the paint industry, photovoltaic cells, photo catalysis, and environmental purification	[18]
Zinc	<100 nm	Rod-like, isometric, hexagonal, etc.	Harnessing a large surface area relative to volume ratio, antimicrobial, anti-corrosive, and unique optical properties like UV scattering, electrical, and photo catalytic activity	Diverse applications as antimicrobial agents, drug delivery vehicles, and guardians of food safety	[19]
Aluminum oxide	10–80 nm	Oriented fibers, spherical-shaped	Distinguished by its fluorescent white powder form, large surface area, and remarkable versatility	Applications across pharmaceuticals, biosensing, anti-cancer therapy, antimicrobial activities, ceramics, packaging materials, and aerospace domains	[20]

electronics, and batteries due to their outstanding resilience and electrical conductivity. AgNPs find applications in textiles, medical equipment, and water treatment technologies because of their renowned antibacterial qualities [21]. The following sections provide insights into the properties and potential impacts of some popular ENMs on the environment and living organisms.

19.2.1 Silver Nanoparticle

The size spectrum of AgNPs may range from 1 to 100 nm and they typically may consist of 20–15,000 Ag atoms. AgNPs find extensive applications in various industries, including residential and commercial sectors, making them one of the most manufactured NPs [22]. These NPs constitute approximately 25% of all consumable NPs [23]. AgNPs exhibit remarkable stability and durability in environmental conditions, particularly in low pH environments and in the presence of nitrous acid. One notable characteristic of AgNPs is their tendency to undergo sulfidation, a natural transition process. Studies have shown that Ag_2S, derived from Ag ions or AgNPs, is the primary species formed in the presence of sulfides, as observed in recent investigations on silver activity in sewage sludge [24–26]. The coexistence of AgNPs and Ag ions in the same environmental settings has been found to have detrimental effects on aquatic organisms [27]. Moreover, sulfidated AgNPs may exhibit reduced hazards compared to their non-sulfidated counterparts. Sulfidation has been shown to influence various biological factors, such as liver oxidative stress, detoxifying enzymes, and brain acetyl cholinesterase activity, potentially reducing the adverse effects on living organisms [28]. In wastewater treatment systems, the majority of AgNPs are converted to Ag_2S through the sulfidation process [29]. Given the widespread use of AgNPs and their potential environmental implications, further research and the development of appropriate management strategies are necessary to understand their environmental fate and address associated risks [30].

19.2.2 Mesoporous SNP

Due to their extensive surface area, simplicity of functionalization, and biocompatibility, SNPs are employed in diversified arenas including drug delivery, medical imaging, diagnostics, etc. Mesoporous SNPs exhibit unique properties that contribute to their wide range of applications. The large internal surface area of mesoporous SNPs, resulting from their precisely controlled pore sizes, enables enhanced reactivity and efficiency in drug loading, catalytic reactions, and adsorption processes. Moreover, their high chemical stability ensures their durability and effectiveness in various environments [31]. These features allow for customized applications of mesoporous SNPs, such as controlled drug release and selective molecule adsorption. Furthermore, the surface of mesoporous SNPs can be easily functionalized with organic or inorganic moieties, offering versatility, and enhancing their effectiveness by attaching targeting ligands, drugs, or catalysts. This functionalization enables targeted and controlled delivery of substances, improving the efficacy and reducing potential adverse effects. Mesoporous SNPs' biocompatibility ensures minimal harm to biological systems, making them suitable for various biomedical applications. Moreover, the excellent stability of mesoporous SNPs under different environmental conditions ensures their longevity in catalytic reactions or environmental remediation processes. The versatility of Mesoporous silica nanoparticles (MSNPs) in terms of shape and size, including spheres, rods, and tubes, further expands their applicability in diverse fields such as electronics, optics, and energy storage.

19.2.3 Titanium Dioxide Nanoparticle

TiO_2 is a highly productive and widely used nanomaterial known for its photoactive properties. It is considered to have relatively low environmental harm, but its consumption in large quantities can help prevent fungal growth. However, metallic NPs like TiO_2 can generate free radicals and peroxides, which can be detrimental to microorganisms. Studies have shown that nano TiO_2 exhibits toxicity toward marine planktonic organisms, although the extent of toxicity depends on the concentration of TiO_2 in the specific environment. The release of ENMs such as TiO_2 NPs into the environment is significantly influenced by wastewater treatment plants. These plants play a crucial role in the dispersal of ENMs, including TiO_2, into the ecosystem. Approximately 85% of total TiO_2 discharges occur through wastewater, with hazardous concentrations often present after the final membrane filtration step [25]. This raises concerns about the potential impacts of TiO_2 NPs on the environment and living organisms. TiO_2 NPs are widely used in various industries, including paint, coatings, and cosmetics, due to their high refractive index and light-scattering properties [32,33]. However, it is important to note that the toxicity of TiO_2 is influenced by multiple factors, including the specific environment and the concentration of TiO_2 NPs present. Further research is necessary to gain a comprehensive understanding of the behavior and potential ecological effects of TiO_2 NPs. It is crucial to develop appropriate strategies for the safe use and disposal of TiO_2 nanomaterials to minimize any adverse effects on environmental and human health. By promoting responsible manufacturing, usage, and disposal practices, we can ensure the sustainable and safe integration of TiO_2 NPs in various applications.

19.2.4 Copper Oxide Nanoparticle

The growing demand for CuNPS in various industries, including agriculture [34], electricity, catalysis, optoelectronics, and antimicrobial compounds, has led to an increase in their synthesis [35]. Among the several copper nanostructured materials, three of interests are copper, cupric oxide, and cuprous oxide NPs, as they exhibit low toxicity to humans. Due to their unique qualities, CuNPs are preferred in certain fields. For instance, the catalyst industry utilizes copper oxide due to its high atom efficiency and ease of product isolation [36]. The interplay between optical and electronic applications is well understood. In electronic applications, band structure and optical absorption are crucial factors. CuNPs have broader band gaps compared to solid copper oxide, resulting in increased optical conductivity at higher photon energies, leading to enhanced absorbance [37]. Furthermore, the size of the crystal affects sensing capabilities. Nano scale copper crystals create electron-depleted regions, enabling the conduction of electricity in response to gas reactions. This property makes CuNPs valuable in the sensing industry [38]. CuO NPs, known as the greatest donors and receivers of electrons among metals, exhibit potent antibacterial activity. They can bind with the cell membrane and interfere with the genetic makeup of bacteria, effectively killing those [39]. Studies have shown that the size of CuNPs plays a significant role in determining their toxicity. Smaller particles, less than a nanometer in size, tend to be more toxic to microbes than larger particles present in bulk materials. While CuNPs are generally considered to have low toxicity to humans, it is crucial to understand their potential environmental impact. The release of CuNPs into ecosystems, whether through intentional use or accidental discharge, can pose risks to aquatic organisms and other living organisms. Factors such as NP size and their interactions with different organisms can influence their toxicity. Further research is needed to evaluate the long-term effects and ecological consequences of CuNPs in the environment [40].

19.3 NANOPARTICLE TRANSFORMATION PROCESSES IN THE ENVIRONMENT

The term "transformation" refers to the structural changes that NPs experience when exposed to their surrounding environments. In recent years, the popularity and widespread use of the ENMs have sparked interest in investigating their transformation processes in the environment. These transformations occur when ENMs interact with environmental factors such as light, chemicals, and biological entities. By understanding these processes, scientists can assess the environmental behavior and potential risks associated with the use of ENMs in agriculture and other sectors. Accurately predicting the fate, transport, and potential ecological effects of ENMs requires a comprehensive understanding of their transformation processes. These transformations can significantly alter the properties and behaviors of ENMs, ultimately influencing their interactions with the environment and living organisms. In the following sections, we will explore some of the key transformation processes observed in AgNPs, CuNPs, TiO_2 NPs, mesoporous SNPs, and ZnO NPs. These transformations can include oxidation, reduction, sulfidation, photo catalysis, functionalization, dissolution, surface complexation, and aggregation (Figure 19.1) [41]. By examining these processes, we aim to gain insights into the behavior, fate, and potential environmental impacts of these popular ENMs

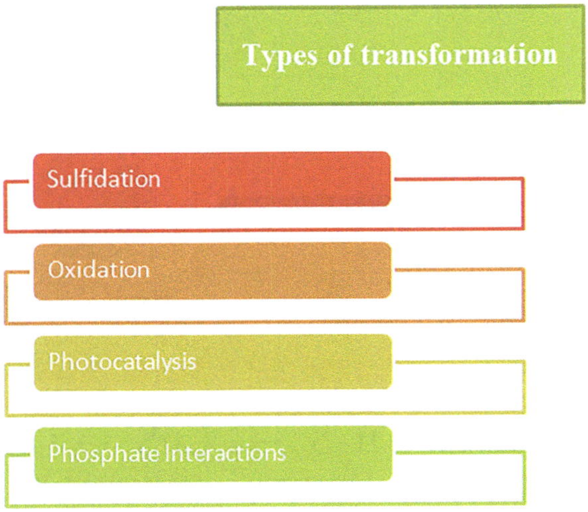

FIGURE 19.1 Common transformation processes of engineered nanomaterials.

in agricultural systems. Understanding the transformation processes of ENMs is crucial for the safe and sustainable integration of Agri nanotechnology into agriculture. By considering the potential risks and benefits associated with these transformations, we can develop appropriate strategies and guidelines for the responsible use of ENMs, minimizing any adverse effects on the environment and ensuring the long-term sustainability of agricultural practices.

19.3.1 Sulfidation

This is a displacement of metal ions with S in low redox conditions, such as sediments and some phases of wastewater treatment. This has a consequence on metal and metal oxide NPs, converting them into degraded products [42]. It was found that AgNPs easily combine with sulfide to produce Ag_2S. Sulfidation leads to reduction of AgNP toxicity. Levard et al. [43] reported that sulfidation of AgNPs showed reduced toxicity to four diverse types of aquatic and terrestrial eukaryotic organisms (*Danio rerio* (zebrafish), *Caenorhabditis elegans* (nematode worm), *Fundulus heteroclitus* (killifish), and the aquatic plant *Lemna minuta* (least duckweed)) [43]. The environment's sulfide and specific nanomaterial molar concentrations determine how quickly a certain nanomaterial sulfidates.

At room temperature, a variety of CuO species, including crystalline CuS (covellite) and amorphous (CuxSy) species are formed by the interaction between CuO and inorganic sulfide (Figure 19.2) [44]. The sulfidation process of CuNPs involves direct solid–fluid sulfidation as the dominant pathway, accompanied by a lower amount of dissolution–precipitation formation of small CuxSy clusters [45]. This transformation affects the stability and reactivity of CuNPs and can have implications for their environmental behavior and toxicity.

Nano TiO_2 undergoes a different transformation process known as photo catalytic sulfidation in the presence of inorganic sulfide and light irradiation. This process leads to the formation of sulfide species on the surface of TiO_2 NPs, altering their properties and reactivity [46]. The sulfidation of TiO_2 NPs can affect their photo catalytic activity and has implications for their fate and behavior in the environment, particularly in the presence of sulfide-rich environments or under light exposure. Similarly, mesoporous SNPs can undergo sulfidation in the presence of inorganic sulfide. The transformation of mesoporous SNPs can lead to the formation of sulfide species within the mesoporous structure, affecting their adsorption properties, drug release behavior, and catalytic activity [47]. The sulfidation process can enhance the reactivity and selectivity of mesoporous SNPs in certain applications, such as gas sensing or catalysis. In the case of ZnO NPs, sulfidation occurs

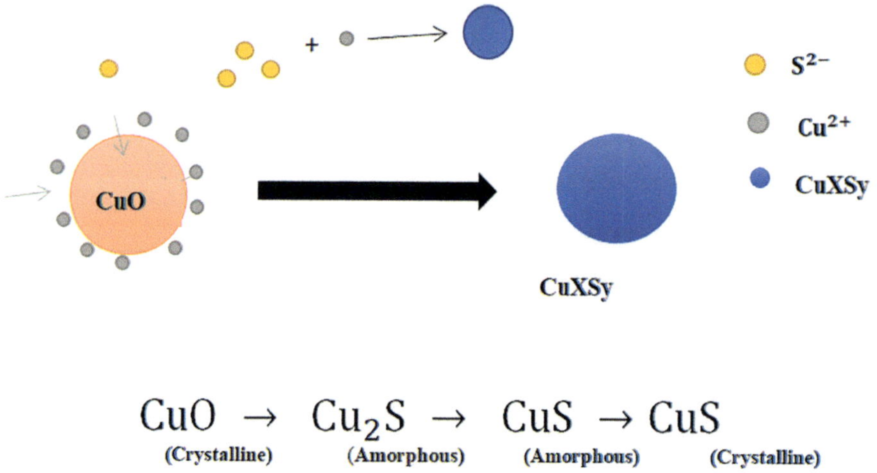

FIGURE 19.2 Diverse CuO species: crystalline covellite and amorphous CuxSy formed via CuO and inorganic sulfide interaction.

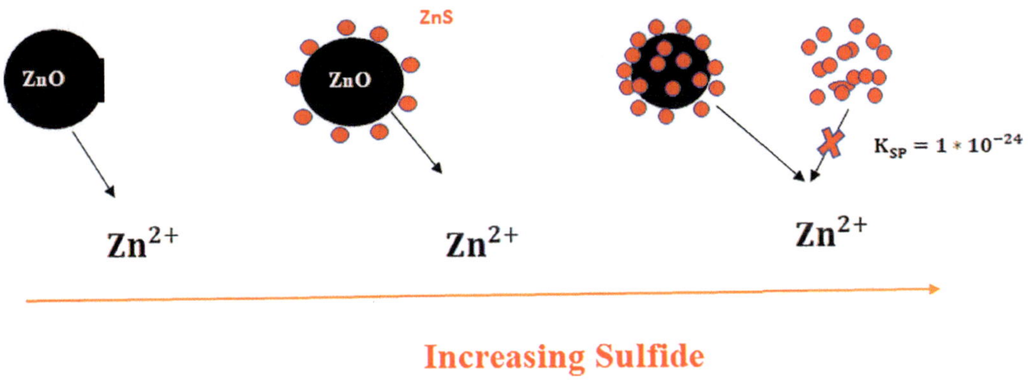

FIGURE 19.3 Sulfidation transformation of ZnO NPs: Complete conversion of ZnO NPs to nano ZnS in the presence of optimal sulfide concentration.

through a mechanism known as dissolution and precipitation in the presence of inorganic sulfide (Figure 19.3). Over a period of 5 days, ZnO NPs can be completely sulfidated into nano zinc sulfide when the appropriate concentration of sulfide is present [48]. This transformation affects the structure and properties of ZnO NPs, influencing their reactivity, solubility, and potential toxicity in the environment.

19.3.2 Oxidation

In the realm of ENMs, oxidation refers to the formation of a relatively insoluble oxide surface coating on NPs, which serves to passivate the surface and prevent further oxidation. This oxidation phenomenon occurs in conditions that promote oxidation, such as freshwater and aerated soil [49]. AgNPs are susceptible to oxidation in seawater, resulting in the release of free silver ions. While these ions possess antimicrobial properties, they can also pose toxicity risks to various aquatic organisms [50]. The development of an oxide layer on the surface of AgNPs significantly influences their stability, reactivity, and ecological implications. Similarly, CuNPs can undergo oxidation in specific environmental settings. This leads to the formation of a copper oxide layer on the surface of CuNPs, influencing their behavior and bioavailability. The oxidation process has the potential to affect the rate of copper ion release and dissolution, which may have implications for the toxicity and environmental fate of these NPs [51]. In the case of ZnO NPs, oxidation results in the creation of a protective oxide layer on their surface. This layer enhances stability and reduces the solubility of ZnO NPs, thereby impacting their reactivity and potential toxicity in the environment. The oxidation process of ZnO NPs plays a significant role in their interactions with organisms and the surrounding environment [52]. Mesoporous SNPs can also undergo oxidation, leading to the formation of an oxide layer on their surface. This alters the surface properties of SNPs, influencing their adsorption capacity, drug delivery behavior, and catalytic activity. The oxidation process has noteworthy effects on the performance and potential applications of mesoporous SNPs across various fields. Overall, the oxidation of AgNPs, CuNPs, ZnO NPs, and mesoporous SNPs has significant implications for their properties, behavior, and potential ecological impacts. It is crucial to comprehend the oxidation processes involved in these nanomaterials to accurately evaluate their environmental fate and ensure their safe and sustainable utilization [53].

19.3.3 Photo Catalysis

Photo catalysis is the process of increasing the ability to absorb light on a substrate using a catalyst. Through the creation of an electron–hole pair, catalysts improve reactions. Substrate and electron interaction results in free radicals, which then react with reactants to generate the final product [54]. Photo catalysis of CuO, AgNPs, mesoporous SNPs, and ZnO NPs in soil and aquatic systems relies on harnessing light energy to facilitate chemical reactions and transform these NPs.

CuNPs exhibit photo catalytic behavior in both soil and aquatic environments. When exposed to light, CuNPs can generate electron–hole pairs that engage in redox reactions. These photo catalytic processes can impact the breakdown of organic pollutants, the alteration of inorganic compounds, and the overall destiny of CuO NPs in the environment. AgNPs also demonstrate photo catalytic properties when illuminated in soil and aquatic systems. Light absorption by AgNPs leads to the production of reactive oxygen species (ROS) and electron–hole pairs, contributing to the degradation of organic pollutants and the antimicrobial effects of AgNPs. These photochemical processes are particularly relevant for environmental remediation and water treatment. Mesoporous SNPs, with their distinctive mesoporous structure, serve as effective photo catalysts in soil and aquatic environments. The presence of light triggers photo excitation of mesoporous SNPs, initiating electron transfer and redox reactions. This photo catalytic activity can be utilized for different purposes such as pollutant degradation, drug delivery, and energy conversion in the environment. ZnO NPs are widely recognized for their remarkable photo catalytic properties. When exposed to light, ZnO NPs generate electron–hole pairs, which actively participate in diverse photochemical reactions. These reactions facilitate the decomposition of organic compounds, the generation of ROS, and the transformation of pollutants in soil and aquatic systems.

Photo catalysis in soil and aquatic systems occurs through the interaction of photo catalytic materials, such as titanium dioxide (TiO_2), with light energy. Here's a general overview of how photo catalysis takes place in both environments:

19.3.3.1 Soil Photo Catalysis

In soil, photo catalysis involves the activation of photo catalytic materials by sunlight or artificial light sources. When sunlight or UV radiation strikes the surface of TiO_2 particles in the soil, the photons provide energy that promotes the movement of electrons from the valence band to the conduction band of TiO_2. This excitation creates electron–hole pairs. The generated electron–hole pairs in TiO_2 can participate in redox reactions, leading to the production of ROS such as hydroxyl radicals (OH) and superoxide ions (O_2^-). These ROS possess strong oxidative properties and can react with various organic compounds present in the soil, including pollutants, organic matter, and contaminants. The ROS produced during photo catalysis in soil initiate oxidation reactions with organic contaminants, breaking them down into smaller, less harmful compounds. The hydroxyl radicals, in particular, play a crucial role in the degradation of pollutants through their strong oxidative nature. This process ultimately leads to the mineralization and removal of pollutants from the soil environment. Aquatic

19.3.3.2 Photo Catalysis in Aquatic System

In aquatic systems, photo catalysis occurs similarly to the soil, but with some variations due to the presence of water. TiO_2 NPs can be suspended in water or attached to surfaces such as sediments or the walls of water treatment systems. When illuminated by light, the photo catalytic process is initiated, and the generated ROS can react with organic pollutants in the water. Hence, photo catalysis of TiO_2 NPs can be utilized in soil remediation, pollutant degradation, and water treatment processes, contributing to environmental purification and remediation.

19.3.3.3 Phosphate Interactions

The transformation of ENMs known as "phosphate interactions" refers to the processes involving the interaction between ENMs and phosphate compounds in the environment. These interactions can occur in various environmental compartments, including soils, sediments, and aquatic systems. The phosphate interactions of ENMs can have significant implications for nutrient cycling, water quality, and overall environmental health.

The presence of natural ligands on natural oxide minerals may have a variety of effects on the transformation of NPs. It was observed that the deposition of TiO_2 NPs decreases the absorbance of phosphate by the soil. A study reveals that the presence of phytate pre-sorbed γ-Al_2O_3 transforms zinc into zinc phytate, and Zn–Al layered double hydroxide precipitates. This transition becomes apparent only when a significant quantity of Zn is present [55]. The primary component of organically dispersed phosphorus in soil and water is speculated to be iron oxide NP, which is extensively prevalent in natural soil. Iron and the phosphate molecule interact quite strongly. It has been revealed that the Oxo-Fe structure in iron oxide NPs behaves as a catalyst for the breakdown of organic phosphate in water. According to the phosphorus cycle, this behavior causes the hydrolysis of phosphate esters and is essential for the transformation of phosphorus [56]. AgNPs are being used in more consumer goods and everyday items, which has resulted in a substantial discharge of these toxic substances into aquatic and terrestrial environments. In the natural environment, AgNPs must interact with many elements that change them and may change their toxicity to ecosystem biodiversity [57]. Exposure to AgNP diminished the standing crop of autotrophs. Aquatic ecosystems with AgNP outflow have been demonstrated to have decreased phytoplankton C:P ratios and poorer phytoplankton development. Ag–P complexes are formed because of an interaction between Ag and P, thereby reducing the degree of severity of the aforementioned effects while intensifying the periphyton's adverse reaction to exposure to AgNP [58,59]. Moreover, the presence of phosphates on ENMs can influence the bioavailability of these nanomaterials to organisms. The absorption of phosphates onto ENMs can alter the mobility and accessibility of these compounds, potentially affecting their uptake by plants, algae, and other organisms. This can have consequences for nutrient cycling and the overall productivity of ecosystems.

Additionally, the phosphate interactions of ENMs can have implications for the behavior and fate of other environmental pollutants. The adsorption of phosphates on ENMs can compete with other chemicals for binding sites, influencing the retention and release of contaminants like heavy metals in soils and aquatic systems. These competitive interactions can modify the bioavailability and toxicity of pollutants, ultimately impacting the stability of the ecosystem. In conclusion, phosphate interactions of ENMs impact bioavailability, pollutant behavior, and ecosystem stability.

19.4 AGING PROCESSES AND ENVIRONMENTAL IMPACT OF NANOPARTICLES: TRANSFORMATION MECHANISMS OF NANOPARTICLES IN THE ENVIRONMENT AND THREAT ASSESSMENT

NPs must endure several aging processes after their discharge into the environment, including chemical change, aggregation, and disaggregation. The main soil transformation mechanisms and the threat posed by NPs are:

a. Dissolution
b. Aggregation
c. Surface chemistry
d. Accumulation in living organisms leading to toxicity

19.4.1 Dissolution

Since every single NP has a unique set of traits, they all display unique patterns of breakdown. ENMs are divided into two groups: soluble and non-soluble, based on their solubility patterns. ENMs are classified into two groups based on their solubility pattern (Figure 19.4).

During the investigation of dissolution, it is observed that certain changes in ENMs affect their solubility behavior, which in turn affects the release rate and toxicity [60]. In contrast to natural water, where carbon NPs are rather insoluble, UV radiation exposure causes the synthesis of C_{60}

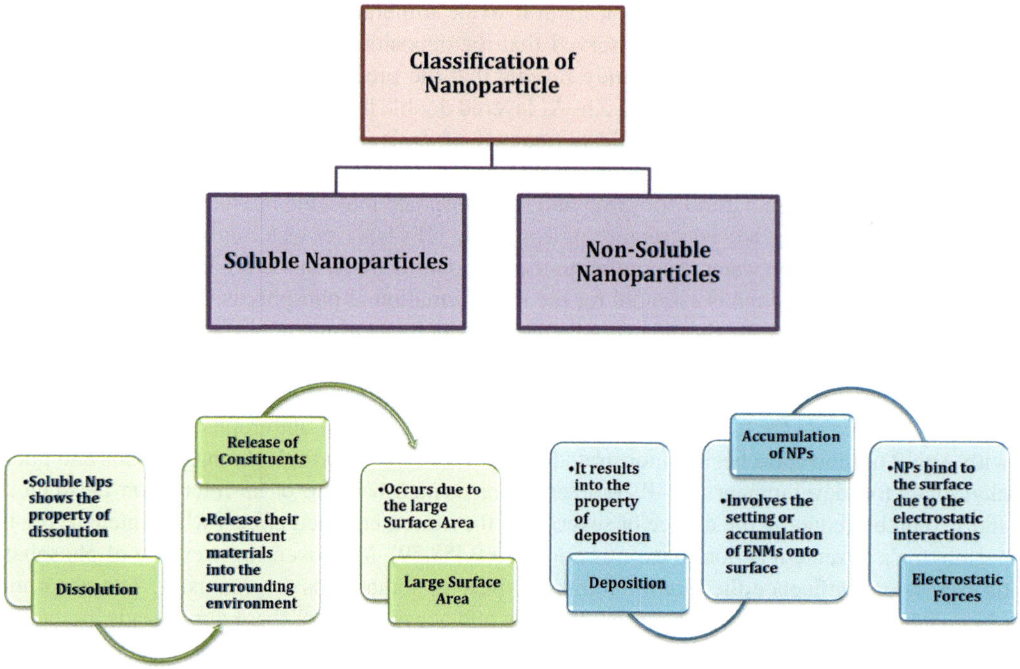

FIGURE 19.4 Classification of engineered nanomaterials based on their solubility.

(a derivative of carbon), which produces highly oxidized fullerene oxide that is soluble in water. Aggregate dissolution is caused by the reaction between C_{60} and dissolved ozone [61,62]. The term dissolution describes a metal ion's propensity to participate in redox and complexation reactions. Since smaller AgNPs have a lower redox potential and more surface atom than larger AgNPs, they tend to dissolve more readily in water [63]. Surface coatings have an enormous effect on the solubility of specific ENMs in the environment. For instance, AgNPs coated with citrate have the propensity to dissolve more slowly than those coated with Polyvinylpyrrolidone (PVP) or tannic acid. Lower solubility is due to the presence of a reducing agent in citric acid which brought reduction of the oxide layer back to zerovalent Ag and the reaction between the Ag ions and carboxylic group of organic acid results in the retention of Ag^+ [64,65]. The dissolution rate of NPs may be impacted by the presence or absence of organic and inorganic ligands in addition to by aerobic and anaerobic conditions [66]. Organic ligands have a capability to change ferric cation into ferrous [67]. Oxalate can induce the formation of ferrous ions, which increases solubility. Different anions also influence how ENMs dissolve. The dissolution rate of some NPs might improve with an increase in ionic strength. While some anions operate as ENM inhibitors, the dissolving rate of some NPs may accelerate with an increase in ionic strength. For instance, oxoanions like phosphate, arsenate, and borate obstruct the dissolution due to the formation of bi- or multinuclear inner sphere surface complexes which is energetically unfavorable [68], whereas Cl, SO_4^{2}, and NO_3 promoted the dissolution.

In the above context, we can conclude that it is the potential of metal ions to form a complex and the presence of ligands and protons which decides whether the ENMs are categorized into soluble or non-soluble [69].

19.4.1.1 Deposition

Deposition is the action directly opposed to dissolution. On the other end of the spectrum, non-soluble ENMs prefer to deposit on the surface whereas soluble particles are more likely to demonstrate the process of dissolution. Deposition indicates the way molecules or atoms accumulate on an identifiable surface or substrate. This phenomenon is triggered by a number of factors, including surface contact, electrostatic interactions, Brownian motion, and gravitational forces [69]. Engineered Nanoparticles (ENPs) get deposited as a result of electrochemical forces and surface contacts working together, however particle surface heterogeneity may have a greater impact on the process. A study demonstrates that compared to the original SiO_2 micro particles, deposition of the nano silicon dioxide microspheres modified by the surface coating of positive charge Poly-(L-lysine) is slow.. However, PLL-modified silicon dioxide microspheres are not depositing on surfaces made of aluminum oxide [70,70]. Particle surfaces in the outside world are always heterogeneously or inconsistently charged and may be partially covered with organic material from the environment or other functional groups. So we can say that particle surface heterogeneity plays important role in deposition of ENPs [71]. The polysaccharide found in the ecosystem's natural biofilm acts as a key point of contact for the ENMs emitted into the environment [72]. Hematite and silica are from those NPs that can interact with the extracellular polymeric elements that make up the biofilm. Electrostatic forces have a significant role in the interaction of NP biofilm, along with other forces including the heterogeneity of surface charges. It has been found that the alginate, dextran sulfate, and dextran-coated surface displays a higher rate of hematite NP deposition than the chitosan surface. This pattern is brought about by the fact that positively charged hematite and strongly negatively charged alginate exhibit electrostatic attraction, whereas hematite and chitosan exhibit repulsion. Alginate, dextran sulfate, and dextran all have relatively low rates of silica NP deposition while chitosan-coated surfaces have higher rates of deposition. Alginate- and dextran sulfate-coated surfaces showed electrostatic repulsion, and it seemed that the variations in average negative charge densities had no impact on the strength of the electrosteric attraction between the silica NPs and polysaccharide-coated surfaces [73].

19.4.2 Aggregation

Aggregation of ENMs refers to the process by which individual nanoscale particles come together to form larger clusters or aggregates. It is a common phenomenon that can occur in various environments where ENMs are present, such as liquid suspensions, aerosols, or even solid surfaces. The aggregation of ENMs can happen through different mechanisms, including Brownian motion, van der Waals forces, electrostatic interactions, and bridging agents. The Derjaguin–Landau–Verwey–Overbeek (DLVO) theory can be applied to explain the aggregation behavior of ENMs in colloidal systems [74]. According to the DLVO theory, the interplay of attractive Van der Waals force, repulsive electrostatic force, total interaction energy, and energy barrier influences the aggregation behavior of particles. Attractive Van der Waals forces promote aggregation, while repulsive electrostatic forces counteract aggregation. The balance between these forces determines the total interaction energy, which, in turn, influences the stability and propensity for aggregation. The energy barrier created by the repulsive forces prevents aggregation unless external influences reduce the barrier height. It's important to note that the DLVO theory provides a simplified framework for understanding ENM aggregation, but it may not fully capture the complexity of real-world systems. Other factors, such as steric effects, surface modifications, specific interactions, and environmental conditions, may also play a role in ENM aggregation and should be considered for a comprehensive understanding of the process [75]. It is important to note that the specific effects of these transformations on NP aggregation can vary depending on the environmental conditions, particle characteristics, and the presence of other substances. The aggregation of AgNPs in soil and aquatic environments can be influenced by various transformations such as coating modifications, aggregation by Natural organic matter (NOM) that occur in these environments [76]. The presence of ions in soil and aquatic environments can affect AgNP aggregation. Certain ions, such as divalent cations (e.g., calcium, magnesium) and multivalent anions (e.g., phosphate, silicate), can form bridges between AgNPs, leading to increased attractive forces and enhanced aggregation. Conversely, the presence of ions with electrostatic repulsion (e.g., monovalent cations) can stabilize AgNPs and inhibit aggregation [77]. AgNPs can undergo redox reactions, particularly in environments with variable oxygen levels or the presence of redox-active substances. Oxidation of AgNPs can lead to the formation of Ag ions, which may then interact with other particles or undergo further transformations. Changes in the redox state of AgNPs can influence their surface charge and aggregation behavior [78]. In natural environments, sulfidation reactions of ZnO NPs can occur through interactions with sulfur-containing compounds present in the environment. One specific example is the reaction between ZnO NPs and hydrogen sulfide (H_2S) gas or sulfide ions (S^{2-}) in water bodies or sediments. This reaction can lead to the formation of ZnS on the surface of ZnO NPs. The sulfidation of ZnO NPs decreases their aggregation propensity due to surface modification, reduced surface energy, and increased stability provided by the ZnS shell formed on the NPs' surface [79].

19.4.3 Surface Chemistry

Interaction between two interfaces, which is also known as surface chemistry, is influenced by the transformations of ENMs that occur in a natural environment. Surface chemistry plays a crucial role in various scientific disciplines and industries, including materials science, catalysis, electrochemistry, colloid and interface science, and biotechnology. Variation in surface chemistry means there is a change in the functional group of surfaces, which results in a change in the surface charge as it depends on the type of the functional group. Sometimes reactions with ROS in the environment may lead to oxidation, and this results in chemical modification by changing the functional group. The ENM with the hydroxyl or amine groups changes into a carbonyl group due to the oxidation. So, these types of reactions have significant implications for their reactivity, stability, and interaction with biological systems, as they can influence their surface chemistry, surface charge, and surface reactivity [36]. For instance, the formation of O-functional groups and the leaching of positively charged mineral phases from pyrogenic/petrogenic carbon NPs (pCNPs) would render their surface charges more negative and reduce their ζ-potential, isoelectric poin (IEP), and point-of-zero charge [80].

19.4.4 Accumulation in Living Organisms Leading to Toxicity

Transformations of ENMs have a significant impact on their toxicity in the environment, underscoring the need to better understand and characterize their potentially harmful effects. Each NP exhibits unique reactions and responses to various transformations, which contribute to their toxicity alongside their beneficial properties. However, the mechanisms underlying their toxicity differ among different types of ENMs. For example, fullerene NPs exert their toxic effects by directly damaging membrane proteins and lipids through oxidative processes, without the production of ROS. In contrast, wide-bandgap semiconductors like TiO_2 or ZnO can generate electron–hole pairs when exposed to UV light [81]. This leads to the reduction/oxidation reactions of nearby species, and certain metal oxides are known to exhibit toxicity due to the production of ROS through photo catalysis. Moreover, the toxic mechanisms of ENMs can also be influenced by light conditions. In the dark, TiO_2 NPs demonstrate toxicity by attaching to surfaces and inducing oxidative stress. However, under UV exposure, their toxicity remains significant, but now it arises from the generation of ROS induced by UV radiation [82,83]. A model for light triggered TiO_2 NP toxicity is depicted in Figure 19.5. The transformations of AgNPs in the environment can significantly impact their toxicity in both soil and aquatic environments. When AgNPs are released into the soil, they can undergo various transformations such as aggregation, dissolution, and sulfidation. Transformation such as aggregation tends to decrease the toxicity of AgNPs due to reduction in available surface area for interactions.

In summary, the transformations of ENMs play a critical role in determining their toxicity in the environment, and understanding the specific mechanisms behind their harmful effects is essential for the comprehensive assessment and management of these nanomaterials. The toxicity of AgNPs

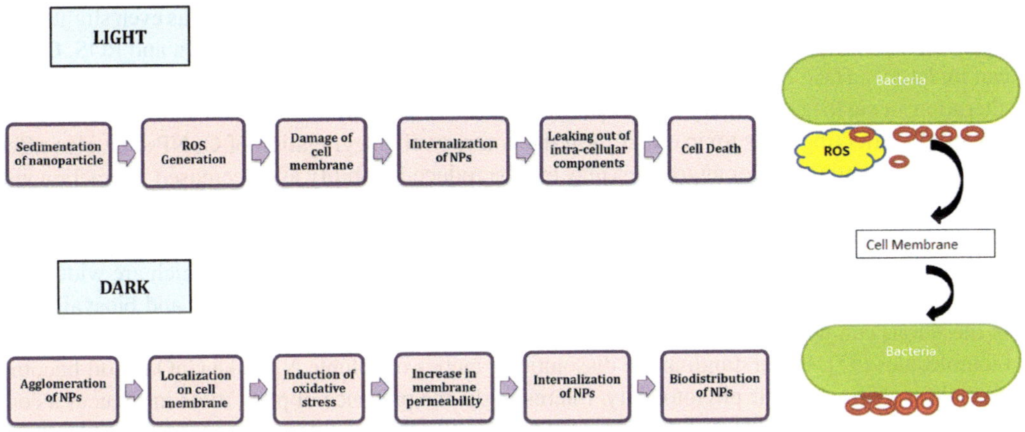

FIGURE 19.5 Interaction between TiO_2 NP and bacteria under light and dark conditions.

in the environment is primarily attributed to the release of silver ions (Ag^+) from AgNPs. These ions can damage cells and inhibit growth [84]. Sulfidation, a rapid process facilitated by natural elements like air and water, occurs more quickly in AgNPs compared to bulk silver and other NPs [85]. Understanding sulfidation processes in both air and water is crucial from an environmental perspective, as AgNPs often end up in sewage pipes, wastewater treatment plant effluent, and bio solids. When AgNPs react with sulfide ions (S^{2-}) in soil and aquatic environments, they form less toxic silver sulfide (Ag_2S) NPs, which are less bioavailable and reactive [86,87]. Aggregation, dissolution, and sulfidation can occur in aquatic environments, influencing the availability, uptake, and toxicity of AgNPs to aquatic organisms such as fish, algae, and benthic organisms. Overall, AgNPs exhibit toxicity toward various organisms, including benthic organisms [88], soil bacteria, algae [89], and *Nitrosamonas europea*, a planktonic ammonia-oxidizing bacteria [90]. Once mesoporous SNPs are released into the environment, they undergo various transformations that can change their physical and chemical properties, ultimately impacting their toxicity. Aggregation of MSNPs influences their availability and uptake by organisms, where larger aggregates may exhibit reduced mobility and decreased interaction with biological systems, potentially reducing their harmful effects [91]. Dissolution of mesoporous SNPs releases SNPs and ions, while surface modification or functionalization can also modify their toxicity by enhancing or diminishing interactions with biological systems [92]. A specific functional group associated with increased toxicity of mesoporous SNPs is the amino group (-NH_2), found on the surface of mesoporous SNPs. The presence of NH_2 groups enhances their interaction with biological systems, potentially leading to cytotoxic effects. For example, NH_2 functionalized mesoporous SNPs have been found to reduce egg fertilization rates and have negative impact on the embryonic development of marine organisms, leading to morphological alterations, increased embryo mortality, and disruptions in mitochondrial function and skeletogenesis. These findings underscore the potential toxic effects of NH_2 functionalized mesoporous SNPs on the marine ecosystem, with implications for the development and reproduction of organisms within it [93]. Furthermore, it should be noted that the degradation of surface modifications over time can have an impact on the toxicity of mesoporous SNPs. It should be noted that the degradation of surface modifications over time can have an impact on the toxicity of mesoporous SNPs [94]. For instance, vinyl-modified SNPs were observed to be internalized by lymphocytes to varying degrees of efficiency and did not exhibit any cytotoxic or genotoxic effects according to various assessment methods. However, these particles did affect the proliferation of lymphocytes, leading to a decrease in the mitotic index value and disruptions in cell cycle progression. In contrast, unmodified SNPs demonstrate cytotoxic and genotoxic properties at high doses, and they also interfere with the cell cycle [95]. This highlights the importance of considering the specific surface modifications of mesoporous SNPs in relation to their potential toxicity, as even slight variations can significantly influence their biological effects. When exposed to oxygen and ROS, CuNPs can undergo surface oxidation, resulting in the formation of copper oxide/hydroxide species. This alteration in surface chemistry can impact the interaction between CuNPs and living organisms, potentially affecting their toxic effects. On the contrary, the aggregation of CuNPs can also affect their toxic potential. Larger aggregates may exhibit reduced mobility and decreased interaction with biological systems, leading to potentially lower toxicity. However, smaller aggregates or individual NPs, with their larger surface area and enhanced reactivity, may pose a greater risk [96]. In water or soil environments, the dissolution of CuNPs can liberate copper ions (Cu^{2+}), which are widely recognized for their toxicity across various organisms. The remarkable reactivity and bioavailability of these ions can instigate oxidative stress, disrupt crucial cellular processes, and inflict damage to biomolecules [97]. Understanding the dissolution kinetics of soluble NPs like CuO in soil becomes crucial for assessing their phytotoxicity. Interestingly, the presence of plant roots introduces a complex dynamic that impacts NP dissolution. Specifically, the presence of roots in the soil exhibits opposing and somewhat compensatory effects on NP dissolution, as observed in rhizosphere soil. The alteration in pH induced by the roots contributes to a reduction in CuNP dissolution [98]. This interplay between roots and CuNPs adds an additional layer of complexity, highlighting the intricate relationship between NPs, soil, and plant systems.

19.5 IMPACT OF SIZE DISTRIBUTION, PH, AND IONIC EFFECT ON NANOPARTICLE TRANSFORMATION IN THE ENVIRONMENT

NPs' physical characteristics, such as their size distribution, and environmental parameters like pH and ionic effect may play a crucial role in NP transformation.

19.5.1 Size Distribution of Nanoparticles

The size of ENMs plays a crucial role in their transformations in soil and aquatic environments. Smaller ENMs have a higher surface-to-volume ratio, increasing their reactivity and interaction with the environment. They exhibit greater mobility, can penetrate deeper into soil, and can be transported over longer distances in aquatic systems. Smaller particles are prone to aggregation, altering their behavior, mobility, and toxicity. Their greater bioavailable potentially leads to greater bioaccumulation and toxic effects. Smaller ENMs also exhibit higher reactivity and undergo faster transformations such as dissolution or reduction. Their smaller size makes them more susceptible to transport, dispersal, and rapid degradation or transformation. Understanding the influence of ENM size is crucial for assessing their environmental fate and potential risks in soil and aquatic environments [99–103]. For example, the smaller-sized AgNPs are found to have increased toxic effects on *C. elegans*. AgNPs with a size range of 10 nm exhibits a higher propensity for aggregation and physical transformation compared to AgNPs of 75 nm size, primarily due to the presence of elevated surface energy in smaller NPs. This phenomenon can be attributed to the surface area differences, which render the smaller particles to be more reactive in nature compared to the larger particles [104,105]. The size-dependent transformations of AgNPs highlight the increased reactivity and faster rates of transformation observed in smaller particles. Similarly, as the size of TiO_2 NPs decreases, several size-dependent transformations can occur. One significant transformation is the phase transition between different crystal structures. TiO_2 exists in various crystalline forms, such as anatase, rutile, and brookite [106]. Smaller NPs have a higher surface energy, which can lead to the stabilization of metastable phases, such as anantase, that are thermodynamically less favorable in bulk materials [107–110]. As the particle size increases, the energy required for the phase transition to more stable phases such as rutile decreases. Therefore, larger TiO_2 NPs are more likely to transform into the rutile phase [111]. The size-dependent transformations of AgNPs, including their increased reactivity and faster rates of transformation, and the phase transition in TiO_2 NPs are examples of how particle size influences the behavior and transformations of nanomaterials. On the contrary, the size-dependent dissolution of TiO_2 NPs in the soil is a distinct transformation process that occurs because of their interaction with environmental factors. The size-dependent dissolution of TiO_2 NPs in soil occurs as a result of their interaction with environmental factors. Smaller TiO_2 NPs, with a larger surface area, are more prone to dissolution compared to larger particles. When these NPs come into contact with soil moisture or other solvents, the dissolution process begins [112]. The dissolution of TiO_2 NPs involves the release of titanium ions from the NP surface into the surrounding soil solution. In acidic soil environments, characterized by low pH, the dissolution of TiO_2 NPs can be enhanced. When smaller TiO_2 NPs get dissolved in acidic soil, Ti ions are released into the soil solution [113,114]. These dissolved Ti ions can then interact with various components in the soil, such as organic matter, clay minerals, or other ions present. They may form complexes or bind to these soil constituents influencing their mobility, bioavailability, and potential impacts on soil organisms and plants [115]. The size-dependent dissolution of TiO_2 NPs in acidic soil highlights the importance of considering both the particle size and soil conditions when evaluating their behavior and potential risks in the environment.

19.5.2 pH

The pH of the environment plays a crucial role in the transformation and interactions of ENMs with organisms. The surface charge and solubility of ENMs are influenced by the pH, which in turn affects their behavior. Rathnayake et al. [116] demonstrated that the transformation of ENMs can vary depending on the concentration of specific functional groups or ions at different pH levels [116]. For example, it is observed that citrate-capped AgNPs are more stable at pH 7 and 9, while Bovine serum albumin (BSA)-capped AgNPs are stable over a wider pH range (2, 3, 7, and 10). The pH-dependent stability of AgNPs can be attributed to the speciation of citric acid and the steric effect of BSA on the surface of the NPs. Surface charge of carbon coated AgNP becomes positive at pH 4 suggesting dependency of surface charge on pH. These results in increase of agglomeration [117]. Change in pH also affects the dissolution rate of AgNP as there is dependency on the protons and dissolved O_2. It is observed that with the increase in pH, release of Ag^+ from AgNP and dissolution is decreased [118,119]. These pH-dependent transformations can be particularly relevant in industrial and municipal wastewaters, where the specific pH conditions can potentially mediate these transformations [120]. Similarly, Loosli and Stoll investigated the effect of pH (3.7, 6.2, and 9.7) on the heteroaggregation of TiO_2 NPs in a latex suspension. They observed that better aggregates were formed at pH 3.7 compared to pH 6.2 and 9.7. This was attributed to the charge neutralization between the negatively charged latex suspension and the positively charged TiO_2 NPs [121]. The pH-dependent aggregation kinetics demonstrates the importance of pH in influencing the interactions between different NPs and their surroundings. Changes in pH can have a significant influence on the transformation of ENMs such as nano CuO. In high pH environments (alkaline conditions), nano CuO tends to undergo dissolution, resulting in the release of copper ions (Cu^{2+}). This dissolution process is enhanced by the presence of hydroxide ions (OH^-) in alkaline solutions [122,123]. On the contrary, in low pH environments (acidic conditions), the dissolution of nano CuO is reduced or inhibited due to the lower concentration of OH^-. The pH also affects the surface reactivity and agglomeration of nano CuO. Change in pH results decrease in size of CuO agglomerates and increase in the stabilization of CuO [124]. In high pH environments, surface hydroxyl groups can form, and this can influence the reactivity and stability of the nanomaterial. Conversely, in low pH environments, surface groups can become protonated, leading to changes in surface charge and reactivity [125]. Furthermore, pH impacts the aggregation behavior of nano CuO. In high pH environments, particle size decreases, which can cause decreased particle aggregation. This aggregation alters the size, shape, and stability of the particles. In contrast, in low pH environments, the stability of nano CuO particles may be enhanced, reducing their tendency to aggregate or precipitate. The oxidation state of copper on the surface of nano CuO is also influenced by pH.

In high pH conditions, the surface of nano CuO may exhibit a higher oxidation state, such as Cu (II) or Cu (III), due to the formation of surface hydroxide or oxide species. In low pH conditions, the surface oxidation state may be lower, such as Cu (I), due to the reduction potential of the acidic environment. Mesoporous SNPs can undergo dissolution in both soil and aquatic environments. In acidic conditions, the low pH can promote the dissolution of silica, resulting in the release of silicate ions. This dissolution process is facilitated by the presence of hydrogen ions (H^+) in the solution. In alkaline conditions, the dissolution of mesoporous SNPs may be reduced due to the lower concentration of H^+ ions. The reactivity of the mesoporous SNPs' surface can be influenced by pH. In acidic conditions, the silanol groups on the surface of mesoporous SNPs can become protonated, leading to changes in surface charge and reactivity. This protonation can affect interactions with surrounding ions, molecules, or organic matter. In alkaline conditions, the deprotonating of silanol groups may occur, altering the surface charge and reactivity. Mesoporous SNPs can exhibit different aggregation behavior in soil and aquatic environments depending on pH. In some cases, at certain pH ranges, the electrostatic interactions between mesoporous SNPs can lead to aggregation and the formation of larger agglomerates. The pH-dependent surface charge and the presence of counter ions can affect the stability and dispersion of mesoporous SNPs. In certain cases, the pH conditions in soil or aquatic environments can facilitate surface modification of mesoporous SNPs. For example, pH-dependent hydrolysis reactions can occur on the surface of mesoporous SNPs, leading to the attachment or release of functional groups. These modifications can impact the properties and behavior of mesoporous SNPs, including their interactions with contaminants or biological entities. For example, there is no effect of pH is observed on bare SNPs but surface coatings such as alumina-coated SNP has shown increase in toxicity level at pH 6. Toxicity of alumina-coated SNP decreases with decrease in pH, whereas it increases with increase in pH. In summary, the pH of the environment has a significant impact on the stability, transformation, and aggregation of ENMs. Understanding these pH-dependent processes is crucial for predicting and managing the environmental fate and behavior of ENMs.

19.5.3 Ionic Effect

The ionic strength of the soil and water system plays a role in the transformation of ENMs. Ionic strength refers to the concentration of ions present in a solution and has an impact on the behavior and fate of ENMs.

In environments with high ionic strength, where there are increased concentrations of ions, the stability and aggregation behavior of ENMs can be affected. The higher ionic strength promotes the aggregation of ENMs by diminishing the electrostatic repulsion between particles. This leads to the formation of larger aggregates or agglomerates. Consequently, the mobility and transport of ENMs in soil and water systems are altered, potentially influencing their distribution and environmental impacts [126]. Numerous studies have indicated that the agglomeration kinetics of Ag, CuO, TiO_2, and ZnO NPs in aquatic and soil suspension systems are accelerated with an increase in ionic strength. Researchers have consistently observed that higher levels of ionic strength promote the aggregation of these ENMs, leading to their increased tendency to form larger agglomerates or clusters. This phenomenon has been consistently observed across various studies conducted in both aquatic and soil environments [127–129]. The critical coagulation concentration (CCC) is the minimum concentration of electrolytes needed for particles to aggregate. It marks the point where repulsive forces between particles are neutralized, allowing attractive forces to dominate and form aggregates. Exceeding the CCC leads to increased aggregation, while staying below it prevents agglomeration. The CCC depends on electrolyte type, concentration, pH, and particle surface properties. Different NPs have varying CCC values, with stronger surface charges or greater electrostatic stabilization resulting in higher CCCs. In a different investigation, it was observed that AgNPs with citrate coating had a lower CCC value (47.6 mmol/L) compared to those with polyvinylpyrrolidone coating (CCC 111.5 mmol/L) in a monovalent NaCl electrolyte [122]. The French group studied the agglomeration kinetics of 4–5 nm TiO_2 NPs and found that they rapidly agglomerated into micro-sized clusters within a few minutes in a 16.5 mmol/L NaCl solution at pH 4.5. When a mixture of mono- and divalent ions was present, divalent cations (Ca^{2+}, Mg^{2+}) were more effective in destabilizing and agglomerating the ENMs due to their strong adsorption on negatively charged surfaces. While the presence of other ions can have a synergistic effect on agglomeration, the dominant ion in the suspension regulates the agglomeration kinetics. Clay minerals, commonly found in soil systems, can either enhance or suppress the agglomeration/stability of ENMs depending on their concentration. Montmorillonite particles, when interacting with Ag and TiO_2, destabilize the system and promote the agglomeration of these ENMs [123]. Understanding the influence of ionic strength on the transformation and agglomeration of ENMs in soil and water systems is essential for assessing their behavior and potential environmental impacts. By investigating the CCC and considering the interactions with various ions and minerals, researchers can advance our knowledge of how ENMs behave and develop strategies for their safe and sustainable use in different environmental contexts.

19.6 EFFECT OF ENGINEERED NANOMATERIALS (ENMS) IN SOIL AND AQUATIC SYSTEM

19.6.1 Bioavailability

The bioavailability of ENMs refers to the proportion of a given ENM that is available and capable of interacting with biological systems, such as living organisms or cells, once it enters the body or the environment. ENMs are intentionally designed and manufactured at the nanoscale, typically ranging from 1 to 100 nm in size. Due to their unique properties at the nanoscale, they exhibit different behaviors compared to their bulk counterparts. When ENMs are introduced into biological systems, their interactions can be influenced by various factors, including their size, shape, surface chemistry, and surface charge. Bioavailability encompasses the processes involved in the uptake, distribution, metabolism, and elimination of ENMs within living organisms. It describes how much of the administered or environmentally released ENMs can reach the target site or exert an effect on biological systems. It is important to understand the bioavailability of ENMs because it influences their potential toxicity, biological effects, and overall risk assessment. The bioavailability of ENMs refers to the proportion of a given ENM that is available and capable of interacting with biological systems, such as living organisms or cells, once it enters the body or the environment. The concentrations of environmental pollutants, specifically ENMs, are directly linked to the accumulation of these substances within the bodies of aquatic organisms, such as fish and micro crustaceans, as well as soil-dwelling creatures like earthworms and other soil insects [124]. When ENMs are introduced into biological systems, their interactions can be influenced by various factors, including their size, shape, surface chemistry, and surface charge. Bioavailability encompasses the processes involved in the uptake, distribution, metabolism, and elimination of ENMs within living organisms [125]. It describes how much of the administered or environmentally released ENMs can reach the target site or exert an effect on biological systems. It is important to understand the bioavailability of ENMs because it influences their potential toxicity, biological effects, and overall risk assessment. The bioavailable species of ENMs are potentially taken up by aquatic and terrestrial biota resulting in bioaccumulation and bio magnification in the food chain. The study conducted by the research group focused on the transformation of AgNPs in soil and sediment within freshwater mesocosms. It was observed that in soil, AgNPs transformed into Ag_2S to a degree of 52%, whereas in subaquatic sediment, the transformation reached 55%. The sulfidation process was found to be more pronounced in aquatic sediment due to the presence of anoxic conditions and favorable geochemical factors [126]. Simultaneously, erosion and runoff were observed to transport AgNPs from terrestrial soil to sediment. Notably, even after significant sulfidation, AgNPs still exhibited potential bioavailability as evidenced by their presence in terrestrial plant biomass, mosquito fish, and chironomids.

However, the duration of Ag's bioavailability and the specific bioavailable species remain uncertain. Transformation products, which could include ions, sulfides, sulfhydryl compounds, or products with unknown compositions, might contribute to explaining the observed effects. The bioavailability and aging of ENMs are influenced by factors such as their size, bulk composition, and surface coating. In a related study it is observed that smaller gold NPs (20 nm) are more bioavailable compared to larger ones (55 nm) due to differences in their aggregation within soil pore water. In another study by the research group, it was found that surface coatings of humic acid, fulvic acid (FA), and alginate increased the bioavailability of zinc NPs compared to coating with tannic acid [127]. It is observed that compared to aged CuNPs and oxide forms of Cu, fresh CuNPs show more sorption capacity of organic acids such as citric and oxalic acid. The concentrations of Cu complexation with organic acids influenced the bioavailability of the ENMs. Furthermore, smaller ENMs demonstrate higher solubility and stability, while aged CuNPs increase in size due to coalescence and coarsening under ambient environmental conditions [128,129]. This increased colloidal stability of ENMs contributes to their potential bioavailability and extended residence time in the aqueous environment. The duration of exposure can also impact bioavailability. Empirical evidence has demonstrated the accelerated leaching of TiO_2 from aged plastic materials. The mobilized particles that are released fall within a size range larger than biotically active TiO_2 NPs. However, modeling studies suggest that the biotically active NPs could originate from TiO_2 pigments present in the environment. While rutile is considered the most prominent polymorph of TiO_2 in non-fibrous plastics, the extent and nature of engineered surface modifications in consumer and environmental plastics remain largely unknown. The presence of surface modifications is expected to significantly influence the photo-oxidative degradation of plastics and the release of fine TiO_2 particles, including potentially nano-sized particles. Consequently, further research is necessary to investigate the effects of surface modification on these processes. In conclusion, understanding the bioavailability of ENMs is crucial for assessing their potential risks and impacts on biological systems. Factors such as ENM size, shape, surface chemistry, and environmental conditions can influence their interactions and transformations. The presence of bioavailable species, even after significant transformations, highlights the importance of considering the fate and effects of ENMs in different environmental compartments. Further research is needed to explore the influence of surface modifications on the behavior and bioavailability of ENMs, as well as their potential impacts on the photo-oxidative degradation of plastics and the release of fine particles.

19.6.2 Toxicity of Specific ENM to Overall Aquatic Ecosystem

ENMs pose a threat to the aquatic ecosystem as they possess distinct characteristics that make them toxic. These characteristics include the ability to generate ROS, undergo bioaccumulation and biomagnification, disrupt biological activity, and directly interact with organisms. In addition, ENMs can indirectly impact the aquatic ecosystem by causing changes in community structure and affecting ecosystem functions. The following are several ways in which ENMs exhibit toxicity in aquatic environments. Aquatic plants play a crucial role in ecosystems and have a significant influence on the behavior and destiny of ENMs. However, the toxic effects of ENMs in aquatic plants have not been extensively studied, and there is a limited amount of literature available on this topic. For instance, Jiang et al. investigated the concentration-dependent toxicity of AgNPs on the aquatic plant *Spirodela polyhriza*. They observed an increase in ROS levels in a dose-dependent manner when the plant was exposed to AgNPs, indicating AgNP-mediated nanotoxicity [129].

19.6.3 ENM Toxicity to Fishes

Water has been identified as a primary pathway for the entry of ENMs into the environment, as they are utilized, degraded, and disposed of, ultimately reaching aquatic ecosystems. Within these ecosystems, both plants and animals are vulnerable to the presence of these contaminants. In a recent study, a team of researchers investigated the uptake, distribution, and elimination of AgNPs in the common carp *Cyprinus carpio*. The toxicokinetic properties of AgNPs were examined in carp exposed for 7 days and depurated for 2 weeks. After the exposure period, various fish tissues (brain, gills, skeletal muscle, liver, gastrointestinal tract, and blood) were collected, digested, and analyzed for total silver concentration. Interestingly, no evidence of silver accumulation in the form of AgNPs was found; instead, silver was observed primarily in its ionic form across all tissues. This study provides valuable insights into the mechanisms of silver accumulation in different tissues of exposed fish, which can aid in assessing the ecological risks associated with dissolved silver ions. From another study, it was found TiO_2 NPs exhibit significant toxicity to aquatic vertebrates such as zebra fish and *Daphnia*. Zebra fish exposed to TiO_2 NPs at parts per billion (ppb) concentrations for 23 days experienced mortality. The presence of TiO_2 NPs combined with artificial solar irradiation led to the generation of dose-dependent ROS. These ROS accumulated in the organs of the zebra fish, resulting in stunted growth, organ pathology, delayed metamorphosis, and DNA damage. The size of citrate stabilized TiO_2 NPs has been discovered to significantly affect their toxicity to zebra fish embryos. ROS was shown to be produced at higher rates by smaller TiO_2 NPs (6, 12, and 15 nm) with larger surface areas. These abnormalities in zebra fish embryos included pericardial edema, yolk sac edema, craniofacial deformity, and opaque yolk. The study came to the conclusion that rather than concentration, the toxicity mechanism of TiO_2 NPs is mostly dependent on their surface area. The assessment of copper toxicity in aquatic organisms is crucial in environments vulnerable to copper pollution, particularly in proximity to sources such as industrial and urban wastewater discharge and atmospheric deposition from copper mining and smelting activities. These sources can lead to elevated copper concentrations in the aquatic ecosystem. When copper levels surpass the normal requirements for growth and development in species, accumulation can occur, resulting in irreversible harm. Research has shown that exposure to low (180 μg/L of Cu^{2+}), medium, and high levels (3,200, 1,000, and 560 μg/L) of Cu^{2+} can induce morphological changes in winter flounder (*Pseudopleuronectes americanus*) fish. Through histological techniques and electron microscope analysis, fatty metamorphosis in the liver, kidney necrosis, damage to hematopoietic tissue, and alterations in gills were observed in groups exposed to high and medium copper levels. Model organisms exposed to low copper levels displayed vacuolated epithelial layers. It suggests that exposure to elevated copper levels in aquatic ecosystems can lead to morphological changes and tissue damage in aquatic organisms, highlighting the importance of assessing copper toxicity in these environments.

19.6.4 Toxic Effect of ENMs on Rhizosphere Activities

The rhizosphere, which encompasses the area surrounding plant roots, contains a diverse community of organisms, root exudates, metabolites, and specific chemicals. These components play a crucial role in facilitating interactions between plants, pathogens, microbes, and the soil, while also serving as ecosystem engineers that impact soil health, productivity, and crop yields. NPs exhibit both positive and negative associations with plants and rhizospheric communities. It is evident that NPs present in the soil can penetrate the underground environment, leading to detrimental effects on soil properties and potentially posing risks to the surrounding ecosystem. NPs have been found to significantly hinder soil micro flora either by exerting toxic effects or altering the availability of toxins. This can indirectly impact the synthesis of organic compounds and create antagonistic relationships. The impact of NPs on microbes is influenced by various factors such as their physicochemical properties, concentration, exposure time, and growth medium. Notably, NPs possess colloidal properties, which make it challenging for microbes to uptake them. Consequently, NPs are believed to exert toxicity by solubilizing ions and invading cells through membrane disruption. Free radicals generated by NPs further impede microbial function by causing damage to cell walls, nuclei, exopolysaccharides suppression, inhibition of biofilm biosynthesis, and lipid peroxidation. Additionally, NPs can lead to membrane leakage of soluble sugars, lipids, proteins, enzyme denaturation, cell lysis, disrupted vesicles, and inhibition of cell respiration.

19.6.5 Toxic Effect on Soil Enzymes

Among the various ecosystems, the soil ecosystem is considered to receive the most significant levels of NPs from all sources. Consequently, it is crucial to understand the effects of NPs on soil enzymatic activities and their potential environmental impact. ENMs have the potential to harm soil enzymes by directly interacting with them, inhibiting their activity, and modifying their structure. They can also trigger oxidative stress, disrupt the natural defense mechanisms against it, and interfere with enzyme functions. Furthermore, ENMs can bring about changes in soil properties and microbial communities, which can have indirect effects on enzyme activity.

19.7 CONCLUSION

Nowadays several ENMs like nano Ag, CuO, ZnO, TiO_2, and SiO_2 are being utilized in all facets of human lives starting from electronic gadgets and agriculture to medical science. Hence, it is very pertinent to study their transformation in the ecosystem as the toxicity of the nanomaterials depends on their final physicochemical characteristics. The process of ENM transformation is a very complex process, and this depends on various environmental factors like pH and ions present in the surrounding environment. The transformation process eventually dictates the dissolution, aggregation, and surface properties of an ENM. Ultimately these properties become crucial for the determination of toxicity of ENMs in soil and aquatic systems. This book chapter serves as a valuable resource for researchers, environmental scientists, policymakers, and stakeholders interested in the field of nanomaterials and their interactions with soil and aquatic environments. The findings and insights provided contribute to our knowledge of the environmental fate and impacts of ENMs, paving the way for informed decision-making and responsible use of these materials in various applications.

REFERENCES

1. Kumar, N., Tripathi, P., & Nara, S. (2018). Gold nanomaterials to plants. *Nanomaterials in Plants, Algae, and Microorganisms*, 1, 195–220. doi:10.1016/b978-0-12-811487-2.00009-8
2. Abbas, Q., Yousaf, B., Amina, Ali, M. U., Munir, M. A. M., El-Naggar, A., & Naushad, M. (2020). Transformation pathways and fate of engineered nanoparticles (ENPs) in distinct interactive environmental compartments: a review. *Environment International*, 138, 105646. doi:10.1016/j.envint.2020.105646
3. Adamczyk, Z., & Weroński, P. (1999). Application of the DLVO theory for particle deposition problems. *Advances in Colloid and Interface Science*, 83(1–3), 137–226. doi:10.1016/s0001-8686(99)00009-3
4. Agler, M. T., Wrenn, B. A., Zinder, S. H., & Angenent, L. T. (2011). Waste to bioproduct conversion with undefined mixed cultures: the carboxylate platform. *Trends in Biotechnology*, 29(2), 70–78. doi:10.1016/j.tibtech.2010.11.006
5. Ahmad Md., I., Bhattacharya, S. S., & Hahn, H. (2009). Structure, thermal stability, and optical properties of boron modified nanocrystalline anatase prepared by chemical vapor synthesis. *Journal of Applied Physics*, 105(11), 113526. doi:10.1063/1.3143029
6. Al-Fakeh, M. S., Samir Osman Mohammed, O., Gassoumi, M., Rabhi, M., & Omer, M. (2021). Characterization, antimicrobial and anticancer properties of palladium nanoparticles biosynthesized optimally using Saudi propolis. *Nano* 11, 2666. doi:10.3390/nano11102666.
7. Alves, D., Santos, C. G., Paixão, M. W., Soares, L. C., Souza, D. de, Rodrigues, O. E. D., & Braga, A. L. (2009). CuO nanoparticles: an efficient and recyclable catalyst for cross-coupling reactions of organic diselenides with aryl boronic acids. *Tetrahedron Letters*, 50(48), 6635–6638. doi:10.1016/j.tetlet.2009.09.052
8. Gupta, A., Maynes, M., & Silver, S. (1998). Effects of halides on plasmid- mediated silver resistance in Escherichia coli. *ASM Journals Applied and Environmental Microbiology*, 64(12), 6635–6638. doi:10.1128/AEM.64.12.5042-5045.1998
9. Turner, A., & Filella, M. (2023) The role of titanium dioxide on the behavior and fate of plastics in the aquatic environment. *Science of the Total Environment*, 869, 161727. doi:10.1016/j.scitotenv.2023.161727
10. Anschutz, A. J., & Penn, R. L. (2005) Reduction of crystalline iron(III) oxyhydroxides using hydroquinone: Influence of phase and particle size. *Geochemical Transactions*, 6, 60–66.

11. Aruguete, D. M., & Hochella, M. F. (2010). Bacteria–nanoparticle interactions and their environmental implications. *Environmental Chemistry*, 7(1), 3. doi:10.1071/en09115
12. Augustine, R., Hasan, A., Primavera, R., Wilson, R. J., Thakor, A. S., & Kevadiya, B. D. (2020). Cellular uptake and retention of nanoparticles: Insights on particle properties and interaction with cellular components. *Materials Today Communications*, 25, 101692. doi:10.1016/j.mtcomm.2020.101692
13. Chambers, B. A., Afrooz, A. R. M. N., Bae, S., Aich, N., Katz, L., Saleh, N. B., et al. (2014). Effects of chloride and ionic strength on physical morphology, dissolution, and bacterial toxicity of silver nanoparticles. *Environmental Science & Technology*, 48, 761–769. doi:10.1021/es403969x
14. Badawy, A. M. E., Luxton, T. P., Silva, R. G., Scheckel, K. G., Suidan, M. T., & Tolaymat, T. M. (2010). Impact of environmental conditions (pH, ionic strength, and electrolyte type) on the surface charge and aggregation of silver nanoparticles suspensions. *Environmental Science & Technology*, 44(4), 1260–1266. doi:10.1021/es902240k
15. Bakshi, M., & Kumar, A. (2021). Copper-based nanoparticles in the soil-plant environment: assessing their applications, interactions, fate and toxicity. *Chemosphere*, 281, 130940. doi:10.1016/j.chemosphere.2021.130940
16. Bar-Ilan, O., Chuang, C. C., Schwahn, D. J., Yang, S., Joshi, S., Pedersen, J. A., Heideman, W. (2013). TiO_2 nanoparticle exposure and illumination during zebra fish development: mortality at parts per billion concentrations. *Environmental Science & Technology*, 47(9), 4726–4733. doi:10.1021/es304514r
17. Barker, L. K., Giska, J. R., Radniecki, T. S., & Semprini, L. (2018). Effects of short- and long-term exposure of silver nanoparticles and silver ions to nitrosomonas europaea biofilms and planktonic cells. *Chemosphere*, 206, 606–614. doi:10.1016/j.chemosphere.2018.05
18. Bernd, G., Fred, K., Barry, P., Ralf, K., Michael, S., Henning, W., Arnim, G., & Fadri Gottschalk, R. (2018). Release and concentrations of engineered nanomaterial in the environment. *Scientific Reports*, 8, 1565.
19. Blaser, S. A., Scheringer, M., MacLeod, M., & Hungerbühler, K. (2008). Estimation of cumulative aquatic exposure and risk due to silver: contribution of nano-functionalized plastics and textiles. *Science of the Total Environment*, 390(2–3), 396–409. doi:10.1016/j.scitotenv.2007.10.0
20. Calderón-Jiménez, B., Johnson, M. E., Montoro Bustos, A. R., Murphy, K. E., Winchester, M. R., & Vega Baudrit, J. R. (2017). Silver nanoparticles: technological advances, societal impacts, and metrological challenges. *Frontiers in Chemistry*, 5, 6. doi:10.3389/fchem.2017.00006
21. Chai, H., Yao, J., Sun, J., Zhang, C., Liu, W., Zhu, M., & Ceccanti, B. (2015). The effect of metal oxide nanoparticles on functional bacteria and metabolic profiles in agricultural soil. *Bulletin of Environmental Contamination and Toxicology*, 94(4), 490–495. doi:10.1007/s00128-015-1485-9
22. Chen, D., Huang, F., Cheng, Y.-B., & Caruso, R. A. (2009). Mesoporous anatase TiO_2 beads with high surface areas and controllable pore sizes: a superior candidate for high-performance dye-sensitized solar cells. *Advanced Materials*, 21(21), 2206–2210. doi:10.1002/adma.200802603
23. Connor, E. E., Mwamuka, J., Gole, A., Murphy, C. J., & Wyatt, M. D. (2005). Gold nanoparticles are taken up by human cells but do not cause acute cytotoxicity. *Small*, 1(3), 325–327. doi:10. .1002/smll.200400093
24. Conway, J. R., Adeleye, A. S., Gardea-Torresdey, J., & Keller, A. A. (2015). Aggregation, dissolution, and transformation of copper nanoparticles in natural waters. *Environmental Science & Technology*, 49(5), 2749–2756. doi:10.1021/es504918q
25. Dwivedi, A. D., Dubey, S. P., Sillanpää, M., Kwon, Y. N., Lee, C., & Varma, R. S. (2015). Fate of engineered nanoparticles: implications in the environment. *Coordination Chemistry Reviews*, 287, 64–78. doi:10.1016/j.ccr.2014.12.014
26. Corsi, I., Cherr, G. N., Lenihan, H. S., Labille, J., Hassellov, M., Canesi, L., Matranga, V. (2014). Common strategies and technologies for the eco safety assessment and design of nanomaterials entering the marine environment. *ACS Nano*, 8(10), 9694–9709. doi:10.1021/nn504684k
27. Cwiertny, D. M., Hunter, G. J., Pettibone, J. M., Scherer, M. M., & Grassian, V. H. (2008). Surface chemistry and dissolution of α-FeOOH nano rods and microrods: environmental implications of size-dependent interactions with oxalate. *The Journal of Physical Chemistry C*, 113(6), 2175–2186. doi:10.1021/jp807336t
28. Zhou, D., Abdel-Fattah, A. I., Keller, A. A., Dalai, S., Pakrashi, S., Kumar, R. S. S., Chandrasekaran, N., & Mukherjee, A. (2012). A comparative cytotoxicity study of TiO_2 nanoparticles under light and dark conditions at low exposure concentrations. *Toxicology Research*, 1(2), 116. doi:10.1039/c2tx00012a
29. Das, P., Metcalfe, C. D., & Xenopoulos, M. A. (2014). Interactive effects of silver nanoparticles and phosphorus on phytoplankton growth in natural waters. *Environmental Science & Technology*, 48(8), 4573–80. doi:10.1021/es405039w

30. Desmau, M., Carboni, A., Le Bars, M., Doelsch, E., Benedetti, M. F., Auffan, M., & Gelabert, A. (2020). How microbial biofilms control the environmental fate of engineered nanoparticles? *Frontiers in Environmental Science*, 8, 82. doi:10.3389/fenvs.2020.00082
31. Du, J., Zhang, Y., Cui, M., Yang, J., Lin, Z., & Zhang, H. (2017). Evidence for negative effects of ZnO nanoparticles on leaf litter decomposition in freshwater ecosystems. *Environmental Science: Nano*, 4(12), 2377–2387. doi:10.1039/c7en00784a
32. Dwivedi, A. D., Dubey, S. P., Sillanpää, M., Kwon, Y. N., Lee, C., & Varma, R. S. (2015). Fate of engineered nanoparticles: implications in the environment. *Coordination Chemistry Reviews*, 287, 64–78. doi:10.1016/j.ccr.2014.12.014
32. Eduok, S., Martin, B., Villa, R., Nocker, A., Jefferson, B., & Coulon, F. (2013). Evaluation of engineered nanoparticle toxic effect on wastewater microorganisms: current status and challenges. *Ecotoxicology and Environmental Safety*, 95, 1–9. doi:10.1016/j.ecoenv.2013.05.022
33. Petersen, E. J., Henry, T. B., Zhao, J., Mac-Cuspie, R. I., Kirschling, T. L., Dobrovolskaia, M. A., Hackley, V., Xing, B., & White, J. C. (2014). Identification and avoidance of potential artifacts and misinterpretations in nanomaterial eco toxicity measurements. *Environmental Science & Technology*, 48(8), 4226–46. doi:10.1021/es4052999
34. Li, F., Liang, Z., Zheng, X., Zhao, W., Wu, M., & Wang, Z. (2015). Toxicity of nano-TiO_2 on algae and the site of reactive oxygen species production. *Aquatic Toxicology*, 158, 1–13.
35. Fortner, J. D., Kim, D.-I., Boyd, A. M., Falkner, J. C., Moran, S., Colvin, V. L., Kim, J.-H. (2007). Reaction of water-stable C_{60} aggregates with ozone. *Environmental Science & Technology*, 41(21), 7497–7502. doi:10.1021/es0708058
36. Montes de Oca-Vásquez, G., Solano-Campos, F., Vega-Baudrit, J. R., López-Mondéjar, R., Odriozola, I., Vera, A., Bastida, F. (2020). Environmentally relevant concentrations of silver nanoparticles diminish soil microbial biomass but do not alter enzyme activities or microbial diversity. *Journal of Hazardous Materials*, 391, 122224. doi:10.1016/j.jhazmat.2020.122224
37. Galaktionova, L., Gavrish, I., & Lebedev, S. (2019). Bio effects of Zn and Cu nanoparticles in soil systems. *Toxicology and Environmental Health Sciences*, 11(4), 259–270. doi:10.1007/s13530-019-0413-5
38. Gao, X., Avellan, A., Laughton, S., Vaidya, R., Rodrigues, S. M., Casman, E. A., & Lowry, G. V. (2018). CuO nanoparticle dissolution and toxicity to wheat (Triticum aestivum) in rhizosphere soil. *Environmental Science & Technology*, 52(5), 2888–2897. doi:10.1021/acs.est.7b05816
39. Gao, X., Spielman-Sun, E., Rodrigues, S. M., Casman, E. A., & Lowry, G. V. (2017). Time and nanoparticle concentration affect the extractability of Cu from CuO NP-amended soil. *Environmental Science & Technology*, 51(4), 2226–2234. doi:10.1021/acs.est.6b04705
40. Garner, K. L., & Keller, A. A. (2014). Emerging patterns for engineered nanomaterials in the environment: a review of fate and toxicity studies. *Journal of Nanoparticle Research*, 16(8), 2. doi:10.1007/s11051-014-2503-2
41. Gottschalk, F., Sonderer, T., Scholz, R. W., & Nowack, B. (2009). Modeled environmental concentrations of engineered nanomaterials (TiO_2, ZnO, Ag, CNT, and Fullerenes) for different regions. *Environmental Science & Technology*, 43(24), 9216–9222. doi:10.1021/es9015553
42. Hall-Stoodley, L., Costerton, J. W., & Stoodley, P. (2004). Bacterial biofilms: from the natural environment to infectious diseases. *Nature Reviews Microbiology*, 2(2), 95–108. doi:10.1038/nrmicro821
43. Levard, C., Hotze, E. M., Colman, B. P., Dale, A. L., Truong, L., Yang, X. Y., & Lowry, G. V. (2013). Sulfidation of silver nanoparticles: natural antidote to their toxicity. *Environmental Science & Technology*, 47(23), 13440–13448. doi:10.1021/es403527n
44. Hendi, A. A., & Rashad, M. (2018). Photo-induced changes in nano-copper oxide for optoelectronic applications. *Physica B: Condensed Matter*, 538, 185–190. doi:10.1016/j.physb.2018.03.035
45. Hendren, C. O., Mesnard, X., Dröge, J., & Wiesner, M. R. (2011). Estimating production data for five engineered nanomaterials as a basis for exposure assessment. *Environmental Science & Technology*, 45(7), 2562–2569. doi:10.1021/es103300g
46. Ho, C.-M., Yau, S. K.-W., Lok, C.-N., So, M.-H., & Che, C.-M. (2010). Oxidative dissolution of silver nanoparticles by biologically relevant oxidants: a kinetic and mechanistic study. *Chemistry - An Asian Journal*, 5(2), 285–293. doi:10.1002/asia.200900387
47. Hotze, E. M., Phenrat, T., & Lowry, G. V. (2010). Nanoparticle aggregation: challenges to understanding transport and reactivity in the environment. *Journal of Environment Quality*, 39(6), 1909. doi:10.2134/jeq2009.0462
48. Mbanga, O., Cukrowska, E., & Gulumian, M. (2022). Dissolution of titanium dioxide nanoparticles in synthetic biological and environmental media to predict their biodurability and persistence. *Toxicology in Vitro*, 84, 105457. doi:10.1016/j.tiv.2022.105457

49. Zhang, H., Yang, Y., Zhang, Q., Tan, Z., & Xing, B. Redox reactions of silver nanoparticles with dissolved oxygen and dissolved organic matter: implications for aggregation and surface transformations. *Environmental Science & Technology*, 48(16), 9004–9012.
50. Huang, X.-L. (2019). Iron oxide nanoparticles: an inorganic phosphatase. *Nano Catalysts*, 89, 82650. doi:10.5772/intechopen.82650
51. Huynh, K.A., Chen, K.L. (2011). Aggregation kinetics of citrate and polyvinylpyrrolidone coated silver nanoparticles in monovalent and divalent electrolyte solutions. *Environmental Science & Technology*, 45, 5564–5571. doi:10.1021/es200157h
52. Park, H.-G., Kim, J., Kang, M. & Yeo, M.-K. (2014). The effect of metal-doped TiO_2 nanoparticles on zebra fish embryogenesis. *Molecular & Cellular Toxicology*, 10, 293–301.
53. Ikuma, K., Madden, A. S., Decho, A. W., & Lau, B. L. T. (2014). Deposition of nanoparticles onto polysaccharide-coated surfaces: implications for nanoparticle–biofilm interactions. Environmental Science: Nano, 1(2), 117–122. doi:10.1039/c3en00075c
54. Iviza, F., Afrasiabi, Z., & Jose, E. (2018). Effects of silver nanoparticles on the activities of soil enzymes involved in carbon and nutrient cycling. *Pedosphere*, 28(2), 209–214. doi:10.1016/s1002-0160(18)60019-0
55. Cañas, J. E., Qi, B., & Maul, J. (2011) Acute and reproductive toxicity of nano-sized metal oxides (ZnO and TiO_2) to earthworms (Eisenia fetida). *Journal of Environmental Monitoring*, 13, 3351–3357.
56. Jang, M.-H., Kim, W.-K., Lee, S.-K., Henry, T. B., & Park, J.-W. (2014). Uptake, tissue distribution, and depuration of total silver in common carp (cyprinus carpio) after aqueous exposure to silver nanoparticles. *Environmental Science & Technology*, 48(19), 11568–11574. doi:10.1021/es5022813
57. John, A., Küpper, M., Manders-Groot, A., Debray, B., Lacome, J.-M., & Kuhlbusch, T. (2017). Emissions and possible environmental implication of engineered nanomaterials (enms) in the atmosphere. *Atmosphere*, 8(12), 84. doi:10.3390/atmos8050084
58. Aschberger, K., Micheletti, C., Sokull-Klüttgen, B., & Christensen, F.M. (2011). Analysis of currently available data for characterizing the risk of engineered nanomaterials to the environment and human health—Lessons learned from four case studies European Commission, Joint Research Centre (JRC), Institute for Health and Consumer Protection (IHCP), Via E. Fermi, 2749, I-21027 Ispra, Italy. doi:10.1016/j.envint.2011.02.005
59. Siddiqi, K. S., Rahman, A., Tajuddin, T., & Husen, A. (2018). Properties of zinc oxide nanoparticles and their activity against microbes. *Nano Review*, 13, 141.
60. Kim, M.-S., Louis, K. M., Pedersen, J. A., Hamers, R. J., Peterson, R. E., & Heideman, W. (2014). Using citrate-functionalized TiO_2 nanoparticles to study the effect of particle size on zebra fish embryo toxicity. *The Analyst*, 139(5), 964. doi:10.1039/c3an01966g
61. Kloepfer, J. A., Mielke, R. E., & Nadeau, J. L. (2005). Uptake of CdSe and CdSe/ZnS quantum dots into bacteria via purine-dependent mechanisms. *Applied and Environmental Microbiology*, 71(5), 2548–2557. doi:10.1128/aem.71.5.2548-2557.2005
62. Kurniawan, T. A., Sillanpää, M. E. T., & Sillanpää, M. (2012). Nanoadsorbents for remediation of aquatic environment: local and practical solutions for global water pollution problems. *Critical Reviews in Environmental Science and Technology*, 42(12), 1233–1295. doi:10.1080/10643389.2011.556553
63. Lankoff, A., Arabski, M., Wegierek-Ciuk, A., Kruszewski, M., Lisowska, H., Banasik-Nowak, A., & Slomkowski, S. (2012). Effect of surface modification of silica nanoparticles on toxicity and cellular uptake by human peripheral blood lymphocytes in vitro. *Nano Toxicology*, 7(3), 235–250. doi:10.3109/17435390.2011.649796
64. Larue, C., Castillo-Michel, H., Sobanska, S., Cécillon, L., Bureau, S., Barthès, V., Sarret, G. (2014). Foliar exposure of the crop Lactuca sativa to silver nanoparticles: evidence for internalization and changes in Ag speciation. *Journal of Hazardous Materials*, 264, 98–106. doi:10.1016/j.jhazmat.2013.10.053
65. Levard, C., Hotze, E. M., Lowry, G. V., & Brown, G. E. (2012). Environmental transformations of silver nanoparticles: impact on stability and toxicity. *Environmental Science & Technology*, 46(13), 6900–6914. doi:10.1021/es2037405
66. Li, K., & Ma, H. (2019). Rotation and retention dynamics of rod-shaped colloids with surface charge heterogeneity in sphere-in-cell porous media model. *Langmuir*, 35(16), 5471–5483. doi:10.1021/acs.langmuir.9b00748
67. Li, W.-R., Xie, X.-B., Shi, Q.-S., Zeng, H.-Y., Ou-Yang, Y.-S., & Chen, Y.-B. (2009). Antibacterial activity and mechanism of silver nanoparticles on Escherichia coli. *Applied Microbiology and Biotechnology*, 85(4), 1115–1122. doi:10.1007/s00253-009-2159-5
68. Li, X., Lenhart, J. J., & Walker, H. W. (2011). Aggregation kinetics and dissolution of coated silver nanoparticles. *Langmuir*, 28(2), 1095–1104. doi:10.1021/la202328n
69. Liu, J., & Hurt, R. H. (2010). Ion release kinetics and particle persistence in aqueous nano-silver colloids. *Environmental Science & Technology*, 44(6), 2169–2175. doi:10.1021/es9035557

70. Lok, C. N., Ho, C. M., Chen, R., He, Q. Y., Yu, W. Y., Sun, H., Tam, P. K. H., Chiu, J. F., Che, C. M. (2007) Silver nanoparticles: partial oxidation and antibacterial activities. *Journal of Biological Inorganic Chemistry*, 12, 527–534.
71. Lombi, E., Donner, E., Tavakkoli, E., Turney, T. W., Naidu, R., Miller, B. W., & Scheckel, K. G. (2012). Fate of zinc oxide nanoparticles during anaerobic digestion of wastewater and post-treatment processing of sewage sludge. *Environmental Science & Technology*, 46(16), 9089–9096. doi:10.1021/es301487s
72. Loosli, F., Le Coustumer, P., & Stoll, S. (2013). TiO_2 nanoparticles aggregation and disaggregation in presence of alginate and Suwannee River humic acids pH and concentration effects on nanoparticle stability. *Water Research*, 47(16), 6052–6063. doi:10.1016/j.watres.2013.07.021
73. Li, M., Pokhrel, S., Jin, X., Madler, L., Damoiseaux, R., Hoek, E. M. V. (2011). Stability, bioavailability, and bacterial toxicity of ZnO and iron-doped ZnO nanoparticles in aquatic media. *Environmental Science & Technology*, 45, 755–761.
74. Ma, R., Stegemeier, J., Levard, C., Dale, J. G., Noack, C. W., Yang, T., Lowry, G. V. (2014). Sulfidation of copper oxide nanoparticles and properties of resulting copper sulfide. *Environmental Science: Nano*, 1(4), 347–357. doi:10.1039/c4en00018h
75. Horie, M., Stowe, M., Tabei, M., Etsushi Kuroda Maurer-Jones, M. A., Gunsolus, I. L., Murphy, C. J., & Haynes, C. L. (2013). Toxicity of engineered nanoparticles in the environment. *Analytical Chemistry*, 85(6), 3036–3049. doi:10.1021/ac303636s
76. McShane, H. V. A., Sunahara, G. I., Whalen, J. K., & Hendershot, W. H. (2014). Differences in soil solution chemistry between soils amended with nanosized CuO or Cu reference materials: implications for nanotoxicity tests. *Environmental Science & Technology*, 48(14), 8135–8142. doi:10.1021/es500141h
77. Mikami, K., Kido, Y., Akaishi, Y., Quitain, A., & Kida, T. (2019). Synthesis of Cu_2O/CuO nano crystals and their application to H_2S sensing. *Sensors*, 19(1), 211. doi:10.3390/s19010211
78. Miller, R. J., Lenihan, H. S., Muller, E. B., Tseng, N., Hanna, S. K., & Keller, A. A. (2010). Impacts of metal oxide nanoparticles on marine phytoplankton. *Environmental Science & Technology*, 44(19), 7329–7334. doi:10.1021/es100247x
79. Mitrano, D.M., & Nowack, B. (2017). The need for a life-cycle based aging paradigm for nanomaterials: importance of real-world test systems to identify realistic particle transformations. *Nanotechnology* 28, 072001. doi:10.1088/1361-6528/28/7/072001
80. Morones, J. R., Elechiguerra, J. L., Camacho, A., Holt, K., Kouri, J. B., Ramírez, J. T., & Yacaman, M. J. (2005). The bactericidal effect of silver nanoparticles. *Nanotechnology*, 16(10), 2346–2353. doi:10.1088/0957-4484/16/10/059
81. Movafeghi, A., Khataee, A., Abedi, M., Tarrahi, R., Dadpour, M., & Vafaei, F. (2018). Effects of TiO_2 nanoparticles on the aquatic plant Spirodela polyrrhiza: Evaluation of growth parameters, pigment contents and antioxidant enzyme activities. *Journal of Environmental Sciences*, 64, 130–138. doi:10.1016/j.jes.2016.12.020
82. Mueller, N. C., & Nowack, B. (2008). Exposure modeling of engineered nanoparticles in the environment. *Environmental Science & Technology*, 42(12), 4447–4453. doi:10.1021/es7029637
83. Tahir, M. B., Sohaib, M., Sagir, M., & Rafique, M. (2022). Role of nanotechnology in photo catalysis. *Encyclopedia of Smart Materials*, 2, 578–589. doi:10.1016/B978-0-12-815732-9.00006-1
84. Mwilu, S. K., El Badawy, A. M., Bradham, K., Nelson, C., Thomas, D., Scheckel, K. G., & Rogers, K. R. (2013). Changes in silver nanoparticles exposed to human synthetic stomach fluid: effects of particle size and surface chemistry. *Science of the Total Environment*, 447, 90–98. doi:10.1016/j.scitotenv.2012.12.036
85. Silva, N., Ramírez, S., Díaz, I., Garcia, A., & Hassan, N. (2019). Easy, quick, and reproducible sonochemical synthesis of CuO nanoparticles. *Materials (Basel)*, 12(5): 804. doi:10.3390/ma12050804
86. Navarro, E., Piccapietra, F., Wagner, B., Marconi, F., Kaegi, R., Odzak, N., & Behra, R. (2008). Toxicity of silver nanoparticles to chlamydomonas reinhardtii. *Environmental Science & Technology*, 42(23), 8959–8964. doi:10.1021/es801785m
87. Nemi, M., Tzong-Rong, G., Boontida, U., Jong-Chin, H., Kelvin, H.-C. C., & Chung-Der, H. Review of copper and copper nanoparticle toxicity in fish. *Nanomaterials (Basel)*, 10(6), 1126. doi:10.3390%2Fnano10061126
88. Norman, B. C., Xenopoulos, M. A., Braun, D., & Frost, P. C. (2015). Phosphorus availability alters the effects of silver nanoparticles on periphyton growth and stoichiometry. *PLoS One*, 10(6), e0129328. doi:10.1371/journal.pone.0129328
89. Odwa M., Ewa C., & Mary, G. (2022). Dissolution of titanium dioxide nanoparticles in synthetic biological and environmental media to predict their biodurabilityand persistence. *Toxicology in Vitro*, 84, 105457.

90. Oukarroum, A., Samadani, M., & Dewez, D. (2014). Influence of pH on the toxicity of silver nanoparticles in the green alga chlamydomonas acidophila. *Water, Air, & Soil Pollution*, 225(8), 2038. doi:10.1007/s11270-014-2038-2
91. Christian, P., von der Kammer, F., Baalousha, M., & Hofmann, T. (2008). Nanoparticles: structure, properties, preparation and behaviour in environmental media. *Ecotoxicology*, 17(5), 326–343.
92. Han, P., Wang, X., Cai, L., Tong, M., & Kim, H. (2014). Transport and retention behaviors of titanium dioxide nanoparticles in iron oxide-coated quartz sand: effects of pH, ionic strength, and humic acid. *Colloids and Surfaces A: Physicochemical and Engineering Aspects*, 454, 119–127, doi:10.1016/j.colsurfa.2014.04.020
93. Palmer, D. A. (2017). The solubility of crystalline cupric oxide in aqueous solution from 25°C to 400°C. *The Journal of Chemical Thermodynamics*, 114, 122–134. doi:10.1016/j.jct.2017.03.012
94. Patra, J. K., & Baek, K.-H. (2014). Green nano biotechnology: factors affecting synthesis and characterization techniques. *Journal of Nanomaterials*, 2014, 1–12. doi:10.1155/2014/417305
95. Oleszczuk, P., Czech, B., Kończak, M., Bogusz, A., Siatecka, A., Godlewska, P., & Wiesner, M. Impact of ZnO and ZnS nanoparticles in sewage sludge-amended soil on bacteria, plant and 3 invertebrates. *Chemosphere*, 237, 124359.
96. Peng, C., Xu, C., Liu, Q., Sun, L., Luo, Y., & Shi, J. (2017). Fate and transformation of CuO nanoparticles in the soil–rice system during the life cycle of rice plants. *Environmental Science & Technology*, 51(9), 4907–4917. doi:10.1021/acs.est.6b05882
97. Peretyazhko, T. S., Zhang, Q., & Colvin, V. L. (2014). Size-controlled dissolution of silver nanoparticles at neutral and acidic ph conditions: kinetics and size changes. *Environmental Science & Technology*, 48(20), 11954–11961. doi:10.1021/es5023202
98. Petersen, E. J., Reipa, V., Watson, S. S., Stanley, D. L., Rabb, S. A., & Nelson, B. C. (2014). DNA damaging potential of photo activated P25 titanium dioxide nanoparticles. *Chemical Research in Toxicology*, 27(10), 1877–1884. doi:10.1021/tx500340v
99. Phenrat, T., Saleh, N., Sirk, K., Tilton, R.D., Lowry, G.V., 2007. Aggregation and sedimentation of aqueous nanoscale zerovalent iron dispersions. *Environmental Science & Technology* 41, 284–290. doi:10.1021/es061349a
100. Piccapietra, F., Sigg, L., & Behra, R. (2011). Colloidal stability of carbonate-coated silver nanoparticles in synthetic and natural freshwater. *Environmental Science & Technology*, 46(2), 818–825. doi:10.1021/es202843h
101. Rajala, J. E., Vehniäinen, E.-R., Väisänen, A., & Kukkonen, J. V. K. (2018). Toxicity of silver nanoparticles to Lumbriculus variegatus is a function of dissolved silver and promoted by low sediment pH. *Environmental Toxicology and Chemistry*, 37(7), 1889–1897. doi:10.1002/etc.4136
102. Ranade, M. R., Navrotsky, A., Zhang, H. Z., Banfield, J. F., Elder, S. H., Zaban, A., & Whitfield, H. J. (2002). Energetics of nano crystalline TiO_2. *Proceedings of the National Academy of Sciences*, 99(Supplement 2), 6476–6481. doi:10.1073/pnas.251534898
103. MacCuspie, R. I. (2011). Colloidal stability of silver nanoparticles in biologically relevant conditions. *Journal of Nanoparticle Research*, 13, 2893–2908.
104. Rosenfeldt, R. R., Seitz, F., Schulz, R., & Bundschuh, M. (2014). Heavy metal uptake and toxicity in the presence of titanium dioxide nanoparticles: a factorial approach using daphnia magna. *Environmental Science & Technology*, 48(12), 6965–6972. doi:10.1021/es405396a
105. Sabyrov, K., Burrows, N. D., & Penn, R. L. (2012). Size-dependent anatase to rutile phase transformation and particle growth. *Chemistry of Materials*, 25(8), 1408–1415. doi:10.1021/cm302129a
106. Sekine, R., Marzouk, E. R., Khaksar, M., Scheckel, K. G., Stegemeier, J. P., Lowry, G. V., & Lombi, E. (2017). Aging of dissolved copper and copper-based nanoparticles in five different soils: short-term kinetics vs. long-term fate. *Journal of Environment Quality*, 46(6), 1198. doi:10.2134/jeq2016.12.0485
107. Shanaz, J., Ismail Bin, Y., Yatimah Binti, A., & Ahmad Farid Bin, A. B. (2017). Reviews of the toxicity behavior of five potential engineered nanomaterials (ENMs) into the aquatic ecosystem. Toxicology Reports, 4, 211–220. doi:10.1016/j.toxrep.2017.04.001
108. Sigmund, G., Jiang, C., Hofmann, T., & Chen, W. (2018). Environmental transformation of natural and engineered carbon nanoparticles and implications for the fate of organic contaminants. *Environmental Science: Nano*, 5, 2500–2518 doi:10.1039/c8en00676h
109. Stefano, T., Simone, A., Francesca, P., Carolina, S., Elisabetta, C., Stefano, L., Marco, F., Luciana, D. (2022). Amino-functionalized mesoporous silica nanoparticles (NH_2-MSiNPs) impair the embryonic development of the sea urchin Paracentrotus lividus. *Environmental Toxicology and Pharmacology*, 95, 103956. doi:10.1016/j.etap.2022.103956

110. Stumm, W. (1997). Reactivity at the mineral-water interface: dissolution and inhibition. *Colloids and Surfaces A: Physicochemical and Engineering Aspects*, 120(1–3), 143–166. doi:10.1016/s0927-7757(96)03866-6
111. Tarn, D., Ashley, C. E., Xue, M., Carnes, E. C., Zink, J. I., & Brinker, C. J. (2013). Mesoporous Silica nanoparticle nano carriers: bio functionality and biocompatibility. *Accounts of Chemical Research*, 46(3), 792–801. doi:10.1021/ar3000986
112. Thiagarajan, V., & Ramasubbu, S. (2021). Fate and behaviour of TiO_2 nanoparticles in the soil: their impact on staple food crops. *Water, Air, & Soil Pollution*, 232(7), 274. doi:10.1007/s11270-021-05219-8
113. Thio, B. J. R., Zhou, D., & Keller, A. A. (2011). Influence of natural organic matter on the aggregation and deposition of titanium dioxide nanoparticles. *Journal of Hazardous Materials*, 189(1–2), 556–563. doi:10.1016/j.jhazmat.2011.02.072
114. Tratnyek, P. G., & Johnson, R. L. (2006). Nanotechnologies for environmental cleanup. *Nano Today*, 1(2), 44–48. doi:10.1016/s1748-0132(06)70048-2
115. Rathnayake, S., Unrine, J. M., Judy, J., Miller, A.-F., Rao, W., & Bertsch, P. M. (2014). Multitechnique investigation of the pH dependence of phosphate induced transformations of ZnO nanoparticles. *Environmental Science & Technology*, 48(9), 4757–4764. doi:10.1021/es404544w
116. Van Hoecke, K., De Schamphelaere, K. A. C., Ramirez–Garcia, S., Van der Meeren, P., Smagghe, G., & Janssen, C. R. (2011). Influence of alumina coating on characteristics and effects of SiO_2 nanoparticles in algal growth inhibition assays at various pH and organic matter contents. *Enviroment International*, 37(6), 1118–1125. doi:10.1016/j.envint.2011.02.009
117. Verena, K., Christian, S., & Sebastian, K. (2021). Bioaccumulation assessment of nanomaterials using freshwater invertebrate species. *Environmental Sciences Europe*, 33, 9.
118. Wu, P., Cui, P., Du, H., Alves, M. E., Liu, C., Zhou, D., & Wang, Y. (2019). Dissolution and transformation of ZnO nano- and microparticles in soil mineral suspensions. *ACS Earth and Space Chemistry*, 3(4), 495–502. doi:10.1021/acsearthspacechem.8b
119. Yang, G., Hu, L., Keiper, T. D., Xiong, P., & Hallinan, D. T. (2016). Gold nanoparticle monolayers with tunable optical and electrical properties. *Langmuir*, 32(16), 4022–4033. doi:10.1021/acs.langmuir.6b00347
120. Yin, L., Cheng, Y., Espinasse, B., Colman, B. P., Auffan, M., Wiesner, M., & Bernhardt, E. S. (2011). More than the ions: the effects of silver nanoparticles on lolium multiflorum. *Environmental Science & Technology*, 45(6), 2360–2367. doi:10.1021/es103995x
121. You, T., Liu, D., Chen, J., Yang, Z., Dou, R., Gao, X., & Wang, L. (2017). Effects of metal oxide nanoparticles on soil enzyme activities and bacterial communities in two different soil types. *Journal of Soils and Sediments*, 18(1), 211–221. doi:10.1007/s11368-017-1716-2
122. Yuan, Y., Karahan, H. E., Yıldırım, C., Wei, L., Birer, Ö., Zhai, S., & Chen, Y. (2016). Smart poisoning of Co/SiO_2 catalysts by sulfidation for chirality-selective synthesis of (9, 8) single-walled carbon nanotubes. *Nanoscale*, 8(40), 17705–17713. doi:10.1039/c6nr05938d
123. Zarrin, H., & Iqbal, A. (2017) Impact of metal oxide nanoparticles on beneficial soil microorganisms and their secondary metabolites. *International Journal of Life- Sciences Scientific Research*, 3, 1020–1030. doi:10.21276/ijlssr.2017.3.3.10
124. Zhang, H., & Banfield, J. F. (1998). Thermodynamic analysis of phase stability of nano crystalline titania. *Journal of Materials Chemistry*, 8(9), 2073–2076. doi:10.1039/a802619j
125. Zhang, H., Huang, Q., Xu, A., & Wu, L. (2016). Spectroscopic probe to contribution of physicochemical transformations in the toxicity of aged ZnO NPs to Chlorella vulgaris: new insight into the variation of toxicity of ZnO NPs under aging process. *Nano Toxicology*, 10(8), 1177–1187. doi:10.1080/17435390.2016.1196252
126. Zhang, W., Yao, Y., Sullivan, N., & Chen, Y. (2011). Modeling the primary size effects of citrate-coated silver nanoparticles on their ion release kinetics. *Environmental Science & Technology*, 45(10), 4422–4428. doi:10.1021/es104205a
127. Zhao, S., Su, X., Wang, Y., Yang, X., Bi, M., He, Q., & Chen, Y. (2019). Copper oxide nanoparticles inhibited denitrifying enzymes and electron transport system activities to influence soil denitrification and N_2O emission. *Chemosphere*, 245, 125394. doi:10.1016/j.chemosphere.2019.12
128. Zhou, W., Liu, Y.-L., Stallworth, A. M., Ye, C., & Lenhart, J. J. (2016). Effects of pH, electrlyte, humic acid, and light exposure on the long-term fate of silver nanoparticles. *Environmental Science & Technology*, 50(22), 12214–12224. doi:10.1021/acs.est.6b03237
129. Jiang, H.-S., Qiu, X.-N., Li, G.-B., Li, W., & Yin, L.-Y. (2014). Silver nanoparticles induced acumulation of reactive oxygen species and alteration of antioxidant systems in the aquatic plant Sprodela polyrhiza. *Environmental Toxicology and Chemistry*, 33(6), 1398–1405. doi:10.1002/etc.2577

20 Ethical, Legal, and Socio-Economic Impacts of Nanotechnology

Shivani, Preeti Thakur, and Atul Thakur
Amity University Haryana

20.1 INTRODUCTION

20.1.1 Definition and Overview of Nanotechnology

An area of science and technology known as nanotechnology focuses on manipulating and controlling matter at the nanoscale or the scale of atoms and molecules. It entails the creation, synthesis, evaluation, and use of materials and gadgets with particular nanoscale features and capabilities. Nanotechnology is a field of science and technology that deals with the manipulation and control of matter at the nanoscale, typically involving materials and structures with dimensions between 1 and 100 nm [1]. It involves the understanding and utilization of unique properties and behaviors exhibited by matter at the nanoscale. At such tiny scales, the properties of materials can differ significantly from their bulk counterparts due to quantum effects and increased surface area.

In comparison to their bulk counterparts, materials exhibit novel and frequently improved characteristics at the nanoscale. This results in various electrical, optical, mechanical, and chemical behaviors due to quantum mechanical phenomena and an enhanced surface-to-volume ratio. Understanding and utilizing these special qualities is the goal of nanotechnology, which strives to produce new substances, tools, and systems with enhanced functionality [2–4]. Numerous industries, including electronics, healthcare, energy, materials science, environmental science, and others use nanotechnology as shown in Figure 20.1. Examples of goods and uses based on nanotechnology include:

- **Electronics:** Nanotechnology has had a significant impact on the electronics industry by enabling the fabrication of smaller, faster, and more efficient devices. It has facilitated the development of nanoscale transistors, memory devices, and sensors, leading to advancements in computing, data storage, and communication technologies.
- **Medicine:** Nanotechnology has immense potential in medicine and healthcare. It enables precise drug delivery systems, targeted therapies, and diagnostic tools. Nanoparticles can be designed to selectively deliver drugs to specific cells or organs, reducing side effects. Nano sensors can detect biomarkers for early disease diagnosis, and nanodevices can assist in tissue engineering and regenerative medicine.
- **Energy**: Nanotechnology offers solutions for renewable energy generation, energy storage, and environmental remediation. Nanomaterials can improve the efficiency of solar cells, enhance battery performance, and enable lightweight and durable materials for energy applications. Nanotechnology also plays a role in pollution detection and water purification.
- **Manufacturing and Industries:** Nanotechnology is transforming various industries by improving manufacturing processes and creating new materials with advanced properties.

DOI: 10.1201/9781003502692-20

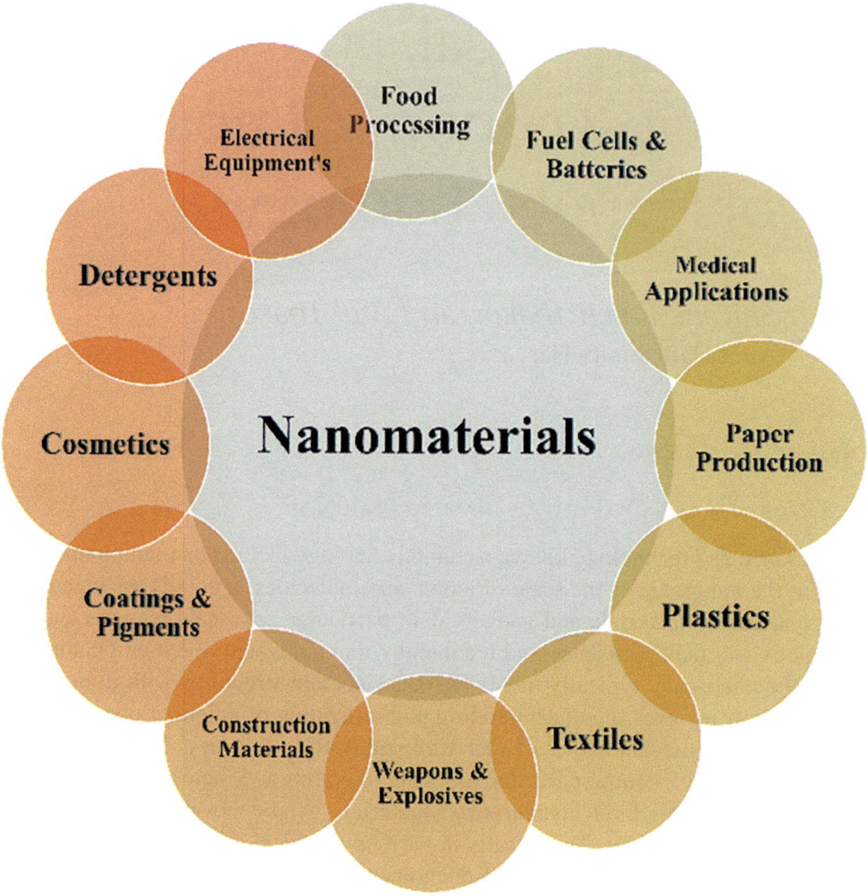

FIGURE 20.1 Applications of nanomaterials.

It enables precise control and manipulation of materials at the atomic and molecular level, leading to improved product quality, increased efficiency, and reduced waste.

- **Environmental science:** Nano catalysts for effective chemical reactions, nanofiltration for the purification of water, and nano sensors for the detection of contaminants.

As with any emerging technology, nanotechnology raises concerns about its potential environmental and health impacts. Researchers and regulatory bodies are actively studying the safety aspects and developing guidelines to ensure responsible development and use of nanomaterials [5–8].

Nanotechnology uses a variety of methods and techniques to work with and create materials at the nanoscale. These include bottom-up procedures, in which smaller building blocks are put together to make larger structures, and top-down approaches, in which larger structures are gradually shrunk [9].

20.2 IMPORTANCE OF ETHICAL, LEGAL, AND SOCIO-ECONOMIC IMPACTS

It is impossible to overestimate the significance of how nanotechnology will affect ethics, law, and society. These factors must be considered to ensure the ethical development, application, and integration of nanotechnology into society. These effects are important for the following main reasons:

a. Ethical significance
   - **Protection of human health and safety:** Ethical considerations assist in making sure that possible concerns connected with nanotechnology, such as exposure to nanomaterials, are properly assessed and managed to safeguard human health and safety.
   - **Environmental responsibility:** By addressing issues regarding the manufacture, usage, and disposal of nanomaterials, ethical talks aim to reduce the environmental impact of nanotechnology. The appropriate and responsible use of nanotechnology is guided by ethical considerations that address concerns including privacy, surveillance, and potential abuse [10].

b. Legal Importance
   - **Regulatory framework:** Creating particular legal frameworks for nanotechnology aids in regulating its creation, application, and commercialization. Setting safety standards, labeling specifications, and guidelines for the responsible handling of nanoparticles are all part of this.
   - **Protection of intellectual property:** Legal considerations make sure that researchers and innovators are fairly compensated and protected for their intellectual property, which encourages more innovation in the field of nanotechnology. Clear legal frameworks address liability issues in the event that nanomaterials or products based on nanotechnology cause any harm. They outline accountability and offer procedures for redress and compensation [11,12].

c. **Social and economic significance:** Nanotechnology has the ability to generate new employment opportunities and stimulate economic growth. Policymakers and stakeholders may design plans to take advantage of the economic potential of nanotechnology and generate long-term employment by having an understanding of the socio-economic implications.
   - **Addressing economic disparities:** To prevent escalating economic inequalities, socio-economic conversations take into account the fair sharing of nanotechnology benefits. When adopting and using nanotechnology, efforts are taken to ensure accessibility, affordability, and diversity [13].
   - **Public perception and trust:** The successful integration of nanotechnology into society depends on public acceptance and trust. To foster understanding, trust, and informed decision-making, socio-economic factors put a strong emphasis on public participation, knowledge, and education [14].

   It supports responsible and sustainable growth, helps to reduce risks, encourages equity, and maximizes the social advantages received from this quickly developing sector by taking into account the ethical, legal, and socio-economic aspects of nanotechnology. It promotes a multifaceted strategy that considers the welfare of each person, the environment, and society at large.

20.3 ETHICAL IMPACTS OF NANOTECHNOLOGY

20.3.1 Ethical Guidelines and Principles

Several standards and principles have been put forth to address the ethical implications of nanotechnology. The following guiding principles for ethical behavior in nanotechnology are:

- **Responsible development:** From conception to disposal, researchers and developers should place a high priority on ethical and environmentally friendly procedures. This entails taking into account potential dangers, reducing environmental effects, and assuring the security of staff, customers, and the general public.

- **Safety and risk assessment:** To detect possible risks related to nanomaterials and uses of nanotechnology, thorough and transparent risk assessments should be carried out. To safeguard both human health and the environment, it is necessary to assess the toxicity, exposure routes, and long-term impacts of manufactured nanoparticles.

A concept called responsible research and innovation (RRI) encourages moral concerns and societal effects in the creation and application of technologies [15,16], including nanotechnology. In the context of nanotechnology, the following are some important components of RRI:

Responsible nanotechnology research entails foreseeing and addressing potential ethical, social, and environmental difficulties early in the development process. This is known as anticipatory governance. The hazards and effects of nanotechnology can be understood and managed by researchers and stakeholders before they become pervasive.

- **Engagement of stakeholders:** For the responsible development of nanotechnology, it is essential to involve stakeholders, including the general public, in the research and innovation process. Considering many viewpoints can aid in identifying and addressing issues, ensuring that nanotechnology is in line with social demands, and boosting public confidence and acceptance [17–19]. By reducing resource consumption, waste production, and environmental effects, nanotechnology research should put sustainability first. This includes taking nanoparticles' life cycles into account and creating environmentally friendly manufacturing procedures. Responsible nanotech research should consider all aspects of the environmental and health dangers that may be posed by nanomaterials. Sharing of data, methodologies, and research outcomes is encouraged by open science, which is promoted by ethical nanotechnology research. Transparency, reproducibility, and cooperation are encouraged by openness, which enables responsible innovation and allows for the expansion of preexisting knowledge.
- **Education and training:** Ethical considerations in nanotechnology research and innovation require teaching and training scientists, engineers, and other stakeholders. To make sure that nanotechnology research is carried out with ethical considerations in mind, it can be helpful to raise awareness of the potential ethical effects and provide guidelines and tools.
- **Technology transfer with ethics:** As part of responsible innovation in nanotechnology, it is important to take the moral ramifications of technology transfer and commercialization into account. Especially in poor nations and marginalized people, this entails addressing concerns about intellectual property rights, equitable distribution, and access to nanotechnology benefits.
- **Continuous evaluation and improvement:** Nanotechnology innovation and responsible research are iterative processes. To find areas for improvement, address new ethical issues, and make sure that responsible practices are incorporated into future research and innovation efforts, regular review, feedback, and learning from prior experiences are crucial. Researchers can promote responsible innovation that takes into account the broader societal implications and ethical considerations of their work by incorporating these ideas into nanotechnology research and development. Nanotechnology innovation and responsible research work to match scientific progress with society's values, requirements, and sustainability objectives [20].

20.3.2 Responsible Research and Accountability

In the field of nanotechnology, accountability and responsibility are crucial ethical factors. The following are some significant ethical implications for accountability and responsibility in nanotechnology:

- **Responsible development:** It is the duty of researchers and developers in the field of nanotechnology to make sure that their work is done responsibly. This includes taking into account the potential drawbacks, advantages, and effects of nanotechnology applications over the course of their whole life cycle. A proactive risk assessment, commitment to ethical standards, and the adoption of suitable actions to reduce harm and increase social benefits are all part of responsible development.
- Information sharing and transparency are essential for ethical nanotechnology research. The work of researchers and developers should be transparently described, including any potential dangers, constraints, and uncertainties. Open communication promotes stakeholder trust and fosters the ability to make well-informed decisions.
- **Ethical resource usage:** Ethical resource usage is a key component of responsible nanotechnology development. Researchers should think about how their study will affect the environment and aim to reduce resource use, waste production, and pollution. This involves supporting environmentally friendly manufacturing techniques, recycling and waste management strategies, and developing nanomaterials with minimal negative effects on the environment.
- **Product safety and consumer protection:** To guarantee consumer safety, products based on nanotechnology must be appropriately created and marketed. Manufacturers should carry out thorough safety evaluations and follow all relevant rules and regulations. For consumers to make educated decisions, transparent labeling and clear communication of potential dangers and safety rules are essential.
- **Ethical resource management:** An essential element of responsible nanotechnology development is ethical resource management. Researchers should consider the environmental impact of their research and work to reduce resource use, waste generation, and pollution. Supporting ecologically friendly manufacturing practices, waste reduction and recycling programs, and the creation of nanomaterials with low environmental impact are all part of this [21,22].
- **Consumer safety and product safety:** Nanotechnology-based products must be adequately developed and promoted to ensure consumer safety. Manufacturers must conduct in-depth safety assessments and abide by all applicable laws and regulations. Transparent labeling and clear explanations of potential risks and safety guidelines are crucial for enabling customers to make informed decisions.
- **Impacts accountability:** Researchers and developers need to take responsibility for any potential effects of their nanotechnology-related work. As part of this, one must accept responsibility for any unforeseen repercussions or negative effects on society, the environment, or human health. When harm happens, proper steps should be taken to treat the effects and stop a repeat of the situation.
- For ethical practices to be promoted, the safety and well-being of people and the environment to be ensured, and the public's confidence in the sector to be upheld, responsible development and accountability in nanotechnology are essential. To promote ethical and sustainable nanotech innovation, it is crucial for academics, developers, policymakers, and stakeholders to respect these values.
- **Regulatory compliance:** Conducting ethical nanotech research entails adhering to all applicable laws and moral principles. Researchers and developers should keep up with the changing regulatory environment and make sure that their work complies with ethical and legal guidelines. This includes getting the required permits, following safety guidelines, and disclosing any potential dangers or negative outcomes [23–25].

20.3.3 Ethical Concerns in Nanotechnology

The study of controlling matter at the nanoscale, or nanotechnology, has the potential to transform multiple sectors and have a significant impact on society. The ethical issues surrounding nanotechnology must be taken into account even though there are numerous promising uses. Following are a few ethical issues raised by applications of nanotechnology:

- **Health and safety:** The potential hazards to human health and safety posed by nanoparticles are a major issue. Concerns regarding the effects of certain nanoparticles on human health and the environment are raised by their toxicity and ability to cross biological barriers. To guarantee that nanoparticles are used safely, extensive study on their toxicity and long-term impacts must be done.
- **Environmental impact:** Novel materials could be introduced into the environment as a result of nanotechnology. These manufactured nanoparticles raise questions about their long-term environmental impact, including their persistence, bioaccumulation, and potential implications on ecosystems. To reduce negative environmental effects, thorough risk analyses and ethical disposal techniques are required.
- **Privacy and surveillance:** Nanotechnology makes it possible to create surveillance equipment and extremely sensitive sensors at the nanoscale. As these technologies evolve, there are worries about potential privacy breaches and civil liberties violations. Nanoscale surveillance technologies must be used responsibly and protected from abuse by appropriate laws and safeguards.
- **Equity and access:** As with any new technology, there is a chance that existing social and economic disparities will get worse. A technical divide and greater marginalization of impoverished populations could result from pricey or exclusive nanotechnology applications. For a just and fair society, it is crucial to guarantee that everyone has access to nanotechnology and its advantages [26].
- Nanotechnology offers enormous potential in the medical field, notably in the areas of targeted medication delivery, diagnostics, and regenerative medicine. Ensuring informed permission, preventing the exploitation of vulnerable groups, and addressing issues of equity in access to cutting-edge medical treatments are only a few of the ethical difficulties raised by the use of nanomedicine [27].
- Nanotechnology has the potential to significantly alter military and defense systems, which has ethical ramifications for military applications. The use of nanotechnology for autonomous weaponry, surveillance, and human capability improvement raises ethical questions. When developing and implementing nanotechnology for military purposes, concerns like proportionality, adherence to international humanitarian law, and unexpected repercussions must be taken into consideration.
- Ethical considerations also apply to the research and development stage of the nanotechnology industry. The integrity of nanotechnology research and its uses depends on ethical behavior in scientific research, openness in disclosing results, and avoiding any conflicts of interest [28].

20.4 LEGAL IMPACTS OF NANOTECHNOLOGY

Due to its innovative features and possible hazards, nanotechnology offers particular difficulties for current laws and regulations. Figure 20.2 depicts some significant legal ramifications of nanotechnology for current laws and rules.

Due to quick development in nanotechnology, there are now regulatory gaps in a number of different jurisdictions. The specific dangers and properties of nanomaterials and applications of nanotechnology may not be sufficiently covered by current laws and regulations. To close these loopholes and guarantee the secure development, manufacture, and application of nanotechnology, governments are trying to update and adopt new legislation. Specialized risk assessment approaches may be needed due to the particular characteristics of nanomaterials. Governments are

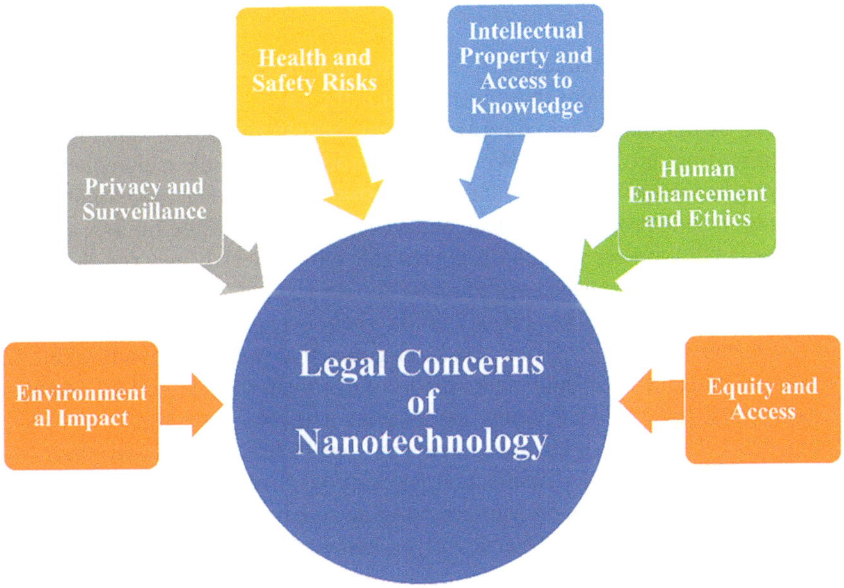

FIGURE 20.2 Legal concerns of nanotechnology.

working to create regulations and standards for risk assessment, labeling, and safety of nanomaterials. In consumer protection and product labeling specific labeling standards may be necessary for products based on nanotechnology to alert customers to the presence of nanoparticles and any potential dangers. To address the unique properties and potential hazards of nanotech products, it may be necessary to amend current consumer protection legislation and labeling regulations. Advances in nanotechnology create questions about intellectual property. To address the special difficulties posed by inventions and discoveries related to nanotechnology, patent laws and regulations may need to be modified [29]. Nanotech creations, such as nanomaterial compositions or nanoscale technologies, may call for particular knowledge and standards for intellectual property protection. Due to the possible environmental effects of nanomaterials and applications of nanotechnology, environmental rules must be evaluated and perhaps modified. The unique properties and possible dangers of nanotechnology, such as the discharge of manufactured nanoparticles into the environment, may need the updating of current legislation. Workers who produce, handle, or use nanomaterials may be at risk for occupational health and safety issues. To address these particular concerns and develop suitable safety measures for workers in businesses connected to nanotechnology, existing occupational health and safety standards may need to be updated [30]. Because nanotechnology is a worldwide field, it is crucial for trade and collaboration that laws and standards in many jurisdictions are consistent. For the safe and responsible use of nanotechnology, as well as to avoid trade restrictions and assure ethical business practices, international cooperation is required. Data protection and privacy issues are raised by uses of nanotechnology that require data collection or surveillance [31]. To secure the security of personal data and private rights, it could be necessary to modify current laws and regulations to address the particular difficulties created by surveillance systems and data gathering techniques based on nanotechnology. Policymakers and regulators are actively striving to update and build rules and regulations that adequately address the unique hazards and ethical considerations involved with this industry as nanotechnology continues to advance. To allow for the safe and ethical development and use of nanotechnology while safeguarding the general public and the environment, it is crucial to make sure that legal frameworks keep up with scientific breakthroughs [32].

20.4.1 Current Laws and Regulations

When opposed to their bulk counterparts, nanomaterials may exhibit various characteristics and behaviors, which may affect how they interact with the environment and pose dangers. Before nanoparticles can be discharged into the environment, certain policies of many nations have mandated specific reporting or evaluation. Workers who produce or handle nanomaterials may run the risk of developing health problems. Regulations governing occupational health and safety may need to be modified to include provisions for nanomaterials, such as exposure limits, safety precautions, and risk analyses [33]. Workers who produce or handle nanomaterials may run the risk of developing health problems. It may be necessary to revise the occupational health and safety legislation to incorporate particular measures for nanomaterials, such as exposure limits, protective measures, and risk assessments. There are various consumer goods that incorporate nanotechnology to educate consumers on the existence and potential risks of nanomaterials in products, regulatory bodies may mandate producers to conduct safety assessments and offer unambiguous labeling for those products. Intellectual property rights, such as patents, may be used to safeguard inventions in nanotechnology. Legal frameworks for intellectual property such as patents and other types of intellectual property can be very important in encouraging research and development in the area. It's significant to remember that the legal framework for nanotechnology is still developing, and various nations may take various methods. International institutions like the International Organization for Standardization (ISO) and the Organization for Economic Cooperation and Development are also striving to create policies and standards for the responsible and safe application of nanotechnology [34]. Governments will probably continue to hone and modify their legal frameworks as the area develops to address the unique issues brought up by nanotechnology, taking into account scientific advancements, risk evaluations, and societal concerns. The use of nanotechnology, however, is already constrained to the extent that it is used to create or improve weapons that are outlawed by current arms control treaties, such as biological weapons, chemical weapons, non-detectable fragments, blinding laser weapons, anti-personnel mines, explosive remnants of war, and, most recently, cluster munitions. The applicable treaty would govern nanotechnology if it were utilized as an enabling technology for the creation of weapons in certain areas. For instance, using nanotechnology, much more potent lasers can be created than those that were previously known. The ability of nanotechnology to create and manipulate molecules with specific properties could result in bio/chemical agents that can cause defined adverse effects, such as temporary incapacitation or death, or multilayered biochemical carriers that can easily control the spread of bio/chemical agents [35].

In international humanitarian law instruments, the broad tenet that "the right of belligerents to adopt means of warfare is not unlimited" has been codified. The Martens Clause must be taken into consideration while interpreting this general norm and other provisions of international humanitarian law [36]. The Martens Clause has gained importance as new technologies increasingly influence the development and sophistication of weapons and delivery systems, something that was not anticipated by the drafters of international humanitarian law instruments, even though "principles of humanity" and "dictates of public conscience" alone may not provide a firm legal basis to prohibit the use of specific weapons.

The use of nanotechnology for military or other purposes is not currently covered by any international treaties [37,38]. It is improbable that a preventive arms control pact to limit or forbid the use of nanotechnology for military purposes will be enacted. Due to the fact that international arms control treaties are often reactive to technical advancements and have a narrow focus, only certain weapons that can be identified by their design, purpose, and attributes are prohibited or regulated.

20.4.2 Legal Challenges in Nanotechnology

To ensure the safe and responsible research, application, and commercialization of nanomaterials and products based on nanotechnology, a number of legal challenges related to nanotechnology must be addressed. Patents and other forms of intellectual property protection are essential for promoting innovation and financial investment in industry. However, the particular difficulties involved in patenting nanoscale ideas, such as figuring out patentability standards and dealing with prior art problems, need considerable thought. Since nanotechnology is a worldwide industry, it is essential to cooperate internationally and harmonize rules and regulations to consistently solve legal issues [39,40]. Collaboration between nations and international organizations can aid in the development of universal standards, best practices, and guiding principles. To address these legal difficulties, it is necessary for governments, regulatory bodies, scientific communities, industrial stakeholders, and legal professionals to work together in a proactive and multifaceted manner [41]. To keep up with technological breakthroughs and changing scientific understanding in the field of nanotechnology, it is crucial to regularly review and adjust regulatory frameworks.

There are three issues with the widespread usage of nanotechnology that relate to the application of international humanitarian law [42]. First, the level of knowledge necessary to decide whether a weapon is legal, to determine whether an attack is excessive, and to take precautions when choosing a target raises concerns about the health and environmental effects of Engineered Nanomaterials (ENMs) and Engineered Nanoparticles (ENPs). Second, aside from situations where the impacts are intended, or may be anticipated, to be widespread, long-lasting, and severe, there is no clear guidance regarding how broadly health and environmental repercussions resulting from armed attacks must be taken into account. This is due to the ongoing argument over how much indirect, long-term effects should be taken into consideration when determining what is unnecessary. When determining what constitutes excessive harm or excessive suffering as well as the excessiveness of an attack, long-term effects should be taken into account. Third, if national regulatory bodies do not acknowledge the significance of more thorough nanotechnology regulation considering wartime scenarios, the principle of differentiation and the obligation to take measures will become increasingly challenging to uphold [43].

20.4.3 Patents and Nanotechnology

Inventions have been demonstrated by the filing of patent applications in the top 50 depositories, which increased from 1,200 in 2000 to over 13,000 in 2008. The United States Patent and Trade Office, the European Patent Office, and the Japan Patent Office are only a few of the overseas patent offices. Per year there are 167 total patents, 64 (or 39% of the total) are owned by government agencies, 45 (27% of the total) by commercial enterprises, and 10 (or 6% of the total) by academic institutions. 37 patents, or 22% of the total, are held by individual creators. The remaining 5% of patents are collaborative patents created in association with businesses from the public or private sectors. The Council for Scientific and Industrial Research and the Defense Research and Development Organization are two top contributors from the government sector [44–46]. The top companies that have industry patents include Ranbaxy Laboratories, Stempeutics Research Private Limited, Panacea Biotech Limited, and Arrow Coated Products Limited. In comparison to industry, academic institutions publish more articles than there are patents [47].

20.4.4 Product Liability and Responsibility

This part is not specifically concerned with these pre-marketing regulations but rather with the civil liability repercussions if N & N-related items hurt the public. Of course, the law, information, and labeling could have an effect on such civil culpability. Hopefully, nanotechnologies will result in new advantages without any negative effects. However, it is possible that some customers may suffer harm as a result of N & N-led advances given the unpredictability and clear policy of not regulating for zero risk [48]. Some product liability theories support dispersing this risk. In light of the current situation's ambiguity regarding the gravity and scope of potential threats, achieving this objective may prove challenging from a legal and economic standpoint. As some products must be deemed

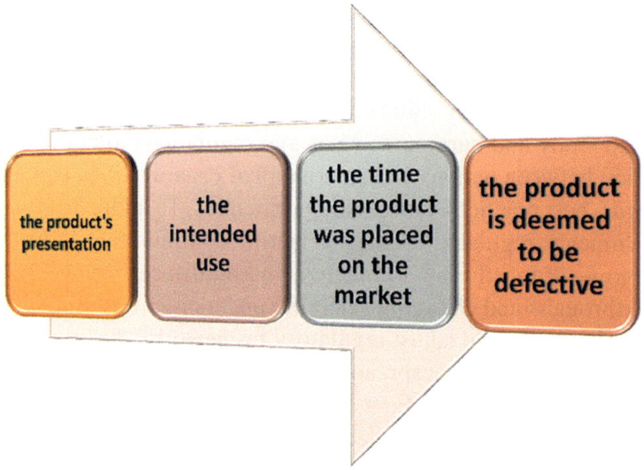

FIGURE 20.3 Safety levels of a product.

"defective" under product liability law or "non-conforming" under European sales legislation, legal issues are likely to arise. There will certainly be discussions concerning the impact of unknown or at least poorly understood dangers on that assessment. Development risk defense is crucial in the area of strict product liability. The primary concern will be how the Products Liability Directive is implemented. A large variety of consumer products contain N & N. The application of N & N is currently most relevant in the fields of food and cosmetics, where the health hazards are clear. The Scientific Committee on Consumer Products has already come to the conclusion that it is essential to reevaluate the safety of sunscreen products in light of the expanding nanoscience research, take into account the influence of physiologically abnormal skin, and possibly take into account the potential impact of mechanical action on skin penetration. It is difficult to apply product responsibility legislation to product problems relating to N & N. The actual status of knowledge in many settings and the court's judgment of the ability to consider knowledge acquired in one context as generic to all N & N will be crucial in practice. Hence, there is a lack of consensus among scientists over whether theories apply equally to soluble and non-soluble, inhaled, and non-inhaled nanoparticles, or even between various types of nanoparticles [49].

1. When a product fails to provide the level of safety that a person has a right to expect, taking into account all relevant factors (Figure 20.3).
2. A product is not considered defective only because a superior product has since entered the market.

Consumer expectations vs. "risk/utility" are two criteria for determining defectiveness that compete in the idea of product responsibility [50]. Although compelling arguments have been made for some role for risk/utility evaluation, at the very least for design errors, the European Directive appears to favor a model based on consumer expectations, as indicated by the reference to "the safety which a person is entitled to expect."

20.5 ENVIRONMENTAL LIABILITY AND REMEDIATION

Numerous recent advancements in material science and nanotechnology have prompted requests for research into the effects of nanomaterials on the environment from people from all walks of life. Over the past ten years, these developments have increased, raising worries about the possible negative environmental effects of nanomaterials. Agriculture, water purification, healthcare, transportation, energy, electronics, and environmental bioremediation are just a few fields where nanotechnology has made significant strides [51]. Thus, there are many opportunities for nanotechnology to raise living standards in a beneficial and sustainable way. However, unless the accompanying economic effect, societal acceptance (including ethical difficulties), environmental impact, safety, and sustainability are thoroughly understood, nanotechnology may not realize its full potential. Agriculture, industrial processes, and a number of regular human activities all naturally emit chemicals and dangerous materials into the environment. These noxious substances disrupt the ecosystem and have detrimental impacts on both people and other living things in the environment when present in high enough concentrations. Numerous environmental remediation technologies have been created to solve the problem of environmental contamination, but bioremediation has proven to be more dependable and successful due to its eco-friendly characteristics [52]. Environmental bioremediation is the process of employing biological agents to address environmental issues related to nanotechnology in various fields of its applications.

Although nanotechnology has enhanced environmental solutions, there are still environmental issues (Figure 20.4) with numerous situations of complexities that call for investigation and multidimensional solutions. Understanding the underlying physical interactions between the ecosystem and nanomaterials is essential to comprehending how nanotechnology affects the environment. Dreher [53] provided a brief overview of the issues arising from manufactured nanomaterial toxicology risk assessment, highlighted the findings and contributions of nanomaterials from companion articles, and put these contributions into various perspectives. It was noted that in the near future, the effects of occupational and public exposure to manufactured nanoparticles will dramatically increase due to the ability of nanomaterials to improve the performance and quality of many public consumer products and the development of medical tests and therapies that will use manufactured nanoparticles. According to Dreher, there are a number of significant risk assessment concerns regarding produced nanoparticles that have been raised by the National Science Foundation and the US Environmental Protection Agency.

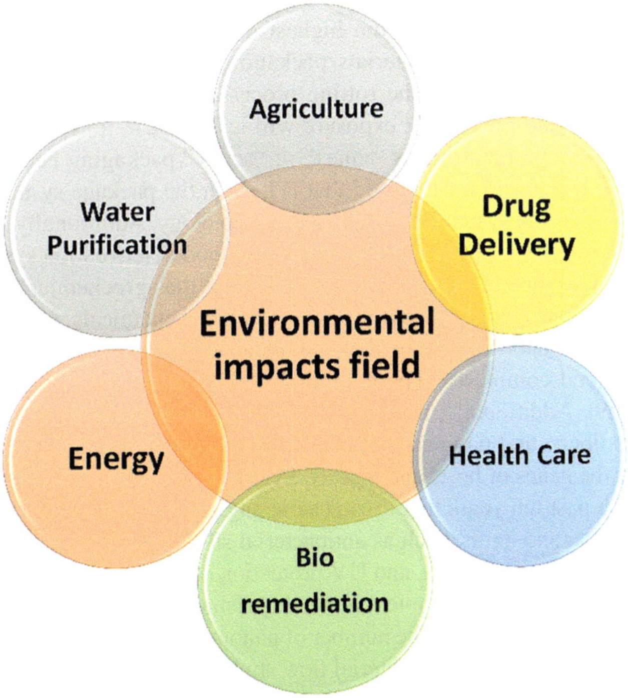

FIGURE 20.4 Environmental impacts of nanotechnology.

- Toxicology of manufactured nanoparticles
- Transport, transformation persistence, environmental, and biological destiny of produced nanoparticles
- Overall sustainability and recyclability of manufactured nanomaterials
- Ability to extrapolate manufactured nanoparticle toxicity using existing particle and fiber toxicological databases

Dreher predicted that, as the quantity and variety of nanoparticles utilized in society increase, so will the risk of exposure. Nanomaterials cause health hazards by evading the body's defensive mechanisms and entering the body through ingestion, skin absorption, or inhalation, which might happen through acute and short-term exposure as well as long-term exposure (such as immersion or polluted air) [54].

20.6 SOCIO-ECONOMIC IMPACTS OF NANOTECHNOLOGY

20.6.1 Food Industry in Nanotechnology

The term "nano food" refers to food produced utilizing nanotechnology. According to news reports, "programmable food" can be heated with a properly calibrated microwave transmitter to change the product's color, taste, and amount of different nutrients. This type of food can include drinks or pre-cooked lasagna that contains nanoparticles. Then there is the idea that milk may be made to taste like cola to encourage more kids to drink it. Processing of green tea has the ability to ten-fold boost selenium absorption [55]. These meals and beverages are being developed by a number of industry heavyweights, including Kraft, Nestlé, and Unilever, thanks to the developing field of nanotechnology [56]. Nanotechnology is also being utilized to change the characteristics of food, such as its shelf life, flavor, texture, and ability to withstand heat. For instance, Unilever has reported advances in the production of low-fat, low-calorie food that maintains its rich, creamy flavor and texture [57]. This technology has been used in a variety of very low-fat ice creams, mayonnaise, and spreads.

The agri-food nanotechnology with the highest commercialization is nano food packaging. Barrier technologies found in nanomaterials packaging extend food's shelf life, durability, and freshness—or at the very least, halt the rotting process. According to DuPont, a nano titanium dioxide plastic addition that lowers UV exposure will lessen harm to food kept in clear packaging [58]. This packaging can be regarded as "smart" or "active" packaging because it functions as an antioxidant and an antibacterial for the food that is kept in the package system. Carbon nanotubes can be included in packaging systems thanks to nanotechnology. Additionally, this carbon nanotube aids in the detection of microbes, harmful proteins, and food deterioration systems.

For proper farm management in the twenty-first century, the agrochemical and information technology industries have considered producing new agricultural chemicals, seeds, etc. in nanodevices. Crop the development of nano pesticides that can alter the alkaline conditions in an insect's stomach through science and commerce. These nano pesticides distribute pesticides more precisely to the intended host [59]. Additionally, nanobiotechnologies are making it possible to create targeted nanoparticles, nano fibers, and nano capsules to transport certain foreign DNA and particular nano chemicals that alter the genes of hosts and insects. Synthetic biology is a new field of technology that was developed in the past ten years that uses genetic engineering, nanotechnology, and informatics approaches. Numerous nano items, such as antibacterial silver products, sunscreens containing zinc oxide and titanium dioxide, medicines, and UV protection coatings, are entering the market, and the number of nanoparticles in the environment is steadily rising [60]. There are now more nanotechnology-based goods available. Naturally, the number of nanomaterials in the waste streams is constantly rising throughout the year. Models for material flow study clearly show that nanoparticles.

Currently, the United States, Australia, New Zealand, South Korea, Taiwan, China, and Israel are using nanoparticles in the food industry as well as other relevant disciplines [61]. It is quite regrettable to suggest that no accurate data has been acquired from the Indian region. It goes without saying that the right new techniques need to be introduced in India to obtain information about the toxicity effect of nanoparticles in the food sector and other various sectors.

20.6.2 Economic Impacts on Various Sector

The discipline of nanotechnology has developed to the point where it has the potential to assist society in reaching the goals of increased productivity and faster economic growth in various industries, including manufacturing, energy, and medicine. Therefore, to inform policy choices and financial investments, quantitative data is required to evaluate the economic impact of nanotechnology. To address societal issues like health and environmental concerns, support industrial competitiveness and economic growth, and expedite the commercialization of new technologies and products, public investment in nanotechnology is crucial [62]. The multipurpose nature of nanotechnology makes it difficult to measure the economic impact; the frequent blending of the impact of many other technologies with that of nanotechnology makes it difficult to understand and determine the precise role of nanotechnology; and difficulties encountered in meaningfully manipulating nanotechnology are just a few of the challenges that could affect the economic impact of nanotechnology: inadequacy of information obtained and difficulty assessing the value of the technology because data is primarily collected through surveys; the complex relationship between science and innovation; difficulties encountered in meaningfully manipulating and cleaning data related to nanotechnology; confidentiality and proprietary which businesses put in their product and services making it difficult to obtain data from industry [63].

Research and development (R&D) are now a national priority for the industrialized world due to the economic implications of nanotechnology. Today, national nanotechnology efforts have been established in over 60 countries, including the United States, Japan, Germany, Taiwan, China, Israel, Australia, and India. Public sector investment for these initiatives was estimated at $US4.6 billion in 2004 alone [64]. Similar to how the private sector has matched government funding, firms like IBM, NEC, Monsanto, and DuPont contend for a piece of the revenues from next-generation scientific advancements. In the developing field of nanotechnology, intellectual property rights (IPR) ownership and enforcement will be more crucial as applications progress from the R&D stage to the commercial market [65]. The requirement for a thorough framework for managing nanotechnology IPRs will be essential to the technology's commercial viability given the growing private sector involvement in nanotechnology R&D.

For government organizations and politicians, it can be challenging to assess the economic impact of nanotechnology and the investment returns that come with it. The economic worth of nanotechnology as depicted in Figure 20.5 is typically assessed in terms of education, employment, research activity, and the commercialization of goods and procedures, all of which differ significantly from one country to the next [66]. Analyzing research by counting the number of articles and papers that have been published is another method for determining the economic impact of nanotechnology. The variety in importance and value that exists in the assessment of the economic impact of research is a disadvantage of this technique. While university degrees, research projects, and outputs in the field of nanotechnology offer a good yardstick for evaluating the worth of academic labor, they are insufficient as economic indicators. Investments in nanotechnology should be evaluated in terms of their impact on export sectors, reduction of manufacturing costs, creation of jobs, development of new goods or services, and establishment of new businesses [67].

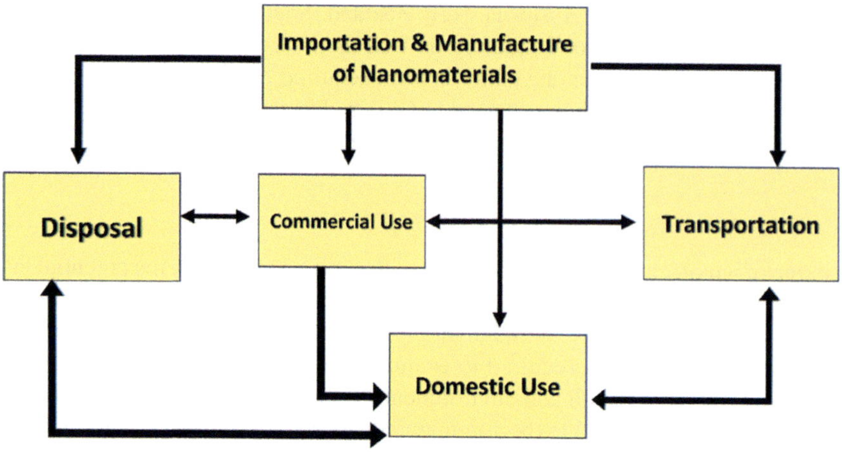

FIGURE 20.5 Assessment of the entire life cycle, from raw material extraction to manufacturing, use, and disposal.

20.7 SOCIAL IMPLICATIONS AND INEQUALITY

Nearly all industries involving material work are anticipated to experience increases in job efficiency as a result of nanotechnology. Major national initiatives should concentrate on how nanotechnology may increase production efficiency, better utilize energy resources, and lessen environmental concerns. Improved healthcare, improved pharmaceuticals, improved agricultural and food production, improved industry and transportation, and increased use of information technologies. By establishing collaborations between venture capitalists, industry, academia, national laboratories, and funding agencies earlier in the R&D processes, a systematic approach should be devised to prepare productive units and to promote synergy in the development of nanotechnology [13]. Major involvement of social scientists, economists, and public advocates to assure the equal distribution of contributions and benefits and to enhance possible applications so they may better serve people and, in turn, succeed in the market. There will be a reduction in input costs in some industries, a significant rise in productivity in others, the creation of whole new industries, and an increase in demand for lower demand for some commodities and others [68]. Real wages will ultimately see a big increase, and an increased level of living, with only a temporary rise in unemployment when labor and capital are transferred from outdated or less valuable uses to new, more valuable uses.

20.7.1 Displacement of Tradition Industries and Workers in Nanotechnology

The development of nanotechnology has the potential to upend established sectors and may result in the loss of some jobs. The use of nanotechnology may necessitate changes to the workforce and industry dynamics, as with any technological advancement. In many industries, more automation and greater efficiency are possible because of nanotechnology. For instance, in some industries, nanoscale manufacturing methods can increase production capacities and decrease the requirement for manual labor. Workers who perform manual or repetitive tasks may lose their jobs as a result of this [69]. Nanotechnology applications frequently necessitate the use of specialist knowledge and abilities. Workers may need to upgrade or reskill when traditional sectors adopt nanotechnology to keep up with the shifting needs. Jobs requiring knowledge in nanotechnology, materials science, nanofabrication, and related fields may become available.

Some established industries may need to be restructured or perhaps become obsolete as a result of the incorporation of nanotechnology. For instance, developments in nanomaterials might affect current production procedures or result in the use of nanomaterials to replace some materials. Finding alternative career prospects or making the transfer to new industries may be difficult for workers in those sectors [70]. Economic repercussions may result from the displacement of existing industries and workers by nanotechnology. Local economies that depend significantly on conventional sectors could endure downturns or other challenges. Governments and authorities must think about ways to assist impacted workers and ease their transition to new industries. Nanotechnology may cause some occupations to go, but it also creates new job prospects. A trained workforce is necessary for the development, manufacture, and commercialization of nanomaterials, nanodevices, and nanoscale technology. Nanotechnology-savvy scientists, engineers, technicians, and other professions are in greater demand, opening up job opportunities. It is critical to offer workers impacted by changes in traditional industries support and training programs to reduce the negative effects of displacement [71]. To assist workers transitioning to new industries, governments, educational institutions, and industry partners can work together to provide retraining programs, skill development efforts, and entrepreneurship support. Hence, it is necessary to remember that nanotechnology is not the only technology that has led to the displacement of conventional industries and people. Throughout history, changes in industries and labor markets have frequently been brought about by technological developments. Proactive planning, spending on education and training, and strong regulations are the keys to ensuring a smooth transition for impacted workers while maximizing the promise of nanotechnology for sustainable development and economic success.

20.8 SUSTAINABLE DEVELOPMENT AND NANOTECHNOLOGY

Nanotechnology and sustainable development are two related topics that have the potential to have a significant impact on our society and environment. A holistic strategy called sustainable development strives to satisfy current needs without endangering the capacity of future generations to satisfy their own. The application of nanotechnology to sustainable development presents both potential and difficulties. Through a variety of uses, nanotechnology can improve energy efficiency. For instance, nanostructured materials can increase the effectiveness of solar cells, increasing the viability of renewable energy sources. By enhancing insulation, lightweight, and fuel economy, nano coatings and nanocomposites can help lower energy usage in buildings and transportation. The creation of sophisticated sensors and environmental parameter monitoring systems is made possible by nanotechnology. Nano sensors can provide real-time data for efficient environmental management by detecting pollutants, diseases, and toxins at low concentrations [72]. To clean up the environment, by removing contaminants from the air, water, and soil, nanomaterials are also investigated. For the treatment and purification of water, nanotechnology provides potential options. Membranes with nanostructure can increase the effectiveness of filtering techniques, eliminating impurities, and enhancing water quality. For the treatment of wastewater, the removal of heavy metals, and the desalination of water, nanomaterials such as nanoparticles and nanofibers are used.

By increasing crop output and minimizing environmental effects, nanotechnology has the potential to change agricultural methods. By increasing the effectiveness of fertilizers, insecticides, and herbicides and reducing their use and environmental release, nanoscale delivery systems can reduce their consumption. Precision soil monitoring, irrigation control, and crop health are other applications for nano sensors and nanomaterials. Nanotechnology's cutting-edge materials and procedures can help with sustainable waste management. The effective separation, recycling, and recovery of valuable resources from waste streams can be accomplished using nanomaterials [73]. Nano catalysts can make it easier to turn waste into beneficial goods like chemicals or biofuels, lowering reliance on fossil fuels. Although nanotechnology has a lot of potential for sustainable development, there are a number of risks and difficulties that could arise when using it.

20.9 CONCLUSION

It is crucial to address the ethical, legal, and socio-economic implications of the development and application of nanotechnology as this field moves forward. Strong frameworks that encourage responsible and inclusive nanotechnology practices developed through multidisciplinary collaborations combining scientists, policymakers, ethicists, legal specialists, and civil society organizations. Nanotechnology raises ethical concerns regarding how this potent technology should be used responsibly. It is necessary to address concerns like human enhancement, privacy, and equitable access to advances in nanotechnology. To guarantee that nanotechnology is created and applied in a way that respects human dignity, fosters social well-being, and minimizes harm, there is a need for ethical standards and laws. Regarding intellectual property rights, product safety laws, and liability frameworks, nanotechnology poses difficulties. Existing laws may need to be updated, or new ones may need to be created, due to the distinctive qualities and traits of nanomaterials. Legal frameworks should strive to find a balance between fostering innovation and guaranteeing public and environmental safety. Nanotechnology has the ability to transform sectors, produce new employment opportunities, and boost economic growth. Advanced materials, better healthcare technologies, and more effective energy systems could all result from it. However, there are also worries about possible job loss, the digital divide, and the concentration of money and power. Additionally, it is important to carefully consider how nanotechnology might affect the environment. The impact of nanomaterials on ecosystems and human health is uncertain. To produce and use nanomaterials sustainably, it is imperative to do extensive risk assessments, create suitable safety measures, and promote these practices.

REFERENCES

1. Santamaria, A. (2012). Historical overview of nanotechnology and nanotoxicology. *Nanotoxicity: Methods and Protocols*, 926, 1–12.
2. De, A., Bose, R., Kumar, A., Mozumdar, S., De, A., Bose, R., & Mozumdar, S. (2014). A brief overview of nanotechnology. In *Targeted Delivery of Pesticides Using Biodegradable Polymeric Nanoparticles*, Springer, New Delhi, 35–36.
3. Thakur, P., & Thakur, A. (2022). Nanomaterials, their types and properties. In *Synthesis and Applications of Nanoparticles*. Springer Nature, Singapore, 19–44.
4. Safarov, T., Kiran, B., Bagirova, M., Allahverdiyev, A. M., & Abamor, E. S. (2019). An overview of nanotechnology-based treatment approaches against Helicobacter Pylori. *Expert Review of Anti-Infective Therapy*, 17(10), 829–840.
5. Punia, P., Bharti, M. K., Chalia, S., Dhar, R., Ravelo, B., Thakur, P., & Thakur, A. (2021). Recent advances in synthesis, characterization, and applications of nanoparticles for contaminated water treatment-a review. *Ceramics International*, 47(2), 1526–1550.
6. Thakur, A., Thakur, P., & Khurana, S. P. (2022). *Synthesis and Applications of Nanoparticles*. Springer, Singapore, 1–17.
7. Kostoff, R. N., Koytcheff, R. G., & Lau, C. G. (2007). Global nanotechnology research literature overview. *Technological Forecasting and Social Change*, 74(9), 1733–1747.
8. Iqbal, P., Preece, J. A., & Mendes, P. M. (2012). Nanotechnology: the “top-down” and “bottom-up” approaches. In *Supramolecular Chemistry: From Molecules to Nanomaterials*. Wiley, Hoboken, NJ.
9. Krishnia, L., Thakur, P., & Thakur, A. (2022). Synthesis of nanoparticles by physical route. In *Synthesis and Applications of Nanoparticles*. Springer Nature, Singapore, 45–59.
10. Sahoo, S. (2010). Socio-ethical issues and nanotechnology development: perspectives from India. In *10th IEEE International Conference on Nanotechnology*. IEEE, New York, 1205–1210.
11. Ludlow, K. (2007). More than science: ethical and socio-legal concerns in nanotechnology regulation in Australia. In *New Global Frontiers in Regulation: The Age of Nanotechnology*. Edward Elgar, Cheltenham, Gloucestershire, 166–186.
12. Zucker, L. G., & Darby, M. R. (2005). Socio-economic impact of nanoscale science: initial results and nanobank. In: Roco, M. C., Bainbridge, W. S. (Eds.) *Nanotechnology: Societal Implications II—Individual Perspectives*. Springer, Dordrecht, The Netherlands.

13. Miller, G., & Scrinis, G. (2011). Nanotechnology and the extension and transformation of inequity. In: Cozzens, S., Wetmore, J. (Eds.) *Nanotechnology and the Challenges of Equity, Equality and Development*. Springer, Dordrecht, The Netherlands, 109–126.
14. Foladori, G., Invernizzi, N., & Záyago, E. (2009). Two dimensions of the ethical problems related to nanotechnology. *NanoEthics*, 3, 121–127.
15. Marchant, G. E., Sylvester, D. J., & Abbott, K. W. (2008). Risk management principles for nanotechnology. *Nanoethics*, 2, 43–60.
16. Baran, A. (2016). Nanotechnology: legal and ethical issues. *Ekonomia i Zarządzanie*, 8(1), 47–54.
17. Spagnolo, A. G., & Daloiso, V. (2009). Outlining ethical issues in nanotechnologies. *Bioethics*, 23(7), 394–402.
18. Weil, V. (2001). Ethical issues in nanotechnology. *Societal Implications of Nanoscience and Nanotechnology*, 193, 28–29.
19. Schummer, J., & Baird, D. (Eds.). (2006). *Nanotechnology Challenges: Implications for Philosophy, Ethics and Society*. World Scientific, Vancouver, BC.
20. McGinn, R. (2010). Ethical responsibilities of nanotechnology researchers: a short guide. *NanoEthics*, 4, 1–12.
21. Lewenstein, B. V. (2006). What counts as a 'social and ethical issue' in nanotechnology? *Nanotechnology Challenges: Implications for Philosophy, Ethics and Society*, 11, 201–216.
22. Allhoff, F., & Lin, P. (Eds.). (2009). *Nanotechnology & Society: Current and Emerging Ethical Issues*. Springer, Dordrecht, The Netherlands.
23. Grunwald, A. (2005). Nanotechnology—a new field of ethical inquiry? *Science and Engineering Ethics*, 11, 187–201.
24. Gardner, J., & Dhai, A. (2014). Nanotechnology and water: ethical and regulatory considerations. *Application of Nanotechnology in Water Research*, 6, 1–20.
25. Dalton-Brown, S. (2012). Global ethics and nanotechnology: a comparison of the nanoethics environments of the EU and China. *NanoEthics*, 6, 137–150.
26. Hunt, G., & Mehta, M. (2007). An invitation to debate: review of nanotechnology: risk, ethics and law. *Nanotechnology Law & Business*, 4(2), 14–21.
27. Engelmann, W., von Hohendorff, R., & da Silva Leal, D. W. (2022). Ethical issues in nanotechnology-II. In *Environmental, Ethical, and Economical Issues of Nanotechnology*. Jenny Stanford Publishing, Singapore, 131–149.
28. Abuelma'atti, M. T. (2009). Nanotechnolgy: benefits, risks and ethical issues. In *2009 5th IEEE GCC Conference & Exhibition*. IEEE, New York, 1–5.
29. Fiedler, F. A., & Reynolds, G. H. (1993). Legal problems of nanotechnology: an overview. *Southern California Interdisciplinary Law Journal*, 3, 593.
30. Feitshans, I. L., & Sabatier, P. (2022). Global health impacts of nanotechnology law: advances in safer-nano regulation. *Materials Today: Proceedings*, 67, 985–994.
31. Boucher, P. M. (2018). *Nanotechnology: Legal Aspects*. CRC Press, Boca Raton, FL.
32. Hodge, G. A., Bowman, D., & Ludlow, K. (Eds.). (2009). *New Global Frontiers in Regulation: The Age of Nanotechnology*. Edward Elgar Publishing, Cheltenham, Gloucestershire.
33. Singh, R., Thakur, P., Thakur, A., Kumar, H., Chawla, P., Rohit, V. J., & Kumar, N. (2021). Colorimetric sensing approaches of surface-modified gold and silver nanoparticles for detection of residual pesticides: a review. *International Journal of Environmental Analytical Chemistry*, 101(15), 3006–3022.
34. Brazell, L. (2012). *Nanotechnology Law: Best Practices*. Kluwer Law International BV, Netherlands.
35. Kai, K. (2012). Nanotechnology and medical robotics; legal and ethical responsibility. *Waseda Bulletin of Comparative Law*, 30, 1–6.
36. Cassotta, S. (2012). Extended producer responsibility in waste regulations in a multilevel global approach: nanotechnology as a case study. *European Energy and Environmental Law Review*, 21(5), 198–219.
37. De Ville, K. A. (2008). Law, regulation and the medical use of nanotechnology. In: Jotterand, F. (Eds.) *Emerging Conceptual, Ethical and Policy Issues in Bio Nanotechnology*. Springer, Dordrecht, 181–200.
38. Abbott, K. W., Marchant, G. E., & Corley, E. A. (2012). Soft law oversight mechanisms for nanotechnology. *Jurimetrics*, 52, 279–312.
39. Dana, D. A. (Ed.). (2011). *The Nanotechnology Challenge: Creating Legal Institutions for Uncertain Risks*. Cambridge University Press, Cambridge.
40. Miernicki, M., Hofmann, T., Eisenberger, I., von der Kammer, F., & Praetorius, A. (2019). Legal and practical challenges in classifying nanomaterials according to regulatory definitions. *Nature Nanotechnology*, 14(3), 208–216.

41. Cassotta, S. (2012). Extended producer responsibility in waste regulations in a multilevel global approach: nanotechnology as a case study. *European Energy and Environmental Law Review*, 21(5), 198–219.
42. Matsuda, M., & Hunt, G. (2005). Nanotechnology and public health. *Nihon Koshu Eisei Zasshi (Japanese Journal of Public Health)*, 52(11), 923–927.
43. Baranowska, N. N. (2018). Public authority liability and the regulation of nanotechnology: a european perspective. *Canadian Journal of Law and Technology*, 16(1), 3.
44. Wernette, R. C., & Nilsen, R. L. (2008). The legal risks of nanotechnologies. *For The Defense*, 50(11), 73–81.
45. Feitshans, I. L., & Sabatier, P. (2022). Global health impacts of nanotechnology law: advances in safer-nano regulation. *Materials Today: Proceedings*, 67, 985–994.
46. Roco, M. C., & Bainbridge, W. S. (2005). Societal implications of nanoscience and nanotechnology: maximizing human benefit. *Journal of Nanoparticle Research*, 7, 1–13.
47. Nasu, H. (2012). Nanotechnology and challenges to international humanitarian law: a preliminary legal assessment. *International Review of the Red Cross*, 94(886), 653–672.
48. Howells, G. (2009). Product liability for nanotechnology. *Journal of Consumer Policy*, 32, 381–391.
49. Lee, R., & Jose, P. D. (2008). Self-interest, self-restraint and corporate responsibility for nanotechnologies: emerging dilemmas for modern managers. *Technology Analysis & Strategic Management*, 20(1), 113–125.
50. Kokotovich, A. E., Kuzma, J., Cummings, C. L., & Grieger, K. (2021). Responsible innovation definitions, practices, and motivations from nanotechnology researchers in food and agriculture. *NanoEthics*, 15, 1–15.
51. Elcock, D. (2007). Potential impacts of nanotechnology on energy transmission applications and needs (No. ANL/EVS/TM/08-3). Argonne National Lab. (ANL), Argonne, IL (United States).
52. Dunphy Guzmán, K. A., Taylor, M. R., & Banfield, J. F. (2006). Environmental risks of nanotechnology: national nanotechnology initiative funding, 2000–2004. *Environmental Science & Technology*, 40, 1401–1407.
53. Theodore, L., & Kunz, R. G. (2005). *Nanotechnology: Environmental Implications and Solutions*. John Wiley & Sons, Hoboken, NJ.
54. Shatkin, J. A. (2017). *Nanotechnology: Health and Environmental Risks*. CRC Press, Boca Raton, FL.
55. Yadav, S. K., Khan, Z. A., & Mishra, B. (2013). Impact of nanotechnology on socio-economic aspects: an overview. *Reviews in Nanoscience and Nanotechnology*, 2(2), 127–142.
56. Bozeman, B., & Youtie, J. (2017). Socio-economic impacts and public value of government-funded research: lessons from four US National Science Foundation initiatives. *Research Policy*, 46(8), 1387–1398.
57. Rashidi, L., & Khosravi-Darani, K. (2011). The applications of nanotechnology in food industry. *Critical Reviews in Food Science and Nutrition*, 51(8), 723–730.
58. Thiruvengadam, M., Rajakumar, G., & Chung, I. M. (2018). Nanotechnology: current uses and future applications in the food industry. *3 Biotech*, 8, 1–13.
59. Chellaram, C., Murugaboopathi, G., John, A. A., Sivakumar, R., Ganesan, S., Krithika, S., & Priya, G. (2014). Significance of nanotechnology in food industry. *APCBEE Procedia*, 8, 109–113.
60. Mohammad, Z. H., Ahmad, F., Ibrahim, S. A., & Zaidi, S. (2022). Application of nanotechnology in different aspects of the food industry. *Discover Food*, 2(1), 12.
61. Amini, S. M., Gilaki, M., & Karchani, M. (2014). Safety of nanotechnology in food industries. *Electronic Physician*, 6(4), 962.
62. Hobson, D. W. (2009). Commercialization of nanotechnology. *Wiley Interdisciplinary Reviews: Nanomedicine and Nanobiotechnology*, 1(2), 189–202.
63. Iavicoli, I., Leso, V., Ricciardi, W., Hodson, L. L., & Hoover, M. D. (2014). Opportunities and challenges of nanotechnology in the green economy. *Environmental Health*, 13(1), 1–11.
64. Morse, J. (2012). Assessing the economic impact of nanotechnology: The role of nanomanufacturing. *NNN Newsletter*, 5(3), 1745.
65. Roco, M. C. (2017). Overview: Affirmation of Nanotechnology between 2000 and 2030. In *Nanotechnology Commercialization: Manufacturing Processes and Products*. John Wiley & Sons, Hoboken, NJ, 1–23.
66. Shapira, P., & Youtie, J. (2015). The economic contributions of nanotechnology to green and sustainable growth. In: Basiuk, V., Basiuk, E. (Eds.) *Green Processes for Nanotechnology: From Inorganic to Bioinspired Nanomaterials*, Springer, Cham, 409–434.

67. Hullmann, A. (2007). Measuring and assessing the development of nanotechnology. *Scientometrics*, 70(3), 739–758.
68. Cozzens, S. E., & Wetmore, J. (Eds.). (2010). *Nanotechnology and the Challenges of Equity, Equality and Development* (Vol. 2). Springer Science & Business Media, Singapore.
69. Scrinis, G., & Lyons, K. (2007). The emerging nano-corporate paradigm: nanotechnology and the transformation of nature, food and agri-food systems. *The International Journal of Sociology of Agriculture and Food*, 15(2), 22–44.
70. Peng, B., Tang, J., Luo, J., Wang, P., Ding, B., & Tam, K. C. (2018). Applications of nanotechnology in oil and gas industry: progress and perspective. *The Canadian Journal of Chemical Engineering*, 96(1), 91–100.
71. Hanus, M. J., & Harris, A. T. (2013). Nanotechnology innovations for the construction industry. *Progress in Materials Science*, 58(7), 1056–1102.
72. Diallo, M. S., Fromer, N. A., & Jhon, M. S. (2014). *Nanotechnology for Sustainable Development: Retrospective and Outlook*. Springer International Publishing, Singapore, 1–16.
73. Helland, A., & Kastenholz, H. (2008). Development of nanotechnology in light of sustainability. *Journal of Cleaner Production*, 16(8–9), 885–888.

Index

9781032820453